Book of Abstracts of the 50th Annual
European Association for Animal Pr...

The EAAP Book of Abstracts is published under the direction of Johan van Arendonk

EAAP - European Association for Animal Production

The European Association for Animal Production wishes to express its appreciation to
Ministero per le Politiche Agricole (Italy), Alimentari e Forestali (Italy),
Associazione Italiana Allevatori (Italy)
for their valuable support of its activities.

Book of Abstracts of the 50th Annual Meeting of the European Association for Animal Production

Zurich, Switzerland, 22-26 August 1999

J.A.M. van Arendonk, Editor-in-chief

A. Hofer, Y. van der Honing, F. Madec, M. Bonneau, D. Pullar, M. Schneeberger, J.A. Fernández, E.W. Bruns

Wageningen 1999

CIP-data Koninklijke Bibliotheek
Den Haag

ISSN 1382-6077
ISBN 90-74134-73-4
NUGI 835

Subject headings:
animal production,
book of abstracts

© **Wageningen Pers, Wageningen
The Netherlands, 1999**

Printed in The Netherlands

All rights reserved.
Nothing from this publication may be reproduced, stored in a computerized system or published in any form or in any manner, including electronic, mechanical, reprographic or photographic, without prior written permission from the publisher, Wageningen Pers, P.O. Box 42, NL-6700 AA Wageningen, The Netherlands.

The individual contributions in this publication and any liabilities arising from them remain the responsibility of the authors.

The designation employed and the presentation of material in this publication do not imply the expression of any option whatsoever on the part of the European Association for Animal Production concerning the legal status of any country, territory, city or area or of its authorities, or concerning the delimitation of its frontiers or boundaries.

In so far as photocopies from this publication are permitted by the Copyright Act 1912, Article 16B and Royal Netherlands Decree of 20 June 1974 (Staatsblad 351) as amended in Royal Netherlands Decree of 23 August 1985 (Staatsblad 471) and by Copyright Act 1912, Article 17, the legally defined copyright fee for any copies should be transferred to the Stichting Reprorecht (P.O. Box 882, 1180 AW Amstelveen, The Netherlands).

For reproduction of parts of this publication in compilations such as anthologies or readers (Copyright Act 1912, Article 16), written permission must be obtained from the publisher.

Preface

The 50th annual meeting of the European Association for Animal Production (EAAP) will be held in Zurich, Switzerland. Fifty years ago, the EAAP was founded in Switzerland and this year we will celebrate its 50th anniversary at the Federal Institute of Technology in Zurich. Livestock researchers and their companions are welcomed to join this special event. As part of the celebrations of the 50th anniversary, a special issue of our international journal "Livestock Production Science" will be published soon which gives an overview of the activities of EAAP and contains papers of a number of young scientists active in our field of interest.

The book of abstract forms the main publication of the scientific contributions to this meeting which covers a wide range of disciplines and livestock species. It contains the full program and abstracts of the invited as well as the contributed papers including posters of all 33 sessions.

The number of abstracts submitted for presentation at this meeting was again large, and it has been a challenge to the different study commissions and chairpersons to put together the scientific program. In addition to the presentations in theatre, there will be large number of poster presentations during the conference. I hope you will all take the time to have a good look at these contributions as well as to the theatre presentations. Several people have been involved in the production of this book of abstracts. In order to improve the quality of the publication, this is the first year in which we moved away from camera-ready production to the use of electronic versions of all material. Mike Jacobs of Wageningen Pers has been responsible for a large proportion of the administrative work and the production of the book of abstracts. The contact persons of the study commissions have been responsible for organizing the scientific programme and communication with the chairpersons and invited people.

The program looks very interesting and I trust we will have a good meeting in Zurich. I hope will find this book of abstracts an useful source of reference as well as a reminder of a good meeting in Zurich during which a large number of people actively involved in animal production have met and exchanged ideas.

Johan van Arendonk

Editor-in-Chief

European Association for Animal Production (EAAP)

President   Dr. h.c. Philipp R. Fürst zu Solms-Lich
Executive Vice-President   Prof. Dr. Jean Boyazoglu

Address   Villa del Ragno,
            Via Nomentana 134, I-00161 Rome, Italy
Phone   +39 06 86329141
Telefax   +39 06 86329263
E-Mail   zoorec@rmnet.it

In cooperation with

The Swiss Organizing Committee of EAAP 1999
Swiss Association for Animal Breeding

Honorary Committee

Minister Pascal Couchepin   Minister of Federal Dept. of Economics
Mayor Josef Estermann   Mayor of the City of Zurich
Executive Councilor Robert Bisig   Head of Dept. of Economics, Canton Zug
Prof. Dr. Olaf Kübler   President of the Swiss Federal Institute of
                                Technology, ETH, Zurich

Members of the Swiss Organizing Committee

Prof. Dr. N. Künzi   Institute of Animal Science, ETH Zurich, Animal Breeding Group
                        Chairman Organizing Committee
Dr. C. Marguerat   Institute of Animal Science, ETH Zurich, Animal Breeding Group
                        General secretariat
Dir. E. Germann   Director: Swiss Simmental + Red & White Cattle Breeders'
                        Federation, Zollikofen
                        Finances
Prof. Dr. J. Morel   Vice Director: Federal Division of Agriculture, Berne
                        Public relations
Dr. J. Schmidlin   Swiss Federation for Small Breeds, Berne
                        Excursions and cattle exhibition
F.-H. Bovet   President: Swiss Association of Beef Breeds, Givrins
                        Excursions and cattle exhibition
Dr. U. Witschi   Swiss Federation for Artificial Insemination, Zollikofen
                        Excursions and cattle exhibition
Dir. M. Zogg   Director: Swiss Brown Cattle Breeders' Federation, Zug
                        Excursions and cattle exhibition
Dir. J. Schletti   Director: Swiss Meat Board, Berne
                        Sponsoring
Prof. Dr. J. Blum   Vet. Med. Faculty, University of Berne
                        Scientific matters

Time Table

Saturday, August 21st
- 09.00 - 18.00   Registration of participants
- 09.00 - 18.00   Symposium I: Teaching Methods in Animal Science
- 09.00 - 12.00   Board meeting
- 10.00 - 16.00   Scientific Programme Committee Meeting
- 16.00 - 18.00   Livestock Production Science (Editorial Board)

Sunday, August 22nd
- 08.30 - 12.00   Session of the Council with Presidents and Secretaries of Study Commissions and with Presidents of Working Groups
- 08.30 - 16.00   Special Satellite Symposium: How to prepare and present scientific papers
- 09.00 - 16.00   Registration of participants
- 13.30 - 16.00   Council Meeting
- 17.00 - 19.00   Ceremony "50 years of EAAP" in the Tonhalle, Zürich
- 19.00 - 23.00   Jubilee Dinner at the Kongresshaus, Zürich

Monday, August 23rd
- 08.00 - 18.00   Registration of participants/Information
- 09.00 - 12.30   Session I of study commissions
- 14.00 - 17.30   Session II of study commissions
- 18.00 - 22.00   Visit of Poster Exhibition / social mix at ETH, main Building

Tuesday, August 24th
- 08.00 - 12.00   Registration of participants/Information
- 08.30 - 12.00   Session III of study commissions
- 13.00 - 22.00   Mid-Conference Tour: Farm Visit/Animal Show/Folklore Evening in Zug

Wednesday, August 25th
- 07.00 - 08.30   at Hotel: Breakfast Meeting: Scientific Programme Committee
- 08.00 - 18.00   Information/Excursions etc.
- 09.00 - 12.30   Session IV of study commissions
- 14.00 - 17.30   Session V of study commissions
- 14.00 - 17.30   Council Meeting
- evening free

Thursday, August 26th
- 08.00 - 10.00   Session VI of study commissions
- 10.30 - 12.30   General Assembly
- 12.30   Council Meeting
- 12.30   Scientific Programme Committee
- 14.00   Departure for Post Conference Tours I + II
- 14.00   Beginning of Symposium III: Biology of the Pancreas in Growing Farm Animals
- 14.00   Beginning of Interbull Meeting at ETH

Scientific programme - Zürich 1999

Study Commissions	Monday, 23 August 09.00 - 12.30 Session I	Monday, 23 August 14.00 - 17.30 Session II	Tuesday, 24 August 08.30 - 12.00 Session III
Genetics (G)	(G* + P + H) Consequences of new technologies (with special emphasis on pigs and horses)	Free Communications/ Business meeting	Advances in statistical methods
Chairperson	S. Neuenschwander (CH)	E. Mäntysaari (FIN) J. van Arendonk (NL)	R. Reents (D)
Animal Nutrition (N)	(N* + C) Nutrition and breeding of cattle to optimize use of grass and forages for milk and beef production	(N + Ph*) Physiological and nutritional aspects of growth and development in new born animals	(N + S* + IGA) Feeding of dairy goats under intensive management
Chairperson	A. Kuipers (NL)	S. Pierzynowski (S)	P. Morand-Fehr (F)
Animal Management and Health (M)	Impact of welfare research and legislation on production, economy and consumer acceptance	(M* + C + OIE) Reproduction technologies (AI, embryo transfer, cryopreservation...) and risks of disease transmission	(M + Ph*) Physiology of suboptimal growth
Chairperson	B. Wechsler (CH)	M. Thibier (F)	J. Blum (CH)
Animal Physiology (Ph)	Cell division and protein synthesis for growth in animal tissues	(Ph* + N) Physiological and nutritional aspects of growth and development in new born animals	(Ph* + M) Physiology of suboptimal growth
Chairperson	N. Oksbjerg (DK)	S. Pierzynowski (S)	J. Blum (CH)
Cattle Production (C)	(C + N*) Nutrition and breeding of cattle to optimize use of grass and forages for milk and beef production	(C + M* + OIE) Reproduction technologies (AI, embryo transfer, cryopreservation...) and risks of disease transmission	Free communications / Business meeting
Chairperson	A. Kuipers (NL)	M. Thibier (F)	S. Gigli (I)
Sheep and Goat Production (S)	Sheep and goat production in wet mountain areas	Free communcations	(S* + N + IGA) Feeding of dairy goats under intensive management
Chairperson	M. Schneeberger (CH)	D. Croston (UK)	P. Morand-Fehr (F)
Pig Production (P)	(P + G* + H) Consequences of new technologies (with special emphasis on pigs and horses)	The role of dietary fibre in pig production	Free communications / Business meeting
Chairperson	S. Neuenschwander (CH)	A. Aumaitre (F)	J.A. Fernández (DK)
Horse Production (H)	(H + G* + P) Consequences of new technologies (with special emphasis on pigs and horses)	a) Free communications b) Business meeting	Horse breeding in Switzerland
Chairperson	S. Neuenschwander (CH)	a) E. Barrey (F) b) E. Bruns (D)	A. Lüth (CH)

Scientific programme - Zürich 1999

Study Commissions	Wednesday, 25 August 09.00 - 12.30 Session IV	Wednesday, 25 August 14.00 - 17.30 Session V	Thursday, 26 August 08.00 - 10.00 Session VI
Genetics (G)	a) Free communications b) (G + N + P*) Quality of meat and fat as affected by genetics and nutrition	(G* + Ph) Biological selection limits and fitness constraints	Free communications/ Ideas box
Chairpersons	a) I. Olesen (N) b) C. Wenk (CH)	A. Visscher (NL)	M. Carabaño (E)
Animal Nutrition (N)	(N + G + P*) Quality of meat and fat as affected by genetics and nutrition	a) Free communications b) Business meeting	(N* + P) 08.00 - 12.00 Future strategies with regard to the use of feed without antibiotic feed additives in pig production
Chairperson	C. Wenk (CH)	a) H. Hagemeister (D) b) Y. van der Honing (NL)	K.B. Pedersen (DK)
Animal Management and Health (M)	(M* + H) Electronic identification	Free communications	Business meeting
Chairperson	R. Geers (B)	O. Szenci (H)	F. Madec (F)
Animal Physiology (Ph)	Business meeting	(Ph + G*) Biological selection limits and fitness constraints	Free communications
Chairperson	M. Bonneau (F)	A. Visscher (NL)	K. Sejrsen (DK)
Cattle Production (C)	Future role of dual purpose cattle	Management tools to improve the physical and financial performance of livestock farms	Free communications / Emerging issues
Chairperson	Ch. Böbner (CH)	R. Palmer (USA)	G. Pollott (UK)
Sheep and Goat Production (S)	Genetic resistance to disease and parasites - alternatives to chemotherapy	Economic, genetic and management aspects of fine fibre production	Business meeting / Emerging issues
Chairperson	M. Stear (UK)	J. Milne (UK)	D. Croston (UK)
Pig Production (P)	(P* + G + N) Quality of meat and fat as affected by genetics and nutrition	Feeding and management of the weaned pig	(P + N*) 08.00 - 12.00 Future strategies with regard to the use of feed without antibiotic feed additives in pig production
Chairperson	C. Wenk (CH)	V. Danielsen (DK)	K.B. Pedersen (DK)
Horse Production (H)	(H + M*) Electronic identification	New developments in reproduction	Free communications
Chairperson	R. Geers (B)	T. Allen (UK)	D. Austbø (N)

51st Annual Meeting of the European Association for Animal Production

The 51st Annual Meeting will take place in The Hague, The Netherlands from 21-24 August 2000.

Addresses of the 51st Annual Meeting of EAAP:

Organizing Committee
c/o S. van der Beek
CR Delta
P.O. Box 454
NL 6800 AL Arnhem
The Netherlands
tel:+31 26 3898 700
fax:+31 26 3898 777
eaap2000@alg.vf.wau.nl

Official Congress Agency
c/o Bernie Brilman PCO BV
Huygenstraat 1
NL 2271 BV Voorburg
The Netherlands
tel:+31 70 387 00 70
fax:+31 70 386 33 72
eaap2000@bbpco.nl

Abstract submission
Wageningen Pers
c/o A.F.M. Jacobs
P.O. Box 42
NL 6700 AA Wageningen
The Netherlands
tel:+31 317 47 65 16
fax:+31 317 42 60 44
eaap2000@wageningenpers.nl

EAAP Congresses coming up

EAAP serie no. 98
32nd ICAR session

Bled, Slovenia
16-20 May, 2000

Secretariat
Groblje 3
Svn - 1230 Domzale
Slovenia

EAAP serie no. 99
Prospects for a sustainable dairy sector in the Mediterranean

Hammamet, Tunesia
26-28 October, 2000

Symposium Secretariat
Office de l'Elevage et des Pâturages
30, rue Alain Savary
TN 1002 TUNIS
Tunesia

Agendas of Study Commissions

– Commission on Animal Genetics (G) — Page XV
– Commission on Animal Nutrition (N) — Page XXIX
– Commission on Animal Management and Health (M) — Page XXXVIII
– Commission on Animal Physiology (Ph) — Page XLVI
– Commission on Cattle Production (C) — Page LII
– Commission on Sheep and Goat Production (S) — Page LXII
– Commission on Pig Production (P) — Page LXXII
– Commission on Horse Production (H) — Page LXXXVI

Commission on Animal Genetics (G)

President
Dr.ir. J.A.M. van Arendonk
Wageningen Agricultural University
Department of Animal Breeding
Marijkeweg 40
6709 PG Wageningen
The Netherlands
Phone:+31 317 483378
Fax:+31 317 483929
E-mail:Johan.VanArendonk@alg.vf.wau.nl

Secretaries
Dr. A. Hofer
Swiss Federal Institute of Technology (ETH)
Institute of Animal Science
ETH Centre/LFW B56
8092 Zürich
SWITZERLAND
Phone:+41 1632 3248
Fax:+41 1632 1260
E-mail:hofer@inw.agrl.ethz.ch

Prof.Dr. L. Fésüs
Research Institute for Animal Breeding & Nutrition
Gesztenyés út 1
2053 Herceghalom
HUNGARY
Phone:+36-23-319082
Fax:+36-23-319082

CONSEQUENCES OF NEW TECHNOLOGIES
(WITH SPECIAL EMPHASIS ON PIGS AND HORSES)

Session 1

Joint session with the commissions on Pig Production and Horse Production (G*+P+H)

Date   23-aug-99
                9:00 - 12:30 hours
Chairperson   S. Neuenschwander (CH)

Genetics

Papers

	GPH1 no.	Page
Is functional genomics possible in farm animals? Archibald, A.L.	1	1
A genetic marker for litter size in landrace based pig lines Southwood, O., T.H. Short, G.S. Plastow and M.F. Rothschild	2	1
Visualisation of differential gene expression among equine tissues using cDNA-AFLP Verini Supplizi, A., K. Cappelli, A. Porceddu, F. De Marchis, C. Moretti, M. Falcinelli, M. Pezotti, D. Casciotti and M. Silvestrelli	3	2
A diagnostic assay discriminating between two major sheep mitochondrial DNA haplotypes suitable for QTL association studies Hiendleder, S., W. Hecht and S.H. Phua	4	2
Genomics and patenting Brenig, B.	5	3
Detection of quantitative trait loci for fatness traits in Dutch Meishan-crossbreds Koning, D-J. de, L.L.G. Janss, A.P. Rattink, M.A.M. Groenen, P.N. de Groot, E.W. Brascamp and J.A.M. van Arendonk	6	3
Status of Genome and QTL mapping in Hohenheim F2 Pig-Families Moser, G., G. Reiner, E. Mueller, P. Beeckmann, G. Yue, M. Dragos, H. Bartenschlager, S. Cepica, A. Stratil, J. Schroeffel and H. Geldermann	7	4
Mapping loci for the degree of spotting, colour dilution, and circumocular pigmentation in the ADR granddaughter design. Reinsch, N., H. Thomsen, N. Xu, C. Looft, E. Kalm, G. Brockmann, S. Grupe, C. Kühn, M. Schwerin, B. Leyhe, S. Hiendleder, G. Erhardt, I. Medjugorac, I. Russ and M. Förster	8	4
Quantifying the impact of introgressing major genes for disease resistance MacKenzie, K. and S.C. Bishop	9	5
Genetic and economic consequences of alternative strategies for introgressing QTL responsible for disease resistance, using different breeding schemes Waaij, E.H. van der and J.A.M. van Arendonk	10	5

Posters

	GPH1 no.	Page
Species-specific splicing of transgenic RNA in the mammary glands of pigs and rabbits Aigner, B., K. Pambalk, M. Renner, W.H. Günzburg, E. Wolf, M. Müller and G. Brem	11	6

Genetics

GPH1 no. Page

The fertilizing ability of spermatozoa differentiated from female primordial germ cells in male chimeric chickens 12 6
Tagami, T., H. Kagami, T. Harumi, Y. Matsubara, H. Hanada and M. Naito

Producing chimeras by transfer of chicken blastodermal cells cultured without a feeder layer 13 7
Matsubara, Y., A. Hirota, H. Sobajima, A. Onishi, H. Kagami, T. Tagami, T. Harumi, M. Sakurai and M. Naito

Inbreeding and homozygosity in a Hungarian Lipizzan horse population 14 7
Mihók, S., B. Pataki, E. Takács, I. Bodó, Z. Egri, B. Bán and D. Ritter

Horse parentage testing using high polymorphic microsatellites: a case study 15 8
Ginja, C., J. Matos, A. Clemente and T. Rangel

RFLP based genetic mapping of the ovine inhibin/activin (*INHA, INHBA, INHBB*) genes identifies microsatellites for QTL association studies 16 8
Hiendleder, S., K.G. Dodds and R. Wassmuth

Molecular genetic studies on the porcine c-myc proto-oncogene regarding phylogeny, associations with performance traits and interactions with the Ryr1-genotype 17 9
Reiner, G., G. Moser, H. Geldermann and V. Dzapo

Determination of $\alpha(1,2)$fucosyltransferase gene function in resistance to oedema disease and post-weaning diarrhoea and in blood group antigen A-0 synthesis 18 9
Meijerink, E., P. Vögeli, S. Neuenschwander, H.U. Bertschinger, R. Fries and G. Stranzinger

Properties of a sperm ligand that are necessary for zona binding 19 10
Jansen, S., B. Marschall, I. Jeneckens and B. Brenig

The genomic structure of the ovine Uroporphyrinogen Decarboxylase gene 20 10
Nezamzadeh, R., A. Krempler and B. Brenig

FREE COMMUNICATIONS / BUSINESS MEETING

Session 2

Date   23-aug-99
                14:00 - 17:30 hours
Chairperson   E. Mäntysaari (Fin) and J. van Arendonk (NL)

Genetics

Papers
G2 no. Page

Implementation of a routine genetic evaluation for longevity based on survival analysis techniques in dairy cattle populations in Switzerland   1   11
Vukasinovic, N., J. Moll and L. Casanova

A proposal for genetic evaluation for functional herd life in Italian Holsteins   2   11
Schneider, P. and F. Miglior

The use of type traits as indicators for survival of dairy cows   3   12
Bünger, A. and H. Swalve

Phenotypic relationship between type and longevity in the French Holstein breed   4   12
Larroque, H. and V. Ducrocq

Genetic analysis of functional longevity in Danish Landrace sows   5   13
Damgaard, L.H., I.R. Korsgaard, J. Jensen and L.G. Christensen

Bayesian analysis of variance components of fertility in Spanish Landrace pigs.   6   13
Varona, L. and J.L. Noguera

Genetic parameters of a random regression model for daily feed intake of performance tested French Landrace and Large White growing pigs   7   14
Schnyder, U., A. Hofer, F. Labroue and N. Künzi

Estimates of direct and maternal genetic covariance functions for early growth of Australian beef cattle   8   14
Meyer, K.

Posters
G2 no. Page

Genetic parameters for individual birth and weaning weight of Large White piglets   9   15
Kaufmann, D., A. Hofer, J.-P. Bidanel and N. Künzi

Simultaneous estimation of the covariance structure for production and reproduction traits in pigs   10   15
Peskovicová, D., J. Wolf, E. Groeneveld and L. Hetényi

Estimates of genetic parameters for reproduction traits at different parities in Dutch Landrace pigs   11   16
Hanenberg, E.H.A.T., E.F. Knol and J.W.M. Merks

Implications of a non-genetic antagonism between maternal effects of two generations in the dispersion structure of the maternal animal model   12   16
Quintanilla, R., M.R. Pujol and J. Piedrafita

Three lactations vs all lactations model   13   17
Samoré, A.B., F. Canavesi and F. Miglior

Genetics

G2 no. Page

A parallel solver for Mixed Model Equations stratified to minimise communication 14 17
Madsen, P. and M. Larsen

An integrated use of statistical packages in organizing, manipulating and analyzing a set of one hundred thirty thousand beef data for practical purposes 15 18
Tzortzios, S. and N. Gitsakis

The (co)variance structure of Sarda ewe's fat and protein content Test Day records 16 18
Carta, A., N.P.P Macciotta, S. Sanna and A. Cappio Borlino

Modelling milk production in extended lactations of dairy cows 17 19
Vargas, B., W.J. Koops and J.A.M. van Arendonk

Genetic parameters of longevity traits in Holstein dairy cattle of Iran 18 19
Heydarpour, M.

Genetic parameter estimation of cow survival in the Israeli dairy cattle population 19 20
Settar, P. and J.I. Weller

The heritability of visual acceptability in South African Merino sheep 20 20
Erasmus, G.J., F.W.C. Neser, J. van Wyk and J.J. Olivier

Genetic parameters for growth traits in Moroccan Timahdit breed and its D'man x Timahdit cross 21 21
El Fadili, M., C. Michaux, J. Detilleux, F. Farnir and P.L. Leroy

Performance, carcass yield and quality of meat in broiler rabbits: a comparison of six breeds 22 21
Skrivanová, V., M. Marounek, E. Tůmová, M. Skřivan, P. Klein and J. Laštovková

The effect of selection for growth rate on carcass composition and meat characteristics of rabbits 23 22
Piles, M., A. Blasco and M. Pla

Estimates of genetic parameters for hip and elbow dysplasia in Finnish Rottweilers 24 22
Mäki, K., M. Ojala and A.E. Liinamo

Lactation and sample test day correlation and repeatability of milk somatic cell count in Hungarian Holstein Fresian cows 25 23
Amin, A.A., and T. Gere

Genetic parameters for production traits in mink 26 23
Hansen, B.K., P. Berg and J. Jensen

The heritability coefficients of some traits in Polar blue fox population 27 24
Socha, S.

Genetics

ADVANCES IN STATISTICAL METHODS

Session 3

Date	24-aug-99
	8:30 - 12:00 hours
Chairperson	R. Reents (D)

Papers

	G3 no.	Page

Statistical methods in animal breeding: wherefrom and whereto?   1   25
Gianola, D.

Random regression models to describe phenotypic variation in weights of beef cows when age and season effects are confounded   2   25
Meyer, K.

Derivation of multiple trait reduced rank random regression (RR) model for the first lactation test day records of milk, protein and fat   3   26
Mäntysaari, E.A.

The effect of incomplete lactation records on covariance function estimates in test day models   4   26
Pool, M.H. and T.H.E. Meuwissen

Accounting for variance heterogeneity in French dairy cattle genetic evaluation   5   27
Robert-Granié, C., B. Bonaïti, D. Boichard and A. Barbat

Recent developments in linkage and association analysis to search for single genes   6   27
Stricker, C.

Robust Bayesian polygene mapping using skewed student-t distributions and finite polygenic models   7   28
Rohr, P. von and I. Hoeschele

Descent graphs in animal pedigrees: an application   8   28
Schelling, M., R.L. Fernando, N. Künzi and C. Stricker

Genetic improvement of laying hens viability using survival analysis techniques   9   29
Ducrocq, V., B. Besbes and M. Protais

Bayesian estimation of parameters of a structural model for genetic covariances between milk yield in five regions of the USA   10   29
Rekaya, R., K. Weigel and D. Gianola

Genetics

Posters

	G3 no.	Page
Multivariate analysis of Gaussian, right censored Gaussian, ordered categorical and binary traits *Korsgaard, I.R., M.S. Lund, D. Sorensen, P. Madsen and D. Gianola*	11	30
Impact of variation of length of individual testing periods on estimation of variance components of a random regression model for feed intake of growing pigs *Schnyder, U., A. Hofer and N. Künzi*	12	30
Direct and maternal (co)variances in multitrait analysis in a multibreed beef cattle herd *Schoeman, S.J. and G.F. Jordaan*	13	31
The breeding value for direct and maternal effect in the early stage of growth of beef cattle *Pribyl, J., K Šeba and J. Pribylová*	14	31
Sire effects on weaning weight of simmental calves *Lukács, P., F. Szabó, L. Vajda, E. Vörösmarthy, P.J. Polgár and Zs. Wagenhoffer*	15	32
Statistical analysis of Dna polimorphism *Bek, Y. and Y. Tahtalỹ*	16	32

FREE COMMUNICATIONS

Session 4

Date 25-aug-99
9:00 - 12:30 hours
Chairperson I. Olesen (N)

Papers

	Ga4 no.	Page
Accomodation and evaluation of ethical, strategic and economic values in animal breeding goals *Olesen, I., B. Gjerde and A. Groen*	1	33
Stochastic simulation of breeding schemes for dairy cattle *Sørensen, M.K., P. Berg, J. Jensen and L.G. Christensen*	2	33
Prediction of rates of inbreeding in selected populations *Bijma, P. and J. Woolliams*	3	34
Efficient use of test capacity in pig breeding *Serenius, T., M.-L. Sevón-Aimonen and A. Mäki-Tanila*	4	34

Genetics

	Ga4 no.	Page
Genetic parameters, inbreeding depression and genetic trend for litter traits in a closed population of Meishan pigs *Lende, T. van der, J.A.M. van Arendonk, V.D. Kremer and O.L.A.M. de Rouw*	5	35
Genetic correlation for backfat thickness and daily gain measured on boars and gilts raised in different environments *Malovrh, S., H. Brandt, P. Glodek and M. Kovac*	6	35
Genetic variation for traits of commercial importance exists in Danish rainbow trout *Henryon, M. and P. Berg*	7	36
Mapping of QTL affecting functional and type traits in the Dutch Holstein population *Schrooten, C., H. Bovenhuis, W. Coppieters and J.A.M. van Arendonk*	8	36
Mixed model analysis of markers in the growth hormone axis and their associations with production traits in Holsteins *Aggrey, S.E., C.Y. Lin, J.F. Hayes, D. Zadworny and U. Kuhnlein*	9	37
Polygenic inheritance of the bovine blood group systems F, J, R and Z determined by loci on different chromosomes *Thomsen, H., N. Reinsch, N. Xu, C. Looft, E. Kalm, S. Grupe, C. Kühn, G. Brockmann, M. Schwerin, B. Leyhe, S. Hiendleder, G. Erhardt, I. Medjugorac, I. Russ and M. Förster*	10	37
Mapping of quantitative trait loci affecting egg quality in chicken *Honkatukia, M.S., M. Tuiskula-Haavisto, J. Vilkki, D-J. de Koning, N. Schulman and A. Mäki-Tanila*	11	38
Analysis of quantitative trait loci in broilers using a Bayesian mixed model *Kaam, J.B.C.H.M. van, M.C.A.M. Bink, M.A.M. Groenen, H. Bovenhuis and J.A.M. van Arendonk*	12	38

Posters

	Ga4 no.	Page
Potential use of information on identified quantitative trait loci (QTLs) in prediction of response to long-term selection *Satoh, M., K. Ishii and T. Furukawa*	13	39
The effects of missing markers and combination of the number of grandsire and his sires on mapping quantitative trait loci in a small dairy cattle population *Togashi, K., N. Yamamoto, O. Sasaki and A. Nakamura*	14	39
Genotypes of bGH and bPRL genes in relationships to milk production *Chrenek, P., J. Huba, M. Oravcová, L. Hetényi, D. Peskovicová and J. Bulla*	15	40

Genetics

	Ga4 no.	Page
Microsatellite typing of ancient parchment and leather Pfeifer, I., J. Burger and B. Brenig	16	40
Isolation of the porcine four and a half LIM domain protein 1 (FHL 1) Krempler, A., S. Kollers, R. Fries, H. Al-Bayati and B. Brenig	17	41
Mapping and exclusion mapping of genomic imprinting effects in mouse F_2-families Mantey, C., N. Reinsch, G. Brockmann and E. Kalm	18	41
Three-layer selection of bulls: a farmers' choice Beek, S. van der and G. de Jong	19	42
Objectives in dairy cattle improvement in Estonia Pärna, E.	20	42
Sire x ecological region interaction in Bonsmara cattle Nephawe, K.A., F.W.C. Neser, C.Z. Roux, H.E. Theron, J. van der Westhuizen and G.J. Erasmus	21	43
Breed differences, heterosis and maternal effect on weaning weight of beef calves Szabó, F., J. Dohy, K. Szentpáli and J. Tari	22	43
Joint genetic evaluation of purebred and crossbred data under dominance Mielenz, N., L. Schüler and E. Groeneveld	23	44
Estimation of genetic parameters in Hanoverian horses including special combining ability Brade, W. and E. Groeneveld	24	44
Empirical evidence for segregation variance in beef cattle Birchmeier, A.N., R.J.C. Cantet, R.L. Fernando and C.A. Morris	25	45
Inbreeding effects on the parameters of the growth funtion of Iberian pigs Rodrigáñez, J., L. Silió, M.A. Toro and M.C. Rodríguez	26	45
A demonstration project of the EU DG XII biotechnology research programme: characterizing genetic variation in the European pig to facilitate the maintenance and exploitation of biodiversity Alderson, L., M.Y. Boscher, C. Chevalet, B. Coudurier, R. Davoli, B. Danell, J.V. Delgado, P. Glodek, M. Groenen, C. Haley, K. Hammond, D. Milan, L. Ollivier and G. Plastow	27	46
Relationships among subpopulations of Iberian Pig using microsatellites Martínez, A.M., A. Rodero and J.L. Vega-Pla	28	46

Genetics

Posters

	Ga4 no.	Page
Genetic variation of two local Romanian pig breeds assessed using DNA markers *Ciobanu, D., A. Nagy, R. Wales and G.S. Plastow*	29	47
Protein polymorphic system of swine bload serum *Ambrosjeva, E.D. and A.A. Novicov*	30	47
Dinamics of genetic structure in zivilsky pigs *Novicov, A.A., N.I. Romaneko, M.S. Semak and V.N. Ananjev*	31	48
Genetic analysis of two cattle populations using DNA microsatellites *Zamorano, M.J., A. Rodero and J.L. Vega-Pla*	32	48
Genetic similarity for erythrocyte antigens in animals of Holstein *Zheltikov, F.I., V.L. Petukhov and V.G. Marenkov*	33	49
Genetic analysis of synthetic multifertility of midfine-wool Sheep *Marzanov, N.S., B.S. Iolchiev, M.R. Nassiry and V.P. Shikalova*	34	49
Frequency of sister chromatid exchanges in Egyptian water buffalo *Ahmed, S.*	35	50
The analysis of The Holmogor Bulls Lines Differentiation by A DNA-fingerprinting *Yakovlev, A.F., N.V. Dementjeva and V.P. Terletsky*	36	50
A new allele of MC1-R shows a 12 bp insertion in Brown Swiss breeds *Kriegesmann, B., I. Pfeifer, B. Dierkes, A. Krempler and B. Brenig*	37	51
Monitoring of genetic variation of Estonian cattle breeds - temporal changes *Viinalass, H. and S Värv*	38	51
Evolution, improvement and conservation of Simmental breed in Ukraine *Ruban, J.D.*	39	52

QUALITY OF MEAT AND FAT AS AFFECTED BY GENETICS AND NUTRITION

Session 4

Joint session with the commissions on Animal Nutrition and Pig Production (P*+G+N)

Genetics

Date   25-aug-99
                 9:00 - 12:30 hours
Chairperson   C. Wenk (CH)

	PGN4	Page	Page
Papers	1 - 11	LXXIX	301 - 306
Posters	12 - 43	LXXX	307 - 322

BIOLOGICAL SELECTION LIMITS AND FITNESS CONSTRAINTS

Session 5

Joint session with the commission on Animal Physiology (G*+Ph)

Date   25-aug-99
                 14:00 - 17:30 hours
Chairperson   A. Visscher (NL)

Papers   GPh5 no. Page

Long term selection for litter size in mice (>100 generations); correlated
responses and biological constraints   1   52
Vangen, O.

Selection limits and fitness constraints in pigs   2   53
Knap, P.W. and P. Luiting

Effects of growth hormone deficiency on mice selected for increased and
decreased body weight and fatness   3   53
Bünger, L. and W.G. Hill

Selection for litter size and its consequences for the allocation of feed resources -
a concept and its implications illustrated by mice selection experiments   4   54
Rauw, W.M., P. Luiting, R. Beilharz, M.W.A. Verstegen and O. Vangen

Relation between voluntary feed intake of sows during lactation and sow and
litter performance   5   54
Apeldoorn, E.J. and J.J. Eissen

Genetic aspects of disease incidence in fattening pigs   6   55
Steenbergen, E.J. van and A.H. Visscher

Genetics

	GPh5 no.	Page
Quantifying genetic contributions to a dairy cattle population using pedigree analysis *Visscher, P.M., S. Brotherstone and T. Roughsedge*	7	55
Genetic parameters of female and male fertility traits in US Holstein cattle *Weigel, K.A.*	8	56

Posters

	GPh5 no.	Page
Analysis of heterosis in stress susceptibility using DNA markers on mouse chromosome 3 *Brunsch, C., M. Starke, P. Reinecke and G. Leuthold*	9	56
Testing the suitability of the used selection index for Japanese quail (Coturnix japonica) *Richtrová, A., J. Hrouz and D. Klecker*	10	57
Spirometric performances in Belgian Blue calves : environmental factors analysis and genetic parameters estimation *Bureau, F., C. Michaux, J. Coghe, Ch. Uystepruyst, M. -L. van de Weerdt, C. Husson, P.L. Leroy and P. Lekeux*	11	57
Selection response after long-term selection for high and low running activity in mice with special consideration of a selection limit *Renne, U. and M. Langhammer*	12	58
Plasma levels of thyroid hormones and cortisol in stressed Pietrain and Duroc pigs *Rosochacki, S., A. Piekarzewska and J. Połoszynowicz*	13	58

FREE COMMUNICATIONS / IDEAS BOX

Session 6

Date	26-aug-99
	8:00 - 10:00 hours
Chairperson	M. Carabaño (E)

Papers

	G6 no.	Page
Genetic parameters for clinical mastitis, somatic cell score, production, udder type traits, and milking ease in French first lactation Holsteins *Rupp, R. and D. Boichard*	1	59

Genetics

	G6 no.	Page
Bayesian analysis of heritability of liability to clinical mastitis in Norwegian Cattle with a threshold model *Heringstad, B., R. Rekaya, D. Gianola, G. Klemetsdal and K.A. Weigel*	2	59
Genetic analysis of conception rate in heifer and lactating dairy cows *Boichard, D., A. Barbat and M. Briend*	3	60
Estimates of genetic parameters for carcass traits in Finnish Ayrshire and Holstein-Friesian *Parkkonen, P., A.-E. Liinamo and M. Ojala*	4	60
Relationship of body weight and carcass quality traits with first lactation milk production in Finnish Ayrshire cows *Liinamo, A.-E. and M. Ojala*	5	61
Relationships between feed intake in different test periods of potential AI-bulls *Wassmuth, R., H. Alps and H.J. Langholz*	6	61
Genetic improvement of feed efficiency of beef cattle *Arthur, P.F., J.A. Archer, R.M. Herd and E.C. Richardson*	7	62
Genetic background of milk coagulation properties in Finnish dairy cows *Ruottinen, O., T. Ikonen and M. Ojala*	8	62

Posters

	G6 no.	Page
Relationships between milk protein genotype and milk production traits in Jersey breed *Sáblíková, L., O. Jandurova and M. Štípková*	9	63
Association between ß-lactoglobulin genotypes and its content in milk *Futerová, J., L. Sáblíková, J. Kopecny, M. Stípková and O. Jandurova*	10	63
Estimates of repeatability for and factors affecting the milk coagulation traits *Tyrisevä, M., T. Ikonen and M. Ojala*	11	64
Frequency and heritability of supernumerary teats in German Holstein cows *Brka, M., N. Reinsch, E. Kalm and N. Junge*	12	64
Relative efficiency of two methods of selection for feed efficiency: residual feed consumption and feed conversion ratio *Campo, J.L. and J. González*	13	65
Study on postweaning growth of heifers of different breeds on the same condition in Hungary *Wagenhoffer, Zs. and F. Szabó*	14	65

Genetics

	G6 no.	Page
Effects of Crossbreeding on Fertility Traits in a Synthetic Dairy Cattle Population *Stenske, M., G. Seeland and O. Distl*	15	66
Physico-chemical and sensorial evaluation of beef from Polish bulls *Węglarz, A., E. Gardzina, P. Zapletal and J. Szarek*	16	66
Relationship between type traits and production in Polish Black and White sire evaluation *Zarnecki, A.S. and W. Jagusiak*	17	67
Identification of genomic regions linked to atresia coli in cattle *Kempers, A., u. Thieven, S. Neander, C. Drögemüller, J. Pohlenz, O. Distl and B. Harlizius*	18	67
Hereditary determination of cattle to mastitis *Kochnev, N.N.*	19	68
The influence of the genetic factors on immune reactivity and natural resistance in cattle *Marenkov, V.G., V.L. Petukhov and Y.A. Krinskiy*	20	68
Significance of Lys-mic genotyping in the genetic improvement of mastitis resistance in dairy cattle *Pareek, C.S., M. Schwerin and K. Walawski*	21	69
Vital study of Bovine granulosa cells during follicular development by differential staining with acridine orange for detection of apoptosis *Proshin, S.N., T.A. Smirnova, T.I. Kuzmina and A.F. Yakovlev*	22	69
Cytogenetic and its relation to fertility in bovine *Mahrous, K.F., T.M. El-Sayed and I.A. Abd El-Hamid*	23	70
The level of somatic genome mutations in farm animals *Kochneva, M.L., V.L. Petukhov, G.N. Korotkova, T.B. Paramonova and S.G. Kulikova*	24	70
Chromosome mutations in cattle: consequence of the Tomsk Siberian chemical plant (SCP) accident *Masun, S.R., M.L. Kochneva, P.L. Petukhov, S.G. Kulikova, B.L. Panov and N.N. Shipilin*	25	71
Somatic chromosome instability of phenotypically healthy and abnormal calves in different ecological areas *Kulikova, S.G., V.L. Petukhov and M.L. Kochneva and G.N. Korotkova*	26	71
Comparative analysis of the immunological and molecular methods to identify the BLV - positive cows at the different stages of viral infection *Koutsenko, N., D. Shayakhmetov, M. Smaragdov, N.V. Dementieva and T. Starozhilova*	27	72

COMMISSION ON ANIMAL NUTRITION (N)

President
Dr. Ir. Y. van der Honing
Institute of Animal Science and Health
P.O. Box 65
8200 AB Lelystad
The Netherlands
Phone:+31 320 237314
Fax:+31 320 237320
E-mail:Y.van.der.Honing@id.dlo.nl

Secretaries
Prof.dr. J. Ramalho Ribeiro
Estação Zootécnica Nacional /EZN
Fonte Boa
P-2000 Vale de Santarém Ribatejo
PORTUGAL
Phone:+351 043 760202/3/4
Fax:+351 043 760540

G.M. Crovetto
Univ. di Milano /Facoltà Agraria
Istituto di Zootecnica Generali
Via Celoria 2
20133 Milano
ITALY
Phone:+2 70 60 01 59
Fax:+2 70 63 80 83
E-mail:enernet@imiucca.csi.it

NUTRITION AND BREEDING OF CATTLE TO OPTIMIZE USE OF GRASS AND FORAGES FOR MILK AND BEEF PRODUCTION

Session 1

Joint session with the commission on Cattle Production (N*+C)

Date	23-aug-99
	9:00 - 12:30 hours
Chairperson	A. Kuipers (NL)

Nutrition

Papers

	NC1 no.	Page
Economic evaluation of different dairy cow types under heavy roughage rations *Erdin, D., R. Schwager-Suter, D. von Euw and N. Künzi*	1	73
Breeding to optimise use of grass and forages for milk production *Veerkamp, R.*	2	73
Feeding high genetic merit dairy cows in grass based systems *Gordon, F.J., C.P. Ferris and A. Cromie*	3	74
Planning and monitoring herbage growth and utilization to feed grazing animals under uncertain conditions *Duru, M.*	4	74
Grazing managements with Belgian Blue growing bulls before an indoors finishing *Dufrasne, I., A. Clinquart, J.-L. Hornick, C. van Eenaeme and L. Istasse*	5	75
Comparison of Friesian and Charolais x Friesian steers in grass-based production systems *Keane, M.G. and E.G. O'Riordan*	6	75
Use of extensive breeding system for beef production in Italy: the Maremmana case *Gigli, S. and M. Iacurto*	7	76
Use of the automated gas production technique to determine the fermentation kinetics of carbohydrate fractions in maize silage *Deaville, E. and I. Givens*	8	76
Long term effects of a large concentratre decrease for dairy cows. *Brocard, V., J. Kerouanton and D. Le Meur*	9	77
Energy supplementation of high altitude grazed dairy cows: effect on pasture intake, blood and milk parameters as well as nitrogen utilization *Berry, N., F. Sutter, M. Bruckmaier, J.W. Blum and M. Kreuzer*	10	77

Posters

	NC1 no.	Page
Evaluation of nutritive value of different meadow communities in Lower Silesia *Wolski, K., P. Nowakowski and A. Szyszkowska*	11	78
Evaluation of milk performance of Red Angus cows in first three lactations *Makulska, J., A. Węglarz and J. Szarek*	12	78

Nutrition

	NC1 no.	Page
Milk yield and milk composition of buffalo and Friesian cows as affected by supplementation of milk plus El-Feel, F.M.R., A.K.I. Abd El-Moty, AA.A. Abd-El-Hakeam, M.A.A. El-Barody and A.A. Baiomy	13	79
Differential redox properties of meat after biocatalist as quality indexes Matthes, H. -D., V. Pastushenko and H. Heinrich	14	79
The interrelation between economically useful characters and biological characters in ayrshire race cows and use their in the selection Proshina, O.V. and Yu.V. Boykov	15	80
Potentionals for spent *Pleurotus ostreatus* compost ensiling Adamović, M., R. Jovanović, G. Grubić, I. Milenković, Lj. Sretenović and Lj. Stoićević	16	80
Wirtschaftlichkeit der Bullen- und Färsenmast mit Kreuzungstieren aus sechs verschiedenen Fleischrassen und bayerischen Fleckviehkühen Kögel, J., M. Pickl and A. Obermaier	17	81
The Hereford animals of Siberian selection. Gamarnic, N.G., V.G. Antropov and N.V. Konskich	18	81
The vinasse utilization in dairy cattle nutrition Valizadeh, R. and A. Ziaei	19	82
Modelling a cow's energy balance using repeated measurements Schwager-Suter, R., C. Stricker, D. Erdin and N. Künzi	20	82
Crossbreeding of dairy cattle in Kenya. An economic evaluation Kahi, A.K., G. Nitter, J.A.M. van Arendonk, W. Thorpe and C.F. Gall	21	83
Adiposity and winter feeding of dairy heifers Marie, M., V. Thénard and C. Bazard	22	83
Testing stability of performance in young sires? Panicke, L., R. Staufenbiel, O. Burkert, E. Fischer and F. Reinhardt	23	84
An observation of the significance of the relationship between the dry matter intake and body weight and between the DMI and milk production (zield) Mudřik, Z., Z. Nemec and B. Hucko	24	84
Indicators influencing economy of beef cattle breeds breeding in Czech Republic Ježková, A.	25	85
Substitution of protein supplement by urea in diet of fatted steers. Vrzalová, D., A. Krása and J. Třinácty	26	85

Nutrition

	NC1 no.	Page
Feed intake and feed utilization of suckling cows during early lactation *Teichmann, S., R.-D. Fahr, G. von Lengerken and F. Mörchen*	27	86
Study on the relations between nutritive and environment factors for growing-fattening steers *Bunghiuz, C., D. Georgescu, L. Rădulescu and F. Beiu*	28	86
Growth ability of Limousin breed and crossbreds of Czech Pied cattle with beef breeds *Nová, V., F. Louda and L. Štolc*	29	87
Slaughter value of bulls of Czech Pied breed, Limousin breed and their crossbreds *Louda, F., V. Nová and L. Štolc*	30	87
Variability of chemical composition and heavy metal content in commodity milk produced in industrial and agricultural regions *Litwińczuk, A., A. Drozd-Janczak, M. Florek and A. Filozof*	31	88
Possibilities of increasing nutrition quality of milk *Toušová, R., F. Louda and L. Stádník*	32	88
Lifetime milk production of purebred and crossbred Russian Black Pied cows *Karlikov, D.V., A.P. Pyzhov and O.G. Tsvetkova*	33	89
Genetic relationship of Romanian Black and White with different Holstein breeds *Florescu, E. and I. Granciu*	34	89
Genetic variability in docility in German Angus and Simmental cattle *Mathiak, H., M. Gauly, K. Hoffmann, R. Beuing, M. Kraus and G. Erhardt*	35	90
The negative influence of intensive farming on the antioxidants property of meat *Matthes, H. -D., V. Pastushenko, H. Heinrich and Z. Holzer*	36	90
Effect of three contrasting calving patterns on the performance of dairy cows *Ryan, G., S. Crosse and M. Rath*	37	91
Production of quality milk for processing in a grass based system *O'brien, B.*	38	91
Research regarding the intake capacity for hay and green forages in young buffaloes and steers *Georgescu, D., F. Beiu, C. Bunghiuz, G. Macarie and H. Macarie*	39	92
Effect of soyoil on the DM disappearance of high fiber diets from nylon bags *Nasserian, A.A.*	40	92

Nutrition

PHYSIOLOGICAL AND NUTRITIONAL ASPECTS OF GROWTH AND DEVELOPMENT IN NEW BORN ANIMALS

Session 2

Joint session with the commission on Animal Physiology (Ph*+N)

Date	23-aug-99		
	14:00 - 17:30 hours		
Chairperson	S. Pierzynowski (S)		

	PhN2	Page	Page
Papers	1 - 10	XLVII	150 - 154
Posters	11 - 16	XLVIII	155 - 157

FEEDING OF DAIRY GOATS UNDER INTENSIVE MANAGEMENT (JOINT SESSION EAAP + IGA)

Session 3

Joint session with the commission on Sheep and Goat Production (S*+N)

Date	24-aug-99		
	8:30 - 12:00 hours		
Chairperson	P. Morand-Fehr (F)		

	SN3	Page	Page
Papers	1 - 6	LXVIII	253 - 256
Posters	7 - 12	LXIX	256 - 259

QUALITY OF MEAT AND FAT AS AFFECTED BY GENETICS AND NUTRITION

Session 4

Joint session with the commissions on Animal Genetics and Pig Production (P*+G+N)

Nutrition

Date	25-aug-99
	9:00 - 12:30 hours
Chairperson	C. Wenk (CH)

	PGN4	Page	Page
Papers	1 - 11	LXXIX	301 - 306
Posters	12 - 43	LXXX	307 - 322

FREE COMMUNICATIONS

Session 5

Date	25-aug-99
	14:00 - 17:30 hours
Chairperson	H. Hagemeister (D)

Papers

N5 no. Page

The nutritive value of palm kernel meal measured in vivo and using rumen fluid and enzymatic techniques 1 93
O'Mara, F.P., F.J. Mulligan, E.J. Cronin, P. Caffrey and M. Rath

Effects of altering hourly nitrogen supply to the rumen on metabolism in growing lambs fed either barley or sugar beet based diets. 2 93
Richardson, J.M., L.A. Sinclair and R.G. Wilkinson

Comparison of wheat and corn for dairy cows 3 94
Daenicke, R., D. Gädeken and P. Lebzien

Prediction of nutrient rumen turnover from animal and dietary information 4 94
Cannas, A. and P.J. van Soest

Effect of casein supply to the abomasum on post-ruminally starch digestibility and glucose metabolism of dairy cows 5 95
Abramson, S.M., A. Arieli, I. Bruckental, S. Zamwel and Z. Shabi

A comparison of the feed value of triticale, winter wheat and winter rye for growing pigs including the effect of variety, soil and nitrogen fertilisation 6 95
Fernández, J.A., H. Jørgensen and A. Just

The effect of feed mixtures on histopathological changes of mucous membrane of digestive tract 7 96
Nemec, Z., M. Doskočil, Z. Mudrik, A. Kodes, B. Hucko and J. Obadálek

Effect of protein concentration in the diets balanced according to ileal digestible amino acids content on protein utilization in lean pigs 8 96
Fandrejewski, H., St. Raj, D. Weremko and G. Skiba

Nutrition

	N5 no.	Page
Substitution of yellow corn by sorghum in layer ration and the effect of metionine and kemzyme supplementaion on hen performance and egg quality *Hashish, S.M. and A. El-Ghamry*	9	97
Influence of fusariotoxins on the growing performance of broilers and turkeys *Leitgeb, R., H. Lew, R. Khidr, W. Wetscherek, J. Böhm and W. Zollitsch*	10	97
A comprehensive approach of the rabbit digestion : consequences of a reduction in dietary fibre supply *Gidenne, T., V. Pinheiro and L. Falcão e Cunha*	11	98
The effect of amino acid supplement and feeding system in gestation on reproductive performance of sows *Schmeiserová, L., J. Třináctý, L. Klapil and K. Šimeček*	12	98

Posters

	N5 no.	Page
Application of an in vitro incubation technique for estimating utilisable amino acids in feedstuffs for ruminants *Lebzien, P., H. Valenta, R. Daenicke and H. Böhme*	13	99
Nylon bag degradability and mobile nylon bag digestibility of crude protein and amino acids in hydrolysed feather meal. *Harazim, J., L. Pavelek, L. Pavelková, J. Třináctý and P. Homolka*	14	99
Effect of rape lecithins on growth, carcass quality and body fat composition of bulls *Wettstein, H.R., F. Sutter, M.R.L. Scheeder and M. Kreuzer*	15	100
The balance of microelements (Fe and Mn) in green mass and hay of natural pastures in the nutrition of sheep *Muratovic, S.R.*	16	100
The effect of a mineral drink on a dry metter intake, a milk yield and a physiological state of dairy cows *Kudrna, V., P. Lang and P. Mlázovská*	17	101
The effects of different ration energy levels on fattening results of lambs *Ružić, D., D. Negovanović, A. Pavlicevic and M.P. Petrović*	18	101
Effect of length of vegetation season of some maize hybrids on their nutritional characteristics *Loucka, R. and E. Machacová*	19	102

Nutrition

	N5 no.	Page
Estimation of usability of intermediate method AIA in research of feeding value green forage *amaranths* for nutrition of ruminants *Łukaszewski, Z., K. Petkov and K. Antczak*	20	102
Critical size of plastic particles passing digestive tract of cows. *Třinácty, J., P. Homolka and M. Šustala*	21	103
Improved equations for predicting the metabolizable energy (ME) content of feeds for ruminants *Palic, D. and M. Brits*	22	103
Effect of donor feeding in the gas production capacity *Borba, A.E.S., P.M.M.O. Gonçalves, C.F.M. Vouzela and A.F.R.S. Borba*	23	104
Determination of maintenance requirement of limiting amino acids for pigs *Kodes, A., D. Vodehnal, J. Heger and B. Hucko*	24	104
N and P reduction from livestock waste *Yano, F., D. Kondo, H. Osako and S. Kato*	25	105
Ileal digestible amino acid requirements of Belgian lean meat-type pigs *Schrijver, R. de and J. vande Ginste*	26	105
Effect of ensiled corn grain on the hens laying *Stekar, J., E. Tkalčic, A. Holcman and M. Kovac*	27	106
Nutritional evaluation of corn-barley diets supplemented with different levels of enzymes preparation for layers and broiler chicks *Hashish, S.M., G.M. El-Mallah and A.A. El-Ghamry*	28	106
Effect of extrusion of rapeseeds on performance of hens *Zeman, L., M. Lichovníková and D. Klecker*	29	107
Effect of sodium and calcium phosphates on egg production and quality *Klecker, D., M. Lichovníková and L. Zeman*	30	107
Productive efficiency and safety of broiler's diets containing wheat treated with fungistatic drugs *Harazim, J., E. Mareček, L. Pavelek, P. Suchý, I. Herzig, P. Suchý and E. Straková*	31	108
Amino acid pattern and biological protein value of spelt and winter wheats *Chrenková, M., Z. Ceresnáková, A. Sommer, Z. Gálová and L. Dahlstedt*	32	108
Antinutritional substances and their effect on the quality of the rape protein *Kodes, A., Z. Nemec and B. Hucko*	33	109

Nutrition

	N5 no.	Page
Fermentation of soybean meal improves zinc and iron availabilities in rats *Hirabayashi, M., T. Matsui and H. Yano*	34	109
The use of N-alkanes as markers for intake and digestibility determination of hay fed to Islandic toelter horses *Mølbak, L., O. Gudmundson and P.D. Møller*	35	110
Determination of fat-soluble vitamins in animal feeds *Wágner, L., J. Csapó, L. Vincze, G. Szüts, G. Kovács and K. Dublecz*	36	110
Relation between ochratoxin A content in cereal grain and mixed meals determined by the Elisa and HPLC methods and attempt to evaluate their usability for monitoring studies *Jarczyk, A., L. Jędrychowski, B. Wróblewska and R. Jędryczko*	37	111
The effect of anthropogenic activity on fodder crop quality in an area of CR with long-term pollutant load *Wittlingerová, Z., A. Kodes and B. Hucko*	38	111
"Flushing method" to improve the reproductive efficiency in rabbit does *Luzi, F., S. Barbieri, C. Lazzaroni, C. Cavani, M. Zecchini and C. Crimella*	39	112

FUTURE STRATEGIES WITH REGARD TO THE USE OF FEED WITHOUT ANTIBIOTIC FEED ADDITIVES IN PIG PRODUCTION

Session 6
Joint session with the commission on Pig Production (P* + N)

Date	26-aug-99	
	8:00 - 11:30 hours	
Chairperson	K.B. Pedersen (DK)	

	PN6	Page	Page
Papers	1 - 8	LXXXIV	330 - 333
Posters	9 - 14	LXXXV	334 - 336

Commission on Animal Management and Health (M)

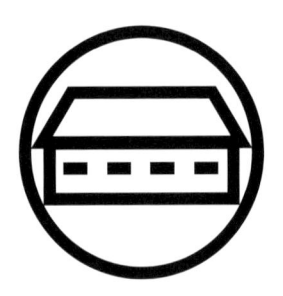

President
Dr.ing. F. Madec
CNEVA Ploufragan
Zoopole - les croix
P.O. Box 53
F-22440 Ploufragan
FRANCE
Phone:+33 296 760130
Fax:+33 296 786861
E-mail:croix.cneva@zoopole.asso.fr

Secretaries
Prof. Dr. O. Szenczi
University of Veterinary Science
István u. 2
1078 Budapest, Pf. 2.
HUNGARY
Phone:+36 1 3426731
Fax:+36 1 3426518
E-mail:OSZENCZI@NS.UNIVET.HU

E. von Borell
Martin Luther Universität
Institute of Animal Breeding & Husbandry
E. Abderhalden Str.
06108 Halle
GERMANY
Phone:+49 345 55 22 332
Fax:+49 345 55 27 105
E-mail:borell@mluitzs1.Landw.uni-halle.de

IMPACT OF WELFARE RESEARCH AND LEGISLATION ON PRODUCTION, ECONOMY AND CONSUMER ACCEPTANCE

Session 1

Date	23-aug-99
	9:00 - 12:30 hours
Chairperson	B. Wechsler (CH)

Management and Health

Papers
	M1 no.	Page

Impact of animal behaviour research on new developments in housing design for farm animals — 1 — 113
Borell, E. von

Relationships between farmers' attitudes, retailers' actions, legislation and welfare research — 2 — 113
Broom, D.M.

Dutch farm animal welfare research in interaction with society, legislation, consumer demand and industry — 3 — 114
Blokhuis, H.J. and G.F.V. van der Peet

The impact of welfare research on legislation and animal housing in Sweden — 4 — 114
Keeling, L.

Direct payments to promote housing systems that are especially adapted to the behavioural needs of farm animals — 5 — 115
Meier, W.

French people's willingness to pay for farm animal welfare — 6 — 115
Latouche, K.

Legislation of farm animal welfare requirements in Slovakia — 7 — 116
Mihina, S., B. Lovas, L. Hetenyi and J. Sokol

Implications of changes to legislative space allowance for performance, aggression and immune competence of growing pigs at different group sizes — 8 — 116
Turner, S.P., S.A. Edwards, M. Ewen and J.A. Rooke

Posters
	M1 no.	Page

Schallanalyse in Ferkelerzeugerbetrieben — 9 — 117
Marquardt, V., D. Schäffer, G. Marx and H. Prange

The prises of the Polish Polar blue foxes skins on international auction — 10 — 117
Socha, S., D. Pomykala, J. Slawon and G. Jezewska

Management and Health

REPRODUCTION TECHNOLOGIES (AI, EMBRYO TRANSFER, CRYOPRESERVATION) AND RISKS OF DISEASE TRANSMISSION (JOINT SESSION EAAP + OIE)

Session 2
Joint session with the commission on Cattle Production (M*+C)

Date	23-aug-99
	14:00 - 17:30 hours
Chairperson	M. Thibier (F)

Papers

	MC2 no.	Page
Prevention of disease transmission by semen in cattle *Wentink, G.H., J.C. Bosch, J.E.D. Vandehoek and Th. van den Berg*	1	118
Prevention of diseases transmission by the use of semen in the porcine AI industry *Leiding, C.*	2	118
Prevention of disease transmission through the transfer in-vivo-derived bovine embryos *Stringfellow, D. and M.D. Givens*	3	119
Embryo transfer in small ruminants: the method of choice for health control in germplasm exchanges *Thibier, M. and B. Guérin*	4	119
The effective control of sanitary risks associated with the production of *in vitro* produced bovine embryos. *Guérin, B., B. Le Guienne and M. Thibier*	5	120
Risks of transmission of spongiform encephalopathies by reproductive technologies in domesticated ruminants *Wrathall, A.E.*	6	120
Application of survival analysis to identify management factors related to the rate of BHV1 seroconversions at Dutch dairy farms *Schaik, G. van, Y. H. Schukken, M. Nielen, A.A. Dijkhuizen and R. Huirne*	7	121

Management and Health
Posters

	MC2 no.	Page
Results of a MAS study in dairy cattle with respect to longevity Freyer, G. and L. Panicke	8	121
The evaluation of mastitis pathogenic agents and their possible influence on consumers health Jemeljanovs, A., J. Bluzmanis, V. Mozgis, V. Jonins and A. Reine	9	122
Microbiological and mycological investigations on reproductive organs of A.I. bulls and cows Jemeljanovs, A.	10	122
Comparison of two staining methods used in morphometry of bull spermatozoa heads by computer-assisted image analysis Yániz, J., F. López-Gatius and P. Santolaria	11	123
Relationship between peripartum climatic conditions and retained placenta in dairy cows Yániz, J. and F. López-Gatius	12	123
Ewe fertility following cervical AI of fresh or frozen semen Boland, M.P., J.P. Hanrahan, P. Duffy and A. Donovan	13	124
Time scheduled insemination of lactating dairy cows Gábor, G.	14	124
Factors influencing conception of cows Stádník, L., F. Louda and R. Toušová	15	125
Cryopreservation in long-term selected lines of laboratory mice Langhammer, M. and U. Renne	16	125
Comparison of viability and acrosome status of boar spermatozoa frozen in mini-or maxi -straws Bali Papp, A., SZ. Nagy, J. Ivancsics, A. Kovacs, J. Dohy and T. Pecsi	17	126
Laparoscopic insemination in sheep Catillo, G., G.M. Terzano, L. Taibi, G. Ficco and G. Noia	18	126
Bovine oocytes recovering by the method of slicking and scaryfication of ovarian cortex Toborowicz, R., J. Żychlińska, J. Buleca and J. Szarek	19	127
In vitro embryos production in cattle Vintila, I., A. Grozea, A. Guler and Gh. Ghişe	20	127

Management and Health

	MC2 no.	Page
Effect of selenium supplementation prepartum on reproductive function of dairy cows at pasture *Erokhin, A.S.*	21	128
MHC and milk production *Kostomakhin, N.M.*	22	128
The immunologic factors in pig production *Akulich, E.G.*	23	129
Research of incident factors and minimization of embryonic mortality of animal at early stages embryogenesis *Karmanova, E.P., I.A.Hakana, M.E. Huobonen and A.E.Bolgov*	24	129
The programmable freezer for use on farm *Moroz, T.A. and A.M. Malinovskiy*	25	130

PHYSIOLOGY OF SUBOPTIMAL GROWTH

Session 3

Joint session with the commission on Animal Physiology (Ph*+M)

Date	24-aug-99
	8:30 - 12:00 hours
Chairperson	J. Blum (CH)

	PhM3	Page	Page
Papers	1 - 6	XLIX	158 - 160
Posters	7 - 9	XLIX	161 - 162

ELECTRONIC IDENTIFICATION

Session 4

Joint session with the commission on Horse Production (M*+H)

Date	25-aug-99
	9:00 - 12:30 hours
Chairperson	R. Geers (B)

Management and Health

Papers

	MH4 no.	Page
The ISO standards for animal Radio Frequency Identification, its present status and future extensions *Jansen, M.B.*	1	130
IDEA project (IDentification Electronique des Animaux): evaluation of the feasibility of a community-wide electronic animal identification system *Ribó, O., M. Cropper, C. Korn, A. Poucet, U. Meloni, M. Cuypers and P. de Winne*	2	131
Electronic identification with passive transponders in veal calves *Lambooy, E., C.E. van 't Klooster, A.C. Smits and C. Pieterse*	3	131
First results about electronic cattle identification in the german part of the european project IDEA (Identification èlectronique des animaux) *Klindtworth, M., G. Wendl, H. Pirkelmann and W. Reimann*	4	132
Comparison of injectable and bolus transponders for the electronic identification of sheep on farm conditions *Caja, G., R. Nehring, C. Conill and O. Ribó*	5	132
The use of electronic identification with horses *Pirkelmann, H.*	6	133
Vergleichende untersuchungen zur kennzeichnung von pferden mit transponder und mit Heißbrand *Pollmann, U.*	7	133
Electronic identification in horses *Søndergaard, E., L.L. Hansen, H. Staun and H. Schougaard*	8	134

Posters

	MH4 no.	Page
An exponential smoothing model in time series analysis of milk electrical conductivity data for the clinical mastitis detection. *Secchiari, P.L., M. Mele and R. Leotta*	9	134

FREE COMMUNICATIONS

Session 5

Date   25-aug-99
        14:00 - 17:30 hours
Chairperson O. Szenci (H)

Management and Health

Papers
M5 no. Page

An emerging problem in the Pig: PMWS (Postweaning Multisystemic Wasting Syndrome)   1   135
Madec, F., E. Eveno, P. Morvan, L. Hamon, D. Mahé, C. Truong, R. Cariolet, E. Albina and A. Jestin

Economic solution of heating the piglets nests   2   135
Stuhec, I., M. Vogrin-Bračič, M. Kovac and S. Malovrh

Pathological changes in the periparturient period induced by fumonisin B1 fed to pregnant sows   3   136
Zomborszky-Kovács, M., Á Tóth, F. Vetési, Á Bata and G. Tornyos

The effects of feeder space allowance and group size on finishing pig welfare   4   136
Spoolder, H., S.A. Edwards and S. Corning

Behaviour of slaughter pigs during separation   5   137
Schäffer, D. and E. von Borell

Production results from sows and their offspring kept outdoors or indoors with respect to their maternal abilities   6   137
Wülbers-Mindermann, M. and B. Algers

Health-control costs and dairy farming systems in western France   7   138
Seegers, H., C. Fourichon, F. Beaudeau and N. Bareille

The meat's thermodynamic parameters correlate with the farming type   8   138
Matthes, H. -D., V. Pastushenko, H. Heinrich and Z. Holzer

Growth and survival of Holstein and Brown calves reared outdoors in individual hutches   9   139
Alpan, O., O. Ertugrul, N. Unal, F. Azeroglu and O. Kaya

Posters
M5 no. Page

Information network - the base for finding economical decisions in the animal production   10   139
Döring, L., H. Saage, R.D. Fahr and G. van Lengerken

Heat production, herbage intake, and heart rate of grazing heifers   11   140
Ando, S. and K. Otsuki

Dominance relationships and mating efficiency in boars kept with gilts in a dynamic service system   12   140
Grigoriadis, D.F., S.A. Edwards and P.R. English

Management and Health

	M5 no.	Page
Sexual courtship of Andalusian stallions in a directed mount *Rodero, E., B. Alcaide, M. Anglada and J.R.B. Sereno*	13	141
Sexual behavior of Andalusian stallions in a directed mount *Rodero, E., J. Cantarero, M. Anglada and J.R.B. Sereno*	14	141
Sexual behavior of Andalusian stallions during their conduction towards a direct mount *Rodero, E., B. Alcaide, C. Fernández Barranquero and J.R.B. Sereno*	15	142
Sexual behavior of the mares directed mount *Rodero, E., J. Cantarero, C. Fernández Barranquero and J.R.B. Sereno*	16	142
Mechanical damage to the body of turkeys raised on different types of floors. *Wójcik, A., J. Sowińska and K. Iwańczuk-Czernik*	17	143
Growth curves of intensively reared ostriches (*Struthio camelus*) in Northern Italy *Sabbioni, A., P. Superchi, A. Bonomi, A. Summer and G. Boidi*	18	143
Vergleich der verkaufsform chinchillapelzen in den jahren 1997/98 *Sulik, M., L. Felska and G. Mieleńczuk*	19	144
Nutzung des internet für effektive forschung in der tierproduktion *Andres, M.*	20	144
Emission burdening of an area in connection with putting an incinerator into operation *Zikova, E. and T. Adamec*	21	145
Effects of selenium on health and productivity of farm animals in Western Pomerania (Poland) *Ramisz, A., J. Malecki and A. Balicka-Ramisz*	22	145
Einfluß der sommerhitze auf ethophysiologische parameter bei Holstein kühen im Holstein Deutschland *Schahidi, R., M. Steinhardt and H.H. Thielscher*	23	146
Effect of type of castration on meat quality in rabbits *Kenawy, M.N.*	24	146

Commission on Animal Physiology (Ph)

President
Dr. M. Bonneau
INRA
Station de Recherches Porcines
35590 Saint-Gilles
FRANCE
Phone:+33 99 285051
Fax:+33 99 285080
E-mail:bonneau@rennes.inra.fr

Secretaries
Dr. T. van der Lende
Department of Animal Breeding
Wageningen Agricultural University
Postbus 338
6700 AM Wageningen
The Netherlands
Phone:+31 317 483972
Fax:+31 317 483962
E-mail:vanderlende@alg.vf.wau.nl

Dr. K. Sejrsen
Danbish Institute of Agricultural Science
Foulum, PO Box 50
8830 Tjele
Denmark
Phone:+45 8999 1513
Fax:+45 8999 1564
E-mail:ks@sh.dk

CELL DIVISION AND PROTEIN SYNTHESIS FOR GROWTH IN ANIMAL TISSUES

Session 1

Date	23-aug-99
	9:00 - 12:30 hours
Chairperson	N. Oksbjerg (DK)

Papers Ph1 no. Page

The importance of muscle cell division in myogenesis and postnatal muscle growth *Rehfeldt, C.*	1	147

Physiology

	Ph1 no.	Page
The importance of cell division in udder development and lactation *Knight, C.H.*	2	147
The components of protein synthesis (intra and intercellular regulation) *Riis, B.*	3	148
Quantitative aspects of protein synthesis in muscle growth *Sève, B.*	4	148
Myogenesis and expression pattern of myogenin and myf-3 in fetal pigs *Christensen, M., N. Oksbjerg, P. Henckel and P.F. Jørgensen*	5	149
Growth and mammary development in gilts provided different energy levels from weaning to puberty *Sørensen, M.T., N. Oksbjerg and M. Vestergaard*	6	149

PHYSIOLOGICAL AND NUTRITIONAL ASPECTS OF GROWTH AND DEVELOPMENT IN NEW BORN ANIMALS

Session 2

Joint session with the commission on Animal Nutrition (Ph*+N)

Date	23-aug-99 14:00 - 17:30 hours
Chairperson	S. Pierzynowski (S)

Papers

	PhN2 no.	Page
Postnatal adaptation of the gastrointestinal tract in neonatal animals: a possible role of milk-borne growth factors *Xu, R.J.*	1	150
Intestinal maturation induced by spermine in young animals *Dandrifosse, G., C. Grandfils, O. Peulen, P. Deloyer and S. Loret*	2	150
Nutritional and hormonal regulation of development in the neonatal pig *Louveau, I., M.J. Dauncey and J. Le Dividich*	3	151
Aspects of GIT motility in relation to the development of digestive functions in neonates *Lærke, H.N., V. Lesniewska, M.S. Hedemann, B.B. Jensen and S.G. Pierzynowski*	4	151

Physiology

	PhN2 no.	Page
The prenatal gastrointestinal tract: hormonal and luminal influences on its functional maturation *Sangild, P.T.*	5	152
Nutrition, metabolism and endocrine changes in neonatal calves *Blum, J.W. and H. Hammon*	6	152
Development of digestive and immunological function in neonates. Role of early nutrition *Kelly, D.*	7	153
Effect of cation-anion difference and EDTA on performance, ruminal fermentation, blood acid-base status and Fe availability in grain-fed calves *Ghodratnama, A., S.L. Scott, J.R. Seoane and G. St-Laurent*	8	153
Contribution to the study of gut hypersensitivity reactions to soyabean proteins in preruminant calves and early-weaned piglets *Dréau, D. and J.-P. Lallès*	9	154
Effect of milk intake level on the development of digestive function in piglets during the first postnatal week *Le Huërou-Luron, I, M.J. Lafuente, F. Thomas, V. Romé and J. Le Dividich*	10	154

Posters

	PhN2 no.	Page
Mid term effect of formula milk on the development of digestive function in piglets *Le Huërou-Luron, I, B. Codjo, F. Thomas, V. Romé and J. Le Dividich*	11	155
Sow metabolism and piglet performance *Lindqvist, A., H. Saloniemi, M. Rundgren and B. Algers*	12	155
Effects of colostrum intake on diarrhoea incidence and growth of neonatal calves *Gutzwiller, A.*	13	156
The dependence between AVP and electrolytes concentration in the blood of calves in neonatal period *Ożgo, M. and W.F. Skrzypczak*	14	156
Cereals as partial replacements of the milk ration in veal calves *Morel, I.*	15	157
Immune response and performance of chickens under different factors *El-Kaiaty, A.M. and F.A. Ragab*	16	157

Physiology

PHYSIOLOGY OF SUBOPTIMAL GROWTH
Session 3
Joint session with the commission on Animal Management and Health (Ph*+M)

Date	24-aug-99
	8:30 - 12:00 hours
Chairperson	J. Blum (CH)

Papers

	PhM3 no.	Page
Genetically caused retarded growth Sellier, P.	1	158
Cytokine-related free radical modulation of reduced growth performance Elsasser, T.H., S. Kahl, T.N. Rumsey and J.W. Blum	2	158
Reduced growth of calves and its reversal by use of anabolic agents Sartin, J.L. and M.A. Shores	3	159
Mechanisms involved in the process of reduced and compensatory growth Hornick, J.-L., C. van Eenaeme, O. Gérard, I. Dufrasne and L. Istasse	4	159
Effects of underfeeding during the weaning period on the growth and metabolism of the piglet Le Dividich, J and B. Sève	5	160
Body weight gain and reduced bovine mammary growth and development: implications for potential milk yields Sejrsen, K., S. Purup and M. Vestergaard	6	160

Posters

	PhM3 no.	Page
Physiological mechanisms involved in the effects of concurrent pregnancy and lactation on foetal growth and mortality in the rabbit Fortun-Lamothe, L., A. Prunier, G. Bolet and F. Lebas	7	161
Comparison of intramuscular fat development in rabbits born from either simultaneously pregnant and lactating does or only pregnant does Gondret, F., L. Fortun-Lamothe and M. Bonneau	8	161
Effect of energy intake on protein deposition in the body of pigs during compensatory growth Fandrejewski, H. and H. St. Raj	9	162

Physiology

BIOLOGICAL SELECTION LIMITS AND FITNESS CONSTRAINTS

Session 5

Joint session with the commission on Animal Genetics (G*+Ph)

Date	25-aug-99
	14:00 - 17:30 hours
Chairperson	A. Visscher (NL)

	GPh5	Page	Page
Papers	1 - 8	XXV	52 - 56
Posters	9 - 13	XXVI	56 - 58

FREE COMMUNICATIONS

Session 6

Date	26-aug-99
	8:00 - 10:00 hours
Chairperson	K. Sejrsen (DK)

Papers
Ph6 no. Page

Interactions between growth of skeletal muscle and IGF-I in pigs of different sex — 1 — 162
Biereder, S., M. Wicke, G. van Lengerken, F. Schneider and W. Kanitz

Effect of divergent selection for body weight on plasma growth hormone concentration in 90 generation mice — 2 — 163
Wirth-Dzięciołowksa, E., B. Reklewska and A. Karaszewska

Plasma growth hormone levels in ewe lambs of two genotypes and association with weight gain and milk yield — 3 — 163
Peclaris, G.M., K. Koutsotolis, E. Nikolaou, G. Kann and A. Mantzios

Effects of Mg^{2+} on prolactin and somatotropin binding to bovine granulosa cells — 4 — 164
Lebedeva, I.Yu., V.A. Lebedev and T.I. Kuzmina

Effect of somatotropin on nuclear status and morphology of bovine cumulus oocytes complexes — 5 — 164
Kuzmina, T.I., T.V. Shelouchina and N.N. Neckrasova

Physiology

	Ph6 no.	Page
Role of prostaglandins in ovulation in the swine *Yamada, Y., H. Kadokawa and Y. Kawai*	6	165
Relationships between follicular fluid composition and follicular/oocyte quality in the mare *Gérard, N., G. Duchamp and M. Magistrini*	7	165
Time of LH peak in oestrus synchronized buffaloes in two different seasons *Borghese, A., V.L. Barile, G. Terzano, A. Galasso, A. Malfatti, O. Barbato and A. Debenedetti*	8	166
Effects of prenatal stress on endocrine and immune responsiveness in neonatal pigs *Kanitz, E., M. Tuchscherer and W. Otten*	9	166
Effects of milking frequency on milk proteins *Sorensen, A., D. Muir and C. Knight*	10	167

Posters

	Ph6 no.	Page
The changes of electrolytes concentration and plasma osmolality in the blood of cows in perinatal period. *Skrzypczak, W.F. and M. Ożgo*	11	167
On immunoprotective effect of cerebral neuropeptides *Kumar, J., A. Karus, T. Schattschneider, M.-A. Kumar, A. Kaljo and Ü. Pavel*	12	168
The role of protein kinase C in realization of effect of prolactin on pig granulosa cells *Denisenko, V.Y. and T.I. Kuzmina*	13	168
Relationship between mitochondrial activity and intracellular stored Ca^{2+} levels in parthenogenetically activated bovine oocytes *Fedoskov, E.D., L.D. Galieva, T.I. Kuzmina and Ju. S. Kirukova*	14	169
Embryo cloning by nuclear transfer: experiences in sheep *Loi, L., S. Boyazoglu, J. Fulka, S. Naitana and P. Cappai*	15	169
Some physiological responses related to dietary magnesium oxide in laying Japanese quail diets *Hamdy, A.M.M.*	16	170

Commission on Cattle Production (C)

President
Dr. C. Thomas
Scottish Agricultural College
Auchincruive
KA6 5HW Ayr
UNITED KINGDOM
Phone:+44 12 92 52 03 31
Fax:+44 12 92 52 12 84
E-mail:C.Thomas@au.sac.ac.uk

Secretaries
Dr. D. Pullar
Meat and Livestock Commission
Winterhill House, Snowdon Drive
P.O. Box 44
MK6 1AX Milton Keynes
UNITED KINGDOM
Phone:+44 1908 844 347
Fax:+44 1908 844 214
E-mail:duncan_pullar@mlc.org.uk

Dr. E. Villa
Associazione Italiana Allevatori
Via G. Tomassetti 9
00161 Rome, ITALY
Phone:+39 06 85 451308
Fax:+39 06 44 293027

NUTRITION AND BREEDING OF CATTLE TO OPTIMIZE USE OF GRASS AND FORAGES FOR MILK AND BEEF PRODUCTION

Session 1

Joint session with the commission on Animal Nutrition (N*+C)

Date	23-aug-99	
	9:00 - 12:30 hours	
Chairperson	A. Kuipers (NL)	

	NC1	Page	Page
Papers	1 - 10	XXX	73 - 77
Posters	11 - 40	XXX	78 - 92

Cattle

REPRODUCTION TECHNOLOGIES (AI, EMBRYO TRANSFER, CRYOPRESERVATION) AND RISKS OF DISEASE TRANSMISSION (JOINT SESSION EAAP + OIE)

Session 2

Joint session with the commission on Animal Management and Health (M*+C)

Date	23-aug-99
	14:00 - 17:30 hours
Chairperson	M. Thibier (F)

	MC2	Page	Page
Papers	1 - 7	XL	118 - 121
Posters	8 - 25	XLI	121 - 130

FREE COMMUNICATIO NS / BUSINESS MEETING

Session 3

Date	24-aug-99
	8:30 - 12:00 hours
Chairperson	S. Gigli (I)

Papers

	C3 no.	Page
Does the milk redoxpotential correlate with the variability of microbes? *Matthes, H. -D., V. Pastushenko and H. Heinrich*	1	171
The effect of feeding intensity and protein supplementation during the last 6 or 12 weeks before parturition on performance of primiparous cows *Mäntysaari, P.*	2	171
Urea milk content in response to different supplementation on grazing dairy cows *González-Rodríguez, A. and O.P. Vázquez Yáñez*	3	172
Changing of milk composition in first ten days of lactation in Black Pied Cattle *Şekerden, Ö., I. Tapkı and M. Şahin*	4	172
The claw horn parameters in Holstein-Friesians in West Siberia *Telezhenko, E.V., V. Petukhov, N.N. Kochnev and A.V. Kosolapikov*	5	173

Cattle

	C3 no.	Page
Comparison of leg injuries and behaviour in dairy cows kept in cubicle systems with straw bedding or soft lying mats *Wechsler, B., J. Schaub, K. Friedli and R. Hauser*	6	173
Comparison of free stall cattle mattresses in a preference test *Sonck, B., V. Vervaeke and J. Daelemans*	7	174
Types estimated by using of linear scoring and measurements of dairy cows *Püski, J., I. Györkös, A. Gáspárdy, S. Bozó and E. Szücs*	8	174
Survey of Albanian Dairy Feed Resources *Kellems, R.O. and R. Balogh*	9	175
The adaptation of brown Swiss and Simmental cattle originated Switzerland to Kazova State Farm in Turkey *Şekerden, Ö.*	10	175
Efficiency of imported Holstein heifers *Urban, F., J. Bouska and M. Štípková*	11	176
Evaluation of lifetime production of top cows in different dairy breeds *Horvai Szabó, M., J. Dohy and G. Holló*	12	176
Effect of growth rate on tenderness of meat from heifer calves *Therkildsen, M. and M. Vestergaard*	13	177
Investigation of a polymorphism in the growth hormone (GH) gene, plasma concentrations of GH and IGF-1 and carcass traits in Friesian bulls *Grochowska, R., A. Lunden, L. Zwierzchowski, M. Snochowski, J. Oprzadek and E. Dymnicki*	14	177
Dynamic developement of the deposition of energy and nutrients in the carcass during growth and meat quality of different cattle breeds *Ender, K., H.J. Papstein, M. Gabel and B. Ender*	15	178
Reducing cross-suckling in calves by the use of a gated feeding stall *Weber, R. and B. Wechsler*	16	178
Meat production in pure and cross-bred Piemontese cattle *Lazzaroni, C., D. Biagini, G. Toscano Pagano and M. Iacurto*	17	179
Effects of a restricted breeding season versus annual repartition of calvings on Belgian Blue suckler beef heifers reproduction performance *Behr, V. de, O. Gérard, M. Diez, J.-L. Hornick and L. Istasse*	18	179

Cattle

Posters

Title	C3 no.	Page
Some aspects regarding water intake of milking buffalo cows *Beiu, F., L. Rădulescu and E. Bolocan*	19	180
Effect of propylene glycol administration on blood metabolites on periparturient Holstein Dairy cows *Yi, Z., R.O. Kellems and L. Roeder*	20	180
Cattle SOD and GPD Isoenzymes, GH, PRL and IGF-1 *Karus, A. and V. Karus*	21	181
Prediction of the main productive and reproductive traits of calves *Shabla, V.P.*	22	181
Method of early prognostication of cattle milk productivity *Lebengartz, Ya. Z.*	23	182
Development of behaviour in Holstein-Friesian cows and calves *Györkös, I., K. Kovács, E. Szücs, G. Gábor, G. Borka, K. Bölcskey and J. Völgyi Csík*	24	182
The problem of Multifetal ability in dairy cattle husbandry *Makeeva, T.V., N.S. Ufimtseva and V.I. Ustinova*	25	183
Twin calvings in Black-and-White cows of West Siberia *Khimich, N.G. and N.N. Nesterenko*	26	183
The relationship between visually scored foot and leg traits and some foot disorders in West Siberian dairy cows *Eryomenko, V.V., E.V. Telezhenko and N.N. Kochnev*	27	184
Effect of housing and climate on the intensity of gorwth in heifers *Györkös, I., K. Kovács, E. Szücs, G. Borka, K. Bölcskey and G. Gábor*	28	184
Growth curve estimation in the Romanian Spotted heifers and Red Holstein crossbreds and some factors of influence *Stanciu, G., L.T. Cziszter and S. Acatincăi*	29	185
Milking routine has an effect on udder health and milk quality of cows *Kiiman, H. and O. Saveli*	30	185
Thermographic study of milker's thermal comfort during milking in milking parlour *Knížková, I., P. Kunc and M. Koubková*	31	186
The changes of teat surface temperature during milking with 45 and 40 kpa *Kunc, P., N. Knizkova and M. Koubková*	32	186

Cattle

	C3 no.	Page
Sanitary conditions of udders and the yield, composition and physicochemical properties of milk from cows imported from France and Germany *Puchajda, Z., M. Czaplicka, E. Radzka and R. Szatkowski*	33	187
Changes in the level of saturated and non-saturated fatty acids in cow milk as a effect of various forms of subclinical mastitis *Gardzina, E., A. Węglarz, J. Makulska and J. Szarek*	34	187
Interaction of daily milk records, parity and stage of lactation with SCC in Holstein-Friesian cows *Dorner, Cs., A. Gáspárdy, L.T. Cziszter and E. Szücs*	35	188
Estimation of relations between fat and protein content in Black and White cows at the first lactation *Gnyp, J., J. Trautman, P. Kowalski and T. Małyska*	36	188
Length of intercalving period and effectiveness of milk production in high milk yielding herds *Zachwieja, A., A. Hibner, J. Juszczak and R. Ziemiński*	37	189
Effects of calving interval and extended lactation length on milk production of Holstein cows raised under Mediterranean conditions *Koç, A., A.E. Okan, H. Akçay and M. Ilaslan*	38	189
Beef meat production with Belgian Blue bulls: comparison of production systems *Behr, V. de, M. Kerrour, J.-L. Hornick, M. Evrard, A. Clinquart, C. van Eenaeme and L. Istasse*	39	190
Genetic comparison of top Holstein bulls and their progeny in Hungary *Jánosa, Á. and J. Dohy*	40	190
The determination of the general and specific breeding values of Holstein-Friesien and Hungarian Simmental species concerning the changes of weight and figure *Polgár, J.P. and F. Szabó*	41	191
Phenotypic correlations between levels of fatty acids in cow milk *Gardzina, E., A. Węglarz, J. Makulska and J. Szarek*	42	191
Protein polymorphism in milk from Black-and-White cows kept in Poland and Lithuania *Litwińczuk, A., Z. Litwińczuk, M. Tumienie and J. Barłowska*	43	192
Genetic trends of the ukrainian Black-and-White dairy breed for productive and reproductive traits *Danshin, V.*	44	192

Cattle

	C3 no.	Page
Effect of genetic and environmental factors on milk production of crossbred cows *Ruban, S.*	45	193
Adaptation of young cows imported from France to the conditions of north-eastern Poland *Puchajda, Z., M. Czaplicka, A. Szymańska and W. Czudy*	46	193
Effect of breeds of the world on breeds of cattle small in number *Saveli, O. and U. Kaasiku*	47	194
Analysis of testing series in the Retinto Beef Cattle Improvement Plan *Álvarez, F., F. Delgado, M. Valera, A. Molina and A. Rodero*	48	194
Evaluation of type classification in Limousin breed *Horvai Szabó, M., J. Dohy and G. Holló*	49	195
Association of growth hormone, λ-casein and β-lactoglobulin gene polymorphism with meat production traits in Friesian bulls. *Oprzadek, J., L. Zwierzchowski and E. Dymnicki*	50	195
Characteristics of browth and reproductive ability in Simmentalized and Hereford cattle of different ecogenesis *Nezavitin, G., A.A. Permyakov and N.B. Zakharov*	51	196
The future model of high quality beef production in herds of milk cows on the way of emploing commercial croosing and transfering embryos of beef breeds *Chmielnik, H. and A. Sawa*	52	196
Carcass variability in seven Spanish beef breeds *Piedrafita, J., R. Quintanilla, C. Sañudo, J.L. Olleta, M.M. Campo, B. Panea, M.A. Oliver, X. Serra, M.D. Garcia-Cachan, R. Cruz-Sagredo, K. Osoro, M. Oliván, M. Espejo and M. Izquierdo*	53	197
Relationships between carcass characteristics and ultrasonography measurements in Belgian Blue double muscle females: preliminary results *Evrard, M., J.-L. Hornick, A. Bielen, V. de Behr and L. Istasse*	54	197
Effect of finishing system on fattty acid composition and tocopherol content of Simmental steers *Schwarz, F.J., M. Timm, C. Augustini, K. Voigt, M. Kirchgeßner and H. Steinhart*	55	198
Effect of castration on the fatty acid profile and the incidence of yellow fat in bovine adipose tissue *Dufey, P.A.*	56	198

Cattle

	C3 no.	Page

Relationship between SEUROP conformation and fat grade and composition of carcasses of Belgian Blue slaughter bulls   57   199
Voorde, G. van de, S. de Smet, M. Seynaeve and D. Demeyer

Relationship between SEUROP-grading, genotype and meat quality in Belgian Blue slaughter bulls   58   199
Smet, S. de, E. Claeys, D. van den Brink, G. van de Voorde and D. Demeyer

Slaughtering value and quality of meat in Polish BW cattle with regard to macro and microelement content   59   200
Litwińczuk, A., Z. Litwińczuk, J. Barłowska and M. Kędzierska

Histological structure of skin tissue of crossbreeds from Black-and-White cows x Italian beef bulls crossed with Aberdeen Angus   60   200
Zapletal, P., J. Szarek and A. Węglarz

The influence of interbreed crossing on the quality of hides in cattle husbandry   61   201
Zakharov, N.B.

Évaluation des intrevalles velage-velage des vaches Limousines dans une élevage hongroise   62   201
Horvai Szabó, M., S. Balika, L. Gulyás, A. Kovács and J. Tözsér

Belgium Blue cattle growth and reproduction as influenced by management   63   202
Gérard, O., M. Evrard, V. de Behr, I. Dufrasne and L. Istasse

Utility crossing using Belgian White Blue and Charolais semen in Hungarian Grey herds   64   202
Bölcskey, K., I. Bárány, I. Bodó, S. Bozó, I. Györkös, A. Lugasi and J. Sárdi

Commercial crossing with Belgian White Blue semen in Hungarian Simmental herds   65   203
Bölcskey, K., I. Bárány, S. Bozó, I. Györkös and J. Sárdi

Comparison of three multiphasic growth models for bulls of Czech Pied Cattle   66   203
Nešetřilová, H. and J. Pulkrábek

Double muscle cattle: Piemontese (PD) x Friesian (FR) and Blanc-Blue-Belge (BBB) x Friesian (FR) young bull crosses in Italy   67   204
Iacurto, M., C. Lazzaroni, S. Failla and S. Gigli

Meat productivity in cattle breeds of different ccogenesis   68   204
Permyakov, A.A., N.B. Zakharov, A.G. Nezavitin and O.A. Ivanova

Influence of calving season and age of dams on rearing results of beef calves   69   205
Litwińczuk, Z., W. Zalewski, P. Jankowski, P. Stanek and B. Kuryło

Cattle

	C3 no.	Page
The distribution of lactate dehydrogenase isocomponents in blood serum of bulls fattening with antidotes *Buleca, J., J. Szarek, D. Magic, J. Pavlík, Z. Vilinská and J. Mattová*	70	205
Effect of slaughtering age on carcass traits of "Tipo cebon" animals with the Rubia Gallega breed *Monserrat, L., A. Varela, J.A. Carballo and L. Sanchez*	71	206
The use of video image analysis in grading of beef quality. *Sakowski, T., M. Słowiński and J. Cytowski*	72	206

FUTURE ROLE OF DUAL PURPOSE CATTLE
Session 4

Date 25-aug-99
9:00 - 12:30 hours
Chairperson Ch. Böbner (CH)

Papers

	C4 no.	Page
Some thoughts about the development of dual purpose breeds-history, presence and future *Averdunk, G.*	1	207
Total merit indices in dual purpose cattle *Sölkner, J., J. Miesenberger, R. Baumung, C. Fuerst and A. Willam*	2	207
Les races à deux fins en France : situation actuelle, principaux systèmes d'exploitation, spécificités de leurs programmes de sélection *Arnaud, F., N. Bloc, J. Pavie and J.L. Reuillon*	3	208
Use of Multiple-trait Across Country Evaluation (MACE) procedures to estimate genetic correlations in dual purpose Austrian Simmental *Druet, T., J. Sölkner and N. Gengler*	4	208
Ecological Total Merit Index for an Austrian dual purpose cattle breed *Baumung, R. and J. Sölkner*	5	209
Future role of Simmental cattle in Austria: views of breeders and decision-makers *Gierzinger, E., A. Willam, J. Sölkner and C. Egger-Danner*	6	209
Relationships of polledness to production traits in the German Fleckvieh population *Lamminger, A., O. Distl, H. Hamann, G. Röhrmoser, E. Rosenberger and H. Kräußlich*	7	210

Cattle

Posters

	C4 no.	Page
Discussing lines of Simental bulls imported to Poland Trautman, J., J. Gnyp, P. Stanek and J. Tarkowski	8	210
Difference in Economy of Czech Spotted and Holstein Breeds Šafus, P. and J. Pribyl	9	211
The effectivity of cows of dual purpose in comparison with dairy breed Kudrna, V., J. Pribyl and P. Lang	10	211
A crossbreeding system for cattle improvement in developing countries Bhuiyan, A.K.F.H., G. Dietl and G. Klautschek	11	212
Milk yield of crosses of Slovak Simmental and Holstein Red and White breeds in dependence on sire's breed Huba, J., M. Oravcová, L. Hetényi, D. Peskovicová and P. Polák	12	212
In vivo carcass value evaluation of Simmental bulls by ultrasonographic method Polák, P., T. Sakowski, K. Słoniewski, L. Hetényi, J. Huba, D. Peskovicová and E.N. Blanco Roa	13	213
Frequency of umbilical hernias in German Fleckvieh and their effect on meat performance Herrmann, R., J. Utz, E. Rosenberger, R. Wanke, K. Doll and O. Distl	14	213
Inheritance and Introgression of Polledness in the German Fleckvieh Population Hamman, H., A. Lamminger and O. Distl	15	214
Comparaison of Holstein and Meuse Rhine Ysel lactation curves with three mathematical models Félix, A., J. Detilleux and P.L. Leroy	16	214

MANAGEMENT TOOLS TO IMPROVE THE PHYSICAL AND FINANCIAL PERFORMANCE OF LIVESTOCK FARMS

Session 5

Date   25-aug-99
                14:00 - 17:30 hours
Chairperson   R. Palmer (USA)

Papers

	C5 no.	Page
Herd health schemes - problems and potential Wassell, T.R., B.R. Wassell and R.J. Esslemont	1	215

Cattle

	C5 no.	Page
Quality monitoring of milk recording *Galesloot, P., G. de Jong and G. van de Boer*	2	215
A dairy farm budget program as a tool for planning and evaluating *Mandersloot, F., A.T.J. van Scheppingen and J.M.A. Nijssen*	3	216
Management systems to improve the production and financial performance of dairy farms *Palmer, R.W.*		359
Modelling lactation curves of Holstein Friesian cows on large scale farms in Malawi *Chagunda, M.G., Z. Guo, E.W. Bruns and C.B.A. Wollny*	4	216
Field performance recording in smallholder dairy production systems in India *Schneider, F., M.R. Goe, C.T. Chacko, M. Wieser and H. Mulder*	5	217
Etude du stress thermique dans les conditions tunisiennes à travers le calcul de l'index température-humidité *Bouraoui, R., A. Madjoub and M. Djemali*	6	217
INFOLEITE - Educational Programs, Research and Sustainability within a Dairy Production System *Guilhermino, M.M.*	7	218

Posters

	C5 no.	Page
Breeding efficiency in evaluation of reproductive performance of dairy cows *Szücs, E., K. Bódis, G. Látits, J. Tözsér, I. Györkös and A. Gáspárdy*	7	218
Application of software programme „Herd hierachy" (for Windows) for beef cattle herds kept on pasture *Dobicki, A. and K. Chudoba*	8	219
Dairy marketing trends in the suburban perimeter of Kenitra city, in the heart of the irrigated zone of Northern Morocco *Sraïri, M.T.*	9	219
The role of the very low somatic cell count on clinical mastitis: a review *Suriyasathaporn, W. and Y.H. Schukken*	10	220
The effect of various milk feeding periods on growth performance of Black Pied calves *Şekerden, Ö. and M. Şahin*	11	220

Commission on Sheep and Goat Production (S)

President
Dr. D. Croston
Meat and Livestock Commission
Winterhill House, Snowdon Drive
P.O. Box 44
MK6 1AX Milton Keynes
UNITED KINGDOM
Phone:+44 1908 844 366
Fax:+44 1908 692 856
E-mail:david_croston@mlc.org.uk

Secretaries
Dr. J. Folch
Unitad de Technologia en Producí
Dep. de Agricultura, Ganaderia y
Monta ana 176
50080 Zaragoza
SPAIN
Phone:+34 76 578338
Fax:+34 76 575501
E-mail:jose@mizar.csic.es

Dr. M. Schneeberger
Swiss Sheep Breeders' Association
P.O. Box 76
3360 Herzogenbuchsee
SWITZERLAND
Phone:+41 62 956 6868
Fax:+41 62 956 6879
E-mail:schneeberger@pop.agri.ch

SHEEP AND GOAT PRODUCTION IN WET MOUNTAIN AREAS

Session 1

Date	23-aug-99
	9:00 - 12:30 hours
Chairperson	M. Schneeberger (CH)

Papers   S1 no. Page

Mountain sheep farming in the British isles   1   221
Waterhouse, A.

Sheep and Goat

	S1 no.	Page
Sheep and goat production in Norwegian wet mountain areas Ådnøy, T., G. Steinheim and L.O. Eik	2	221
Sheep and goat production in wet mountain areas of Switzerland Lüchinger-Wüest, R., A. Zaugg and M. Schneeberger	3	222
Sheep production systems under the hard conditions in alpine regions in Austria Ringdorfer, F.	4	222
Attempts of describing the influence of some climatic conditions on the effects of sheep breeding production in mountainous regions of Poland Niznikowski, R. and J. Ciuruś	5	223
Sheep raising in wet mountain areas of Czech Republic : a three-level system Matlova, V.	6	223
Alpine lambs out of optimal reproduction and grazing techniques Emler, K., C. Marguerat, H. Leuenberger and N. Künzi	7	224

Posters

	S1 no.	Page
Use of N-alkane technique to determine herbage intake of sheep from natural mountain pasture in Sudetes. Nowakowski, P., K. Aniołowski, K. Wolski and A. Cwikla	8	224
Small ruminants grazing as a management tool for controling bush invasion of natural mountain pastures Cwikla, A., J. Gawęcki, W. Łuczak and P. Nowakowski	9	225

FREE COMMUNICATIONS

Session 2

Date 23-aug-99
      14:00 - 17:30 hours
Chairperson M.L. Puntila (Fin)

Papers
Milk production

	S2 no.	Page
The effect of Nigella Sativa supplementation on sheep milk composition and cheese manufacture Abd El-Razek, S.T. and M.A.A. El-Barody	1	225

Sheep and Goat

Posters
Sheep breeding

	S2 no.	Page
Comparison of matings using Moroccan Timahdit and D'man purebreds, first and terminal crosses. 1. Ewe productivity, lambs survival and growth performances *El Fadili, M., C. Michaux, J. Detilleux and P.L. Leroy*	2	226
Carcass, meat and fat quality of Suffolk and Charmoise sired crossbred lambs compared with pure-breds of three common types of German breeds *Scheeder, M.R.L., C. Sürie and H.J. Langholz*	3	226
Mapping quantitative trait loci causing the muscular hypertrophy of Belgian Texel sheep *Marcq, F., J.M. Elsen, J. Bouix, F. Eychenne, M. Georges and P.L. Leroy*	4	227
Repeatability of the breeding value of growth traits in the Merino sheep. *Sierra, A.C., F. Delgado, A. Rodero, A. Molina, F. Barajas and C. Barba*	5	227
Sire breed effects on glucose tolerance and plasma metabolite levels in Merino cross lambs *Bray, A.R., D. O'Connell and S.R. Young*	6	228
Réponse à la sélection d'un élevage ovin à viande dans les conditions du semi-aride *Bedhiaf, S., A. Ben Gara, M. Ben Hamouda and M. Djemali*	7	228
Genetic evaluation of sheep to improve carcass quality in Ireland *Murphy, O.J., E.E. Wall, E.J. Crosby, D.L. Kelleher and V. Olori*	8	229
Using of French Alpine and Nubian breeds to improving performance of Egyptian local Baladi *Sallam, M.T., K.M. Marzouk, A.A. El Hakaam and M.M. Abd Alla*	9	229
Phenotypic and genotypic variability of slaughter indexes in interbreeding *Mirzabekov, S.Sh. and M.A. Yermekov*	10	230
Genotypic differences of southern Kazakh merinoes of Merkensky intrabreed type *Mirzabekov, S.Sh.*	11	230

Management

Role of small holder in improvement of management of small ruminant in Egypt *Al-Khbeer, A.M.S. and K.M. Marzouk*	12	231
Improvement of sheep and goat raising - development project for the Republic of Mali, West Africa *Fantova, M. and V. Matlova*	13	231

Sheep and Goat

	S2 no.	Page
A new lamb feeding method Filya, I. and A. Karabulut	14	232

Milk production

Observations on the appearance of mastitis in milked Merino ewes and in their crossbreds with prolific breeds Pakulski, T., B. Borys and M. Osikowski	15	232
Milk fat composition in goats representing different genetic variants of α S1 casein Reklewska, B., Z. Ryniewicz, M. Góralczyk, A. Karaszewska and K. Zdziarski	16	233
The growth and reproductive characteristics and milk yield of Karakaş sheep in rural farm conditions Gökdal, Ö., M. Bingöl, A. Çivi, Y. Aşkın and F. Cengiz	17	233
Milk protein polymorphism in Portuguese sheep breeds: αS1-casein and β-lactoglobulin Ramos, A.M., P. Russo-Almeida, A. Martins, F. Simões, J. Matos, A. Clemente and T. Rangel-Figueiredo	18	234
The influence of different lymphocytes subpopulations in milk on the health state of udder in sheep Świderek, W.P., A. Winnicka, Wł Kluciński and K.M. Charon	19	234
The milk production of Finnish landrace sheep and their crosses with meat-wool breeds Nassiry, M.R., V.P. Shekalova and E.A. Karasov	20	235
The effect of ration's maize gluten meal -in substitution of soy bean meal- on yield and composition of ewes' milk in early lactation Liamadis, D. and Ch. Milis	21	235
Correlation between the udder health state, its dimensions and milk productivity in the milking hybrids ewes F_1 East Friesian x Polish Merino. Mroczkowski, S., B. Borys and D. Piwczyński	22	236

Meat production and quality aspects

Competitiveness of lamb meat quality in Hungary Molnár, B.	23	236
A study of performance traits of Charolaise sheep imported from France to Poland Czarniawska-Zajaç, S. and W. Szczepański	24	237

Sheep and Goat

	S2 no.	Page
The influence of weaning and movement to a slaughter-house on the level of cortisol, glucose and haematocrit in blood of lambs *Sowińska, J., H. Brzostowski and Z. Tański*	25	237
The influence of age at castration on carcass quality in sheep *Süss, R., U.E. Mahrous and E. von Borell*	26	238
Effect of castration method and age on behaviour and cortisol in sheep *Borell, E. von, U.E. Mahrous and R. Süss*	27	238
Relationship between SEUROP grading, composition and value of lamb carcasses *Voorde, G. van de, S. de Smet and J. Depuydt*	28	239
Ovine omental and subcutaneous adipocytes display depot and breed differences in culture and in vivo *Soret, B., A. Arana, P. Eguinoa, J.A. Mendizabal and A. Purroy*	29	239
An investigation of fatty acid composition and meat quality in lambs from different breed and production system backgrounds *Kurt, E., J.D. Wood, M. Enser, G.R. Nute, L.A. Sinclair and R.G. Wilkinson*	30	240
Quality of lamb's meat and fat as affected by crossing of Polish Merino with prolific and meat breeds *Janicki, B., B. Borys, M. Osikowski and E. Siminski*	31	240
Growth, carcass and fleece quality traits in Swedish sheep breeding *Näsholm, A.*	32	241
Field testing of progeny of breeding rams on growth and slaughter value *Shaker Momani, M., I. Šáda, F. Vohradsky and Ghassan Agil*	33	241
Conformation data of different Tsigai types in Hungary *Gáspárdy, A., F. Eszes, L. Jávorka and T. Keszthelyi*	34	242
The carcass traits of Finnish Landrace sheep and their crosses with Meat - wool breeds *Nassiry, M.R. and V.P. Shekalova*	35	242
The weaning weight and measurement traits of Finnish Landrace and their crosses with Meat-wool breeds *Nassiry, M.R. and V.P. Shekalova*	36	243
The growth and the slaughter value of Pomeranian ewes and their crossbreds with meat rams *Tański, Z., H. Brzostowski and J. Sowińska*	37	243

Sheep and Goat

	S2 no.	Page
Bestimmung der körperzusammensetzung bei lebenden Merinoschafen mit hilfe des computertomographen *Pászthy, G. and A. Lengyel*	38	244
The meat quality of Pomeranian and Ile de France rams and their crossbreds *Brzostowski, H., Z. Tański and J. Sowińska*	39	244
Effect of muscle location and sex on the evaluation of meat quality in sheep *Heylen, K., R. Suess and G. von Lengerken*	40	245
Meat production characteristics of three goat genotypes kept under intensive (stable) and extensive (biotope) feeding conditions *Haumann, P., H. Snell and E.S. Tawfik*	41	245
Contemporary orientation of utility type of breeding of sheep in Czech Republic *Štolc, L., L. Nohejlová, V. Dřevo and V. Nová*	42	246

Nutrition

Hay and silage in the feeding of pregnant and lactating ewes *Sormunen-Cristian, R.*	43	246
Effect of selenium and zinc supplement on fattened lambs *Kuchtík, J., G. Chládek, V. Koutník and M. Hošek*	44	247
Nutrient consumption and body mass development of suckling goat kids of the production genotypes of milk, meat and fibre *Snell, H. and E.S. Tawfik*	45	247
Effect of weaning age and ration energy level on performance, carcass quality and growth rate of quadrolocular stomach of fattening lambs *Liamadis, D., S. Milioudis, A. Hatzikas and Ch. Milis*	46	248
Effects of dietary by-pass glucose on plasma metabolite concentrations in lambs *Abbas, S., T. Matsui and H. Yano*	47	248

Reproductive physiology

Conception rate and embryonic mortality in Booraoola * German Mutton Merinos depending on the time of mating in relation to ovulation *Kaulfuß, K.-H. and S. Moritz*	48	249
Synchronization of oestrus in goats: Dose effect of progestagen *Greyling, J.P.C. and M. van der Nest*	49	249
An attempt to find an optimum time for mating in Anglo Nubian goats in and outside breeding season basen on luteinizing hormone (LH) concentration *Udała, J., B. Blaszczyk, M. Baran, K. Romanowicz-Barcikowska and B. Barcikowski*	50	250

Sheep and Goat

	S2 no.	Page
Der Einfluß von Selenzusatz in monatlichen Abständen auf die Spermaqualität von Schafböcken im Verlauf des Jahres *Seremak, B., J. Udała and B. Lasota*	51	250
Dynamics of estradiol and progesterone blood concentration changes in Anglo-Nubian goats with stimulated ovulation in and outside breeding season *Udała, J. and B. Błaszczyk*	52	251
The prolactin blood level in Anglo-Nubian goats with synchronised oesrtrus in and outside breeding season *Błaszczyk, B. and J. Udała*	53	251
Performances Zootechniques á la Jeunesse Ovin, (l' anné deuxieme), en Fonction de Fertilisation *Scurtu, I.*	54	252
Nutritional evaluation of complex phosphate (Ca, Mg and Na) in growing lambs *Meschy, F.*	55	252
Mean values of milk yield and its components in Awassi x Barki crossbred ewes *El Shahat, A.A. and Y.S. Ghsnem*	56	253

FEEDING OF DAIRY GOATS UNDER INTENSIVE MANAGEMENT (JOINT SESSION EAAP + IGA)

Session 3

Joint session with the commission on Animal Nutrition (S*+N)

Date	24-aug-99
	8:30 - 12:00 hours
Chairperson	P. Morand-Fehr (F)

Papers

	SN3 no.	Page
The energy and protein requirements of lactating goats - the AFRC report *Sutton, J.*	1	253
Recent progress in the assessment of mineral requirements of goats *Meschy, F.*	2	254
Concentrate supplies of dairy goats on pasture *Fedele, V. and Y. Le Frileux*	3	254

Sheep and Goat

	SN3 no.	Page
Influence of the intensity of forage production on nitrogen nutrition of dairy goats *Daccord,*	4	255
Relationship between feeding and goat cheese quality: present days problems *Morand-Fehr, P., R. Rubino and Y le Frileux*	5	255
Effects of recombinant bovine somatotropin administration to lactating goats *Kyriakou, K., S. Chadio, G.P. Zervas, C. Goulas and J. Menegatos*	6	256

Posters

	SN3 no.	Page
The influence of stimulating level of feeding Carpathians goats on their milk production *Tafta, V. and C. Neacsu*	7	256
Effect of rumen-protected methionine and lysine in diets with different protein content on N balance of lactating goats *Rapetti, L., L. Bava, A. Sandrucci, A. Tamburini, G. Galassi and M.G. Crovetto*	8	257
Milk performance and milk composition of goats under conditions of restriction as well as realimentation of energy and nutrient supply *Mantzke, V., H. Münchow and S. Dündar*	9	257
Investigations of rumen and intermediary metabolism of goats under conditions of restriction as well as realimentation of energy and nutrient supply *Münchow, H., V. Mantzke and S. Dündar*	10	258
Effect of physiological stage on the feeding behaviour of dairy goats fed at the trough *Abijaoudé, J.A., P. Morand-Fehr, J. Tessier and Ph. Schmidely*	11	258
Use of Decoquinate to improve the growth of female goats and milk production of one year old goats *Morand-Fehr, P., A. Richard, J. Tessier and J. Hervieu*	12	259

GENETIC RESISTANCE TO DISEASE AND PARASITES - ALTERNATIVES TO CHEMOTHERAPY

Session 4

Date	25-aug-99
	9:00 - 12:30 hours
Chairperson	M. Stear (UK)

Sheep and Goat

Papers

	S4 no.	Page
Genetic resistance in sheep to scrapie, Salmonella and gastrointestinal nematodes *Elsen, J.M., F. Lantier, J. Bouix, J. Vu Tien Khang, C. Moreno and L. Gruner*	1	259
Selecting sheep for resistance to gastrointestinal nematode parasites *Bishop, S.C. and M.J. Stear*	2	260
Effect of expected degree of parasitism on the economic value of genetic host resistance to internal parasites *Amer, P.R., S. Eady and J.C. McEwan*	3	260
Evidence for breed differences in resistance to nematode parasitism *Hanrahan, J.P. and B.A. Crowley*	4	261
Genetic parameters for Eimeria resistance followed natural infections in Merinoland lambs *Reeg, K.J., M. Gauly, C. Bauer, R. Beuing, M. Kraus and G. Erhardt*	5	261
Genetic parameters of gastrointestinal nematodes resistance in German Rhön sheep *Gauly, M., H. Mathiak, K. Hoffmann and G. Erhardt*	6	262
The relation between nematode parasite infection and frequency of lymphocytes subpopulations in blood of the Heath Sheep lambs *Charon, K.M., R. Rutkowski, B. Moskwa, A. Winnicka and W.P. Świderek*	7	262

ECONOMIC, GENETIC AND MANAGEMENT ASPECTS OF FINE FIBRE PRODUCTION

Session 5

Date 25-aug-99
14:00 - 17:30 hours
Chairperson J. Milne (UK)

Papers

	S5 no.	Page
The genetic improvement of Angora goats in France *Allain, D. and J.M. Roguet*	1	263
Genetic improvement of cashmere goats *Bishop, S.C. and A.J.F. Russel*	2	263
Cashmere production on Norwegian goats *Eik, L.O., T. Ådnøy and N. Standal*	3	264

Sheep and Goat

	S5 no.	Page
Genetic parameters for wool traits in Finnsheep Puntila, M.-J., A. Nylander and E-M Nuutila	4	264
Main aspects of fine wool production in Hungary Kukovics, S. and A. Jávor	5	265
The production, wool trade and wool quality in Poland Radzik-Rant, A., D. Sztych and R. Niznikowski	6	265
Speciality wool production in Europe Merchant, M. and A.J.F. Russel	7	266

Posters

	S5 no.	Page
Economic analysis of hair goat breeding farms in Antalya province in Mediterranean region in Turkey Dellal, I. and A. Erkus	8	266
Histomorphological characteristics of the skin structure of grey blueKarakul lambs from Talassky sheep stud Dunayeva, G.S.	9	267

Commission on Pig Production (P)

President
Dr. J.A. Fernández
Danish Institute of Animal Science (DIAS)
Research Centre Foulum
P.O. Box 39
8830 Tjele
DENMARK
Phone:+45 89 99 1374
Fax:+45 89 99 1919
Josea.fernandez@agrsci.dk

Secretaries
Dr. S.M. Rillo
Fundacion para la Investigación
en Biologia Animal
c/Tormenta 4
28220 Majadahonda (Madrid)
SPAIN
Phone:+34 1 636 02 68
Fax:+34 1 637 53 13

Prof.Dr. J. Falkowski
University Agriculture and Technology
Inst. Animal Nutrition & Feed Management
10-718 Olsztyn
POLAND
Phone:+48 89 5234 859
Fax:+48 89 5233 341
E-mail:Falk@moskit.art.olsztyn.pl

CONSEQUENCES OF NEW TECHNOLOGIES
(WITH SPECIAL EMPHASIS ON PIGS AND HORSES)

Session 1

Joint session with the commissions on Animal Genetics and Horse Production (G*+P+H)

Date	23-aug-99
	9:00 - 12:30 hours
Chairperson	S. Neuenschwander (CH)

	GPH1 no.	Page	Pig Page
papers	1 - 10	XVI	1 - 5
posters	11- 20	XVI	6 - 10

THE ROLE OF DIETARY FIBRE IN PIG PRODUCTION
Session 2

Date   23-aug-99
                14:00 - 17:30 hours
Chairperson   A. Aumaitre (F)

Papers   P2 no. Page

The nutritional significance of "dietary fibre" analysis   1   268
Bach Knudsen, K.E.

Role of dietary fibre in the digestive physiology of the pig   2   268
Wenk, C.

Effect of dietary fibre on the energy value of feeds for pigs   3   269
Noblet, J. and G. Le Goff

Effect of dietary fibre on ileal digestibiliy and endogenous nitrogen losses   4   269
Souffrant, W.B.

Effect of dietary fibre on the behaviour and health of the restricted fed sow   5   270
Meunier-Salaün, M.C., S. Robert and S.A. Edwards

Dietary fibre for pregnant sows; - effect on performance and behaviour   6   270
Danielsen, V. and M. Vestergaard

Sugar beet pulp silage as dietary fermentable carbohydrate source for
group-housed sows: effects on physical activity and energy metabolism   7   271
Rijnen, M.M.J.A., J.W. Schrama, M.J.W. Heetkamp, M.W.A. Verstegen and J. Haaksma

Ileal amino acid digestibility of high fibrous oil cake meals in growing pig   8   271
Février, C., Y. Lechevestrier and Y. Jaguelin-Peyraud

The use of sugar beet-pulp in the diet of heavy pigs   9   272
Scipioni, R.

The effects of sugar beet pulp silage added with vinasse in heavy pigs feeding   10   272
Martelli, G., L. Sardi, A. Mordenti, P. Parsini and R. Scipioni

The performance response of growing and finishing pigs fed differing proportions   11   273
Chadd, S. and D.J.A. Cole

Pig

	P2 no.	Page
Influence of energy supply on growth characteristics in pigs and consequences for growth modelling *Quiniou, N., J. Noblet, J.-Y. Dourmad and J. van Milgen*	12	273

Posters

	P2 no.	Page
Ad libitum feeding of loose housed pregnant sows with fibre rich diets *Sørensen, G. and B.N. Fisker*	13	274
Effect of fibre rich diets for loose housed pregnant sows *Fisker, B.N. and G. Sørensen*	14	274
Non-starch polysaccharides from soybean hulls and beet pulp in combination with Ca-sulphate affect digestion and utilisation of nutrients and manure quality in pigs *Mroz, Z., A.W. Jongbloed, K. Vreman, J. Th. M. van Diepen and Y. van der Honing*	15	275
Ammonia emmision from finishing pigs fed 15 percent pelleted sugar beet pulp *Sloth, N.M. and H.B. Rom*	16	275
Impact of de-inking sludge used for bedding on aluminium, copper and HAP concentrations in blood, and in liver, fat, meat and urine of pigs. *Beauchamp, C.J., R. Boulanger and G. St-Laurent*	17	276
Leistungsstabilisierung beim absetzferkel durch einsatz von nahrungsfasern *Münchow, H., L. Hasselmann and V. Mantzke*	18	276
The use of expeller copra meal in grower and finisher pig diets *McKeon, M.P., M.G. Dore, A. Leek and J.V. O'Doherty*	19	277
Lipoic acid and vitamin C used in rations of pregnant sows *Nosenko, N.A.*	20	277
The efficacy of feeding fattening pigs with meal prepared from discarded peanuts *Koczanowski, J., J.B. Pyś, W. Migdal, C. Klocek, A. Gardzińska and A. Siuta*	21	278

FREE COMMUNICATIONS / BUSINESS MEETING

Session 3

Date 24-aug-99
 8:30 - 12:00 hours
Chairpersons P. Glodek (D) / J.A. Fernández (DK) / L. den Hartog (NL)

Pig

Papers
Breeding

	P3 no.	Page
Fattening and meat qualities of different purebred and crossbred pigs raised in Lithuania *Remeikiene, J.*	1	278
Comparison of different pig combinations by using data from Piglog 105 *Tänavots, A. and T. Kaart*	2	279
Heterosis effect demonstrated as an increase of the testes size and an improvement of the semen traits of cross-breed boars between Duroc and Pietrain breeds *Czarnecki, R., M. Rózycki, J. Udala, M. Kawęcka, M. Kamyczek and A. Pietruszka*	3	279
Monitoring of hybrid pig performance in commercial herd thorough the field testation. *Šprysl, M., R. Stupka, J. Èítek and M. Pour*	4	280
Risk factors and genetic variance components of pre-weaning mortality in piglets *Röhe, R. and E. Kalm*	5	280
Productive characterization of varieties of the Iberian pig branch *Barba, C., J.V. Delgado and E. Dieguez*	6	281

Posters
Breeding

	P3 no.	Page
The evaluation of the level of utility traits and breeding steps in pigs by means of correlation analysis *Fiedler, J., L. Houška, J. Pavlík and J. Pulkrábek*	7	281
Relationship between the breeding value of boars and fertility of their sisters *Czarnecki, R., J. Owsianny, M. Rózicki, M. Kawęcka, B. Delikator and K. Dziadek*	8	282
Perinatal mortality in the pig in relation to genetic merit for piglet vitality *Leenhouwers, J.I., E.F. Knol and T. van der Lende*	9	282
Reproductive characteristics of different types of Large White boars bred in Siberia *Kharchenko, P.G.*	10	283
Field testation - a source of enhancing of hybrid pig performance in commercial herds *Stupka, R., M. Šprysl, J. Èítek and M. Pour*	11	283
Genetic correlations between weight of tissues from different carcass cuts of pigs *Orzechowska, B. and M. Róžycki*	12	284

Pig

	P3 no.	Page
The age dynamics on selected hematological and immunological parameters of different breed types of pigs *Buleca, J., J. Szarek, I. Mikula, L. Tkáčiková and J. Mattová*	13	284
The profitable Iberian pigs crossed effect at reproductive and maternal ability traits *Benito, J., C. Vázquez Cisneros, J.L. Ferrera, C. Menaya and J.M. García Casco*	14	285
The genetic differences in the pigs reaction to ultrasound treatment *Nikolaeva, T.N., V.L. Petykhov and O.S. Korotkevich*	15	285

Papers
Nutrition / Body composition

	P3 no.	Page
Feeding strategies for meeting lysine requirements of the grower-finisher pig *Leek, A. and J.V. O'Doherty*	16	286
Relationships between *in vivo* conformation measurements and the composition of growing and finishing pigs *Abrutat, D.J., C.P. Schofield, J.D. Wood, A.R. Frost and R.P. White*	17	286

Posters
Nutrition / Body composition

	P3 no.	Page
Low phytic acid barley for growing pigs: pig performance and bone strength. *Veum, T.L., D.R. Ledoux and V. Raboy*	18	287
Low Phytic Acid Barley for Growing Pigs: Calcium and Phosphorus Balance. *Veum, T.L., D.R. Ledoux and V. Raboy*	19	287
The performance of large white and local Malawian pigs fed rations based on cowpeas (vigna unguiculata), soyabeans (glycinemax), or pigeon peas (cajanus cajan) *Simoongwe, V., J.P. Mtimuni, R.K.D. Phoya and C.B.A. Wollny*	20	288
Performance of local pigs under village conditions in the rural areas of Malawi *Mulume, C.G., C.B.A. Wollny, J.W. Banda and R.K.D. Phoya*	21	288
Suitability of crambe cake and crambe meal in pig feeding *Kampf, D., H. Böhme and R. Daenicke*	22	289
Choice and phase feeding methods of growing pigs: use of diets containing yellow lupin (*Lupinus luteus* L.) *Falkowski, J., W. Kozera and D. Bugnacka*	23	289

Pig

	P3 no.	Page
The effect of rapeseed oil and tallow supplement in a diet on the reproductive performance in sows *Paschma, J.-M.*	24	290
The influence of feeding methods on the results of fattening in crossbred gilts and castrates *Koczanowski, J., W. Migdal and C. Klocek*	25	290
Efficiency of various balanced protein-vitamin supplements to pig ration *Bekenev, V.A. and V.G. Pilnikov*	26	291
Evaluation of the fat score by NIR-spectroscopy *Schwörer, D., D. Lorenz and A. Rebsamen*	27	290
A comparison between different methods to measure lean meat content in carcasses of pigs *Eilart, K. and A. Põldvere*	28	292

Papers
Management / Health

	P3 no.	Page
Piglet mortality in a research herd - causes and effect of parity of the sow *Stern, S. and I. Wigren*	29	292
Influence of extended photoperiod during lactation on behaviour and performance in domestic sows and piglets *Deligeorgis, S. and P.N. Nikokyris*	30	293

Posters
Management / Health

	P3 no.	Page
Influence of the number of piglets born on the composition of sow's colostrum milked immediately after parturition *Csapó, J., Zs. Csapó-Kiss, Z. Házas, P. Horn and T. Németh*	31	293
Behaviour and performance of sows and piglets in crates and a Thorstensson system *Bradshaw, R.H. and D.M. Broom*	32	294
Sows behaviour after regrouping *Zhuchaev, K. and V. Koshel*	33	294
Inducing of exploratory activity in sows after regrouping *Koshel, V. and K. Zhuchaev*	34	295

Pig

	P3 no.	Page
The fattening pigs welfare in various housing system *Walczak, J.*	35	295
Influence of tropical climate and season on energy balance and chemical body composition in young growing pigs *Rinaldo, D., G. Saminadin, G. Gravillon and J. Le Dividich*	36	296
The change of piglets' biochemical indices under laser influence *Korotkevich, O.S.*	37	296
Effect of ultrasound influence on pigs' biologically active points (BAPs) *Sebezhko, O.I., O.S. Korotkevich, D.A. Odnoshevsky and G.N. Korotkova*	38	297
Stress resistance of Siberian North pigs *Kuznetsov, V.G., O.S. Korotkevich and V.L. Petukhov*	39	297
The relationships between foot measurements and body weight in piglets *Kotomin, K.N., V.L. Petukhov, T.N. Nikolaeva and V.G. Kuznetsov*	40	298

Posters
Reproduction

	P3 no.	Page
The investigation into relationship between cholesterol, sex steroids concentrations and fertility in gilts *Klocek, C. and J. Koczanowski and W. Migdal*	41	298
Sexual behavior of Polish Landrace sows and their actual fertility *Stasiak, A. and A. Walkiewicz and P. Kamyk*	42	299
Reproduktionsleistungen von zuchtsauen unter praxisbedingungen bei zusätzlicher b-carotin-zufuhr *Mantzke, V., H. Münchow and H. Fechner*	43	299
Comparison of the semen quality in pure-bred hybrid boars based on ORT test *Udala, J., R. Czarnecki, M. Kawęcka, M. Rózycki and M. Kamyczek*	44	300
Relationship between the growth rate and the thickness of the back fat of young boars and the size of their testes as well as the parameters of their libido and sperm *Czarnecki, R., M. Rózicki, M. Kawęcka, B. Delikator, K. Dziadek and A. Pietruszka*	45	300
Relationship between the sexual activity of young boars and the traits of their semen. *Czarnecki, R., M. Kawęcka, M. Rózycki, B. Delikator and K. Dziadek*	46	301

QUALITY OF MEAT AND FAT AS AFFECTED BY GENETICS AND NUTRITION

Session 4

Joint session with the commissions on Animal Genetics and Animal Nutrition (P*+G+N)

Date	25-aug-99
	9:00 - 12:30 hours
Chairperson	C. Wenk (CH)

Papers

	PGN4 no.	Page
What is pork quality? Andersen, H.J.	1	301
Influence of genetics on pork quality Vries, A.G. de, L. Faucitano, A. Sosnicki and G.S. Plastow	2	302
Genetic parameters of meat quality traits in station tested pigs in France Tribout, T. and J.-P. Bidanel	3	302
Dissection of genetic background underlying meat quality traits in swine Szyda, J. and S. Lien and E. Grindflek	4	303
Skeletal muscle fibres as factors for pork quality Karlsson, A.H., X. Fernandez and R.E. Klont	5	303
Selection progress of intramuscular fat in Swiss pig production Schwörer, D., A. Hofer, D. Lorenz and A. Rebsamen	6	304
Nutritional and genetic influences on meat and fat quality in pigs Ender, K., K. Nürnberg, G. Kuhn and U. Küchenmeister	7	304
Food waste products in diets for growing-finishing pigs Kjos, N.-P. and M. Øverland	8	305
Conservation and development of the bísaro pig. Characterization and zootechnical evaluation of the breed for production and genetic management Santos e Silva, J., J. Ventura, P. Albano and J.S. Pires da Costa	9	305
Meat consumption and health Zimmermann, M.	10	306

Pig

Papers

	PGN4 no.	Page

Who eats meat: factors effecting pork consumption in Europe and the United States — 11 — 306
Jamison, W.

Posters
Genetics and pork quality

	PGN4 no.	Page

Breed effect on meat quality of Belgian Landrace, Duroc and their reciprocal crossbred pigs — 12 — 307
Michalska, G., J. Nowachowicz, B. Rak and W. Kapelanski

Genetic and energy effects on pig meat quality — 13 — 307
Gajic, Z. and V. Isakov

Genetic parameters for fattening traits in the Belgian Piétrain population — 14 — 308
Geysen, D., S. Janssens and W. Vandepitte

Genetic parameters for colour traits and pH and correlations to production traits — 15 — 308
Pedersen, B. and S. Andersen

Gene effects on pork quality

Halothane gene effect on carcass and meat quality by use of Duroc x Pietrain boars — 16 — 309
Busk, H., A. Karlsson and S.H. Hertel

In vivo and post mortem changes of muscle phosphorus metabolites in pigs of different malignant hyperthermia genotype — 17 — 309
Henning, M., U. Baulain, G. Kohn and R. Lahucky

Interactive effects of the *HAL* and *RN* major genes on carcass quality traits in pigs — 18 — 310
Roy, P. le, C. Moreno, J.M. Elsen, J.C. Caritez, Y. Billon and H. Lagant

Meat quantity to meat quality relations when the RYR1 gene effect is eliminated — 19 — 310
Kortz, J., W. Kapelanski, S. Grajewska, J. Kuryl, M. Bocian and A. Rybarczyk

Correlations between growth rate, slaughter yield and meat quality traits after the elimination of RYR1 gene effect — 20 — 311
Kapelanski, W., J. Kortz, J. Kuryl, T. Karamucki and M. Bocian

Detection of the RYR1 locus variants and its implications in the Portuguese pig breeding programs — 21 — 311
Ramos, A.M., F. Simões, J. Matos, A. Clemente and T. Rangel-Figueiredo

Pig

	PGN4 no.	Page
Effect of RYR 1 gene on meat quality in pigs of Large White, Landrace and Czech Meat Pig breeds *Becková, R. and P. David*	22	312
Development of a highly accurate DNA-test for the *RN* gene in the pig *Looft, C., D. Milan, J.T. Jeon, S. Paul, C. Rogel Gaillard, V. Rey, A. Tornsten, N. Reinsch, M. Yerle, V. Amarger, A. Robic, P. le Roy, E. Kalm, P. Chardon and L. Andersson*	23	312
Performances of the Piétrain ReHal, the new stress negative Piétrain line. *Leroy, P.L. and V. Verleyen*	24	313
Effect of RN⁻ gene on the growth rate and carcass quality in crossbreeding of large white sows with P-76 boars *Koćwin-Podsiadła, M., S. Kaczorek, W. Przybylski and E. Krzęcio*	25	313
Breed and slaughter weight effects on meat quality traits in hal- pig populations *Puigvert, X., J. Tibau, J. Soler, M. Gispert and A. Diestre*	26	314
Genotypic and allelic frequencies of the RYR1 locus in the Manchado de Jabugo pig breed *Ramos, A.M., J.V. Delgado, J. Matos, C. Barba, F. Simões, R. Sereno, A. Clemente, T. Rangel-Figueiredo and M. Cumberas*	27	314

Variation of meat quality in pig breeds

Intramuscular fat content in some native German pig breeds *Baulain, U., P. Köhler, E. Kallweit and W. Brade*	28	315
Fatty acid and cholesterol composition of the lard of different genotypes of swine *Csapó, J., F. Húsvéth, Zs. Csapó-Kiss, P. Horn, Z. Házas and É. Varga-Visi*	29	315
The effect of paternal breed on meat quality of progeny of Hampshire, Duroc and Polish Large White boars *Nowachowicz, J., G. Michalska, B. Rak and W. Kapelanski*	30	316
Comparison of several pig breeds in fattening and meat quality in some experimental conditions of a Czech region. *Adamec, T., B. Naděje, J. Laštovková and M. Koucký*	31	316

Criteria of pork quality

Early detection of breed differences in fat distribution in pigs measured by CT *Kolstad, N.*	32	317
Fat score, an index value for fat quality in pigs - its ability to predict properties of backfat differing in fatty acid composition *Gläser, K., M.R.L. Scheeder and C. Wenk*	33	317

Pig

	PGN4 no.	Page
Meat quality with reference to EUROP carcass grading system Kapelanski, W., B. Rak, J. Kapelanska and H. Zurawski	34	318

Pig production and pork quality

The evaluation of pork meat productivity Dementjev, V.N., I.I. Gudilin and V.G. Pilnikov	35	318
The correlations between the fattening and slaughtering performance in pigs Pietruszka, A., R. Czarnecki and E. Jacyno	36	319
Comparison of fat supplements of different fatty acid profile with growing-finishing swine Gundel, J., A. Hermán, M. Szelényi and G. Agárdi	37	319
The pork meat quality in pigs with a different intenzity of nitrogen substances retention Cechová, M., V. Prokop, K. Dřimalová, V. Mikule and Z. Tvrdoň	38	320

Influence of nutrition on pork quality

The effect of dietary Ca-fatty acid salts of linseed oil on cholesterol content in *longissimus dorsi* muscle of finishing pigs* Barowicz, T.	39	320
Soybean oil, sex, slaughter weight, cross-breeding - influence on fattening performance and carcass traits of pigs Kratz, R., E. Schulz, G. Flachowsky and P. Glodek	40	321
Transfer of vitamin E supplements from feed into pig tissues Flachowsky, G., H. Rosenbauer, A. Berk, H. Vemmer and R. Daenicke	41	321
Einfluß der Fütterungsintensität in der Ferkelaufzucht auf die Mast- und Schlachtleistung Wetscherek, W., S. Bickel, H. Huber, F. Lettner and F. Gaheis	42	322
Effect of healing herb (symhytum peregrinum) fed to pigs on meat quality traits Bee, G., J. Seewer Lötscher and P.A. Dufey	43	322

FEEDING AND MANAGEMENT OF THE WEANED PIG
Session 5

Date	25-aug-99
	14:00 - 17:30 hours
Chairperson	V. Danielsen (DK)

Pig

Papers

	P5 no.	Page
Effects of management during the suckling period on post-weaning performance of pigs *Edwards, S.A. and J.A. Rooke*	1	323
Development of the gastrointestinal ecosystem during the weaning period. Need for feed additives? *Jensen, B.B.*	2	323
Nutritional requirements of weaned pigs and how to meet the demands by appropriate diet formulations *Sève, B. and J. Le Dividich*	3	324
Strategies and methods for allocation of food and water in the post-weaning period *Brooks, P.*	4	324
Environmental requirements of the weaned pigs *Dividich, J. Le*	5	325
Housing of weaners - meeting their environmental demands *Pedersen, B.K.*	6	325
Individual feed intake and performance of group-housed weanling pigs *Bruininx, E., C.M.C. van der Peet-Schwering, J.W. Schrama, P.C. Vesseur, L.A. den Hartog, H. Everts and A.C. Beynen*	7	326
Feeding weaned pigs pellets or meal? Effects on performance, water intake and eating behaviour *Laitat, M., M. Vandenheede, A. Désiron, B. Canart and B. Nicks*	8	326
Restricted feeding for prevention of *E. coli* associated diarrhoea in weaned pigs *Jørgensen, L., M. Johansen and C.F. Hansen*	9	327

Posters

	P5 no.	Page
Body composition changes in piglets at weaning in response to nutritional modification of sow milk composition and effects on post-weaning performance *Jones, G., S.A. Edwards, S. Traver, S. Jagger and S. Hoste*	10	327
Effects of sow nutrition and environmental enrichment during the suckling period on post weaning performance of pigs *Foster, E., S.A. Edwards, F.M. Davidson and J. Duncan*	11	328
Effects of L-carnitine with different lysine levels on growth and nutrient digestibility in pigs weaned at 21 days of age *Cho, W.T., S.W. Kim, In K. Han, K.N. Heo and J. Odle*	12	328

Pig

	P5 no.	Page
Effect of FIX-A-TOX on piglets after weaning *Mikule, V., M. Cechová and Z. Tvrdoň*	13	329
Differentiation of piglets on fear of human *Papshev, S., K. Zhuchaev and M. Baruskova*	14	329

FUTURE STRATEGIES WITH REGARD TO THE USE OF FEED WITHOUT ANTIBIOTIC FEED ADDITIVES IN PIG PRODUCTION

Session 6

Joint session with the commission on Animal Nutrition

Date 26-aug-99
8:00 - 11:30 hours
Chairperson K.B. Pedersen (DK)

Papers

	PN6 no.	Page
Selective pressure by antibiotic use in food animals *Witte, W.*	1	330
Experience with stop for the use of antimicrobial growth promoters in pig production and adjustment of management *Jorsal, S.E.*	2	330
Nutritional and gastrointestinal effects of organic acis supplementation in young pigs *Roth, F.X. and M. Kirchgeßner*	3	331
Current value of alternatives to antimicrobial growth promoters with special emphasis on anti secretory factor *Göransson, L. and S. Lange*	4	331
Comparison of fumaric acid, calcium formate and mineral levels in diets for newly weaned pigs *Lawlor, P., B. Lynch and P. Caffrey*	5	332
Commercial acid products for piglets from 7-30 kg BW *Callesen, J., H. Maribo and L. Jørgensen*	6	332
Investigations on the effects of dietary nucleotides in weaned piglets *Zomborszky-Kovács, M., S. Tuboly, P. Soós, G. Tornyos and E. Wolf-Táskai*	7	333

Pig

	PN6 no.	Page
Experiences of the voluntary ban of growth promoters for pigs in Denmark Kjeldsen, N.J., C.F. Hansen and A.O. Pedersen	8	333

Posters

	PN6 no.	Page
Effect of creep feeding, dietary fumaric acid and whey level on post-weaning pig performance Lawlor, P., B. Lynch and P. Caffrey	9	334
Effect of different spray dried plasmas on growth, ileal digestibility and health of early weaned pigs challenged with *E. Coli* k88 Bosi, P., In-Kyu. Han, S. Perini, L. Casini, D. Creston, C. Gremokolini and S. Mattuzzi	10	334
Formi™TLHS - an alternative to antibiotic growth promoters Overland, M., S.H. Steien, G. Gotterbarm, T. Granli and W. Close	11	335
Effects of passive protection with spray dried egg protein (PROTIMAX®) against post-weaning diarrhoea and growth check in piglets Krasucki, W., E.R. Grela, J. Matras and Z. Mroz	12	335
The effect of the inclusion of YeaSacc^{1026} at different levels of the concentrate diet on calf performance Fallon, R.J. and B. Early	13	336
Inclusion of organic chromium in the calf milk replacer: effects on immunological responses of healthy calves and calves with respiratory disease Early, B. and R.J. Fallon	14	336

Commission on Horse Production (H)

President
Prof. Dr. E.W. Bruns
Georg-August-Universität Göttingen
Institut für Tierzucht und Haustiergenetik
Albrecht-Thaerweg 3
37075 Göttingen
GERMANY
Phone:+49 551 395608/ +49
Fax:+49 551 395587
E-mail:ebruns@gwdg.de

Secretaries
Dr. F. Habe
University of Ljubljana
Biotechnical Faculty
Groblje 3
1230 Domzale
SLOVENIA
Phone+386 61 711 988
Fax:+386 61 721 005
E-mail:franc.habe@bfeo.uni-lj.sl

E. Olsson
Swedish University of Agricultural Sciences
Dept. Animal Breeding & Genetics
P.O. Box 7023
750 07 Uppsala 7
SWEDEN
Phone:+46-(0)18-672789
Fax:+46-(0)18-672648
E-mail:elisabeth.gerber@hgen.slu.se

CONSEQUENCES OF NEW TECHNOLOGIES
(WITH SPECIAL EMPHASIS ON PIGS AND HORSES)

Session 1

Joint session with the commissions on Animal Genetics and Pig Production (G*+P+H)

Date	23-aug-99
	9:00 - 12:30 hours
Chairperson	S. Neuenschwander (CH)

	GPH1 no.	Page	Horse Page
papers	1 - 10	XVI	1 - 5
posters	11- 20	XVI	6 - 10

FREE COMMUNICATIONS
Session 2

Date   23-aug-99
               14:00 - 17:30 hours
Chairperson   E. Barrey (F)

Papers

	H2 no.	Page
Genetic parameters of morphofunctional traits in Andalusian horse Molina, A., M. Valera, R. Dos Santos and A. Rodero	1	337
Genetic study of the Equine population of the Lusitanian Horse Valera, M., L. Esteves, M.M. Oom and A. Molina	2	337
The immunogenetic analysis of the Russian trotter mares Chechushkova, M.A., V.V. Sivtunova, M.L. Kochneva and V.L. Petukhov	3	338
Damping characteristics of a shock absorbing steel-polyurethane horseshoe Weishaupt, M.A., E. Mumprecht, C. Lenzlinger and H. Inglin	4	338
The overall performance and prospects of use of Silesian stallions from Ksią Depot in European breeding scheme of heavy warmblood horses Jodkowska, E., E. Walkowicz and H. Geringer de Oedenberg	5	339
The share of Arabian and Thoroughbred horses blood in the Małopolska horses breeding. Kulisa, M., M. Pieszka and G. Ciuraszkiewicz	6	339
Evaluation of horse population of Shagya-Arab breed bred in Czech Republic Navrátil, J.	7	340

Posters

	H2 no.	Page
The variation of Russian Trotter coat colors Sivtunova, V.V., M.A. Chechushkova and M.L. Kochneva	8	340
Percentage distribution of the main whey proteins in milk of Italian Saddle Horse nursing mares during the first two lactation months Martuzzi, F., A. Tirelli, A. Summer, A.L. Catalano and P. Mariani	9	341

Horse

Posters

	H2 no.	Page
The relationships between horse gray coat color and prenatal vitality of Orlov trotters *Knyazev, S.P., N.V. Guturova, S. Nikitin, O.I. Staroverova and R.M. Dubrovskaya*	10	341
Tyrosinase activity and ocular characteristics in Asinara white donkeys *Pinna, W., P. Todde, G.M. Cosseddu, G. Moniello and A. Solarino*	11	342
Genetic analysis of pedigree in the Andalusian Horse *Valera, M., A. Molina and A. Rodero*	12	342

HORSE BREEDING IN SWITZERLAND

Session 3

Date   24-aug-99
                08:30 - 12:00 hours
Chairperson   A. Lüth (CH)

Papers

	H3 no.	Page
Pferdezucht und -haltung in der Schweiz *Schatzmann, R.*	1	343
Die Freibergerzucht in der Schweiz - Population, Zuchtziele und Selektion *Lüth, A.*	2	343
Der Fremdblutanteil beim Freibergerpferd *Binder, H. and P.-A. Juillerat*	3	344
Die Warmblutzucht in der Schweiz - Population, Zuchtziele und Selektion *Lüth, A.*	4	344
Der Feldtest beim CH-Warmblutpferd *Egli, L.*	5	345
Die Haflingerzucht in der Schweiz - Population, Zuchtziele und Selektion *Lüth, A.*	6	345

Horse

ELECTRONIC IDENTIFICATION

Session 4

Joint session with the commission on Animal Management and Health (M*+H)

Date	25-aug-99
	9:00 - 12:30 hours
Chairperson	R. Geers (B)

	MH4	Page	Page
Papers	1 - 8	XLIII	130 - 134
Posters	9	XLIII	134

NEW DEVELOPMENTS IN REPRODUCTION

Session 5

Date	25-aug-99
	14:00 - 17:30 hours
Chairperson	T. Allen (UK)

Papers

H5 no. Page

Effect of insemination timing and dose on pregnancy rates in mares bred with frozen-thawed equine semen 1 346
Knaap, J.H., J.L. Tremoleda, A. van Buiten and B. Colenbrander

Studies on possibilities of stallion semen selection for deep freezing and artificial insemination (AI) 2 346
Kosiniak-Kamysz, K. and Z. Podstawski

Successful low-dose insemination by hysterscopy in the mare 3 347
Morris, L. H-A., R.H.F. Hunter and W.R. Allen

Variation in fertility of warmblood stallions 4 347
Dohms, T. and E.W. Bruns

The gestation length of the Andalusian and Arabian Horse 5 348
Blesa, F., M. Valera, M. Vinuesa and A. Molina

Genetic determination of the period between consecutive foalings in East Bulgarian mares 6 348
Sabeva, I.I.

Horse

	H5 no.	Page
Influence of maternal size on fetal and postnatal development in the horse Allen, W.R., F. Stewart, C. Turnbull, M. Ball, A. Fowden, J. Ousey and P.D. Rossdale	7	349
Influence of maternal age on placental size and function in the mare Wilsher, S. and W.R. Allen	8	349
Superovulation treatment is associated with endogenous LH depression in mares Briant, C. and D. Guillaume	9	350
Treatment of endometrosis in infertile mares by intra-uterine infusion of kerosene Bracher, V., A. Neuschaefer and W.R. Allen	10	350
Immunisation of recurrently aborting mares with stallion lymphocytes Mathias, S. and W.R. Allen	11	351
Maternal recognition of pregnancy in the mare Stout, T.A.E. and W.R. Allen	12	351

FREE COMMUNICATIONS
Session 6

Date	26-aug-99
	8:00 - 10:00 hours
Chairperson	D. Austbø (N)

Papers

	H6 no.	Page
How to breed and train a showjumper? Barneveld, A., R. van Weeren and J.H. Knaap	1	352
A new type of early performance test: gait and conformation measurements in 3 years old horses Barrey, E., M. Holmström, S. Biau, D. Poirel and B. Langlois	2	352
Influence of early exercise on the locomotor system Barneveld, A., R. van Weeren and J.H. Knaap	3	353
The rider effect in a genetic evaluation of showjumping performance of horses. Janssens, S., D. Geysen and W. Vandepitte	4	353

Posters

	H6 no.	Page
Genetic correlations in performance test results among Haflinger mares and stallion *Zeiler, M., H. Hamann and O. Distl*	5	354
Les trends genetiques des caracterès utilitaires des étalons demi-sang dans les stations d'entraînement polonaises *Kapron, M., M. Łukaszewicz and G. Zięba*	6	354
Corrélation entre les indices choisis de l'appréciation de l'entraînement des chevaux de course et la vitesse et distance du galop préparatoire *Kapron, M., I. Janczarek and H. Kaproń*	7	355
Relationship between the racing performance of pure-bred Arabian horses and their coat colour *Pikula, R., M. Smugala, D. Gronet and W. Grzesiak*	8	355
Characteristics of racing performance of Thoroughbreds in Korea *Lee, K.J. and K.D. Park*	9	356
Racing performances of Anglo-Arab-Sardinian horses analysed by mixed linear models *Rassu, S.P.G., N.P.P Macciotta, A. Cappio Borlino, s. Delogu and G. Enne*	10	356
A human encounter test to assess young horses' reaction to humans *Søndergaard, E., J. Ladewig and C.C. Krohn*	11	357
Individual differences in temperamental traits of horses *Visser, E.K., H.J. Blokhuis, J. Knaap and A. Barneveld*	12	357
Behaviour performance of half bred horses tested on race track *Geringer, H.K. and J. Kasprzak*	13	358

ANIMAL GENETICS [G]

Paper GPH1.1

Is functional genomics possible in farm animals?
Alan L. Archibald, Roslin Institute, Roslin, Midlothian EH25 9PS, Scotland, UK

Functional genomics is concerned with functional analyses on a genome-wide scale and with determining function for sequenced genomes. I will seek to address the question 'is functional genomics possible in farm animals (in the absence of genome sequence data)?' I will argue that the farm animal genome projects are already engaged in functional genomics studies. Quantitative trait locus (QTL) mapping experiments are concerned with scanning the genome for chromosomal regions influencing the trait of interest. Thus, QTL-mapping projects link function and the genome. New technologies such as microarrays are creating opportunities for functional genomics studies in a wide range of organisims including farm animals. Microarray technology will also create chances to integrate physiology and whole animal studies with molecular genetics and genomics to dissect the control of biologically and economically important traits. In microorganisms such as yeast and bacteria it is possible to study the function of specific sequences or genes by true reverse genetics in which mutations are created and their phenotypes assessed. In higher animals such reverse genetics is currently only possible in mice. However, nuclear transfer technology may be creating opportunities for direct genetic manipulation in sheep and other farmed animals. I will identify the tools, resources and capabilities necessary for functional genomics in farm animals and I will discuss the implications for animal production.

Paper GPH1.2

A genetic marker for litter size in landrace based pig lines
O.I. Southwood[1], *T.H. Short*[2], *G.S. Plastow*[3] *and M.F. Rothschild*[4]. *PIC W. Europe*[1] *and Group*[3], *Fyfield Wick, Abingdon, Oxfordshire, UK, OX13 5NA,* [2]*PIC USA, Franklin, KY 42135 USA,* [4]*Department of Animal Science, Iowa State University, Ames IA 50011, USA*

The number of genetic markers that are becoming available for use in selection is increasing. They are particularly useful for traits, such as litter size, which are difficult to improve using traditional methods. A DNA marker for the Estrogen Receptor (ESR) gene has previously been shown to have large effects on litter size in Large White or Meishan based lines. Here we report results from a large validation trial with a DNA marker within the Prolactin Receptor (PRLR) gene, which appears to have effects similar to ESR but in lines of Landrace origin. Litter size was recorded for 2615 litters from 5 PIC Landrace based sows from the USA and Europe and genotyped for PRLR. The additive effect of one copy of the beneficial allele differed by line and ranged between +0.10 to +0.8 for total numbers of piglets born and +0.10 to +0.9 for numbers born alive. There was a tendency for piglets, nursed by sows with the beneficial allele, to be heavier at weaning and there was no evidence of adverse effects on carcase quality. Future research will focus on the joint effects of ESR and PRLR in crossbred sows with both Large White and Landrace backgrounds.

ANIMAL GENETICS   [G]

Paper GPH1.3

Visualisation of differential gene expression among equine tissues using cDNA-AFLP
A. Verini Supplizi*[1], K. Cappelli [1], A. Porceddu [2], F. De Marchis [2], C. Moretti [2], M. Falcinelli [2], D. Casciotti [1], M. Pezzotti [2] and M. Silvestrelli [1] . [1] Centro di studio del cavallo sportivo - Facoltà di Medicina Veterinaria - Perugia - Italy. [2] Istituto di miglioramento genetico vegetale - Facoltà di Agraria - Perugia - Italy.

Many biological processes such as cellular growth, organogenesis and response to environmental stimuli are mediated by programs of differential gene expression. The identification of the subsets of differentially expressed genes is a fundamental prerequisite in understanding the molecular regulation of these processes. A novel approach for achieving a RNA fingerprinting, called cDNA AFLP, has been developed (Bachem et al., 1996). The technique is based on the use of highly stringent PCR conditions facilitated by the addition of double stranded adaptors at the ends of cDNA restriction fragments. Adaptors' sequences are used as annealing sites for specific primers during PCR amplification. The addition of one or more bases at the 3' end of the primers will selectively reduce the number of amplified fragments, since selective primers will be extended only in case they find a perfect match to complementary sequence of the cDNA fragment. To visualise changes in gene expression, a system of RNA fingerprinting, based on cDNA AFLP, was optimised to be applied to different equine tissues. Several transcript profiles of horse liver, cartilage, muscle and lymphocytes were analysed and tissue specific bands were identified. The possibility to visualise variation in the rate of transcription of a gene and the low amount of material required for the analysis, render this technique of particular interest to study physiological and pathological events at molecular level.

Paper GPH1.4

A diagnostic assay discriminating between two major sheep mitochondrial DNA haplotypes suitable for QTL association studies
S. Hiendleder*[1], S.H. Phua[2], W. Hecht[3]. [1]Department of Animal Breeding and Genetics, Justus-Liebig-University, Ludwigstr. 21B, 35390 Giessen, Germany, [2]AgResearch Mol. Biol. Unit, Dep. Biochem., University of Otago, PO Box 56, Dunedin, New Zealand, [3]Vet. Genet. Cytogenet., Justus-Liebig-University, Hofmannstr. 10, D-35392 Giessen.

Variation in maternally inherited mitochondrial DNA (mtDNA) has been reported to affect performance traits in livestock. In sheep, two major mtDNA haplogroups have been de-scribed in independent studies. Sequencing data obtained from a representative of each group have revealed an average sequence divergence of ~ 0.5% between both mt genomes. Extensive sequence variation was observed in the mitochondrial control region, and in rRNA and protein coding genes, including replacement substitutions. *Hin*fI RFLP data from a large sample set (n = 239) comprising mitochondrial genomes of both groups indicated an ancient mutation separating both mtDNA types. A completely determined sheep mtDNA sequence was used to assign this mutation to the COI gene and to develop a PCR based assay discriminating between the two phylogenetic branches. The haplogroup specificity of the mutation was further investigated in 28 randomly selected individuals of 6 different breeds. Those were unequivocally assigned to the respective groups on the basis of the developed test and their complete control region sequences. The presented assay provides a rapid and economic means of discriminating between both major domestic sheep mtDNAs in QTL association studies between mtDNA haplotype and performance traits.

ANIMAL GENETICS [G]

Paper GPH1.5

Genomics and patenting
B. Brenig. Institute of Veterinary Medicine, University of Göttingen, Groner Landstraße 2, 37073 Göttingen, Germany.

Genome analysis in livestock species is rapidly developing. For the more important livestock species, i.e. cattle, sheep, goat, and pig, the number of isolated and characterized genes as well as markers for QTL mapping, is growing exponentially and through comparative mapping additional data become available. Currently, not all of these genes can directly be linked to economical important traits. However, in the future our understanding of gene expression and function will improve and therefore at least to some of these genes specific functions will be assigned. On the other hand, genes and markers have already been identified that are of economical interest, e.g. the porcine ryanodine receptor 1 gene. In these cases patenting seems to be necessary, whereas the patenting of undefined DNA sequences is generally not possible. Most of the patents filed in the field of animal breeding and genetics covering DNA-based diagnosis of disorders are within section C, class 12, subclass Q and group 1/68 according to the International Patent Classification (IPC). C12Q 1/68 covers inventions in measuring or testing processes involving nucleic acids. According to the current U.S. classification these inventions are within class 435/6. For obtaining a patent it is not required to contact a patent attorney or patent agent. However, most universities collaborate with innovation or research foundations that undertake the patenting. Finally of course the question remains, whether breeders or breeding organization will be able to pay licence fees or royalties to patent owners for an increasing number of patents.

Paper GPH1.6

Detection of quantitative trait loci for fatness traits in Dutch Meishan-crossbreds
D.J. de Koning, L.L.G. Janss[2], A.P. Rattink, M.A.M. Groenen, P.N. de Groot, E.W. Brascamp and J.A.M. van Arendonk, Wageningen Institute of Animal Sciences, PO Box 338, NL 6700 AH Wageningen, The Netherlands, [2]Institute for Animal Science and Health, PO Box 65, NL-8200 AL Lelystad*

In an experimental cross between Meishan en Dutch Large White and Landrace lines, 1200 F2 animals and their parents were typed for molecular markers covering the entire porcine genome. Associations were studied between these markers and two fatness traits: intramuscular fat content and backfat thickness. Association analyses were performed using interval mapping by regression under two genetic models: 1) An outbred line cross model where the founder lines were assumed to be fixed for different QTL alleles. 2) A half-sib model where a unique allele substitution effect was fitted within each of the 38 half-sib families. Both approaches revealed for backfat thickness a highly significant QTL on chromosome 7 and suggestive evidence for a QTL at chromosome 2. Furthermore suggestive QTLs affecting backfat thickness were detected on chromo-somes 1 and 6 under the line cross model. For intramuscular fat content the line cross approach showed suggestive evidence for QTLs on chromosomes 2, 4 and 6 whereas the half-sib analysis showed suggestive linkage for chromosomes 4 and 7. The nature of the QTL effects and assumptions underlying both models could explain discrepancies between the findings under the two models.

ANIMAL GENETICS   [G]

Paper GPH1.7

Status of Genome and QTL mapping in Hohenheim F2 Pig-Families
G. Moser[1], G. Reiner*[1], E. Mueller[1], P. Beeckmann[1], G. Yue[1], M. Dragos[1], H. Bartenschlager[1], S. Cepica[2], A. Stratil[2], J. Schroeffel[2], H. Geldermann[1]. [1]Department of Animal Breeding and Biotechnology, University of Hohenheim, Stuttgart, Germany, [2]Institute of Animal Physiology and Genetics, Academy of Science of the Czech Republic, Libechov, Czech Republik.

Three informative F_2 families based on Wild Boar x Pietrain, Meishan boar x Pietrain and Wild Boar x Meishan, each with more than 300 animals, were genotyped for evenly spaced marker loci and recorded for more than 100 quantitative traits. Linkage and QTL mapping data for 8 chromosomes are presented. QTLs were mapped for performance traits of growth, carcass- and meat quality. The QTLs are located on different chromosomes and influenced by families. Larger effects were found on chromosomes 6 and 7. Up to 60 % of the phenotypic F2 variance for meat quality traits was associated with chromosome 6. Candidate genes are proposed for some of the QTL intervals. The subsequent QTL mapping used a combined strategy of genome-wide marker mapping with a positional candidate gene approach in order to identify genes which are significant for breeding.

Paper GPH1.8

Mapping loci for the degree of spotting, colour dilution, and circumocular pigmentation in the ADR granddaughter design.
N. Reinsch[1], H. Thomsen[1], N. Xu[1], C. Looft[1], E. Kalm[1], G. Brockmann[2], S. Grupe[2], C. Kühn[2], M. Schwerin[2], B. Leyhe[3], S. Hiendleder[3], G. Erhardt[3], I. Medjugorac[4], I. Russ[4], M. Förster[4], B. Brenig[5], J. Blümel[6], R. Reents[7], G. Averdunk[8], J. Duda[9]. [1]Institut für Tierzucht und Tierhaltung der Christian-Albrechts-Universität, 24098 Kiel, [2]Forschungsinstitut für die Biologie landwirtschaftlicher Nutztiere, Forschungsbereich Molekularbiologie, 18196 Dummerstorf, [3]Institut für Tierzucht und Haustiergenetik der Justus-Liebig-Universität, 35390 Gießen, [4]Institut für Tierzucht der Ludwig-Maximilians-Universität, 80539 München, [5]Tierärztliches Institut der Georg-August-Universität, 37073 Göttingen, [6]Institut für die Fortpflanzung landwirtschaftlicher Nutztiere, 16321 Schoenow, [7]Vereinigte Informationssysteme Tierhaltung w.V., 27283 Verden/Aller, [8]Bayerische Landesanstalt für Tierzucht, 85586 Grub, [9]LKV Bayern, 80048 München; Germany

The proportion of unpigmented coat on the trunk was determined from photographs of 665 bulls distributed over 18 Holstein and 3 Simmental half-sib families from the ADR granddaughter design. A QTL scan over all chromosomes was performed by fitting a multiple marker regression model. On chromosome 6 a QTL for the proportion of white coat with large effects (experimentwise error probability < 0.0001) was found. Data from progeny of Simmental sires were also matched with marker genotypes. The results from the first QTL scan could be confirmed. Additinally loci for the ocurrence of yellow coat and for circumocular pigmentation could be mapped. The results are discussed with respect to resistance to solar radiation, heat stress, photosensitisation, and eye cancer.

ANIMAL GENETICS [G]

Paper GPH1.9

Quantifying the impact of introgressing major genes for disease resistance
K. MacKenzie and S.C. Bishop*. *Roslin Institute (Edinburgh), Roslin, Midlothian EH25 9PS, UK*

A genetic epidemiological model (GEM) is presented for investigating introgression strategies for major disease resistance genes and the impact of these strategies on the epidemiology of the disease to which the gene confers resistance. Fundamental to the model is Ro, the basic reproductive ratio. Ro is the expected number of secondary infections caused by a single infection. If Ro is greater than one, there will be an epidemic. The aim of any introgression strategy is to reduce Ro to less than one, should the infectious agent be present. As introgression proceeds the number of animals infected, should an epidemic occur, shows a curvilinear decrease. Initial decreases are small, then accelerate as the gene frequency increases. The gene frequency required to reduce Ro below one is a function of the initial Ro of the disease, i.e. the infectiousness of the disease in an unselected population. It is not necessary to take the gene to fixation. For moderately infectious diseases (e.g. Ro = 3), approximately two-thirds of the population is required to be resistant and for highly infectious diseases (e.g. Ro = 10) approximately 90% of the population should be resistant to prevent an epidemic from occurring. These results have important practical implications when utilising disease-resistance genes.

Paper GPH1.10

Genetic and economic consequences of alternative strategies for introgressing QTL responsible for disease resistance, using different breeding schemes
E.H. van der Waaij*[1], J.A.M. van Arendonk[1]. *[1] Animal Breeding and Genetics Group, Wageningen Institute of Animal Sciences, Wageningen Agricultural University, PO Box 338, 6700 AH Wageningen, The Netherlands.*

Genetic and economic consequences of different strategies for introgression of QTL responsible for disease resistance, followed by phenotypic selection for production after fixation, were compared using a deterministic approach. Four introgression strategies were considered, differing in number of generations of backcrossing (0, 1, 3 or 7). Application of zero generations of backcrossing implies immediate fixation of the QTL by creating an F_2 generation. Breeding schemes differed in initial breed difference for production level (0.1, 1, 2.5 or 5 phenotypic standard deviations) and female reproductive capacity (4 or 10 offspring per dam). With small initial breed difference, more gain can be made from selection for production within the donor population compared to introgression. With larger breed differences, application of seven backcross generations was genetically most favourable. However, taking economic consequences into account seven generations of backcrossing was the most unattractive option due to high numbers of animals needed until fixation of the QTL (lower production level compared to recipient breed and larger number of genotypings to be done). The optimal number of backcross generations from economical point of view was one or three, depending on the breeding scheme.

ANIMAL GENETICS   [G]

Poster GPH1.11

Species-specific splicing of transgenic RNA in the mammary glands of pigs and rabbits

B. Aigner[1]*, K. Pambalk[2], M. Renner[2], W.H. Günzburg[2], E. Wolf[3], M. Müller[1] and G. Brem[1].
[1]Institut für Tierzucht und Genetik, Veterinärmedizinische Universität Wien, Veterinärplatz 1, A-1210 Wien, Austria (Fax: +43-1-25077-5690), [2]Institut für Virologie, Veterinärmedizinische Universität Wien, Austria, [3] Institut für Molekulare Tierzucht und Haustiergenetik, LMU München, Germany.

Transgenic livestock is used for different biotechnological purposes, e.g. gene farming. These techniques require the precise processing of pre-mRNA transcribed from genomic constructs. We describe the appearance of different alternative mRNA splice patterns of a gene construct in which a mutant human growth hormone (hGH-N) gene is transcriptionally controlled by 2.5 kb of mouse whey acidic protein (WAP2) regulatory sequences in the mammary gland of different livestock species. Compared to the transcription products in transgenic mice harbouring the same gene construct and to cell transfection experiments, expression analysis in transgenic pigs and rabbits revealed different mRNA splice patterns with regard to the proportion of the processed transcripts. In pigs, most of the hGH transgene mRNA products were represented by two closely related forms both lacking a major part of exon 3. In rabbits, the majority of the processed transgenic hGH transcripts consisted of a shorter form lacking the complete exons 3 and 4. Sequence analysis of the transgenes suggests that the species-specific hGH mRNA patterns may be caused by species- and tissue-specific differences in *trans*-acting splice factors.

Poster GPH1.12

The fertilizing ability of spermatozoa differentiated from female primordial germ cells in male chimeric chickens

T. Tagami*[1], H. Kagami[2], T. Harumi[1], Y. Matsubara[1], H. Hanada[1] and M. Naito[1]
[1]Laboratory of Genetic Engineering, National Institute of Animal Industry, Tsukuba Norindanchi, Ibaraki 305-0901, Japan, [2]Department of Animal Science and Biotechnology, School of Veterinary Medicine, Azabu University, Kanagawa, 299-8501, Japan

We have proved the female primordial germ cells (PGCs) could differentiate into W chromosome-bearing (W-bearing) spermatozoa in male gonads of germline chimeric chicken. If W-bearing spermatozoa could fertilize ova, the sex ratio of female to male may well be biased, and this would greatly benefit for poultry production. The purpose of the present study was to investigate whether the spermatozoa differentiated from female PGCs have the fertilizing ability or not.
Germline chimeric chickens were generated by a recipient embryo and the different-sexed PGCs. Sexing of donor and recipient cells was carried out by PCR analysis. The manipulated embryos were cultured ex vivo to hatching and reared to sexual maturity. The fact that the W chromosome specific sequences were detected in each semen sample of 4 male chimeric chickens by the PCR analysis, making it evident that the female PGCs could settle in the gonads and could start to differentiate in the seminiferous tubules of these chimeras. These 4 germline chimeras were then test mated to assess donor contribution to their germline. The donor-derived offspring, however, could not be obtained. It is suggested that the spermatozoa differentiated from female PGCs in male gonads can hardly fertilize ova in avian species.

ANIMAL GENETICS   [G]

Poster GPH1.13

Producing chimeras by transfer of chicken blastodermal cells cultured without a feeder layer

Y. Matsubara[1], A. Hirota[2], H. Sobajima[3], A. Onishi[1], H. kagami[4], T. Tagami[1], T. Harumi[1], M. Sakurai[5], M. Naito[1], [1]National Institute of Animal Industry, Tsukuba Norin-kenkyu-danchi P.O.Box 5,Ibaraki 305-0901, Japan,[2] Saitama-ken Animal Research Center, Sugahiro 784, Konan-cho,Osato-gun,Saitama 360-0102,Japan,[3]Gifu Prefectural Poultry Experimental Station, Hakuma 2672-1, Seki,Gifu 501-3924, Azabu University, Kanagawa, 299-8501, Japan,[5] National Institute of Animal Health,3-1-1 kannondai, Tsukuba, Ibaraki 305-0856, Japan.

The use of transgenic animals is gaining considerable attention in genetic improvement of farm animals and development of biomedical products. In this study, chicken blastodermal cells were cultured in order to develop a method for producing transgenic chickens efficiently. Chicken blastodermal cells obtained from the central disc of the area pellucida (stage X), which are considered to contain primordial germ cells or their precursor cells, increased by about 7 times in culture for 3 days without a feeder layer. These cultured cells showed alkaline phosphatase activity for 2 to 5 days in culture (Matsubara et al.,1997). It was also found that these cultured chicken blastodermal cells contained the SSEA-I and/or EMA-I- positive cells. We obtained two chimeric chickens, judged by the feather color, by transfer of 2 day-cultured Barred Plymouth Rock donor cells to White Leghorn recipients. Now, possibility that these females are germline chimeras is under assessing.

Poster GPH1.14

Inbreeding and homozygosity in a Hungarian Lipizzan horse population

S. Mihók[1], B. Pataki[2], E. Takács[2], I. Bodó*[1], Z. Egri[3], B. Bán [2] and D. Ritter[2]. [1] Debrecen University of Agriculture H-4032 Debrecen Böszörményi út 138, [2] National Institute for Agricultural Control H-1024 Budapest, Keleti K. u 24, [3] Hungarian Association of Lipizzan Breeders. H-3348 Szilvásvárad Egri u. 1.

The most important blood groups and polimorphisms of 74 Lipizzan horses were the basis for the calculation of homozygosity grade and for its comparison with Wright coefficient based upon 5 generation pedigrees. The whole population was divided into a non inbred and an inbred group over $F_x=0.03$. The population sizes of the whole group, of the non inbred and of the inbred groups were 74, 19 and 25 respectively. The homozygosity grades of the inbred group based upon the D blood group, transferrin and serum albumin were 0.24, 0.29, 0.70, and the same figures for the non inbred group were 0.18, 0.25, 0.53 respectively. Thus, in some cases the inbreeding rates in five generation pedigrees correspond well to the homozygosity rate, in spite of the small population size. The comparison with the homozygosity of Thoroughbred and Standard bred yearlings does not proved this ascertainment, rather the decisive influence of sires can be stated. The inbreeding within the 5 pedigree generations was based upon 24 horses, they occurred 100 times in the pedigrees of the 49 animals. The correct expressing of the homozygosity structure of different populations needs more investigation. The DNA should be taken into consideration as well.

ANIMAL GENETICS   [G]

Poster GPH1.15

Horse parentage testing using high polymorphic microsatellites: a case study
C. Ginja*[1]; J. Matos[2]; A. Clemente[2]; T. Rangel[1]. [1]UTAD. Apartado 202, 5001 Vila Real-codex, [2]INETI/Bioquímica II./Paço do Lumiar, 1699 Lisboa-codex

Horse interspecific hybrids (mules) are known to be sterile. A fertile female hybrid was analysed for the determination of the putative parents of a male descendent.

Parentage testing is being widely performed with microsatellite analysis, by studying the PCR amplification profiles of VNTRs regions. A list of specific primers is available for genetic variability analysis on horse breeds which have also been used on parentage testing. The aim of this study was to analyse several horse microsatellite markers for their application to hybrids, donkeys and horses.

Paternity tests were performed on four equines: putative mother, two putative fathers, male descendent. Unrelated animals were used as control. DNA was extracted from blood samples and PCR amplified with the primers: HMS2, HMS7 [1] HTG7, HTG8, HTG10 [2]. [α^{35}S]dATP was incorporated in PCR reactions. Amplification products were analysed by electrophoresis and autoradiograph detection.

All microsatellites tested were successfully amplified. However, most did not present enough variability in the present case that would allow correct parentage determination. HTG10 has produced high level of non-specific amplification. Paternity was determined by profile analysis of the PCR amplification with HTG7. The technique was adequate for hybrids, donkeys and horses. This work suggests that further screening with other markers.

Poster GPH1.16

RFLP based genetic mapping of the ovine inhibin/activin (*INHA*, *INHBA*, *INHBB*) genes identifies microsatellites for QTL association studies
S. Hiendleder[1], K.G. Dodds[2], R. Wassmuth*[1]. [1]Department of Animal Breeding and Genetics, Justus-Liebig-University, Ludwigstr. 21B, D-35390 Giessen, Germany, [2]AgResearch Molecular Biology Unit, Department of Biochemistry, University of Otago,
PO Box 56, Dunedin, New Zealand.

Inhibins/activins are dimeric glycoprotein hormones involved in the male and female reproductive axis, e.g. the control of FSH secretion. As members of the transforming
growth factor β superfamily, inhibin/activin genes have an important role in embryonic growth and development, and *INHBA* has recently been identified as a candidate gene for testicle size in mice. We have typed *Taq*I Southern blot RFLPs for all three genes in the AgResearch International Mapping Flock, comprising 9 large full-sib families with grand-parent information. Linkage analyses placed *INHA* on chromosome 2 and *INHBA* on chro-mosome 4, confirming previous chromosomal assignments by fluorescence *in situ* hybrid-ization. *INHA* was linked to 3 microsatellites at a distance of 0 cM with LOD scores of
3.6, 6.6 and 11.7. *INHBA* was placed in a linkage group with two microsatellites at a dis-tance of 4 and 9 cM, with LOD scores of 10.7 and 6.5. No significant LOD scores were obtained for the *INHBB* gene, but a tentative assignment could be made. The presented mapping data identifies microsatellites suitable for QTL association studies with the *INHA* and *INHBA* genes, and provides genetic mapping information for a type I anchor locus, *INHBA*, in a major farm animal species.

ANIMAL GENETICS [G]

Poster GPH1.17

Molecular genetic studies on the porcine c-myc proto-oncogene regarding phylogeny, associations with performance traits and interactions with the Ryr1-genotype

G.Reiner[*,1], G.Moser[1], H. Geldermann[1], V.Dzapo[2]. [1]Department of Animal Breeding and Biotechnology, University of Hohenheim, Stuttgart, Germany, [2]Institute of Animal Breeding and Genetics, University of Giessen, Giessen, Germany.

Overcoming of genetic antagonisms in animal breeding needs insight into the basic molecular mechanisms. From their physiological role in regulation between proliferation and differentiation, proto-oncogenes might be interesting candidate genes to be involved in this. We studied c-myc, the prototype of the proto-oncogenes, essential in e.g. myogenesis and adipogenesis. c-myc was isolated and sequenced in a range of 7730 bp. By screening a pig/rodent cell hybrid panel, the gene was localized on SSC4p13, preciced to the region between SW495 and SW835 by genetic mapping. Seven polymorphisms were detected and PCR systems were established for mass screening. Haplotypes were used to evaluate the phylogeny of the porcine c-myc locus. The association studies showed two points of interactions. An association was evident between c-myc and carcass-quality-traits, namely the fat content. c-myc based variance reached up to 4 %. Gene interactions between c-myc and Ryr1 have been found, being congruent with the picture of a physiological epistasis between both genes. From the AA to the BB-genotype in c-myc, MHS-based variance decreased for 57 % on the average of meat associated traits and for 47 % for fat-associated traits. Simultaneously, the reduction in differences between the homozygous MHS genotypes were 72 % and 65 % respectively.

Poster GPH1.18

Determination of $\alpha(1,2)$fucosyltransferase gene function in resistance to oedema disease and post-weaning diarrhoea and in blood group antigen A-0 synthesis

E. Meijerink[1], P. Vögeli[1], S. Neuenschwander[*,1], H.U. Bertschinger[2], R. Fries[3], G. Stranzinger[1]. [1]Institute of Animal Science, Swiss Federal Institute of Technology, ETH-Zentrum, CH-8092 Zurich, Switzerland, [2]Institute of Veterinary Bacteriology, University of Zurich, CH-8057 Zurich, Switzerland, [3]Chair of Animal Breeding, Technical University Munich, D-85350 Freising-Weihenstephan, Germany.

The porcine $\alpha(1,2)$ fucosyltransferase genes (*FUT1* and *FUT2*) have been studied for their involvement in the genetically determined resistance and susceptibility to oedema disease and postweaning diarrhoea caused by colonisation of *Escherichia coli* F18 bacteria in the small intestine. Based on a mutation in the *FUT1* gene, we have developed a molecular genetic test for the identification of animals susceptible to colonisation with *E. coli* F18 bacteria. The distribution of the *FUT1* alleles, determining resistance and susceptibility, predicts less than 10% *E. coli* F18 resistant animals in commercial pig breeds.

The close linkage of the blood group A-0 inhibitor locus *S* and the locus determining *E. coli* F18 colonisation suggested that these loci may represent the same $\alpha(1,2)$ fucosyltransferase gene. However, whereas *E. coli* F18 colonisation seems to be controlled from the *FUT1* gene, the *S*-locus represents the *FUT2* gene. Results from our molecular genetic studies on the *FUT2* gene, the blood group A-transferase gene and the $\alpha(1,3)$fucosyltransferase (*FUT3*) gene fit to the epistatic relationship which is known for the blood group loci *EAA* and *S*, and are in agreement with the serum antigen A-0 structure.

ANIMAL GENETICS   [G]

Poster GPH1.19

Properties of a sperm ligand that are necessary for zona binding
Stephan Jansen*, Brit Marschall, Ingo Jenneckens and Bertram Brenig Institute of Veterinary Medicine, Univ. of Goettingen, Groner Lanstrasse 2, 37073 Göttingen, Germany

In many of the steps that mediate gamete binding, especially during their initial contact, carbohydrate-carbohydrate interactions play a major role. However, after the acrosome reaction has taken place and has led to an exposure of secondary ligands, another type of biochemical interaction becomes important. In here, the protease acrosin exhibits high binding activity to zona pellucida. Our investigations have shown that the affinity is based on the interaction of positively charged amino acids with negatively charged sulphate groups of the zona pellucida glycoproteins. We have pinpointed the epitopes responsible for binding and have proposed a three dimensional model of proacrosin to explain its structure-function relation. The binding site comprises six loops that are exposed from the molecule's surface. The loops contain positively charged amino acids and surround the proteolytic centre, which is formed by the amino acids His 70, Asp 124 and Ser 222 in boar proacrosin. Both the epitopes are independent properties of the molecule, however in a physiological respect, they act together to support sperm penetration. Although proacrosin has been shown to be not essential for fertilisation, the presence of this mechanism must have granted an evolutionary advantage.

Poster GPH1.20

The genomic structure of the ovine Uroporphyrinogen Decarboxylase gene
R. Nezamzadeh*, A. Krempler, B. Brenig; Institute of Veterinary Medicine, University of Göttingen, Groner Landstrasse 2, 37073 Göttingen

Uroporphyrinogen decarboxylase (UROD) is an enzyme of the heme biosynthetic pathway causing erythropoetic and hepatic porphyria. In domesticated animals the erythropoetic porphyria, an autosomal recessive disease, is of importance. It was first described in cattle and sheep and goes ahead with light sensitivity and photodermatosis. In females the defect also leads to reduced fertility by inactivation of the ovaries and reduced estrus.
The genomic structure of the UROD gene has been solved in man and cDNA sequences are known from human, mouse and rat. They show a high degree of homology that enabled us to construct PCR primers for the amplification of a genomic region in sheep. This product was used as a probe to screen a genomic sheep library. We were able to isolate several overlapping phages that harboured the whole genomic region of UROD. According to the human genomic organisation we found 10 exons spanning 4500 base pairs. Based on these sequences we designed five primer pairs to amplify the coding regions of the gene. On screening healthy and sick animals we could find no differences in length. Therefore we can exclude bigger insertion or deletion events as a reason for misfunction of the protein. Further investigation of the PCR products should now elucidate any point mutations that may cause the described phenotype.

ANIMAL GENETICS [G] Paper G2.1

Implementation of a routine genetic evaluation for longevity based on survival analysis techniques in dairy cattle populations in Switzerland
N. Vukasinovic*[1], J. Moll[2], and L. Casanova[2]. [1]Institute of Animal Science, Swiss Federal Institute of Technology, 8092 Zurich, Switzerland, [2]Swiss Brown Cattle Breeders' Association, 6300 Zug, Switzerland.

A genetic evaluation of sires based on survival analysis of functional longevity of their daughters has been implemented in the populations of Braunvieh, Simmental, and Holstein cattle in Switzerland. A Weibull mixed survival model was used to estimate heritabilities and breeding values of sires, using data on cows calved from April 1, 1980 onward. Data on Braunvieh and Simmental cows included over 1 million records, data on Holstein cows comprised around 250 000 records. The data contained approximately 20% of censored records. Besides the random sire effect, the model included effects of herd-year-season, age at first calving, parity, stage of lactation, alpine pasturing (Braunvieh and Simmental), and relative milk yield and relative fat and protein percentage within herd to account for culling for production. Heritability of functional longevity, estimated on a sample including approximately 150 000 animals, were .18, .20, and .18 for Braunvieh, Simmental, and Holstein, respectively. Breeding values were estimated for all sires having at least 6 daughters or 3 granddaughters in the data. Breeding values of sires can be expressed in genetic standard deviations or in days of functional productive life and published in sire catalogs along with breeding values for production traits.

Paper G2.2

A proposal for genetic evaluation for functional herd life in Italian Holsteins
P. Schneider and F. Miglior. ANAFI, Via Bergamo 292, 26100 Cremona, Italy.

An analysis of the productive life on Italian Holstein has been performed by means of Survival Analysis. The data set consisted of 13,186,886 lactation records that has been edited out to obtain lifetime records from 1978 to 1998. A mixed Weibull model was fitted. The probability of being culled (hazard) was defined as a product of a baseline hazard function and explanatory variables. The model allowed the inclusion of censored records and the correction of the length of productive life for voluntary culling due to low production. The following effects that affect the culling process were included: the fixed effects of age at first calving, interaction stage of lactation- lactation number, milk production within herd, change in herd size, year of calving, and random effects of sire and herd-year. The impact of each effect on functional herd life was evaluated to determine the main factors which influenced the culling process. Estimated Transmitting Abilities expressed as relative culling rate were obtained for each sire. The development of a national genetic evaluation for Functional herd life is underway.

ANIMAL GENETICS [G]

Paper G2.3

The use of type traits as indicators for survival of dairy cows
A. Bünger and H.H. Swalve*. Institute of Animal Husbandry and Genetics, University of Göttingen, Albrecht-Thaer-Weg 3, D-37075 Göttingen, Germany.

In a survival analysis, type traits can be included as covariates to evaluate their use as predictors for survival. Ideally, data from all cows in a population should be used. Whereas data on the length of productive life (LPL) of individual cows can be retrieved from milk recording data, for type traits this requires that all cows in the population are scored for type at least once. A data set from the Osnabrueck region in North-Western Germany, for which this requirement was fulfilled in recent years, was used. Data consisted of 169,733 cows with information on LPL for calving years 1980 to 1996 (data set 1) and of 39,233 cows with information on LPL and type for calving years 1992 to 1996 (data set 2). The model included stage of lactation, relative production within herd, change of herd size and year-season as time dependent effects, age at calving as a time independent effect, and herd-year-season and sire as random effects.

For data set 2, the scores for 15 linear type traits were included as corrected phenotypic values, estimated breeding values and residuals from a previous analysis. The results indicate a moderate heritability of 0.22 (data set 1). Almost all type traits analyzed exceeded a 0.001 level of significance in their effect on survival with strongest effects for udder depth, fore udder attachment, and front teat placement.

Paper G2.4

Phenotypic relationship between type and longevity in the French Holstein breed
H. Larroque*, V. Ducrocq. Institut National de la Recherche Agronomique, Station de Génétique Quantitative et Appliquée, 78352 Jouy-en-Josas Cedex, France.

Survival analysis was used to examine the impact of type traits on length of productive life of 361385 Holstein cows in 3407 herds. Type information was available for only 45% of the cows. A reference Weibull regression model was defined, including all fixed effects of the French routine genetic evaluation for functional longevity and an indicator variable for availability of type information. The analysis consisted in looking at the effect of the addition to this reference model of each type trait separately, depending on : 1) the population considered (registered *vs* non-registered herds) 2) the presence or absence of a correction for milk production traits (functional *vs* true longevity) and 3) the inclusion of type traits as corrected phenotypic scores *vs* estimated genetic values.

In all cases (but even more so in registered herds), cows with a bad udder (especially unsatisfactory udder-hocks distance) had a much higher culling risk. Body size traits had a influence on longevity in registered herds only, suggesting the existence of voluntary culling on size in those herds. Feet and legs traits -at least those presently recorded- did not influence longevity. Correction for milk production further increased the importance of udder traits whereas the influence of body size disappeared in registered herds. These results were generally more pronounced when type variables were included in the model as estimated breeding values.

ANIMAL GENETICS   [G]   Paper G2.5

Genetic analysis of functional longevity in Danish Landrace sows
L.H. Damgaard[*1], I.R. Korsgaard[1], J. Jensen[1] and L.G. Christensen[2]. [1]Department of Animal Breeding and Genetics, Research Centre Foulum, P.O. Box 50, DK 8830 Tjele, Denmark. [2]Department of Animal Science & Animal Health, The Royal Veterinary & Agricultural University, Grønnegaardsvej 3, DK-1870 Frederiksberg C, Denmark,*

The objective of this study was to estimate genetic parameters of functional longevity in Danish Landrace sows. Functional longevity of a sow was defined as the ability of the sow to remain alive irrespective of production, and was accessed as the time from first farrowing until culling corrected for production. Production was defined as number of piglets born and number of days open in earlier parities. The data was provided from the Danish Pig Breeding database (DanBreed, Danske Slagterier) and included 26,775 sows from 80 multiplier herds. Only sows with first farrowing after January 1. 1995 were included, and the end of the study period was December 1. 1998. Data was analysed using a sire model for survival data. This enabled right censored records to be included in the analysis, i.e. records of sows still alive at the end of the study period, as well as records of sows sold to herds not included in the analysis (36 %). The model developed and the corresponding genetic parameters will be presented.

Paper G2.6

Bayesian analysis of variance components of fertility in Spanish Landrace pigs.
Varona, L.(*), Noguera, J. L. Area de Producción Animal. Centro UdL-IRTA. C/ Rovira Roure, 28198. Lleida (SPAIN).

Fertility is a threshold trait with two categories (1=Pregnant; 0=Non Pregnant). A bayesian analysis with a binary threshold animal model has been performed with 18695 IA mating records from 1987 to 1997, concerning 4085 sows and 187 boars. The pedigree included all individuals genetically linked with data. The threshold response was modeled under a probit model. The model for liability includes paternal and maternal genetic effects, paternal and maternal permanent environmental effects, 54 herd-year-season effects and 7 order of parity effects. Posterior Mean (and Standard deviation) for paternal heritability was 0.028 (0.014), and 0.038 (0.014) for maternal heritability, and 0.072 (0.017) and 0.028 (0.009) for ratios of paternal and maternal permanent environmental effects. Posterior mean (and standard deviation) of genetic correlation between paternal and maternal genetic effects was -0.513 (0.264). Differences between parities suggest that fertility is greater in second parity than in the first one, and decrease in subsequent ones.

ANIMAL GENETICS   [G] Paper G2.7

Genetic parameters of a random regression model for daily feed intake of performance tested French Landrace and Large White growing pigs
U. Schnyder[1]*, A. Hofer[1], F. Labroue[2] and N. Künzi[1]. [1]*Institute of Animal Science, Swiss Federal Institute of Technology (ETH), CH-8092 Zürich, Switzerland*, [2]*Institut technique du porc, La Motte au Vicomte, BP 3, 35651 Le Rheu cedex, France.*

Daily feed intake data of 1285 French Landrace (FL, 1044 boars and 241 castrates) and 2425 Large White (LW, 2039 boars and 386 castrates) growing pigs was recorded with electronic feed dispensers in three French central testing stations in years 1992-94. Group housed pigs fed ad lib were performance tested from 35 to 95 (entire males), respectively 100 kg (castrates) live body weight. A second order polynomial in days on test with fixed regressions for sex and batch, random regressions for additive genetic, pen, litter and individual permanent environmental effects and a constant residual variance was used. Variance components were estimated from weekly means of daily feed intake by means of a Bayesian analysis using Gibbs sampling. After a burn-in of 40000 rounds, determined by the coupling chain method, 60000 samples were used to calculate posterior means of (co)-variances. As effective sample size for genetic parameters was very low (between 5 and 30 samples) results are only preliminary. Heritability estimates of regression parameters were 0.54 (FL) and 0.18 (LW) for intercept, 0.10 (FL&LW) for linear, 0.09 (FL) and 0.07 (LW) for quadratic term. For 11 weekly means of daily feed intake, heritabilities increased with time on test from 0.12 to 0.38 (FL) and 0.09 to 0.23 (LW). Genetic eigenfunctions reveal that altering the shape of the feed intake curve by selection is difficult.

Paper G2.8

Estimates of direct and maternal genetic covariance functions for early growth of Australian beef cattle
K. Meyer. *Animal Genetics and Breeding Unit, University of New England, Armidale, NSW 2351, Australia.*

Data from a selection experiment with monthly weighing of calves were utilised to examine the scope for modelling early growth of beef cattle with a random regression model. Up to 9 weights till weaning were available for all calves, and weights up to 18 months for animals remaining in the herd. Estimates of covariance functions for direct and maternal genetic effects as well as permanent environmental effects due to animals and dams were obtained by restricted maximum likelihood, fitting regressions on orthogonal (Legendre) polynomials of age. Orders of polynomial fit up to 5 (quartic regression) were considered, forcing estimated covariance matrices of random regression coefficients to have reduced rank where applicable. Temporary environmental effects (measurement errors) were assumed to be independently distributed but have heterogeneous variances. Numerous, computationally demanding analyses were required to determine the most appropriate model of analysis. Covariance matrices for ages in the data were constructed, and estimated variance components and correlations compared to previous estimates treating weights at different ages as separate traits. Implications for use of random regression models in genetic evaluation of beef cattle for growth are examined.

ANIMAL GENETICS   [G]   Poster G2.9

Genetic parameters for individual birth and weaning weight of Large White piglets
D. Kaufmann[1], *A. Hofer*[1], *J.P. Bidanel*[2] *and N. Künzi*[1]. [1]*Institute of Animal Science, Swiss Federal Institute of Technology (ETH), 8092 Zürich, Switzerland,* [2]*INRA, SGQA, 78352 Jouy en Josas Cedex, France.*

Data from a French experimental herd recorded between 1990 and 1997 were used to estimate genetic parameters for individual birth and weaning weight of Large White piglets using restricted maximum likelihood (REML) methodology applied to a multivarate animal model. Besides fixed effects the model included random common environment of litter (birth litter and nurse litter for birth and weaning weight, respectively), direct and maternal additive genetic effects. The data consisted of weight observations from 18151 animals for birth weight and from 15360 animals for weaning weight with 5% of animals transferred to a nurse. Estimates of direct and maternal heritability and proportion of the common environmental variance for birth weight were 0.02, 0.21 and 0.12, respectively. The corresponding values for weaning weight were 0.09, 0.16 and 0.23. The direct and the maternal genetic correlations between birth and weaning weight were positive. Weak positive (negative) genetic correlations between direct effects for both traits and maternal effects for birth (weaning) weight were found. A single trait model for weaning weight with the maternal additive genetic effect of the nurse instead of the mother resulted in direct and maternal heritability estimates of 0.06 and 0.31, respectively.

Poster G2.10

Simultaneous estimation of the covariance structure for production and reproduction traits in pigs
D. Peškovičová[1]*, J. Wolf*[2]*, E. Groeneveld*[3]*, L. Hetényi*[1]. [1]*Research Institute of Animal Production, Nitra, Slovakia.* [2]*Research Institute of Animal Production, Prague, Czech Republic*[3]. *Institute od Animal Husbandry and Animal Behaviour, Mariensee, Germany.*

Knowledge about the relationship between reproduction and production traits is needed in order to design selection criteria in pig breeding. Data from field test - average daily gain and backfat thickness (ADGFT, BFFT) and station test - daily gain, valuable cuts and backfat thickness (ADGST, VC, BFST) together with number born alive in the first litter (NBA1) and number born alive in the second and higher litters (NBA2+) as a trait with repeated measurements were used to estimate the complex covariance structure. The REML estimates of the 7 trait animal model were computed using the VCE package. Heritability estimates agree very well with those calculated in separate analyses of production and reproduction traits. Low genetic correlations were found between field and litter size traits (from 0.00 to -0.12). The highest unfavourable genetic correlations were estimated between NBA1 and ADGST (-0.16±0.07) and between ADGST and NBA2+ (-0.28±0.07), while genetic correlations between VC and litter size traits were positive (0.14±0.04). The design of the model was discussed more in detail. The covariance matrices will be used in the national genetic evaluation in Slovakia.

ANIMAL GENETICS   [G]

Poster G2.11

Estimates of genetic parameters for reproduction traits at different parities in Dutch Landrace pigs

E.H.A.T. Hanenberg*, E.F. Knol and J.W.M. Merks. IPG, Institute for Pig Genetics BV, P.O. Box 43, 6440 AA Beuningen, The Netherlands.

Data from Stamboek Pigs were used to estimate genetic parameters for reproduction traits in the first 6 parities of Dutch Landrace sows. Analyses were performed with AIREML using a model with equal design and herd-year-season of first parity as fixed effect.
Complete data of 14.739 sows were available. Heritabilities of the traits total number of born pigs (TNB) and number of pigs born alive (NBA) were similar for the different parities (± 0.10), except the heritability for the second parity which was significantly lower (0.04). In additional analyses a higher heritability in the second parity was found when corrections were made for number of weaned pigs in the first parity and interval weaning to conception. Heritability increased with parity number for number of pigs born death (NBD) from 0.02 to 0.09 and for mothering ability (MA, defined as the percentage of weaned pigs) from 0.02 to 0.05. Heritability decreased with parity number for interval weaning to first insemination (INT) from 0.08 to 0.01. More phenotypic and genetic variance was found in first and second parity. Genetic correlations between parities were high for all traits. Undesired correlations were found between TNB and NBD (0.29) and TNB and MA (-0.30).

Poster G2.12

Implications of a non-genetic antagonism between maternal effects of two generations in the dispersion structure of the maternal animal model

R. Quintanilla*[1], M.R. Pujol[2], J. Piedrafita[2]. [1]Departamento de Producción Agraria, Universidad Pública de Navarra, 31006 Pamplona, SPAIN. [2]Departament de Patologia i de Producció Animals, Universitat Autònoma de Barcelona, 08193 Bellaterra, SPAIN.

In the context of the maternal animal model (MAM), the influence of maternal effects (ME) received during suckling on the maternal ability of females has been modelled through a regression coefficient (β) of the maternal permanent environmental effects provided over the whole ME received. This model implies a recurrent relationship with the ME of all previous generations that has been developed analytically through infinite series. According to the expected phenotypic (co)variances, an antagonism between ME of two generations would always reduce the dam-offspring covariance, but the covariance between other maternal relatives and the total variance can be increased or reduced depending upon the magnitude of β and the relationship between genetic and residual maternal variances. Expected biases in the (co)variance components estimated under MAM would affect specially maternal variances, both genetic and permanent environmental, whereas estimates of direct-maternal genetic covariance could be unbiased if they were obtained from some relatives. Results obtained in an analysis carried out on simulated populations with different values of β confirmed these theoretical expectations.

ANIMAL GENETICS   [G]   Poster G2.13

Three lactations vs all lactations model
F. Canavesi*, A.B. Samorè, F.A. Miglior. ANAFI, Via Bergamo 292, 26100 Cremona, Italy.

The official genetic evaluation model in Italy includes all lactations. In several countries around the world usually only the first three lactations are used for genetic evaluation purposes. The aim of the present study was to compare results from an all lactations model with results from a three lactations model. Data for February 1999 official genetic evaluation were used for the analysis. About 7.5 millions lactations from about 3 millions of cows calving from 1983 were used for the all lactations model. Setting a maximum of three lactations reduced the size of the data set to about 6 millions of records.
Results from the two models were compared at the National level and at the International level. Impact on genetic trend, variability of proofs over time and ranking both at national and international level were investigated.

Poster G2.14

A parallel solver for Mixed Model Equations stratified to minimise communication
P. Madsen[*1] and M. Larsen[2]. [1]Danish Institute of Agricultural Sciences, Department of Animal Breeding and Genetics Research Centre Foulum, P. O. Box 50, DK-8830 Tjele, Denmark. [2]UNI·C Danish Computing Centre for Research and Education, DTU, Bld. 304, DK-2800 Lyngby, Denmark.

Attempts to parallelize a solver for Mixed Model Equations (MME) based on "iteration on data" and a combination of Gauss-Seidel and Jacobi iteration, have shown that a parallel speedup could be achieved, but communication among the processors could limit the parallel speedup substantially. Based on an analysis of amount and pattern of communication, a new implementation has been programmed. It is inspired by the idea's for incorporating external information into MME's, and is characterised by a split of the MME system into equations within a processor (local equations) and equations across processors (global equations). Local equations only receive contributions from data on the processor in question, while the global equations receive contributions from data on more than one processor. During iteration, only global equations generate communication. Global equations would typically be equations for fixed effects other than the one solved by Gauss-Siedel, animal equations for sires and dams of sires and equations for animals in herds, that are distributed on more than one processor. In order to keep the number of global equations as low as possible, all data from a herd should always be on the same processor. Test runs on models of dimensions between 1.2 and 30.0 mill. equations has showed scalability close to linearity and in some cases super linear scalability, primarily due to avoidance of IO-operations and better utilisation of cache memory.

ANIMAL GENETICS [G]

Poster G2.15

An integrated use of statistical packages in organizing, manipulating and analyzing a set of one hundred thirty thousand beef data for practical purposes
S.I. Tzortzios* and N. Gitsakis. Lab. of Biometry, Faculty of Agriculture-Crop and Animal Production, University of Thessalia. 38334 Volos, Greece.

The importance of huge amounts of field data in various research purposes of scientific and practical interest is undoubtedly recognized. However, such a significant research material is associated with serious problems in organizing and manipulating it in the proper way. Among the first problems the researcher expects to encounter is the computer platform which is the mostly suitable for the creation of the proper data file in order to avoid difficulties in: (a) certain needs in data manipulation according to specific scientific requirements, and (b) possible needs for transforming it to other more efficient (for specific statistical purposes) statistical packages.

In this study an effort was undertaken to handle a set of one hundred and thirty five thousand beef data retrieved from a national beef cattle recording scheme (MLC) in the form of a microfish. A failure in a satisfactory transfer of the data from the photocopies (coming from the microfish) to the computer, through the OCR-scanner, in electronic form, a data file of 135 variables and more than 1,000 cases was organized by punching it in an excel sheet. The original information referring to various herds, sires, cows, calves, management systems and certain productive traits, was manipulated properly in order to become compatible to other statistical packages, such as Systat and SPSS, which leads to a more efficient statistical analysis of specific scientific and practical purposes, mainly referring to the field of genetics and beef husbandry.

Poster G2.16

The (co)variance structure of Sarda ewe's fat and protein content Test Day records
A. Carta[1], N.P.P. Macciotta*[2], S.R. Sanna[1], A. Cappio-Borlino[2]. [1]Istituto Zootecnico e Caseario per la Sardegna, Loc. Bonassai, 07040-Olmedo, Italia. [2]Dipartimento di Scienze Zootecniche, Università di Sassari, Via De Nicola 12, 07100-Sassari, Italia.

The analysis of test day yields (TDY) with mixed linear models usually assumes that TDYs at different DIM intervals have the same variance and that the correlation between all pairs of TDYs, within each lactation, is the same. This work evaluated the suitability of this compound symmetry (CS) structure to fit the (co)variance pattern of TD fat and protein contents of Sarda ewes. With this aim, 1178 milk TDYs were analysed by a linear mixed model including a random factor associated with each lactation and the test date and the age by DIM interval interaction, as fixed effects. The actual (co)variances were obtained by fitting data to the linear model without any imposed covariance pattern (UN). Then CS and an order 1 autoregressive (AR(1)) models were used, being that measures close in time tend to be more correlated than measures far apart in time. Repeatability estimates obtained with the CS models were 0.37 and 0.51, respectively for F% and P%. The AR(1) analysis resulted in correlations between pairs of records at different DIM intervals which ranged between 0.43 for a lag of 1 to 0.33 for a lag of 6 in the case of F% and between 0.67 for a lag of 1 to 0.28 for a lag of 6 in the case of P%.

ANIMAL GENETICS   [G]   Poster G2.17

Modelling milk production in extended lactations of dairy cows
B. Vargas[1,2], W.J. Koops[2] and J.A.M. van Arendonk[2]. [1]Escuela de Medicina Veterinaria, Universidad Nacional de Costa Rica, P.O. Box 304-3000, Heredia, Costa Rica,[2]Wageningen Institute of Animal Sciences, P.O. Box 338, 6700 AH Wageningen, The Netherlands.*

Nine mathematical models were compared on their accuracy to fit daily milk yields in standard and extended lactations of Holstein-Friesian cows in Costa Rica. Lactations were classified in 26 classes according to parity (first and later), lactation length (9 to 17 mo) and days open (1 to 10 mo). The diphasic model and lactation persistency model resulted in the best goodness of fit as measured by adjusted R^2, residual standard deviation and Durbin-Watson coefficient. All other models showed lower accuracy and positively correlated residuals. Diphasic model showed the best prediction of productions in normal and extended lactations. It is concluded that the diphasic model provides accurate estimates of milk yield for standard and extended lactations. Days open was found to have a negative effect on milk production for cows with a given length of lactation. The total, rather than the 305-d production is important for the economic efficiency of dairy cattle production and extended lactations should, therefore, be given more attention in research and extension.

Poster G2.18

Genetic parameters of longevity traits in Holstein dairy cattle of Iran
M.Heydarpour, Department of Animal Sciences, College of Agriculture, Ferdowsi University of Mashhad, P.O.Box: 91775-1163,Mashhad-Iran

Longevity in dairy cattle is generally defined as the length of time a cow remains productive in the herd. Longevity is a measure of the success of the cow to survive both voluntary and involuntary culling. Various measures are used for longevity. In this study, lifetime measures including total milk production, number of lactation, number of days in lactation, herd life, and length of productive life were analysed. Also, analysed were stayability measures to 36,48,60, or 72 months of age and to 12,24,36, or 48 months of productive life.Measures of longevity were analysed after correction for milk production during first lactation. Data on 5600 cows from 1983 to 1996 at Karaj Milk Improvement Research Centre (KMIRC) were used to calculate means for longevity traits per year of birth.Heritabilities of longevity traits were estimated for cows born in 1983,1990 and 1996 after data were edited to require at least 10 daughters per sire and 4 cows per herd. Phenotypic and genetic correlations were estimated for the 1996 data. Heritability estimates differed between years of birth. Heritability estimates for all stayability traits were low. Genetic correlations among the productive life traits were higher than those among the total life traits. Stayability to 19 months of productive life is suggested to be included in a selection program if sire evaluation for longevity were desired.

ANIMAL GENETICS   [G]

Poster G2.19

Genetic parameter estimation of cow survival in the Israeli dairy cattle population
Petek Settar[*1], Joel. I. Weller[2]*
[1]*Biometry and Genetics Department, Ege University, Izmir 35100, Turkey*
[2]*Institute of Animal Sciences A. R. O., The Volcani Center, Bet Dagan 50250, Israel*

The linear model method of VanRaden and Klaaskate to analyse herd life was expanded. Information on conception and protein yield was included in the estimation of predicted herd life of Israeli Holsteins. Variance components were estimated by a multitrait animal model. Heritability was slightly higher for herd life than for number of lactations, but genetic correlations were close to unity. The expected herd life of pregnant cows was 420 d greater than open cows. Each kg increase in protein yield increased expected herd life by 9.5 d. Heritability of expected herd life increased from 0.11 for cows 6 mo after first calving to 0.14 for cows 3 yr from first calving. The genetic correlation of expected and actual herdlife increased from 0.87 for records cut after 6 mo to 0.99 for records cut 3 yr after first calving. Phenotypic correlations increased from 0.61 to 0.94.

Poster G2.20

The heritability of visual acceptability in South African Merino sheep
G.J. Erasmus[*1], *F.W.C. Neser*[1], *J.B. Van Wyk*[1] *& J.J. Olivier*[2]. [1]*Dept Animal Science, University of the OFS, Box 339, Bloemfontein 9300. South Africa.* [2]*ARC, Animal Improvement Institute, Middelburg. EC. 5900. South Africa*

Selection on production and reproduction in South African Merino sheep is always combined with selection on visual appraisal and will in all probability remain so for many years to come. Heritabilities for visual acceptability were estimated in two parent Merino studs to ascertain whether visual acceptability can be improved genetically with more selection pressure eventually being applied to production and reproduction A threshold model was fitted utilizing the procedure suggested by Gianolla & Foulley (1985). Effects included were sex, birth status, age of dam and year of birth. Acceptability of an animal was defined as still being in the stud at 18 months of age, meaning that such an animal would either be retained or sold for breeding purposes. This provided one threshold and two classes (acceptable and not acceptable). Records from the two studs used were 3011 progeny of 31 sires and 6495 of 70 sires respectively. Error variance was taken as unity and heritability estimated as $h^2 = 4\sigma^2_s / 1+ \sigma^2_s$ where σ^2_s is the sire variance. Heritability estimates were 0.210 and 0.448. The results suggest that selection improvement for visual acceptability is possible. Supplying breeding values for this trait for AI sires should be seriously considered in National evaluation programmes.

ANIMAL GENETICS [G]

Poster G2.21

Genetic parameters for growth traits in Moroccan Timahdit breed and its D'man x Timahdit cross
M. El Fadili*[1,2], C. Michaux[2], J. Detilleux[2], F. Farnir[2], P.L. Leroy[2]. [1]Département de Zootechnie, INRA, BP-415, Rabat, Morocco, [2]Département de génétique, ULg, B- 4000, Liège, Belgium.

Records on 522 lambs of Timahdit (T) and 722 lambs of D'man x Timahdit (DT), born from 1992 to 1997 and managed under similar conditions were used to estimate genetic parameters for lamb weights at birth (BW), at 30 (W30), at 70 (W70) and at 90 (W90) days of age, and for average daily gains from 10 to 30 (ADG1), 30 to 70 (ADG2) and 30 to 90 days of age (ADG3). Estimates of heritability (h^2) and genetic correlation (r_G) were obtained with MTDFREML program. The animal model included fixed effects of year, sex, birth type, rearing type, and ewe age. The DT lambs had weights and gains similar to T lambs. The h^2 were larger in T than in DT lambs for all traits, except for BW, that was larger in DT crosses (0.73 vs 0.43). The h^2 estimates for postnatal weights ranged from 0.39 to 0.46 in T and from 0.17 to 0.40 in DT. The h^2 estimates for ADG's ranged from 0.26 to 0.38 in T and from 0.03 to 0.12 in DT. The r_G among weights, ranged from 0.74 to 0.91 in T and from 0.84 to 0.94 in DT. The r_G of ADG1 with W70 and W90 were 0.96 and 0.75 in DT, and 0.84 and 0.68 in T. Similarly, high positive r_G were observed between ADG2 and ADG3. It is concluded that selection for improvement of lamb performances could be based on early weights and daily gains.

Poster G2.22

Performance, carcass yield and quality of meat in broiler rabbits: a comparison of six breeds
Vira Skřivanová[1], M. Marounek[2], Eva Tůmová[3], M. Skřivan[3], P. Klein, Martina Kuboukova[1], Jana Laštovková[1]. [1]Research Institute of Animal Production, 104 00 Prague 10 - Uhříněves, Czech Republic, [2]Institute of Animal Physiology and Genetics, 104 00 Prague 10 - Uhříněves, Czech Republic, [3]Czech University of Agriculture, 161 00 Prague 6, Czech Republic.

The performance, feed conversion, slaughter parameters and composition of meat were investigated in 48 rabbits between the ages of 30 and 80 days. Six genotypes were compared: New Zealand White (NZW), Californian White (CW), New Zealand White x Californian White (N x C), Hyla 2000 (Hy), Zika (Zi), Cunistar (Cu). The NZW rabbits had the lowest weight gain (25.4 g/day) and the highest consumption of feed per 1 kg of gain (4.56 kg). The highest gain was observed in the N x C breed (34.5 g/day) and the best feed conversion in the CW breed (2.98 kg/kg). In the NZW rabbits the highest dressing percentage (66.1 %) was found. The lowest dressing percentage (59.6 %) was found in the CW breed. In all breeds, the musculus longissimus dorsi contained more protein and less fat, cholesterol and hydroxyproline than muscles of the hindleg. Carcass of the CW rabbits had the lowest fat and cholesterol contents (7.1 - 28.8 and 0.31 - 0.54 g/kg, respectively). The highest fat contents in the musculus longissimus dorsi and hindleg muscles were found in the Zi and NZW rabbits, respectively. In other quality parameters, differences among breeds were not statistically significant.

ANIMAL GENETICS   [G]   Poster G2.23

The effect of selection for growth rate on carcass composition and meat characteristics of rabbits
M.Piles, A. Blasco* and M. Pla. Departamento de Ciencia Animal, Universidad Politécnica de Valencia, P.O.box 22012, 46071 Valencia, Spain

In order to asses the effect of selection for growth rate in carcass composition and meat quality two groups of rabbits belonging to different generations of selection were compared using a bayesian approach. Embryos belonging to the generations 3rd and 4th of selection were frozen and thawed to be contemporary with the animals of the 10th generation of selection. The control group (C), formed from the offspring of these embryos, was contemporary of the offspring of the generation 11th of selection, chosen at random, who constituted the selected group (S). 65 animals of the group S and 66 of the group C were slaughtered when they reached approximately the Spanish commercial liveweight of 2 kg (at 9 weeks of life). Carcasses were retailed and measured according to the norms of the World Rabbit Scientific Association. An animal model including group, sex and live weight as covariate was used for comparisons. The lower degree of maturity at slaughter weight of the animals belonging to the group S, did not lead to appreciable modifications in most of the parameters of carcass composition and meat quality. Only it is noticeable a lower content in dissectible fat of the animals belonging to group S respective to group C as well as a higher content in fat of the hind leg. Females showed a higher dissectible fat content than males.

Poster G2.24

Estimates of genetic parameters for hip and elbow dysplasia in Finnish Rottweilers
K. Mäki*, A.-E. Liinamo and M. Ojala. Department of Animal Breeding, P.O. Box 28, FIN-00014 Helsinki University, Finland.

Data from 2 764 rottweilers born in 1987 - 1996 were analyzed with a REML procedure using a mixed linear animal model to obtain variance component estimates for hip and elbow dysplasia. Hip joints were scored as normal, borderline and slight, moderate and severe hip dysplasia, with numbers from 0 to 5, respectively. Elbow joints were scored so that 0 represented dogs having normal or borderline elbow joints and scores 1, 2 and 3 dogs with a slight, moderate and severe elbow dysplasia, respectively. The mean for the hip scores was 1.07 and for elbow scores 0.60. Estimates of heritability for hip and elbow dysplasia were 0.58±0.04 and 0.31±0.04, with a genetic correlation 0.37±0.08 between the traits. A genetic improvement of about one genetic standard deviation was observed in both traits during a period of ten years.

ANIMAL GENETICS [G]

Poster G2.25

Lactation and sample test day correlation and repeatability of milk somatic cell count in Hungarian Holstein Fresian cows
A.A. Amin*[1] and T. Gere[2]. [1]Suez Canal University, Faculty of Agriculture, Department of Animal Production, Ismialia - Egypt. [2]Godollo University, College of Agriculture, Department of Animal Breeding, H-3200 Gyongyos, Hungary.

Somatic cell count (SCC) and daily milk yield (DY), fat (F%), protein (P%) and lactose percentage (Lac%) were the essential traits studied. The statistical mixed model included cows as a random effect and genetic groups, parity age of calving within parity and herd effect (farm, year and month of calving) as fixed effects. Estimates of lactational repeatability were ranged from 0.05 to 0.50 and were lower than estimates of overall sample test day from 0.24 to 0.59 for all studied traits except for SCC (0.50 vs. 0.42). Lactational and sample test day correlations between all studied traits were ranged from (-0.11 to -0.30), (0.05 to 0.75) and (-0.18 to -0.43), (0.06 to 0.65) respectively. Regression coefficient of SCC on sample test milk yield was -0.95 ± 0.02 and highly significant. On the other hand regression of SCC on milk yield composition was negative only for Lac% ($-0.09\%\pm0.02$) and positive for both F% and P% (0.03 ± 0.002 and 0.03 ± 0.001, respectively). Selection for shortage and homogenous age at calving and reduce SCC could be possible and would be advisable to decrease milk production disorders. Genetic restriction through to selection on age at calving may cause slow response of selection.

Poster G2.26

Genetic parameters for production traits in mink
B.K.Hansen*[1], P. Berg[2] & J. Jensen[3]. Danish Institute of Animal Sciences, Department of Animal Breeding and Genetics, Research Centre Foulum, P.O. Box 50, DK-8830 Tjele, Denmark.

Breeding values for mink are predicted on a within farm basis, because there is few known genetic links between farms under the current breeding system.
The aim of this paper is to develop models for within farm genetic evaluation. The individual farms included are so large, that genetic parameters can be estimated for each farm individually. The data were from black and brown mink colour types from 10 farms. All farms included data from more than 2 generations. Altogether records on litter size from 57984 litters, including 334769 kits. Furthermore, body weight and pelt quality graded in November from 92296 male and 97389 female kits were analysed. Variance components were estimated in univariate Animal Models by an Average Information REML algorithm. The heritability estimates of the direct additive effect (h_a^2) per farm were on body weight between 0.36 to 0.51, for pelt quality between 0.16 to 0.34, and for litter size 0.05 to 0.15, respectively. Common environmental (c_e^2) effects of litter size between repeated litters of the same dam were between 0.04 and 0.15. The repeat-ability on litter size was between 0.10 to 0.24. Including common environmental effects is necessary to obtain unbiased predictions of breeding values.
The within farm average parameters will be compared with heritability estimates from a joint analysis of all data.

ANIMAL GENETICS [G]   Poster G2.27

The heritability coefficients of some traits in Polar blue fox population
S. Socha. Department of Animal Breeding, Agriculture Faculty, Agricultural and Pedagogic University, 08-110 Siedlce, ul. B. Prusa 12. Poland.

The animal model was applied for evaluation of heritability coefficients of body size and fur quality traits of Polar blue fox population (*Alopex logopus* L.). Data perfor-mance evaluation of 15000 animals during 7 years.

Heritability coefficients for traits were as follows: animal's size 0.288; colour intensity 0.555; colour clarity 0.296; fur density 0.415; hair length 0.461; general appearance 0.779; total number of scores 0.423.

In comparison with earlier evaluated heritabilities coefficients for Polar blue fox popu-lation values differ slightly. Some values were lower while the other greater. It can be supposed, that values evaluated in animal model described genetic variability the most precisely.

ANIMAL GENETICS [G]

Paper G3.1

Statistical methods in animal breeding: wherefrom and whereto?
Daniel Gianola. Department of Animal Sciences, Department of Biostatistics and Medical Informatics, Department of Dairy Science. University of Wisconsin, Madison, WI 53706, USA.

Development and applications of statistical ideas in animal breeding are reviewed, from least-squares to likelihood-based and Bayesian methods. The evolution of specifications from early linear models to complex longitudinal, nonlinear, hierarchies is discussed. The problem of taking selection into account for inference continues: alternative constructs to Henderson's treatment are needed, e.g., in selection for intermediate optima. Potential contributions of missing data theory to the analysis of selection experiments with unknown externalities are mentioned. Prediction of nonlinear merit is revisited, highlighting how the problem can be tackled using, e.g., Markov chain Monte Carlo. Sampling methods for calculating probabilities of ordered events are illustrated. Potential future areas of work, excluding statistical genomics (covered by another speaker), are suggested. These include: search for better functional forms and distributional assumptions, robust methods of inference, criteria for model criticism, and the flexibility offered by sampling/resampling methods. Potential dangers of making strong high-dimensional assumptions, e.g., as in multiple-trait analysis, and the role of structural models for reducing dimensionality are emphasized.

Paper G3.2

Random regression models to describe phenotypic variation in weights of beef cows when age and season effects are confounded
K. Meyer. Animal Genetics and Breeding Unit, University of New England, Armidale, NSW 2351, Australia.

Monthly weight records of beef cows in a selection experiment in Western Australia were analysed fitting a range of random regression models. Records from 19 to 84 months of age were considered, yielding up to 62 'repeated' records per animal. Data were characterised by strong seasonal variation in weights due to a Mediterranean climate with early spring rains, subsequent almost complete drought from late spring to early autumn and resulting winter dearth. Short mating periods resulted in the bulk of calves born in a 2 months period each year, and thus confounding between age and season. Fluctuations in variances corresponded only partially to those in weight (scale effects) with coefficients of variations highest at lowest weights. Analyses fitted regression coefficients for animals only, i.e. did not attempt to separate genetic and permanent environmental effects, and ignored relationships between animals. Curves fitted included orthogonal polynomials of age up to an order of fit of 20, segmented quadratic polynomials with nodes every 6 or 12 months, and low-order Fourier series approximations in conjunction with orthogonal polynomials. Temporary environmental, 'measurement error' variances were considered independently but heterogeneously distributed. A large number of parameters were required to model seasonal variation in weights adequately. Implications for genetic analyses of such data and estimation of breeding values for mature weight are considered.

ANIMAL GENETICS [G]

Paper G3.3

Derivation of multiple trait reduced rank random regression (RR) model for the first lactation test day records of milk, protein and fat

E.A. Mäntysaari, Animal Breeding, Animal Production Research, Agricultural Research Centre MTT, FIN-31600 Jokioinen, Finland.

A two step approach of fitting covariance functions to estimated genetic (**G**) and residual (**R**) covariance matrices between predetermined intervals of lactation was used. Lactation was divided into 5 parts (5-20, 31-60, 121-150, 211-240 and 301-330 days in milk) leading into vcv matrices of size 15x15 ordered by parts within traits. Estimates based on data from 23,072 Ayrshire cows were composed from 22 different REML runs having 5 traits each. To assure positive definite matrices, separate runs were compiled together using a procedure similar to maximization step in EM algorithm.

First, the scaled **R*** was partitioned into $P+I\sigma^2_e$, σ^2_e being the smallest eigenvalue of 3 diagonal blocks of **R***. Second, covariance functions $G^* = \phi K_a \phi'$ and $P = \phi K_p \phi'$ were fit, where $\phi = I_3 \otimes \psi$ and ψ is 5x5 matrix of Legendere polynomials for average days in milk in 5 intervals. In breeding value prediction the K_a and K_p could be seen as genetic and permanent environment vcv matrices of 15 random regressions corresponding to covariables in ϕ. Examination of K_a and K_p suggested that 90% of variation in coefficients was explained by 6 largest eigenvalues and corresponding eigenvectors L_a (L_p). Thus, the number of random regressions for both the random effects were reduced by using ϕL_a and ϕL_p as covariables. Proposed RR model leads into equal heritabilities and correlations for individual test days as the multitrait model.

Paper G3.4

The effect of incomplete lactation records on covariance function estimates in test day models

M.H. Pool* and T.H.E. Meuwissen. Department of Animal Breeding and Genetics, DLO Institute for Animal Science and Health, P.O. Box 65, 8200 AB Lelystad, The Netherlands.

The effect of incomplete lactation records, on the estimated covariance components for covariance functions, was investigated in order to reduce the number of parameters to be estimated per animal, i.e. reducing the order of fit needed to model the variance-covariance matrix of test day records over time when using Legendre polynomials within a test day model. A data set of weekly milk recordings and a reduced data set with only complete lactation records (last test day measured after day 285) were used. The covariance function coefficients were estimated from the reduced data set. For analyses of goodness of fit the original data set was used. Previous results showed that at least a 5th order polynomial was needed.

Use of complete lactations to estimate model parameters improved the goodness of fit. Residual variance was slightly lower and the correlation structure modeled resembled the observed correlation structure in the data better. Overall, the goodness of fit improvement with the order of fit, which became negligible after an order of fit of four. It was concluded that the order of fit could be reduced form a 5th to a 4th order Legendre polynomial within a test day model.

ANIMAL GENETICS   [G] Paper G3.5

Accounting for variance heterogeneity in French dairy cattle genetic evaluation
C.Robert-Granié[*1], *B.Bonaïti*[2], *D. Boichard*[1] *and A. Barbat*[1]. *Institut National de la Recherche Agronomique,* [1]*Station de Génétique Quantitative et Appliquée,* [2]*Direction Informatique, 78352 Jouy-en-Josas Cedex, France.*

A linear mixed model assuming heterogeneous residual variances and known constant variance ratios was applied to the analysis of milk, fat, and protein yields, and fat and protein contents in the French Holstein, Montbeliarde, and Normande dairy cattle populations. The method was based on a log-linear model for the residual variances. This log-linear model included a region-year fixed effect and a herd-year random effect with a within-herd autocorrelation. The estimates of this correlation coefficient varied from 0.64 to 0.92, according to the trait and the breed. For yields, residual standard deviation doubled over the last 20 years, whereas differences across regions were more limited and decreased over time. This model was compared with the homogeneous model and with the multiplicative model of Meuwissen *et al* (1996). Both heteroskedastic models provided very similar results. Accounting for heterogeneous variances had important consequences on cows ranking. Consequences on AI bull ranking were very limited, although the method affected estimated genetic trend and within birth year EBV variability.

Paper G3.6

Recent developments in linkage and association analysis to search for single genes
C. Stricker, Institute of Animal Science, ETH Zentrum CLU, 8092 Zurich, Switzerland

Identifying single genes underlying the inheritance of economically important traits often involves statistical modeling of multi-factorial phenotypes in conjunction with genotypes at polymorphic marker loci. The traits may have been observed in especially designed experiments or in complex pedigrees. There may be prior information available about the inheritance of the trait. This leads to so-called candidate gene approaches with only certain individuals being genotyped at specific loci. On the other hand, if there is no such prior information, a global search of the whole genome for DNA-regions containing quantitative trait loci is carried out. Usually relying on linkage analysis, such global searches study a set of polymorphic loci that we have no reason to believe are functionally related to the observed trait. Association analysis on the other hand is mainly used in candidate gene approaches. Several different statistical methods are involved in association or linkage analysis. Model-based methods include pedigree analysis by Maximum Likelihood, linear regression or Markov Chain Monte Carlo principles. Model-free approaches include e.g. sib pair analysis and transmission-disequilibrium tests. This paper gives an overview over these methods and their respective statistical properties. Situations favoring certain approaches are discussed and illustrated by examples.

ANIMAL GENETICS [G]   Paper G3.7

Robust Bayesian polygene mapping using skewed student-t distributions and finite polygenic models
P. von Rohr*, and I. Hoeschele. Department of Dairy Science, Virginia Polytechnic Institute and State University, Litton Reaves Hall, Blacksburg, VA 24061, USA.

In most QTL mapping studies polygenic effects and residuals are assumed to follow normal distributions. Deviations from this assumption may lead to false detection of QTL. However, the practice of transforming phenotypic data prior to analysis may lead to a loss of power of QTL identification. To improve the robustness of Bayesian QTL mapping for continuous traits, the normal distribution for residuals is replaced with a skewed Student-t distribution. The latter distribution is able to account for both heavy tails and skewness, and both components are each controlled by a single parameter. Because the assumption of a multivariate Student-t distribution for the polygenic effects does not allow the estimation of the degrees-of-freedom parameter for this distribution, polygenic effects are fitted with a Finite Polygenic Model with allele frequency fixed at .5 or allowed to deviate from this value. The robust Bayesian analysis is being evaluated with simulated data sets, one conforming with the normality assumption and others deviating from this assumption. Skewness and heavy tails are introduced to the simulated data using an inverse Box-Cox transformation applied to a Student-t distribution. In a different approach skewed and kurtosed residuals are drawn from chi-squared and Student-t distributions, respectively.

Paper G3.8

Descent graphs in animal pedigrees: an application
M. Schelling*[1], R.L. Fernando[2], N. Künzi[1], C. Stricker[1]. [1]*Institute of Animal Science, ETH Zurich, CLU, 8092 Zurich, Switzerland,* [2]*Departement of Animal Science, Iowa State University, Ames IA 50011, USA*

In linkage analysis, peeling based algorithms provide exact likelihood calculation for multi-allelic pedigrees. However, these algorithms become inefficient for large and complex pedigrees. On the other hand, the Gibbs Sampler can handle arbitrary complex pedigrees. In multi-allelic situations however, the Gibbs Sampler may reveal a reducible Markov Chain that gets stuck in subspaces before the chain has converged to the true posterior distribution. Thus in spite of its advantages, the Gibbs Sampler cannot be applied to multi-allelic situations in general.
Combining the advantages of peeling and Markov Chain Monte Carlo methods, Sobel and Lange (1996) partitioned the calculations into two parts using the concept of descent graphs. A descent graph specifies the path of gene flow in a pedigree disregarding the particular founder alleles travelling on a specific path. The likelihood of a descent graph is calculated deterministically, it is used in a Metropolis-Hastings Sampler to generate a Markov chain of descent graphs. Specific transition rules to move from one graph to another are used to maintain irreducibility.
A modified form of this approach of Sobel and Lange is applied to a simulated example. Its efficiency and accuracy is compared with peeling based methods.

ANIMAL GENETICS [G] Paper G3.9

Genetic improvement of laying hens viability using survival analysis techniques
Ducrocq*[1], B. Besbes[2] and M. Protais[3]. [1]Station de Génétique Quantitative et Appliquée, INRA, 78352 Jouy-en-Josas, France, [2]ISA-Hubbard, Centre de Sélection, BP27, 35220 Chateaubourg, France, [3]ISA-Hubbard,Le Foeil, 22800 Quintin, France.

Survival of about 8 generations of a large strain of laying hens was analysed separating the rearing period (RP) from the production period (PP), after hens were housed. For RP (respectively PP), 97.8% (resp., 94.1%) of the 109160 (resp., 100665) female records were censored after on average 106d (resp., 313d). A Cox proportional hazards model stratified by flock (=season) and including a hatch-within-flock fixed effect seemed to reasonably fit the RP data. For PP, this model could be further simplified to a (non-stratified) Weibull model. Extending these models to frailty (mixed) sire-dam models, the sire genetic variances were estimated at 0.261 ± 0.026 and 0.088 ± 0.010 for RP and PP, respectively (corresponding to heritabilities on the log scale of 0.48 and 0.19). Non-additive genetic effects could not be detected, at least for PP. Selection was simulated by evaluating all sires and dams, after excluding all records from the last generation. Then, actual parents of this last generation were distributed into 4 groups according to their own pedigree index. Raw survivor curves of the progeny of extreme parental groups substantially differed (e.g., by more than 1.5% 300d after housing for PP), suggesting that selection based on genetic solutions from the frailty models could be very efficient, despite the very large proportion of censored records.

Paper G3.10

Bayesian estimation of parameters of a structural model for genetic covariances between milk yield in five regions of the USA
R. Rekaya, K. Weigel and D. Gianola. Department of Dairy Science, University of Wisconsin, 1675 Observatory Drive, Madison, WI. 53706, USA

Precise estimation of genetic covariances is difficult in multi-trait models because information is typically insufficient. A structural model explaining potential patterns in genetic covariances with fewer parameters may give more precise inferences. The objective was to describe 10 genetic correlations using a 3-parameter model. Data were first lactation yields of daughters of AI sires in 5 regions (USA). After edits ($\geq$10 records per sire; days in milk=DIM > 275) there were 3,465,334 lactations representing 43,755 sires. A multiple-trait model posited production in each region as a different trait; effects were herd-year-season, age at calving, DIM and sire. In the structural model, the genetic covariance between regions i and j was : Covariance$(i,j) = \beta_1 + \beta_2 G(i,j) + \beta_3 M(i,j)$, where G=genetic similarity (proportion of common sires) and M=management similarity (ratio of concentrate per 1000 kg milk between regions). A Bayesian analysis via Gibbs sampling led to the posterior distributions of the b's and of the genetic correlations. Heritabilities (.21 - .25) agreed with estimates from the standard multivariate analysis but genetic correlations were lower (.89 - .95). Genetic correlations increased with G and decreased with M.

ANIMAL GENETICS [G]   Poster G3.11

Multivariate analysis of Gaussian, right censored Gaussian, ordered categorical and binary traits
I.R. Korsgaard[1], M.S. Lund[1], D. Sorensen[1], D. Gianola[2], P.Madsen[1], J. Jensen[1]. [1]Danish Institute of Animal Sciences, Department of Animal Breeding and Genetics, Research Centre Foulum, P.O. Box 50, DK-8830 Tjele, Denmark, [2]Department of Meat and Animal Sciences, University of Wisconsin-Madison, WI 53706-1284, USA

A fully Bayesian analysis using Gibbs sampling and data augmentation in a multivariate model of Gaussian, right censored Gaussian, ordered categorical and binary traits is outlined. Threshold models are used for the ordered categorical and binary traits. In the threshold model, the existence of an underlying Gaussian distribution is assumed. The observed categorical value is determined via a grouping defined by thresholds on this underlying scale. In this presentation the Gibbs sampler is outlined and strategies for implementation are reviewed. These include joint sampling of the location parameters; efficient sampling from a multivariate truncated normal distribution, the fully conditional posterior distribution of augmented data; and sampling from the conditional inverse Wishart distribution, the fully conditional posterior distribution of the residual covariance matrix between traits. Allowances are made for unequal models, unknown covariance matrices and missing data.

Poster G3.12

Impact of variation of length of individual testing periods on estimation of variance components of a random regression model for feed intake of growing pigs
U. Schnyder*, A. Hofer and N. Künzi. Institute of Animal Science, Swiss Federal Institute of Technology (ETH), CH-8092 Zürich, Switzerland.

A simulation study was conducted to assess the influence of differences in length of individual testing periods on estimates of variance components of a random regression model for daily feed intake of growing pigs performance tested from 30 to 100 kg live weight. A second order polynomial in days on test with fixed regressions for sex, random regressions for additive genetic and permanent environmental effects and a constant residual variance was used to simulate 20 sets of feed intake data. Variance components were estimated for the same model by means of a Bayesian analysis using Gibbs sampling and REML for two versions of the data: full data sets with 18 weekly means of feed intake per animal and reduced data sets with individual length of testing periods.
No important differences between estimates from full and reduced data occurred. Gibbs posterior means were more accurate for (co)variances of intercept and linear regression parameters and test day variances in first six weeks on test, while REML estimates were more accurate for most (co)variances of quadratic regression parameters and test day variances in the second part of the testing period. Compared to full data, effective sample size of Gibbs samples from reduced data has decreased to 18% for residual variance and increased up to five times for other (co)variances. This is due to confounding of parameter estimates and thus effective sample size should be interpreted with caution.

ANIMAL GENETICS [G]

Poster G3.13

Direct and maternal (co)variances in multitrait analysis in a multibreed beef cattle herd
S.J. Schoeman* and G.F. Jordaan. Department of Animal Sciences, Faculty of Agriculture, University of Stellenbosch, Stellenbosch, 7600, Republic of South Africa.

Estimates of (co)variance components were obtained for growth and efficiency traits with the REML 4.5.2 package of Groeneveld in a multibreed synthetic beef cattle herd. Components were estimated simultaneously fitting four alternative seven-trait models.
Direct heritabilities vary according to the model fitted but were high for preweaning traits and lower for postweaning traits. Applying a model which also included maternal heritabilities, the direct-maternal covariance and the permanent environmental effect, direct heritabilities of weaning weight, weaning index, preweaning Kleiber ratio, preweaning relative growth rate, dam efficiency, postweaning Kleiber ratio and postweaning relative growth rate were 0.57, 0.40, 0.35, 0.71, 0.54, 0.16 and 0.13, respectively. Partitioning of phenotypic variances follows an erratic pattern between models, which may be data related. In general, maternal heritabilities correspond to those reported in the literature and varied from 0.03 for postweaning Kleiber ratio to 0.45 for dam efficiency. Direct-maternal correlations were negative and varied from -0.31 to -0.58 for weaning weight related traits, but was -0.77 for dam efficiency. Moderate negative direct-maternal covariances for postweaning efficiency traits were accepted as weaning weight "carry over" effects.

Poster G3.14

The breeding value for direct and maternal effect in the early stage of growth of beef cattle
J. Přibyl[1], K. Šeba[2], J. Přibylová*[1]. [1]Res. Inst. Anim. Prod. Uhříněves - Praha 10, P.O.Box. 1, 104 01 Czech Republic, [2]ČSCHMS Praha, Czech Republic.

The complete database of Charolais breed in the Czech Republic in the years 1992-97 (4362 records of production with pedigree information) with birth weight (BW) and weight at the age of 120 days (GW) is evaluated with the animal model. Effects of a herd-year-season (HYS), age of mother, the sex of calf, breed composition (if crossbred), direct genetics effect (DGE), maternal effect (ME), and mothers permanent environment (MPE) are considered in the evaluation. HYS is the most significant factor, which covers 64 % of variability for BW. and 47 % for GW. All systematic environmental effects cover 72 % of variability for BW and 52 % for GW. The standard deviation for BW for observations (records) is 6.64 kg, random residuum from the model 2.12, DGE 0.68, ME 0.55 and MPE 0.66 kg. Figures of the same for GW are 28.80, 11.70, 3.95, 2.98 and 3.75 kg respectively. The reliability of breeding values in an average corresponds with these values and are 0.37 (DGE of BW), 0.25 (ME of BW), 0.40 (DGE of GW) and 0.24 (ME of GW). Correlation between BW and GW are r= 0.33 for DGE, 0.18 for ME, 0.21 for MPE, and 0.20 for residual effect of model. Correlation of corresponding records of body weights is r= 0.22.

ANIMAL GENETICS   [G] Poster G3.15

Sire effects on weaning weight of simmental calves
P. Lukács, F. Szabó, L. Vajda, E. Vörösmarthy P., P. J. Polgár, Zs. Wagenhoffer.
Pannon University of Agricultural Sciences, Georgikon Faculty, Keszthely Deák F. u. 16. H-8360

Weaning weight data of 2576 calves from 26 sires of Hungarian Simmental breed were collected between 1981 and 1998 in a farm in Hungary and analysed. Pure-bred calves were born in herd from late February to middle of April. Spring born calves were weaned autumn in September, October and November. Weaning weight data were adjusted to 205 days of age. Data processing were carried out with use of General Linear Model by SPSS 7.5 software.

The results indicate significant sire effect ($P<0.05$) on 205-day weight. Based on the Homogenous Subset Test (Waller-Duncan) 11 homogenous groups were found (group 1: 224-231 kg, group 2: 227-238 kg, group 3: 227-240 kg, group 4: 239-252, group 5: 240-254, group 6: 245-256, group 7 : 251-261, group 8: 254-267, group 9: 255-268 kg, group 10: 260-271 kg, group 11: 266-277 kg). The heritability value calculated on the basis of within and between variance components of sire groups was 0.40.

Poster G3.16

Statistical analysis of Dna polimorphism
Y. Bek, Y. Tahtaý. Department of Animal Science, Faculty of Agriculture, University of Çukurova, 01330 Balcalý, Adana, Türkiye

The relationship between the two estimates of genetic variation or Dna polimorphism at the Dna level, namely the number of segregating sites and the avarege number of nücleotide diferences estimated from pairwase comperasion, is investigated.

A large amount of genetic variation can be maintained in natural populations. In order to understand the mechanism maintinig genetic variation, we must first estimate the amount genetic variation. The amount of polimorphism is determined by many factors such as mutation, natural selection, poulation size, poluation structure, migration, random genetic drift, etc. In this paper two measures for estimating the amount of Dna polimorphism have been examined, these measures are the avarege number of pairwise nücleotide differences and number of segregating sites among a sample of Dna sequences.

ANIMAL GENETICS [G]

Paper Ga4.1

Accomodation and evaluation of ethical, strategic and economic values in animal breeding goals
I. Olesen[*1], B. Gjerde[1] and A.F. Groen[2]. [1]AKVAFORSK, Institute of Aquaculture Research Ltd., P.O. Box 5010, N-1432 Ås, Norway. [2]Animal Breeding and Genetics Group, Wageningen Institute of Animal Sciences, P.O. Box 338, NL-6700 AH Wageningen, The Netherlands.*

A method to accommodate and evaluate noneconomic values of traits in the animal breeding objective and overall selection response is presented and considered.

The traits' values may be split in ethical values (EtV) of e.g. improved environment and animal welfare and values of other long term strategic noneconomic priorities (StV) in addition to economic values on a shorter term (EcV). Correspondingly, we will obtain a genetic gain of ethical value and a long term strategic in addition to economic genetic gain. This give the following breeding goal (considering two traits, Y_1 and Y_2): $H = [EtV_1 \times Y_1 + StV_1 \times Y_1 + EcV_1 \times Y_1] + [EtV_2 \times Y_2 + StV_2 \times Y_2 + EcV_2 \times Y_2]$. The value of the ethical gain is $EtV_1 \times \Delta G_1 + EtV_2 \times \Delta G_2$ and likewise for the strategic and economic gain. The total genetic gain is a sum of the ethical genetic, long term strategic and economic genetic gain.

The method is also illustrated by an example of different ethical, strategic and economic weighing of traits in an animal breeding objective.

Paper Ga4.2

Stochastic simulation of breeding schemes for dairy dairy cattle
M.K. Sørensen[*1], P. Berg[1], J. Jensen[1] & L.G. Christensen[2]. [1]Danish Institute of Agricultural Sciences, Department of Animal Breeding and Genetics, P.O. Box 50, DK-8830 Tjele, Denmark. [2]Department of Animal Science and Animal Health, The Royal Veterinary and Agricultural University. Grønnegårdsvej 2, DK-1870 Frb. C, Denmark.*

Total economic merit in dairy cattle includes several traits. Despite of this, most studies of dairy cattle breeding schemes have assumed a single trait breeding goal, i.e. milk production. In this paper, dairy cattle breeding schemes assuming a multi trait breeding goal were studied, using stochastic simulation.

The breeding schemes consisted of a nucleus with 1,500 females and a breeding population of 18,500 females. Bulls were tested in the commercial population, which were not simulated. However, progeny test results of bulls were simulated as daughter yield deviations. All 10 traits in the current Danish breeding goal were included in the model. The schemes studied included open vs. closed nuclei, selection for total merit vs. milk production, use of early predictors for milk production, different daughter group sizes, improved reproduction technology and use of factorial mating strategies. All simulations covered a 25-year period. Genetic gain were from 14 to 25 EURO per cow per year. Highest gains were optained when selections were for total merit and early predictors were used. For different schemes inbreeding increased from 1,0% up to 1,5% per year. The reduction of the genetic variance of the breeding goal was 30-35% from start to year 25.

ANIMAL GENETICS [G]

Paper Ga4.3

Prediction of rates of inbreeding in selected populations
P. Bijma[*1] *and J.A. Woolliams*[2]. [1]*Animal Breeding and Genetics Group, Wageningen Agricultural University, P.O. Box 338, 6700 AH Wageningen, The Netherlands*, [2]*Roslin Institute (Edinburgh), Roslin, Midlothian EH25 9PS, UK.*

Rates of inbreeding (DF) were predicted for mass selection with discrete or overlapping generations and for index and BLUP selection with discrete generations, based on the concept of long term genetic contributions and assuming the infinitesimal model. The general equation is:
$$E[\Delta F] = \tfrac{1}{2}\sum_s n_s E[u_{i,s}^2] + \tfrac{1}{8}\sum_s n_s \delta_s$$
where n_s is the number of parents in category s (e.g. sexes), $u_{i,s}$ is the expected genetic contribution of individual i in category s conditional on its selective advantage (e.g. its own plus its mate breeding value) and δ_s is a correction for deviations of the variance of family size from Poisson variances. For the special case of mass selection with discrete generations the equation reduces to:
$$E(\Delta F) = \frac{1}{8N_m} + \frac{1}{8N_f} + \frac{i^2 h^2}{4(1+\kappa h^2)^2} \left[(1-\kappa_m h^2)(1-\frac{1}{N_m})(\frac{1}{2N_m}+\frac{1}{2N_f}) + (1-\kappa_f h^2)(1-\frac{1}{N_f})\frac{1}{N_f} \right]$$
where i and k are selection intensity and variance reduction coefficient. Due to the Bulmer effect ΔF has a maximum for $h_{max}^2 \approx 0.6$ with mass selection in discrete generations. With overlapping generations h_{max}^2 depended on the distribution of parents over age classes. With index and BLUP selection, ΔF increased very rapidly with selection intensity and less than halved when the number of parents was doubled.

Paper Ga4.4

Efficient use of test capacity in pig breeding
T. Serenius[*1], *M.-L. Sevón-Aimonen*[1] *and A. Mäki-Tanila*[1]. [1]*Agricultural Recearch Centre, Animal Production Research, 31600 Jokioinen.*

Given the same test capacity, combined sib and individual testing was compared with progeny testing in selecting boars for artificial insemination. The required parameters were determined from the Finnish national breeding programme. Predictions of selection response for daily gain ($h^2 = 0.35$) and meat quality ($h^2 = 0.10$) were created and the risk of a breeding scheme was assessed with the rate of inbreeding, which was limited to 0.5 % per year. Both were analytically derived.Completely new possibilities could be seen, when castrates are replaced by boars in the sib/progeny scheme. Combined individual and sib testing gave better genetic gain than progeny testing for both traits, for the existing annual testing capacity of 1650 pigs/breed. This was due to individual and sib testings having shorter generation interval and higher selection intensity. The effect of selection on the rate of inbreeding was stronger in individual and sib testing, because of strong correlation between estimated breeding values amongst full- and half-sibs. Nevertheless the annual rate of inbreeding is of the same order in both these methods, because the number of selected animals is much bigger in individual and sib testing. The genetic gain of meat quality was 47 % better in individual and sib testing, when the annual rate of inbreeding was restricted to 0.5 %. Respectively, the superiority in daily gain was 127 %.

ANIMAL GENETICS   [G]

Paper Ga4.5

Genetic parameters, inbreeding depression and genetic trend for litter traits in a closed population of Meishan pigs
V.D. Kremer[*1], T. van der Lende[2], O.L.A.M. de Rouw[2], J.A.M. van Arendonk[2]. [1]PIC Group, Roslin Institute, Roslin EH25 9PS, Scotland, UK, [2]Animal Breeding and Genetics group, Wageningen Institute of Animal Sciences, Wageningen Agricultural University, PO Box 338, NL 6700 AH Wageningen, The Netherlands

An animal model including a random permanent environmental effect has been used for the analysis of seven traits related to litter size (LS) and birth weight (BW) for 452 litters in a pure-bred population of Meishan pigs. For LS, heritabilities were: 0.14 (NBA = Number of piglets Born Alive), 0.12 (TNB = Total Number of piglets Born) and 0.12 (TNB+M = TNB plus Mummies). For BW, heritabilities were: 0.25 (AW_{NBA} = Average Weight NBA), 0.24 (AW_{TNB} = Average Weight TNB), 0.25 (TW_{NBA} = Total Weight NBA) and 0.33 (TW_{TNB} = Total Weight TNB). Permanent environmental effects were important for LS and AW (variance ratios close to heritabilities) but not for TW. For LS, inbreeding depressions were: 0.98 (NBA), -1.00 (TNB) and -1.29 (TNB+M) piglets per 10% inbreeding. For BW, inbreeding depressions were: -8 (AW_{NBA}), -17 (AW_{TNB}), -1098 (TW_{NBA}) and -1015 (TW_{TNB}) grams per 10% inbreeding. Genetic and environmental correlations ranged from 0.09 through 0.16 and 0.41 through 0.65, respectively, for LS×AW; from 0.60 through 0.84 and 0.66 through 0.92, respectively, for LS×TW; from 0.63 through 0.67 and 0.09 through 0.21, respectively, for AW×TW.

Paper Ga4.6

Genetic correlation for backfat thickness and daily gain measured on boars and gilts raised in different environments
Špela Malovrh[*1], H. Brandt[2], P. Glodek[2] and Milena Kovač[1]. [1]University of Ljubljana, Biotechnical Faculty, Department of Animal Science, SI-1230 Domžale, Slovenia. [2]University of Göttingen, Institute of Animal Breeding and Genetics of Domestic Animals, D-37075 Göttingen, Germany.

Performance data from test-station for boars and on-farm test for gilts for ultrasonic backfat thickness (BF) and live weight daily gain (LDG) were analysed to estimate the magnitude of genetic-environment interaction (GxEI). GxEI was estimated as genetic correlation in multivariate animal model approach. The dataset consisted of 4952 records for boars and 13165 for gilts for three breeds: Swedish Landrace (SL), Large White (LW) and German Landrace (GL) in years 1990 - 1997 for the nucleus herd on farm Ptuj in Slovenia. Altogether pedigree file contained 20472 animals. Separate analyses were performed for each breed using REML method in VCE 4. Estimated genetic correlations between corresponding traits in station and on-farm tests were high (above 0.9), except for LW (0.50 and 0.44 for BF and LDG, respectively). The heritability estimates were 0.11 - 0.35 in on-farm test, 0.23 - 0.40 in the station test for BF and for LDG 0.14 - 0.23 on-farm and 0.13 - 0.31 in the station. Common litter environment variance presents 6 - 23 % of the phenotypic variance for BF and 7 - 25 % for LDG.

ANIMAL GENETICS [G]

Paper Ga4.7

Genetic variation for traits of commercial importance exists in Danish rainbow trout
M. Henryon[1] and P. Berg[1] .[1]Danish Institute of Animal Sciences, Department of Animal Breeding and Genetics, Research Centre Foulum, P.O. Box 50, DK-8830 Tjele, Denmark.

The objective of this study was to establish that genetic variation for growth rate, food conversion efficiency, survival, and disease resistance exists in Danish rainbow trout. Twenty-five sires and 25 dams were mated by a partly factorial design. Each sire was mated with two dams, and each dam was mated with two sires, resulting in 50 full-sib families. The families were reared in separate tanks, and the fish were assessed for growth rate, food conversion efficiency, survival, and disease resistance. REML estimates of the additive genetic variation in growth rate and food conversion efficiency were obtained by fitting a linear animal model, while the additive genetic variation in survival and disease resistance was estimated by fitting a log-normal frailty function in a reduced animal model. REML estimates of genetic correlations between all traits were obtained by fitting bivariate linear reduced animal models. There was very little additive genetic variation detected for survival. However, a low to moderate amount of additive genetic variation was detected for growth rate (additive genetic coefficient of variation = 18.5%), food conversion efficiency (7.2%), and disease resistance (additive genetic variance on the log-normal scale = 0.26, which corresponds to $h^2 = 15\%$), and there were no adverse genetic correlations between any of the traits. These results highlight the potential to genetically improve Danish rainbow trout for traits of commercial importance.

Paper Ga4.8

Mapping of QTL affecting functional and type traits in the Dutch Holstein population
C. Schrooten*[1,2], H. Bovenhuis[1], W. Coppieters[3] and J.A.M. van Arendonk[1]. [1]Department of Animal Breeding and Genetics, Wageningen Agricultural University, P.O. Box 338, 6700 AH Wageningen, The Netherlands, [2]Holland Genetics, P.O. Box 5073, 6802 EB Arnhem, The Netherlands, [3]Department of Genetics, Faculty of Veterinary Medicine, University of Liège (B43), 20 Bd de Colonster, 4000-Liège, Belgium

Recently, several efforts have been undertaken to detect genes affecting quantitative traits (QTL) in dairy cattle. So far, focus has been mainly on milk production traits, but selection in most breeding programs is for a combination of several traits. The aim of this study is to detect QTL affecting non-production traits.
The granddaughter design used in the present study consisted of 22 grandsires with in total 930 sons. Animals were genotyped for 288 microsatellite markers covering the whole genome. Breeding values were available for 18 conformation traits, 2 fertility traits, 4 calving ease traits, 2 workability traits and one udder health trait. The data was analysed using an across-family multimarker regression approach. Significance thresholds were determined using a permutation test.
Of the 783 trait by chromosome tests that were carried out, 76 were significant at the 5% chromosomewise significance level and 7 were significant at the 5% genomewise significance level. Highest significance levels were obtained for a QTL affecting body size and dairy character located on chromosome 6.

ANIMAL GENETICS [G] Paper Ga4.9

Mixed model analysis of markers in the growth hormone axis and their associations with production traits in Holsteins
S.E. Aggrey, C.Y. Lin, J.F. Hayes*, D. Zadworny and U. Kuhnlein. Department of Animal Science, McGill University, Ste. Anne de Bellevue, QC, Canada H9X 3V9.

Marker genotypes (RFLPs) in the growth hormone (GH) and GH-receptor (GHR) genes were investigated for associations with milk, fat and protein lactation yields in Holsteins. The marker data were obtained on 294 progeny tested bulls. There were three genotypes at each of the two markers. Daughter yield deviations (deregressed bull proofs), DYDs, were obtained from the Canadian Dairy Network for milk fat and protein lactation yields for the 294 genotyped bulls. The statistical model to estimate associations of the marker genotypes with the DYDs included the marker genotypes as fixed effects, inbreeding as a covariate and additive genetic effects of bulls and residual as random effects. The additive genetic relationship matrix among the bulls was included when fitting the model and the covariance matrix of the residuals was diagonal with diagonal elements inversely proportional to the number of daughters of a particular bull. There was a significant difference (4.3 kg ETA) between the homozygote genotypes in GH for fat. There were significant differences between the homozygotes of the GHR in fat (9.47 kg ETA) and protein (5.0 kg ETA). Also, the differences between the best homozygote and the heterozygote in GHR were significant for fat (6.46 kg ETA) and protein (2.54 kg ETA).

Paper Ga4.10

Polygenic inheritance of the bovine blood group systems F, J, R and Z determined by loci on different chromosomes
H. Thomsen[1]*, N. Reinsch[1], N. Xu[1], C. Looft[1], E. Kalm[1], S. Grupe[1], C. Kühn[2], G. Brockmann[2], M. Schwerin[2], B. Leyhe[3], S. Hiendleder[3], G. Erhardt[3], I. Medjugorac[4], I. Russ[4], M. Förster[4], B. Brenig[5], J. Blümel[6], [1]Institut für Tierzucht und Tierhaltung der Christian-Albrechts-Universität, 24098 Kiel, [2]Forschungsinstitut für die Biologie landwirtschaftlicher Nutztiere, Forschungsbereich Molekularbiologie, 18196 Dummerstorf, [3]Institut für Tierzucht und Haustiergenetik der Justus-Liebig-Universität, 35390 Giessen, [4]Institut für Tierzucht der Ludwig-Maximilians-Universität, 80539 München, [5]Tierärztliches Institut der Georg-August-Universität, 37073 Göttingen, [6]Institut für die Fortpflanzung landwirtschaftlicher Nutztiere, 16321 Schoenow, Germany

An attempt was made to assign the bovine erythrocyte antigens to their corresponding chromosomes by linkage analysis. In total 9591 genotypes of 22 grandsires with about 1575 sires from the ADR mapping project were determined for the erythrocyte antigens according to standard procedures in the blood typing laboratories during testing of paternity. Linkage analysis was performed to determine associations between the 247 microsatellite markers and 7 SSCP markers covering the 29 autosomes and the pseudoautosomal region of the sex chromosome. Erythrocyte antigens A, B, C, L and S showed significant association to a single chromosome and could definitely be mapped. But for the blood group systems F, J, R' and Z significant associations were detected to markers on various chromosomes. The appearance of a single blood group system is therefore either dependent on the existence of other blood group systems or due to an interaction between different loci on various chromosomes. As it is shown in humans and in pigs these bovine blood group systems also seem to be influenced by several loci throughout the genome.

ANIMAL GENETICS   [G]

Paper Ga4.11

Mapping of quantitative trait loci affecting egg quality in chicken

M. Honkatukia*[1], M. Tuiskula-Haavisto[1], J. Vilkki[1], D.-J. de Koning[2], N. Schulman[1] and A. Mäki-Tanila[1]. [1]Animal Production Research, 31600 Jokioinen, Finland, [2]Department of Animal Breeding, 6700 AH Wageningen, The Netherlands

Egg shell strength and egg white properties are two major traits affecting egg stability. These traits are quantitative, results of interaction between quantitative trait loci (QTL) and environmental effects. The aim of this study was to use molecular genetics methods to localize QTL affecting egg quality traits, and to find suitable markers to breeding. Our mapping population, a cross of two extreme egg laying lines, was segregating for these traits. The F_2 generation consisted of 320 hens. The F_2 hens were scored for egg quality and production traits at given periods. The egg white quality was measured as Haugh-units. The measured traits included also shell strength as specific gravity, egg production, egg weight and effect of storage. We have chosen a set of microsatellite markers from 14 linkage groups. The mapping of QTL's was performed with multiple marker regression. Empirical significance threshold values were calculated using permutation test. Mapped 97 markers covered approximately 2/3 of chicken genome. We have found at 1% chromosomewise significance level QTL areas affecting egg quality, egg production, body weight and sexual maturity. Extreme F_2 individuals were backcrossed with the opposite grand-parental lines to create two backcross population for further fine-mapping.

Paper Ga4.12

Analysis of quantitative trait loci in broilers using a Bayesian mixed model

J.B.C.H.M. van Kaam*[1], M.C.A.M. Bink[2], M.A.M. Groenen[1], H. Bovenhuis[1], and J.A.M. van Arendonk[1]. [1]Animal Breeding and Genetics Group, Wageningen Institute of Animal Sciences, Wageningen Agricultural University, P.O. Box 338, 6700 AH, Wageningen, The Netherlands, [2] DLO Centre for Biometry Wageningen (CPRO-DLO), 6700 AA, Wageningen, The Netherlands and DLO Institute for Animal Science and Health (ID-DLO), 8200 AB Lelystad, The Netherlands.

QTL analyses were undertaken in a population consisting of 10 full sib families of a cross between two broiler lines. Microsatellite genotypes were determined on generation one and two. Observations on generation three animals were collected in two experiments. In initial genome scans, using interval mapping by regression, QTLs for body weight, growth and feed intake traits were found in a feed efficiency experiment. In the carcass experiment QTLs for carcass percentage and meat colour were detected.
A detailed Bayesian analysis is undertaken on chromosomal regions were QTLs were found. Advantages of the Bayesian model in comparison with the regression analysis are that normally distributed random polygenic and QTL components are modelled and variances are estimated for all random terms in the model. Furthermore, individual trait values are used instead of averages and mate correction is no longer necessary, because all genetic relations are taken into account through relationship matrices. Markov Chain Monte Carlo methods and simulated tempering were applied to obtain solutions. The results added valuable information to the previous results.

ANIMAL GENETICS   [G]   Poster Ga4.13

Potential use of information on identified quantitative trait loci (QTLs) in prediction of response to long-term selection
M. Satoh, K. Ishii* and T. Furukawa. *Department of Animal Breeding and Genetics, National Institute of Animal Industry, Norin-kenkyudanchi P.O.Box 5, Tsukuba 305-0901, Japan*

A deterministic model for predicting response to selection when major QTLs are segregating is presented. Assuming equal polygenic effects, genetic value of gametes was obtained by using hypergeometric distribution. The change in genetic properties resulting from long-term selection were predicted using the procedures of repeated phenotypic selection, segregation and recouping of gametes. Reduction in genetic variance and skewness in genotypic distribution over generations was considered. The effects of identified QTLs were assumed known without error. Information on the identified QTLs was used as a trait with a heritability of 1.0 in a selection index. Genetic parameters in the index were adjusted to true values every generation. Response to selection over multiple generations obtained from genotypic selection was compared with selection using phenotypic performance. Genotypic selection produced higher cumulated genetic gain in the first 4-6 generations of selection than phenotypic selection. However, this advantage receded and was eventually lost in subsequent generations. In the long-term, selection using genotypic information on the identified QTLs produced lower response than phenotypic selection using a wide range of parameters, despite the use of true parameters in the former.

Poster Ga4.14

The effects of missing markers and combination of the number of grandsire and his sires on mapping quantitative trait loci in a small dairy cattle population
K.Togashi*, N.Yamamoto, O.Sasaki and A.Nakamura. *Hokkaido National Agricultural Experiment Station, Sapporo, Japan*

Residual maximum likelihood by Grignola et al(1996) was used to assess the effects of missing markers and the numbers of grandsire and his sires on mapping quantitative trait loci(QTL) in a small dairy cattle population. QTL detection was carried out in a specific chromosome with length of 90cM. DNA markers with 8 alleles were determined at 10 cM intervals. The factors are: 1. the rate of missing markers: 0%, 40% and 60%, 2. the number of sires per grandsire when the total number of daughters with markers is the same: 4 and 10, 3. addition of daughters marker records to a small sized granddaughter design(10-10, 10-6; numbers of grandsires and his sires): 30 daughters per sire and no addition. Likelihood ratio became lower with the increase of missing markers. Its ratio, however, is still higher compared to the case where number of daughters is given as the same as the product of the number of daughters per sire and the rate of existing markers. It indicated that more daughters should be given despite missing markers. Likelihood ratio became higher with the increase of the number of sires per grandsire, which would be due to more correct marker phase of grandsire being estimated. Addition of daughters marker records remarkably increased the likelihood compared to the design without daughters marker records.

ANIMAL GENETICS [G]

Poster Ga4.15

Genotypes of bGH and bPRL genes in relationships to milk production
P. Chrenek, J. Huba, M. Oravcová, L. Hetényi, D. Peškovièová, J. Bulla.. Research Institute of Animal Production, 949 92 Nitra, Slovak Republic.*

The genotypes of bovine growth hormone (bGH) and bovine prolactin (bPRL) genes were detected in 107 cows of the Brown Swiss breed using polymerase chain reaction (PCR-RFLP).
The frequencies of genotypes were the following: in bGH - VV - 12.14 %, LL - 11.21%, LV - 76.65 % and in bPRL - AA - 45.79 %, BB - 11.22 %, AB - 42.99 % respectively. In order to evaluate the relationships between polymorphism and milk production, the first three lactations of cows were taken.
The highest fat (4.77%) and protein (3.55%) content were obtained in cows with LL genotype of bGH. Dairy cows with VV genotype achieved only 4.41% and 3.37% fat and protein percentage in milk. Genotype of bGH was not associated with milk, fat and protein production.
No significant differences in bPRL genotypes were found.

Poster Ga4.16

Microsatellite typing of ancient parchment and leather
I. Pfeiffer[1], J. Burger[2], B. Brenig[1]; [1]Institute of Veterinary Medicine, University of Göttingen, Groner Landstr. 2, 37073 Göttingen, Germany; [2]Department of Historical Anthrophology and Human Ecology, Bürgerstr. 50, 37073 Göttingen, Germany*

Ancient parchment and leather represent a widespread archaeological and historico-cultural collection of source material. Up to the 14th/15th century, parchment had been the dominating material for inscriptions in Europe. Due to this characteristics leather can be used broadly, e.g. book covers or even as a wall covering.
Contrary to the mechanical handling of the animal skin for the manufacture of parchment, the tanning of skin is supposed to result in poor conservation of DNA.
While ancient DNA found in parchment could be isolated regularly and consistently, similar results for leather could only be achived in some cases.
Classing two pieces of parchment with each other is a classical question in graphology. The finding that a fragment indicates the same individual genetical fingerprint as a scroll, can support or disprove a graphological hypothesis.
Beyond this, microsatellite frequencies provide information on the source e.g. the population structure of cattle in the middle ages.

ANIMAL GENETICS   [G]   Poster Ga4.17

Isolation of the porcine four and a half LIM domain protein 1 (FHL 1)
A. Krempler[1], S. Kollers[2], R. Fries[2], H. Al-Bayati[1], B. Brenig[1]*; [1]Institute of Veterinary Medicine, Universitiy of Göttingen, Groner Landstrasse 2, 37073 Göttingen, Germany, [2]Institute for Animal Breeding and Genetics, Technical University Munich, Weihenstephan, 85350 Freising, Germany

Zinc finger proteins are known to play a crucial role in transcriptional regulation and protein-protein interaction. One class of zinc finger proteins, the LIM domain proteins, posses a highly conserved double zinc finger domain. They have been implicated in cell differentiation, regulation, and development.

On screening a porcine skeletal muscle cDNA library we isolated a highly abudant transcript with four LIM domains and a GATA 1 zinc finger. The open reading frame coded for a protein of 280 amino acids. Comparison of the amino acid sequence with available data from human and mouse revealed a high degree of conservation within FHL 1. Most of the amino acid substitutions occured in the first LIM domain and the spacer between the first and the second LIM domain. The fourth LIM domain is entirely conserved among these species. On basis of the cDNA sequence we were able to isolate a genomic fragment from a phage library as well as a PAC clone encompassing the whole genomic region. Sequence analysis showed 6 exons spanning 14 kb. The start codon is situated in the second exon, the stop codon at the beginning of the sixth exon, followed by a 1300 bp untranslated region. The phage insert was used for chromosomal localization by FISH analysis. In agreement with the results from the human localization of the FHL 1 gene, the porcine FHL 1 gene was assigned to chromosome X.

Poster Ga4.18

Mapping and exclusion mapping of genomic imprinting effects in mouse F_2-families
Christine Mantey[1], N. Reinsch[1], G. Brockmann[2], E. Kalm[1]. [1]Institut für Tierzucht und Tierhaltung der Christian-Albrechts-Universität;24098 Kiel, Germany [2]Forschungsinstitut für die Biologie landwirtschaftlicher Nutztiere, Forschungsbereich Molekularbiologie;18196 Dummerstorf, Germany

In order to detect Imprinted Quantitative Trait Loci (IQTL) we used an F_2 intercross design (crossing of mouse outbred line DU6 and DUKs: selected for high body weight, control line resp.). Quantitative measurements for all individuals of pedigree (F_2= 341 offspring) contain the traits body weight, abdominal fat and the weight of liver, kidney and spleen.

Using a multi-marker-regression approach an additive , dominance and an imprinting effect were fitted to the model. The significance threshold values were determined empirically based on a permutation test.

With 89 available informative marker we could exclude imprinting effects for an already published QTL (body weight) on mouse chromosome 11 (February 1999).

The next step is to screen the whole genome for mapping, exclusion mapping of imprinting effects resp.; results considering imprinting effects scattered on the entire genome will be presented.

ANIMAL GENETICS   [G]

Poster Ga4.19

Three-layer selection of bulls: a farmers' choice
S. van der Beek* and G. de Jong, NRS, P.O.Box 454, 6800 AL Arnhem, The Netherlands.

Dairy farmers wanting to select a bull are faced with an ever-increasing number of selection criteria. Sensible use of the available information becomes increasingly difficult. We propose to reform the presentation of the Dutch genetic evaluation results. We define three layers of presentation. At the base are all available selection criteria, expressed in (bio)logical units. At the second layer, traits that logically belong to each other will be grouped. The primary second layer trait will be milk-production-efficiency: it combines kg milk, kg fat, kg protein and live weight. Further, there will be second layer traits for functional herd life, udder health and reproduction. At the third layer, all second layer traits are combined in one economic index. The economic index provides a clear and objective ranking of sires. For some farmers this will be sufficient. The second layer traits provide information on the technical parameters determining farm income. Many farmers steer their management at this level. The base level traits provide farmers with detailed information on the genetic quality of sires. The group of farmers that want to fine tune the genetic makeup of their stock will have to resort to this base level. This system provides the farmers clarity, objectivity, and flexibility.

Poster Ga4.20

Objectives in dairy cattle improvement in Estonia
E.Pärna. Department of Animal Breeding, Institute of Animal Science, Estonian Agricultural University, 1 Kreutzwaldi St., 51014 Tartu, Estonia.

Information on values of milk components is combined with genetic parameters and management information to construct tools for maximizing improvement in overall genetic value of the population. Response to selection, and hence achieving breeding goals, is largely dependent on the genetic and phenotypic parameters and economic weights used.
Producing fat in Estonia under two differing quota levels and no quota there was no difference in economic value between the two quota levels and a 4 fold difference when no quota was imposed. As an alternative desired gains indexes have been applied to avoid the task of a breeder determining the economic values of traits and rather simply to describe the changes desired in each trait. Developing the optimal combination of desired changes is considered for the Estonian Holstein.

ANIMAL GENETICS [G]

Poster Ga4.21

Sire x ecological region interaction in Bonsmara cattle
K.A. Nephawe[1], F.W.C. Neser*[2], C.Z. Roux[3], H.E.Theron[1], J. van der Westhuizen[1] & G.J. Erasmus[2]. [1]ARC Animal Improvement Institute, Private Bag X2, Irene, 1672, Republic of South Africa [2]Department of Animal Science, University of the Orange Free State, P.O. Box 339 Bloemfontein, 9300, Republic of South Africa [3] Department of Genetics, University of Pretoria, Pretoria, 0002, Republic of South Africa

The possible interaction between sire and four ecological regions in which Bosmara cattle are mainly found in South Africa was investigated. Birth and weaning weight records of 43 628 registered Bonsmara calves born between 1976 and 1997 from 18 herds were available. Restricted Maximum Likelihood (REML) procedures were used in the analyses. Genetic correlations and expected correlated response to selection were estimated assuming the same trait (eg weaning weight) to be a different trait when measured in each of four different regions. The results indicate that, for the purpose of genetic evaluation, the three Bushveld regions need not be separated, but that it would be advisable to consider the Highveld region as a separate environment. Selection of bulls bred on the Highveld for the purpose of genetic improvement in the Bushveld regions is likely to be less effective than selection of bulls bred anywhere in the Bushveld and conversely. The results have also, once again, confirmed that the interaction between sire and contemporary group (HYS) is usually more important than between sire and any designated region

Poster Ga4.22

Breed differences, heterosis and maternal effect on weaning weight of beef calves
F. Szabó[1], J. Dohy[2], K. Szentpáli[3] J. Tari[3]. [1]Pannon University of Agricultural Sciences, Georgikon Faculty Keszthely, Deák F. u. 16. H-8361, Hungary, [2]Gödöllö University of Agricultural Sciences, Hungary, [3]Mezöfalva Agricultural Ltd., Hungary

Weaning weight of straightbred Hungarian Simmental (HS), Hereford (HE), and Angus(A) and that of reciprocal crossbred HSxHE and HExA F_1 calves were evaluated in Hungary based on more than 600 animals. Calves were born in spring season, kept and creep fed on pasture in summer, weaned in fall at the age of about 6 months. Weaning weight were adjusted to 205-day of age. Least Squares and Maximum Likelihood Computer Program (Harvey 1990) were used for data analyses.
The average weaning weight of straightbred male and female calves were as follows: 219.1 kg (HS), 196.3 kg (HE), 215.2 kg (A) ($P<0.05$). Significant maternal effect ($P<0.05$) were found in the weaning weight of calves from both reciprocal crossings. 11% positive effect in case of HSxHE reciprocal crossbred calves due to HS dams and 8% positive effect in HExA reciprocal crossbred calves due to A dams. Heterosis effect were also found in the results of both F_1 calf groups. That was 7.8% in HSxHE F_1, and that of 5.5% in case of HExA F_1 calves ($P<0.05$). It seemed that the bigger the genetic distance between the breeds, the higher the maternal and heterosis effect in the weaning weight of their crossbred progeny.

ANIMAL GENETICS   [G]

Poster Ga4.23

Joint genetic evaluation of purebred and crossbred data under dominance
N. Mielenz[1], L. Schüler*[1] and E. Groeneveld[2]. [1]Institute of Animal Breeding and Animal Husbandry with Veterinary Clinic, University of Halle, 06108 Halle, [2]Institute of Animal Husbandry and Animal Behaviour Mariensee, FAL, 31535 Neustadt, Germany

Currently available software packages are not able to handle the joint genetic evaluation of purebred and crossbred data correctly. Instead, the same traits originating in the two purebreds and the crossbred are treated as three separate traits and analysed in a three trait animal model. Using selection index theory it can be demonstrated that this procedure contains model violations even if only additive genetic effects are considered. The effectiveness of the correct model which allows different contributions of the parental lines to the genetic variance of the cross including animal specific dominance effects is assessed on the basis of different selection indices. The sources of information chosen reflect those available in typical chicken breeding programs. The genetic parameters required for index construction - in particular the dominance variances - were derived using a two locus model. If the purebred contribute extremely differently to the crossbred variance large losses in efficiency can be observed. With contributions of the sire line of 11, 19, 37 and 40% to the additive genetic variance of the cross corresponding losses in efficiency of selection where 13.1, 6.5, 0.6 and 0.2% respectively. Not considering the dominance variance - depending on its proportion relative to the total genetic variance - lead to a moderate loss of efficiency from 0.1 to 2.6%.

Poster Ga4.24

Estimation of genetic parameters in Hanoverian horses including special combining ability
W. Brade*[1] and E. Groeneveld[2]. [1]Chamber of Agriculture, Johannssenstr. 10, 30159 Hannover, Germany, [2]Institute of Animal Husbandry and Animal Behaviour, FAL, 31535 Neustadt, Germany.

Young Hanoverian mares at the 3 to 4 years are put through a station or field performance test (one-day test).
Traits recorded are walk, trat, gallop, ride ability and free jumping. Animal model with and without special combining ability were investigated for a dataset of 9233 mares with 20025 pedigree records. The special combining ability was estimated as an additional sire * maternal grandsire interaction. The respective variance component (in its genetic interpretation as dominance variance) accounted on average for around 22% of the total variation.
The heritabilities on the basis of the animal component amounted to 0,23 ....0,41.

ANIMAL GENETICS [G]

Poster Ga4.25

Empirical evidence for segregation variance in beef cattle
A.N. Birchmeier[1], R.J.C.Cantet*[1], R.L. Fernando[2], C.A. Morris[3]. [1]Departamento de Zootecnia, Universidad de Buenos Aires, Av San Martín 4453, 1417 Buenos Aires, Argentina. [2]Department of Animal Science, Iowa State University, USA. [3] AgResearch Ruakura Agricultural Research Centre, PB 3123, Hamilton, New Zealand.

Recent developments on the theory of multibreed evaluations require estimating segregation variance, the additive genetic variance in $F2$ individuals over that existing in $F1$'s. No estimates of this parameter have been previously reported. The goal was to obtain REML estimates of the additive variances plus segregation variance, in a composite population. Data used for estimation consisted on 4,082 birth weights, out of 4,989 animals (including foundation parents) from 27 different breed-types crosses between Angus and Hereford. Records were collected from a diallel experiment of the Ruakura Centre, New Zealand. Breed-types included were purebreds, F1, F2, F3, F4, F5, and different backcrosses. The animal model included fixed effects of sex, birth year, age of dam, date of birth, 8 linear parameters representing the mean genetic effects, and random breeding values. REML estimates were calculated using a first-derivatives based algorithm and a specially written program. REML asymptotic standard errors were calculated by the inverse of the information matrix. After 400 iterations, estimates of the variances (kg^2) were 7.49 ± 0.89 (Angus), 10.25 ± 1.12 (Hereford), 0.87 ± 0.84 (segregation), and 7.94 ± 0.06 (error). Despite the similarity between the two British breeds, these results suggest that the segregation variance may not be null.

Poster Ga4.26

Inbreeding effects on the parameters of the growth funtion of Iberian pigs
J. Rodrigañez, L.Silió, M.A.Toro* and M.C. Rodriguez. Departamento de Mejora Genética y Biotecnología, INIA, Carretera La Coruña km.7, 28040 Madrid, Spain

The growth performance of 104 animals coming from matings of 56 sires with their sisters was compared to that of 125 animals coming from the same sires mated to unrelated dams. The animals belong to three strains of Iberian pigs with a different history of previous inbreeding. The individual recorded data were seven measures of weight between 50 and 240 days of age. The statistical model assumed the two parameters that describe the lineal growth function (a = intercept, b = slope) as different traits and the analysis was carried out in a bayesian framework via Gibbs sampling.
The means of the posterior distribution of heritabilities and genetic correlation were 0.455 (h_a^2), 0.434 (h_b^2) and 0.942 (r_{ab}). The inbreeding depression, as a decrease of the mean per 10 per cent increase of the inbreeding coefficient, was -1.70, -1.14 and -2.68 kg for weight at 120 days (a) and -0.016, -0.010 and -0.031 kg/d for daily growth (b) in each one of the three strains. A more de conventional analysis also lead to similar conclusions.

ANIMAL GENETICS   [G]

Poster Ga4.27

A demonstration project of the EU DG XII biotechnology research programme: characterizing genetic variation in the European pig to facilitate the maintenance and exploitation of biodiversity
Pig Biodiversity Project (L. Alderson, M.Y. Boscher, C. Chevalet, B. Coudurier, R. Davoli, B. Danell, J.V. Delgado, P. Glodek, M. Groenen, C. Haley, K. Hammond, D. Milan, L. Ollivier* and G. Plastow). Coordination, INRA-SGQA, 78352 Jouy- en- Josas cedex, France.

Europe contains a large part of the world genetic diversity of the pig species. A two year demonstration project was initiated in October 1998 under EC Framework IV, which aims to evaluate the extent of pig genetic diversity in Europe. The emphasis in the project is on genetic diversity at the between-breed level, and the design is based on the FAO guidelines. Genetic diversity is estimated by sampling 50 individuals from about 60 different breeds (including a range of rare breeds) and commercial lines from 16 different countries. Variation at DNA level will be evaluated by standard marker technologies, namely Simple Sequence Repeat (microsatellites) and the amplification of subsets of fragment length polymorphism using AFLP. Full information on genotype frequencies in each breed as well as on pairwise genetic distances between breeds will be disseminated to a wide range of users via the world wide web and through previously developed databases. The involvement of FAO will ensure that the lessons learnt in this demonstration project are applied to projects on other species around the world.

Poster Ga4.28

Relationships among subpopulations of Iberian Pig using microsatellites
A.M. Martinez[1], A. Rodero[1], J.L. Vega-Pla*[2]. [1]Unidad de Veterinaria. Departamento de Genética. Universidad de Córdoba. Avenida Medina Azahara s/n. Córdoba. Spain. [2]Laboratorio de Grupos Sanguíneos. Servicio de Cría Caballar. Apartado Oficial Sucursal 2. 14071 Córdoba. Spain.

The Iberian Pig is the most emblematic autochthonous Spanish pig breed and there is a great interest by the conservation of its genetic resources.
This breed is traditionally classified in different subpopulations and some of them are on the verge of extinction. The primary goal of this work is to elucidate and clarify the genetic relationships among these subpopulations using molecular markers. Presently the DNA microsatellites provide an interesting tool to define population structures.
The samples were taken from 220 animals representing 9 subpopulations of Iberian Pig.
Microsatellites used in this research (25) are enclosed within those recommended by the ISAG-FAO advisory group for diversity studies in pig. Hardy-Weimberg equilibrium, genetic diversity, and distance using the method proposed by Nei in 1972 have been studied also a phylogenetic tree by UPGMA analysis has been built. Results have shown as some subpopulations are very well defined such as Manchado de Jabugo (Jabugo Spotted), Torbiscal and Negro Lampiño (Black Hairless). The subpopulations Retinto Extremeño (Extremadura Red), Silvela and Entrepelado (Black Hairy) are very closed from the genetic point of view, but they are differentiated of the others mentioned above.

ANIMAL GENETICS [G] Poster Ga4.29

Genetic variation of two local Romanian pig breeds assessed using DNA markers
D. Ciobanu[1]*, A. Nagy[2], R. Wales[3] and G.S. Plastow[3]. [1]*University of Agricultural Sciences and Veterinary Medicine, Animal Genetics Unit, Manastur 3, 3400 Cluj-Napoca, Romania, [2]Agricultural Research Station of Turda, Romania, [3]PIC International Group, Fyfield Wick, Abingdon, Oxon, OX13 5NA, UK.*

Animals from two local Romanian breeds Mangalitsa and Bazna were analysed for variation at a number of genetic loci using PCR based DNA tests. Polymorphism was assessed at loci involved in disease resistance, coat colour, meat quality and prolifacacy. These breeds are part of a conservation programme in Romania and this information will be used to aid the preservation of a representative gene pool of these breeds, an irreplacable resource for future development in the pig industry.

Poster Ga4.30

Protein polymorphic system of swine bloard serum
E.D. Ambrosjeva, A.A. Novikov.All-Russian Institute of Animal Breeding, Box LesnyePolyany, Pushkino district, Moscow region, 141212, Russia.

Ceruloplasmin (Cp), carbonic anhydrase (Ca), albumin (Alb), vitamin D - binding protein (Gc), haptoglobin (Hp), $S\alpha_2$-macroglobulin ($S\alpha_2$), prealbumin (Pa), transferrin (Tf), and three loci of postalbumin zone (Po-1, Pi-2, Po-2) were studied in blood serum of Large-White pigs. The method of vertical electrophoresis in 10-12,5% polyacrylamid gel with using of gel buffer 0,05M tris-HCl pH 7,2 and electrode buffer - 0,005M tris-glycin pH 8,5 were applied in this work. Polymorphism was detected on loci Pa, Tf, $S\alpha_2$, Po-1, Pi-2, Po-2. The frequency of alleles of polymorphic loci was PaA 0,517, PaB 0,483; TfA 0,295, TfB 0,705; Po-1A 0,175, Po-1B 0,144, Po-1C 0,367, Po-1D 0,271, Po-1E 0,043; Pi-2A 0,309, Pi-2B 0,623, Pi-2C 0,068; Po-2A 0,458, Po-2B 0,542.
The loci Po-1, Pi-2, Po-2 were early determined by the method of two-dimensional electrophoresis (Juneja et al., 1983), that limited their analysis in wide-scale researches. Used of us method allowes to devide these loci by means of ordinary electrophoresis and to identify all polymorphic loci of blood serum on the same gel. On the one hand this simplifies populational researches and the another hand this allowes to use not less fife markers of blood serum for genetic expertise.

ANIMAL GENETICS   [G] Poster Ga4.31

Dinamics of genetic structure in zivilsky pigs
A.A. Novikov, N.I. Romanenko, M.S. Semak, V.N. Ananjev. All-Russian Institute of Animal Breeding, Box LesnyePolyany, Pushkino district, Moscow region, 141212, Russia.

The analysis of genetic structure in zivilsky pigs population was carried out in period from 1985 to 1998 years. Genetic diversity, the level of homozygosity, the genetic equilbration of herd on blood groups for 11 loci were studied.
The maintenance of allelofund was detected with change of the friquency for some alleles (lowing of quantity for E^{bdg} and rising for E^{bdf}, K^{bf}). The significant change of herd structure on genotypes of blood groups of locus E was shown.
The complete disappearance of homozygous genotypes aeg/aeg, bdg/bdg and heterozygous aeg/deg and reduction of gene frequency E deg/deg were happened.
The homozygous genotypes Ebdf/bdf with the frequency of appearance 0,09 was appeared.
Some reduction of homozygosity and the breach of equilibration for systemes E, G, K, L were detected. The mentioned changes of herd genetic structure did not concern marker alleles.

Poster Ga4.32

Genetic analysis of two cattle populations using DNA microsatellites
M.J. Zamorano[1], A. Rodero[1], J.L. Vega-Pla[2]. [1]Unidad de Veterinaria. Departamento de Genética. Universidad de Córdoba. Avenida Medina Azahara s/n. Córdoba. Spain. [2]Laboratorio de Grupos Sanguíneos. Servicio de Cría Caballar. Apartado Oficial Sucursal 2. 14071 Córdoba. Spain.*

Genotype data from eight microsatellites typed in two cattle populations samples (Argentinean Creole cattle and two flocks of Berrenda en Negro cattle) are used to asset a primary study of the genetic structure of them. The main goal is the interest to know more about near extinction and rare breeds.
Mean heterozygosity over all loci tested was 58% in Berrenda en Negro cattle and 62% in the Argentinean Creole cattle. Genetic distances between two breeds and between two Berrenda en Negro flocks were estimated. We found great differences between two samples of Berrenda en Negro due a long reproductive isolation of flocks. This information could be interesting to study bottleneck effect in future resources conservation plans. When genetic distance between breeds is estimated the value is too much elevated we have considerate by historical data. No crossbreeding for five centuries and the other breeds influence have produced a great genetic divergence.

ANIMAL GENETICS [G]

Poster Ga4.33

Genetic similarity for erythrocyte antigens in animals of Holstein
A.I.Zheltikov, V.L. Petukhov, V.G.Marenkov. *Chair of Animal Breeding and Genetics, Novosibirsk Agrarian University, 160 Dobrolubov Str.,630039 Novosibirsk, Russia*

In Western Siberia the breeding procedures are carried out to produce Holstenized Black-and-White cattle well adapted to local conditions. To reach the goal Holstein pure and crossbred sires of different lines are used. To identify the genetic similarity between animals of Black-and-White and Holstein lines the frequency of 48 erythrocytes antigens of 9 genetic system were studied and the coefficients of genetic similarity between lines were accounted. The lines are significantly different in some antigens and are not in other ones. The greatest and least difference in the lines was established for antigens A_2, O', J'_2, K, W, V, J and for the frequency of antigens T_2, Y'_2, E'_2, B", F, H', U", respectively. The coefficients of the genetic similarity in Black-and-White, Holstein lines and between them are 0.88- 0.95, 0.911- 0.937 and 0.820 -0.927, respectively, that indicates weak differentiation of the lines resulted from the great number of crosses and great remoteness of offsprings of its lineage pioneer (5-10 generations).

Poster Ga4.34

Genetic analysis of synthetic multifertility of midfine-wool Sheep
N.S. Marzanov[1], B.S. Iolchiev[1], M.R. Nassiry[2] and V.P. Shikalova[1], *[1]All-Union Research Institute of Animal Husbandry RF-142012 P.O.Dubrovitsy, Podolsk District, Russia, [2] Department of Sheep Breeding, Timiryazev Agriculture Academy, Timiryazevskaya st. 49 127550, Moscow, Russia*

The population was established on the basis of Finnish Landrace dams and Texel, Lincoln and Romney-marsh rams. In this study, genetic variability were based upon seven blood group (BG) systems (A, B, C, D, R, M, I) and seven milk protein (MP)loci (α-La, β-La, Ig, αs1-Cn, β-Cn, K-Cn, αs2-Cn)were examined for 198 and 44 animal samples blood and milk respectively.Four systems out of 14 systems were monomorphic: one within BG (I) and 3 within MP (α-La, β-Casein, αs2-Casein). The highest number of alleles were shown in the B and C systems of BG and αs1-Casein. And in B and C systems there are 7 and 3 alleles respectively and 4 are in αs1-Casein loci. In the genetic structural analysis, the highest frequency of two alleles were observed in M blood group system(M^a - 0.58, M^- - 0.42) and the lowest frequency was Ca antigen (0.0074).The average level of homozygosity and effective number of alleles (Ne) of the seven systems of BG and 7 loci of MP were 59.4 %· 1.84 and 73.4 % and 1.65 respectively.For this population, β-La, Ig, αs1-Cn and K-Cn systems can be used for genetical markers. The average fat, protein and dry matter contents of the β-La loci was significantly higher in AA and AB genotypes than BB genotype (P < 0.001).
Therefore at the first time in carrying out general genetical analysis of synthetic multifertility of midfine wool population by 14 BG and MP systems.These results of investigation shows genetical image of the sheep population.

ANIMAL GENETICS   [G]

Poster Ga4.35

Frequency of sister chromatid exchanges in Egyptian water buffalo
Sahar Ahmed*, National Research Centre, Department of Cell Biology, 12622 Dokki, Cairo, Egypt.

Forty healthy buffalo of both sexes (22 females and 18 males) were used to study sister chromatid exchanges (SCEs) frequencies in Egyptian water buffalo breeds. 24 animals were of the Beheri breed from the north of Egypt and 16 animals were of the Saidi breed from the south of Egypt. The mean values of SCEs per cell varied between 5.7 ± 1.9 and 10.6 ± 3.8 with total mean value of 8.09 ± 1.1. No statistical differences ($P<0.05$) were found between SCEs mean values in male (7.96 ± 1) and female (8.19 ± 1.1) cells in the total sample. SCEs in Beheri breed was recorded as 8.3 ± 1.1 compared to 7.76 ± 0.8 in Saidi breed. The difference between the two breeds was found statistically non-significant. The influence of sex on SCEs frequencies in the two breeds was not detected. The results reached by the present study are considered a criterion of SCEs in Egyptian water buffalo breeds detecting the stability of chromosome in proportion to some pathological conditions, mutational events and risks of environment.

Poster Ga4.36

The analysis of the Holmogor Bulls lines differentiation by A DNA-fingerprinting
A.F. Yakovlev, N.V. Dementjeva, V.P. Terletsky. Institute of animal genetics and breeding, St-Peterburg, 189620, Moskovskoje sh.,55-a.Russia.

Through the analysis of DNA-fingerprints investigated a degree of differentiation 4 lines of the Holmogor bulls. For comparison of a animals genotype among themselves used index of similarity (BS),and also parameters of genetic similarity of populations (Is) and genetic distance by Gilbert(Gd).Compared two pairs lines of the bulls: Hlopchatnic (Ch) and Komelek(K),Zvetok(C) and Nailuchsij (N).Average the bends frequency made 11.4. Factor of similarity for a lines to Ch and K -0.38,for lines C and N -0.41,factor Cd accordingly 0.98 and 0.89. Through the statistical analysis of the received data it was not possible to show authentic genetic distinctions between investigated lines of bulls. Probably, it is necessary for registration of buuls lines to have the preise proofs of their differentiation on the basis of genetic markers.

ANIMAL GENETICS [G]

Poster Ga4.37

A new allele of MC1-R shows a 12 bp insertion in Brown Swiss breeds
B. Kriegesmann, I. Pfeiffer, B. Dierkes, A. Krempler, B. Brenig*, Institute of Veterinary Medicine, University of Göttingen, Groner Landstraße 2, 37073 Göttingen

The coat colour of domesticated animals has always been a marker for race identity. Two loci, the extension (E) and the agouti (A) are responsible for the distribution of the hairpigments (LIT). Recent investigations have shown that the E locus is encoded by different alleles of the G-protein coupled melanocyte-stimulating hormone receptor (MC1-R).
On sequencing the MC1-R genomic region in Brown Swiss cattle we found a new allele with a 12 bp nucleotide in-frame duplication. This insertion will lead to the addition of four amino acids within the third intracellular loop of the receptor. We screened 13 animals for the excistence of this longer allele and found two heterozygous carrier-animals. As no differences in coat colour can be observed in these animals, we conclude that any functional effects of this insertion will probably become evident when present in both alleles. This view is underlined by the fact that similar duplications of several basepairs in coding regions have been described in human receptors. In these cases the mutations lead to misfunction or reduced activity of the proteins.

Poster Ga4.38

Monitoring of genetic variation of Estonian cattle breeds - temporal changes
H. Viinalass* and S. Värv. Institute of Animal Science, Estonian Agricultural University, Kreutzwaldi 1, 51014 Tartu, Estonia

The genetic variation of Estonian cattle breeds (Estonian Holstein, Estonian Red and Estonian Native Cattle) was investigated and the temporal changes in the gene pool were analysed. The data from routine blood typing and gene pool investigations from 1975 to 1998 were considered in the analysis. The monitoring based on erythrocyte antigen systems. For comparability of data over the period, the EAB system as most informative has the main emphasis. The interbreed and intrabreed genetic variation was studied. Temporal changes in gene frequencies of polymorphic loci were determined by variance analysis over the monitoring period. Statistically significant changes in allele frequencies over periods of monitoring were fixed. The temporal changes have taken place as a result of intensive using of new breeding material for Estonian Holsteins (Holsteins); for Estonian Red (Swiss blooded Danish Red, Angeln and Holstein); for Estonian Native Cattle (West-Finnish, Jersey, Brown Swiss), in consequence of which the frequency of the previous most frequent EAB alleles has decreased, but however, the position of EAB marker alleles of investigated Estonian breeds occurred to be stable over the time. The number and distribution of new EAB alleles has rapidly increased.

ANIMAL GENETICS [G]

Poster Ga4.39

Evolution, improvement and conservation of Simmental breed in Ukraine
Y.D. Ruban, Kharkov Zooveterinary Institute, p/o Malaya Danilovka, Dergachovsky Distrikt, Kharkov region, 312050, Ukraine.

Simmental breed of cattle took the second place in the herd structure /32,2% in 1969/ in the former Soviet Union, including the second place in Ukraine /39,2% in 1969/ following the Red Steppe Breed.
The process of crossing Simmental cattle with Red Poll Holstein breed and some other dairy breeds started at the end of 70s.(late 70s). The above process resulted in the creation and approbation of Ukrainian dairy Red Poll breed in 1992. The new breed differs from the others by its dairy type and high milking qualities: on the average 5000 - 6000 kg per a cow. At present the work has been done on preservation and improvement of Simmental breed of the mixed (dairy - beef) type and on the development of Simmental breed of beef type. Method of crossing with imported breeds of Simmental origin (American, Canadian, German, and Austrian selection) as well as native breeds of cattle have been used.
The beef cattle evolved has high meat qualities: daily weight gains are 1100 - 1200 g per a bull - carf the live weight og a bull at the age of 15 months is 510 - 560 kg; 18 months - 600 - 660 kg, slaughter output is 60 - 62 %. The cattle mentioned above has good health and strong constitution.

Paper GPh5.1

Long term selection for litter size in mice (>100 generations); correlated responses and biological constraints
O. Vangen. Department of Animal Science, Agricultural University of Norway, P.O.Box 5025, 1432 Aas, Norway.*

Generally, results from selection experiments and from practical breeding programmes indicate that there is a great potential for further improvement of production traits in farm animals. However, plateauing is reported in most long term selection experiments and unwanted side effects of selection are found in breeding programmes. Unwanted side effects are generally due to threshold values for important biological functions, optimum values of body composition and genetic correlations changing with level of selected trait(s). Experiences from 111 generations of selection for litter size in mice are indicating large potentials for genetic progress as well as genetic and biological constraints to selection. Through different selection strategies and randomisation periods it has been possible to increase litter size (total no. born) by 4.6 standard deviations (from 11 to 23.2 born/litter) in the high line. Results show that lifetime performance have been improved, however, less than 20 percent of the progress in litter size. Larger genetic progress was obtained when selecting in optimal maternal environments versus more constrained maternal environments. Correlated responses are presented in mature size, growth, residual feed intake and behavioural traits. Results are discussed in connection to biological limits and fitness constraints.

ANIMAL GENETICS [G]

Paper GPh5.2

Selection limits and fitness constraints in pigs
P.W. Knap* and P. Luiting. PIC Group, Roslin Institute, Roslin EH25 9PS, UK.

For a trait under selection, selection limits may be caused by (i) genetic variation of that trait becoming depleted, by (ii) fitness becoming compromised, and by (iii) environmental constraints becoming limiting for the genotype's expression. This paper deals with these three factors with regard to selection limits in pigs, paying some attention to other species.

There is currently little evidence for the depletion of genetic variation in pigs.

In highly productive pig genotypes, fitness may become compromised directly, in the form of impaired development of the skeletal system and of the reproductive endocrine system, or as a result of mutations with antagonistic effects such as at the *Hal* gene. Fitness may also become compromised indirectly, when production-related processes come to demand so many resources from the organism that functions such as immune response and coping with other stressors, or lactation, get in a resource-limited situation. This leads to loss of adaptive capacity to cope with intensive conditions (loss of "robustness").

In highly productive pig genotypes, environmental constraints to the allocation of sufficient resources towards the production-related processes may cause genotype x environment interactions on the expression of the genotype. This "environmental sensitivity" leads to the need to adapt the environment to match the needs of the genotype. When this fails, selection limits become apparent.

Paper GPh5.3

Effects of growth hormone deficiency on mice selected for increased and decreased body weight and fatness
Lutz Bünger*, William G. Hill, Institute of Cell, Animal and Population Biology, University of Edinburgh, EH9 3JT, UK

To elucidate the involvement of growth hormone (GH) in the genetic change produced by long-term selection in growth and fatness a study on over 900 mice was undertaken. Lines used in this study had been selected for more than 50 generations for high (PH) and low (PL) body weight at 10 weeks and for high (F) and low fat content (L) at 14 weeks, producing a 3-fold difference in body weight and a 5-fold difference in fat content. GH deficiency was achieved by repeated backcrossing into each line a recessive mutated gene (*lit*), which has a defective GH releasing factor receptor. In the absence of GH, the P-lines still differ in body weights (21d to 98d): e.g. at 98d homozygous *lit/lit*: PH=24.2g; PL=10.0g, wildtype (wt): PH=57.4g PL=18.7g). The effect of the GH deficiency on body weight was very much larger in the PH than in the PL, but on the log scale the difference was much smaller, but still significant. This indicates that changes in the GH-system contribute only a small part of the selection response in growth. GH-deficiency increased fat percentage in all lines (e.g. males at 98 days, *lit/lit*: PH=12.4; PL=11.7, F=26.5, L=5.5; wt: PH=6.1, PL=5.1, F=22.2, L=3.4) with significant line x genotype interactions and higher effects in males than in females. The interactions between the effects of the *lit* gene and the genetic background were, however, relatively small and indicate that other loci contributed to the selection response.

ANIMAL GENETICS [G]

Paper GPh5.4

Selection for litter size and its consequences for the allocation of feed resources - a concept and its implications illustrated by mice selection experiments

W.M. Rauw*[1], P. Luiting[2], R.G. Beilharz[3], M.W.A. Verstegen[4], O. Vangen[1]. [1]Department of Animal Science, Agricultural University of Norway, P.O. Box 5025, 1432 Ås, Norway, [2]Roslin Institute (Edinburgh), Roslin, Midlothian, EH25 9PS, United Kingdom, [3]Department of Animal Production, Institute of Land and Food Resources, University of Melbourne, Parkville, VIC 3052, Australia, [4]Animal Nutrition Group, Wageningen Institute of Animal Science, P.O. Box 338, 6700 AH Wageningen, The Netherlands

The present study describes undesirable effects of selection for increased litter size in mice for energy allocation patterns, in relation to reproductive performance and pup development. Lactating females of the selection line reallocate more buffer resources, otherwise available to other processes, towards lactation than lactating females of the control line. Furthermore, they mobilise body stores for a longer period of time. This means that selected females produce more offspring but at a greater cost to their own metabolism. This process was insufficient to supply offspring with adequate resources, resulting in reduced pup development and increased pre-weaning mortality rates. To generate improved production without meeting concomitant energy requirements will merely result in trade-offs. Increased genetic gains will be compromised in the long-term if the short-term focus is on a single production trait only.

Paper GPh5.5

Relation between voluntary feed intake of sows during lactation and sow and litter performance

E. J. Apeldoorn*[1] and J. J. Eissen[2]. [1]IPG, Institute for Pig Genetics BV, P.O. Box 43, 6440 AA Beuningen, The Netherlands. [2]Animal Breeding and Genetics Group, Wageningen Agricultural University, P.O. Box 338, 6700 AH Wageningen, The Netherlands.

Selection on litter size has resulted in increased energy requirements of lactating sows, whereas daily voluntary feed intake (VFI) may have decreased due to selection on leanness. An experiment was set up to investigate wether VFI of lactating sows is sufficient to raise large litters without loosing too much body reserves. Therefore, litter size of 297 first parity sows was standardised to a number varying from 8 to 14 piglets within three days after farrowing. Backfat thickness, body and litter weight of each sow were measured at ten days after farrowing (day10) and at weaning (day28). Average VFI was measured for the intermediate period. VFI of sows linearly increased with increasing litter size. However, this increase did not fully compensate the extra need for energy due to a larger litter, indicated by the linear decrease in body weight and backfat of sows with increasing litter size. Moreover, a larger litter was linearly associated with a reduced piglet growth. VFI, corrected for litter size, showed a positive correlation with weaning weight of sow and piglets, and with piglet growth. A higher VFI also reduced weight and backfat losses of the sow. Therefore, selection for a higher VFI during lactation is recommended to improve sow and litter performance.

ANIMAL GENETICS [G]   Paper GPh5.6

Genetic aspects of disease incidence in fattening pigs
E.J. van Steenbergen[*,1] and A.H. Visscher[2]. [1]IPG, Institute for Pig Genetics BV, P.O. Box 43, 6640 AA Beuningen, The Netherlands, [2]Institute for Animal Science and Health, P.O. Box 65, 8200 AB Lelystad, The Netherlands.

Disease incidence plays a relevant role in both animal welfare and economics of pig production. Investigations in possibilities on selection for a lower disease incidence have mainly been focussed on selection on immunological parameters. Those parameters are reported to be heritable and are supposed to have a close relation to disease incidence. However clinical signs of diseases on live pigs or their carcasses are the most direct and clear indicators for animal health. From 1996 to 1998 on two commercial farms clinical signs of diseases during the fattening period on over 15,000 pigs were recorded using electronic ear tags and data-loggers. Pigs were identified directly after weaning with an accurate registration of both sire and dam. In the slaughterhouse all carcasses were routinely investigated for clinical signs of diseases, mainly concentrated on lung, heart and liver. Carcass identification is based on the electronic ear tag. Preliminary analyses show significant (χ^2-test, F=.002) differences between progeny groups with one or more clinical signs of diseases during the fattening period. Data will be further investigated with emphasis on possibilities of selection

---

                                                                                       Paper GPh5.7

Quantifying genetic contributions to a dairy cattle population using pedigree analysis
T. Roughsedge*, S. Brotherstone, P.M. Visscher. IERM, University of Edinburgh, West Mains Road, Edinburgh, EH9 3JG, Scotland.

A variety of techniques were employed in the analysis of the change in genetic diversity of the UK Holstein-Friesian population over the last thirty years using the Holstein Friesian Society database (which has a base year of 1960). The parameters estimated were average inbreeding coefficient, average degree of relationship between cows, and two measures of genetic diversity, founder equivalent and founder ancestor number. The cow population was seen to change in founder origin from 96% British Friesian in 1967 to 24% British Friesian and 76% North American Holstein in 1997. The change in origin was seen to affect the rate of increase in inbreeding and to a lesser extent relationship, however the measures of genetic diversity were largely unaffected by the Holstein importation. In 1997 average relationship between cows had reached 1.34%, average inbreeding coefficient was 0.4% and the founder equivalent and founder ancestor number had converged at 93. The average inbreeding coefficient was seen to fall from 0.74 in 1982 to 0.38 in 1992 and to remain fairly constant up to 1997. The maternal structure of the cow population born in 1997 was also analysed. It was found that 93% of the cows were in maternal families of only 1-4 cows and only 0.5% of cows were in maternal families with more than 100 members, where a maternal family is a group of cows related only by maternal lineage.

ANIMAL GENETICS [G]

Paper GPh5.8

Genetic parameters of female and male fertility traits in US Holstein cattle
K.A. Weigel, Department of Dairy Science, University of Wisconsin, 1675 Observatory Drive, Madison, WI, 53706, USA.

Data were from first inseminations of Holstein cows in California (CA) and Minnesota (MN) that were bred between April 15 and October 15, 1998. Fertility variables included 56-day non-return rate (NR56) and 77-day veterinary-confirmed pregnancy rate (CP77). After editing, NR56 data consisted of 50,944 records in CA and 26,072 records in MN, and CP77 data consisted of 36,218 records in CA and 16,964 records in MN. Variance components were estimated with a single-trait animal model, assuming normality. Fixed effects included month of insemination, days in milk, age of cow, age-adjusted peak milk yield, previous days dry, previous calving difficulty, and semen price. Random effects included animal additive genetic effect, mating sire and herd-season class. In California, estimated heritability was 0.8% for NR56 and 1.6% for CP77. In Minnesota, estimated heritability was 2.7% for NR56 and 3.1% for CP77. The proportion of phenotypic variance explained by mating sire was 0.1% in CA and 0.3% in MN for both fertility traits. Genetic selection for female fertility seems possible, particularly if traits based on veterinary pregnancy check data are used, but the potential for differentiating among bulls based on male fertility traits is doubtful.

Poster GPh5.9

Analysis of heterosis in stress susceptibility using DNA markers on mouse chromosome 3
C. Brunsch, M. Starke, P. Reinecke*, G. Leuthold. Humboldt-University of Berlin, Faculty of Agriculture and Horticulture, Institute of Animal Sciences, Department of Breeding Biology and Molecular Animal Breeding, Philippstr. 13, 10115 Berlin, Germany.

An intercross of the mouse inbred lines C57BL/6J and Balb/cJ has been carried out to produce a F_2 generation with altogether 861 female animals. The open field test was used to characterize stress susceptibility. Relative to the stress susceptibility a high and a low performance group (25 and 45 animals respectively) has been constituted from the F_2 animals.
The analysis of even distributed DNA markers followed on the animals of the performance groups under consideration of loci, which are known as be significant for stress susceptibility. Dependent on the estimated dominance degree the microsatellites were statistical checked taking as a basis the overdominance respectively dominance model of heterosis. Based on the presented results the conclusion can be drawn, that the region 22,0 cM (D3Mit306) on chromosome 3 (Hsp86-ps2 locus) is associated with heterosis on stress susceptibility. The here presented results concern the investigations on the mouse chromosome 3.

ANIMAL GENETICS   [G]   Poster GPh5.10

Testing the suitability of the used selection index for Japanese quail (Coturnix japonica)
A. Richtrová, J. Hrouz, D. Klecker. *Department of Breeding of Farm Animals, Mendel University of Agriculture and Forestry Brno, Zemědělská 1, 613 00 Brno, Czech Republic*

Basing on individual performance tests of laying lines of Japanese quali (Coturnix japonica) a selection index was constructed using the following selectin parmeters: egg weight, number of eggs and total egg mass.
In a shortened laying test lasting 6 months, 3 different laying lines (03,07 and 20) counting 90 females per line were evaluated. The intesity of selection based on the selection index was 0,75; 0,80 and 0,80 for the 03, 07 and 20 lines, respectively. The applicability of the selection index was tested in the daughter population. In the individual performance tests the parameters of egg yield were studied during the 10-month laying period.
Results showed that the same selection criteria are suitable for poultry and for the Japanese quail.

Poster GPh5.11

Spirometric performances in Belgian Blue calves : environmental factors analysis and genetic parameters estimation
F. Bureau[1], C. Michaux*[2], J. Coghe[1], Ch. Uystepruyst[1], M.-L. Van de Weerdt[1], C. Husson[1], P. L. Leroy[2], and P. Lekeux[1]. *[1]Department of Large Animals Clinical Sciences, [2] Department of Quantitative Genetics, Faculty of Veterinary Medicine, University of Liege, Belgium*

In a clinical trial to detect possible relationships between respiratory capacity and resistance to respiratory disease in bovine, spirometry has been investigated in 734 Belgian Blue calves (15 to 297 days of age) sired by 20 IA bulls. Spirometric variables (SV) were measured after lobeline administration : average ventilation recorded during the 15 seconds of maximal ventilatory changes, vital capacity, maximal expiratory peak flow, maximal inspiratory peak flow and ventilatory reserve. Analysis of environmental factors showed that age of calves, herd, sex and vaccination status had significant effects on SV. A sire model including the pedigree for sires and multiple trait derivative-free REML procedure were used to estimate genetic parameters for SV as well as for body weight and muscling score. Heritability for SV ranged from .44 ($\pm$ .16) to .28 ($\pm$ .11). Genetic correlations and enviromental correlations among SV ranged from .76 to .98 and from .69 to .80 respectively. Genetic correlations of SV with body weight ranged from .25 to .56 and with muscling score from .21 to .76, and environmental correlations of SV with body weight from .44 to .70 and with muscling score from.09 to .25.

ANIMAL GENETICS [G] Poster GPh5.12

Selection response after long-term selection for high and low running activity in mice with special consideration of a selection limit
U. Renne and M. Langhammer. Research Institute for the Biology of Farm Animals, Department for Population Biology and Breeding Research; Wilhelm-Stahl-Allee 2, 18196 Dummerstorf, Germany.

Mice of the Fzt:DU outbred strain were selected for high and low running activity in a computer controlled treadmill (DU-hLB; DU-nLB). Besides the selection lines an unselected control line was maintained (DU-K). Per line and generation 80 mice pairs were mated.

The running performance in line DU-hLB was increased after 58 generations (gen) from 1020 m to 3600 m, in line DU-nLB was obtained a selection response from -428 m (startvalue: 933 m, endvalue: 505 m). That is an increase/decrease of 72% and 46%, respectively. Expressed in units of the phenotypic and genetic standard deviations it is 4.9 times s_p and 8.3 times s_g (DU-hLB) and 0.8 times s_p and 3.4 times s_g (DU-nLB). Heritability coefficient for running activity at start was 0.597/0.222, decreasing to 0.0006/0.0025 (DU-hLB/DU-nLB) at last generation. In high direction a more effective selection was possible than in low one.

The data were analysed by a modified exponential model. The coefficient of determination for the curve was 0.74/0.23 (DU-hLB: gen 0-37/gen 47-58) and 0.29/0.49 (DU-nLB: gen. 0-37/gen. 47-58). The maximal slope per generation runs to 120m (DU-hLB) and -65m (DU-nLB). Theoretical selection limits (function values) can be expected at a mean running activity of about 2600 m /124 m (DU-hLB/DU-nLB).

Poster GPh5.13

Plasma levels of thyroid hormones and cortisol in stressed Pietrain and Duroc pigs
S.J. Rosochacki[1], *A.B. Piekarzewska*[2], *and J. Połoszynowicz*[1]. [1]*Department of Molecular Cytogenetics,* [2]*Department of Behavioral Physiology, Institute of Genetics and Animal Breeding, Polish Academy of Sciences, Jastrzębiec, 05-551 Mroków, Poland.*

The studies were carried out on a total 50 Duroc and 40 Pietrain of both sexes, 8 months old, pigs. Immobilization stress -IMMS- was done by taping all four limbs to metal mounts in a prone position for 5, 15, 30 and 60 min. The blood was taken in all provided points. Classified by DNA test, Pietrain pigs were 90.6% nn and thus susceptible to stress, in contrast Duroc pigs were 84.7% of Nn and NN, and thus not stress susceptible. The purpose of this study was to investigate the relationship between differences in plasma hormone levels of pigs characterized by different susceptibility to stress and the influence of stress duration. The level of cortisol was similar in P (266 nmol/l) and D (284 nmol/l) pigs and increased after killing by 73% and 54%, respectively. Plasma level of tyroxine was 58 nmol/l in P and 65 nmol/l in D pigs and did not changed during the time of IMMS, the level of T_3 was 0.42 nmol/l in P and 0.50 nmol/l in D and was higher by 23% in D pigs as compare to P ones; the average rT_3 level was 1.34 nmol/l in P and 0.84 nmol/l in D pigs being about 44% higher in P than D pigs. Taking up our results together it is obvious, that acute immobilization stress influence the substantial changes in measured hormones in pigs.

ANIMAL GENETICS   [G]   Paper G6.1

Genetic parameters for clinical mastitis, somatic cell score, production, udder type traits, and milking ease in French first lactation Holsteins
R. Rupp*, D. Boichard. Station de Génétique Quantitative et Appliquée, INRA, 78352 Jouy-en-Josas cedex, France.

Genetic parameters were estimated by restricted maximum likelihood with an animal model on first lactation data of 29,284 French Holstein cows for clinical mastitis, lactation somatic cell score, milking ease, production, and nine udder type traits. The heritability was low for clinical mastitis (0.024), moderate for lactation somatic cell score (0.17) and milking ease (0.17), and ranged from 0.17 to 0.30 for type traits. A high (0.72) but lower than unity genetic correlation was found between clinical mastitis and lactation somatic cell score and indicated that both traits were genetically favorably associated. The antagonism with production was stronger for clinical mastitis than for lactation somatic cell score with genetic correlations of 0.45 and 0.15, respectively. Lactation somatic cell score and clinical mastitis were favorably associated with udder depth, foreudder attachment, and udder balance, with genetic correlations ranging from -0.29 to -0.46, whereas low correlations were found with the other udder type traits. Milking ease was found to be unfavorably correlated with lactation somatic cell score but not with clinical mastitis (genetic correlation of 0.44 and 0.06, respectively).

Paper G6.2

Bayesian analysis of heritability of liability to clinical mastitis in Norwegian Cattle with a threshold model
B. Heringstad*[1], R. Rekaya[2], D. Gianola[2], G. Klemetsdal[1] and K.A. Weigel[2].
[1] Department of Animal Science, Agricultural University of Norway, P.O.Box 5025, N-1432 Ås, Norway. [2] Department of Dairy Science, University of Wisconsin-Madison, Wisconsin 53706, USA.

Clinical mastitis records (presence/absence from 30d pre-partum to 120d post-partum) in first lactation cows were used to infer heritability with a threshold model (age-season, herd and sire effects). Cows without mastitis culled before 120d do not have an equal chance of contracting mastitis; including/excluding their records leads to bias. To assess sampling effects, 4 data sets were created: a) cows still in herd up to 120d, b) as in (a), plus mastitic cows culled before 120d; c) all cows, whether culled or not. d) As in (c), but with the model having a probability of mastitis depending on the time to culling date. Set (d) had 12,871 daughters from 257 young sires in 1131 herds. Sampling attained at least 2% precision for the posterior median of heritability. Posterior means (standard deviations) were: a) 0.18 (0.04); b) 0.20 (0.04); c) 0.19 (0.04) and d) 0.19 (0.04). Effects of sampling bias seem negligible, but (d) is theoretically appealing.

ANIMAL GENETICS [G]

Paper G6.3

Genetic analysis of conception rate in heifer and lactating dairy cows
D. Boichard*, A. Barbat, M. Briend. Station de Génétique Quantitative et Appliquée, INRA, 78352 Jouy-en-Josas cedex, France.

Genetic parameters of conception rate (CR) of heifers and lactating cows were estimated by REML in the three major French breeds. The most likely result (0/1) after each first insemination (AI) was determined from AI and calving dates. This trait was analyzed with a linear model including the fixed effects of herd year, month year of AI, age within parity, and the random effect of sires. The data set included 1910, 433, and 385 sires with at least 30 heifers inseminated in 1995 or 1996 in the Holstein, Normande, and Montbéliarde breeds, respectively. The estimates of CR heritability in first lactation were 1.3, 1.2 and 1.1%, respectively. The corresponding estimates in heifer were 1.0, 2.9, and 6.2%. The estimates of genetic correlation between CR in heifer and in first lactation was high in Normande (.85) and Montbéliarde (.89) but lower in Holstein (.52), whereas the residual correlations were close to zero. The estimates of genetic correlation between CR in first and second lactation were one. The estimate of genetic correlation between CR in first lactation and calving first insemination interval was low in Normande (-.02) and Montbéliarde (-.06) but slightly higher (-.11) in Holstein. The genetic correlations between CR in first lactation and 100-d milk, fat, and protein reached -.32, -.29 and -.25 in Holstein, -.11, -.08, and -.11 in Normande, and -.32, -.35, and -.35 in Montbéliarde.

Paper G6.4

Estimates of genetic parameters for carcass traits in Finnish Ayrshire and Holstein-Friesian
P. Parkkonen*, A.-E. Liinamo and M. Ojala. University of Helsinki, Department of Animal Science, P. O. Box 28, FIN - 00014 Helsinki University, Finland.

The aim of this study was to estimate genetic parameters for slaughter weight, and carcass fleshiness and fatness in Finnish Ayrshire and Holstein-Friesian bulls and heifers. Animal model, sire model and sire maternal grandsire model were tested for their suitability to evaluate young sires in progeny test. There were 38 188 records on animals slaughtered during a period of 20 months. Effects of month of slaughter, age at slaughter, sex and breed were statistically significant, and herd accounted for about 20 to 57 % of the total variation in the data. Estimates of heritability were 0.07 to 0.14 for slaughter weight, 0.16 to 0.31 for fleshiness and 0.08 to 0.16 for fatness, within herd heritabilities were 0.15 to 0.29, 0.29 to 39 and 0.12 to 0.29, respectively in the different breed by sex data sets. There was a positive genetic correlation between slaughter weight and fleshiness, 0.38 to 0.66, whereas fatness was not genetically correlated with the other traits studied. All within herd correlations were high, 0.55 to 0.93, and phenotypic and environmental correlations were also high or moderate. Sire model and sire maternal grandsire model were preferred to animal model due to computational requirements, and sire maternal grandsire model to sire model due to the possibility to include the sire path of maternal pedigree.

ANIMAL GENETICS   [G] Paper G6.5

Relationship of body weight and carcass quality traits with first lactation milk production in Finnish Ayrshire cows
A.-E. Liinamo* and M. Ojala. *Department of Animal Science, P.O.Box 28, FIN-00014 Helsinki University, Finland.*

Relationships of body weight and carcass quality traits with first lactation milk production traits were estimated from a field data set of 28 362 Finnish Ayrshire cows, using REML methodology and animal model. Studied body weight traits included heifer and mature live weight, estimated based on heart girth circumference as part of normal milk recording system, and carcass weight recorded in slaughterhouse. Additional carcass quality measures included carcass fleshiness and fatness as recorded in slaughterhouse. Milk production traits included first lactation 305-d milk, fat and protein yield, and fat and protein percentage. Genetic correlations between carcass weight and yield traits were almost zero, while the genetic correlations between live weight and yield traits were of somewhat higher magnitude. This suggests a negative correlation between yield traits and killing out percentage, which might be caused by difference in digestive tract weight and feed intake capacity. The genetic correlations between carcass quality and yield traits were negative and low to moderate. Selection for high yield alone seems thus to lead into reduction in body condition, which in turn might lead into more negative energy balance especially in early lactation.

Paper G6.6

Relationships between feed intake in different test periods of potential AI-bulls
R. Wassmuth*[1], H. Alps[2] and H.-J. Langholz[1]. [1] *Institute of Animal Husbandry and Genetics, University of Göttingen, Albrecht-Thaer-Weg 3, D-37075 Göttingen, Germany,* [2] *Hessian Institute for Animal Production, Neu-Ulrichstein, D-35315 Homberg/Ohm, Germany.*

Selection for milk production has negative correlated effects on disease incidence and fertility. An improvement of disease resistance of cows is achieveable by breeding for feed intake of potential AI-bulls. In many European countries, station testing of potential AI-bulls is well established and feed intake could be included in the breeding goal. Feed intake measurements considerably increase testing costs. Therefore, the relationship between feed intake in different test periods was investigated.
Between 1993 and 1998, feed intake was measured on potential AI-bulls of the breeds German Friesian (219) and German Red and White (55) at the test station in Neu-Ulrichstein/Hessia. The test period lasted from the 112th to the 312th day of life. The bulls were fed ad libitum with maize silage and received fixed amounts of concentrates according to their age. The intake of maize silage was measured automatically and daily records were available for each bull. The test period of 200 days was divided into four successive part-periods of 50 days and feed intake of each bull was calculated in the whole period as well as in the part-periods. The genetic correlations between feed intake in different test periods were estimated using REML under an animal-model.

ANIMAL GENETICS [G]

Paper G6.7

Genetic improvement of feed efficiency of beef cattle
P.F. Arthur*, J.A. Archer, R.M. Herd and E.C. Richardson. NSW Agriculture, Agricultural Research Centre, Trangie, NSW 2823 Australia.

Providing feed to animals is a major production cost and thus needs greater attention in animal genetic improvement programs. The objective of this study was to examine the genetic variation in postweaning feed efficiency in beef cattle and to assess the potential for its use to improve whole herd profitability. Feed intake and growth data on 1416 weaned Angus, Hereford, Poll Hereford and Shorthorn bulls and heifers from 8 tests were used in this study. After a 21-day adjustment period, the animals were fed *ad libitum*, a pelleted diet consisting of 70% lucerne hay and 30% wheat for 70 days. Residual feed intake (RFI) was used as the measure of feed efficiency and was calculated as the residual from the regression of feed intake against mid-weight$^{0.73}$ and growth rate. REML estimate for heritability of RFI was 0.43 ± 0.06, and its genetic correlations with growth traits were very low. Following one generation of divergent selection, the progeny of high efficiency parents consumed less feed for no reduction in postweaning growth performance. Preliminary results indicate no significant phenotypic correlation between postweaning RFI and mature cow weight and milk production, but a significant correlation with mature cow feed intake and RFI. High efficiency weaners consumed less feed as mature cows.

Paper G6.8

Genetic background of milk coagulation properties in Finnish dairy cows
O.Ruottinen*, T.Ikonen and M.Ojala. Department of Animal Science, University of Helsinki, P.O.Box 28, 00014 Helsinki University, Finland.

A research project consisting of two parts is in progress in order to improve milk coagulation properties in Finnish dairy cattle. In the first part, variation of milk coagulation traits was studied and the heritabilities for these traits were estimated using a sample of 789 Finnish Ayrshire (FAy) and 86 Finnish Friesian (FFr) cows from 51 herds. Large variation was found in milk coagulation properties of the cows. The heritability estimate for the milk coagulation time was 0.22 (s.e. 0.05) and for the curd firmness 0.40 (s.e. 0.04). Eight percent of the FAy cows produced milk that did not coagulate in 30 minutes. One third of these cows were daughters of two closely related sires, which indicates that also genetic factors may, in part, cause noncoagulation of milk. The second part of the study is in progress. Using a data set of 90 FAy sires with at least 30 daughters for each, the genetic background of milk coagulation properties will be studied more thoroughly. An effort will be made to locate genes that possibly have a substantial effect on very poor coagulation or noncoagulation of milk. In addition, genetic correlations between milk coagulation traits and milk production traits will be estimated.

ANIMAL GENETICS [G]

Poster G6.9

Relationships between milk protein genotype and milk production traits in Jersey breed
L. Sáblíková, O.M. Jandurová and M. Štípková. *Research Institute of Animal Production, Department of Molecular Genetics 104 00 Prague 1O - Uhřínìves, Přátelství 815, Czech Republic*

Records from the first and second lactation of 60 Jersey cows were used to investigate associations among milk, fat and protein production and κ-Cn and β-Lg genotypes. All animals were typed for A and B alleles of both milk protein genes using PCR and RFLP methods. The frequencies of alleles determined in our experiment confirmed the expected high frequency of κ-Cn B allele. Well known finding, that B allele of κ-Cn increases significantly protein yield, was clearly confirmed.. Homozygote animals BB produced less milk and more proteins. On the other hand, homozygote constitution of β-Lg AA increases milk production as well as protein content. Generally, the benefits of including the genotyping of milk protein genes to the breeding programmes depends on the strategy of dairy farming. The positive selection of κ-Cn BB Jersey cows is recommended when the quality of milk protein is important and milk is mostly used for the cheese production.
Supported by NAZV Mze ČR grant no. EP 0960006204.

Poster G6.10

Association between β-lactoglobulin genotypes and its content in milk
J.Futerová[1], L.Sáblíková[1], J.Kopečny[2], M.Štípková[1] and O.M. Jandurová[1]. [1]*Research Institute of Animal Production, 104 01 Praha - Uhřínìves, Czech Republic,* [2]*Institute of Animal Physiology and Genetics, Academy of Sciences, Czech Republic.*

Effect of genotypes on β-lactoglobulin content in milk was observed in 65 Holstein cows. Milk samples were collected monthly during standard lactation. The content of β-Lg (g / l) was determined in separated whey by using HPLC method. β-Lg genotypes were detected by polyacrylamide gel electrophoresis. Effect of different genotypes of β-Lg (AA, AB, BB) and alleles (A, B) on content of β-Lg in milk was analysed. The reletionship between β-Lg (g / l) and the total milk protein content was investigated. Obtained data were analysed statistically.
The highest production of β-Lg was detected in cows with AA genotype (5,68 g / l), followed by AB (4,83 g / l) and BB (3,54 g / l) genotypes. The influence of alleles A and B on β-Lg content was also tested. In homozygote animals allele A had higher expression than allele B (A=2,84 g/l, B=1,77 g/l). In heterozygote animals allele A showed dominant effect over allele B.
Supported by NAZV Mze ČR grant no. P 0960006204.

ANIMAL GENETICS   [G]   Poster G6.11

Estimates of repeatability for and factors affecting the milk coagulation traits
A.-M.Tyrisevä*, T.Ikonen and M.Ojala. Department of Animal Science, University of Helsinki, P.O.Box 28, 00014 Helsinki University, Finland.

A total of 979 individual milk samples from 83 Finnish Ayrshire cows in one herd were collected monthly during a period of two years to study the effects of environmental factors, milk composition and pH on milk renneting time and curd firmness, and to estimate the repeatability of the traits. About 10 % of the milk samples did not coagulate at all and one fourth of the cows repeatably produced poorly coagulating milk. Protein content and pH were the major factors affecting milk coagulation properties. Renneting time and curd firmness were at their best at the beginning and at the end of the lactation when protein and fat contents were at their highest. Increased milk pH at the end of the lactation had, however, an unfavourable effect on milk coagulation traits. SCC higher than 600 000 SCC/ml had a detrimental effect on renneting time. Coagulation properties declined with increasing age of cows. The estimate of repeatability for renneting time was 0.56 and for curd firmness 0.65. A large variation was observed between the coagulation measurements of some cows within a lactation. In conclusion, more than one measurements are needed during a lactation to give a reliable estimate of the milk coagulation ability of an individual cow.

Poster G6.12

Frequency and heritability of supernumerary teats in German Holstein cows
M. Brka*, Reinsch, N., Junge, W., Kalm, E. Institut für Tierzucht und Tierhaltung der Christian-Albrechts-Universität, 24098 Kiel.

Supernumerary teats in cows are undesirable since they represent a possible way of mastitis infection in the udder. Besides, milking with milking pails may be hampered. The mode of inheritance of this trait is still unknown but it is believed that the occurrence of this trait is based on polygenic inheritance.
The number of supernumerary teats in German Holsteins was studied in two herds with in total 628 animals, consisting of 452 females and 176 males during one year. Age of calves or heifers was from approximately four weeks to one year. Supernumerary teats were found in 18% of examened animals. The supernumerary teats trait was analysed as a qualitative threshold character with two classes in an animal model with herd as fixed effect. A Bayesian analysis was carried out using a Gibbs Sampling approach. An estimate for heritability of supernumerary teats was 0.21 with a standard error of 0.15. Data from the routine exterieur examination of test bull daugters in the Simmental population contains information about the occurrence of supernumerary teats and shall be used to confirm our observations.

ANIMAL GENETICS [G]   Poster G6.13

Relative efficiency of two methods of selection for feed efficiency: residual feed consumption and feed conversion ratio
J.L. Campo* and J. González. Departamento de Genética, Instituto Nacional de Investigación Agraria y Alimentaria, Apartado 8111, 28080 Madrid, Spain.

Two methods of selection for feed efficiency were compared in *Tribolium*. The traits measured were feed consumption (x_1) and weight gain (x_2) from 7 to 14 days, egg mass from 28 to 30 days (x_3), and initial weight at 7 days (x_4). Selection was intended to decrease the sum of the ratios (x_1/x_2) + (x_1/x_3). Residual feed consumption, used as selection criterion in line RE, was defined as x_1 less the genotypic regression of x_1 on x_2, x_3 and x_4; this regression is equivalent to a selection index restricted to hold x_2, x_3 and x_4 constant. Direct selection was used in line RA. A control line was kept in each of three replicates. Responses observed for the selection objective did not differ between lines, although line RE had the greatest response (-75.41 ± 19.18 vs -56.27 ± 14.25); the observed responses were less than expected. When residual feed consumption was considered as the selection objective, mean responses differed significantly (-46.88 ± 5.06 vs -14.57 ± 9.94); the proportion of realized versus predicted response was 83% in line RE and 49% in line RA. The selection response for feed consumption was significant in either RE (-48.67 ± 3.02) or RA (-26.90 ± 6.38) lines; the realized response for feed consumption was less than the expected response (79%) in line RE and higher than that expected (113%) in line RA. Neither the RE nor the RA lines showed significant responses for weight gain, egg mass or initial weight.

Poster G6.14

Study on postweaning growth of heifers of different breeds on the same condition in Hungary
Zs. Wagenhoffer, F. Szabó, J.P. Polgar, Pannon University, Georgikon Faculty, Keszthely, Hungary

Postweaning growth of heifers of Hereford, Aberdeen Angus, Red Angus and Hungarian Simmental (n=22) were evaluated and compared on the same condition in a one year period (1998) in Hungary. The experimental beef herd were kept outdoor around the year on peat bog soil pasture in Keszthely. Calves were born in spring, beginning of March 1997, creep feeding from July to October and weaned in October. The heifers were kept in free-stall house and were fed by concentrated feed and meadow hay. Weaning weights were adjusted to 205-day. The 205-day weight and the following weight data were processed and standardized with the SPSS 7.5 software. The result of 205-day weight shows significant difference (P<0.05) between Hereford and Hungarian Simmental. No significant differences were found between Hereford and Angus, and between Hungarian Simmental and Angus breeds. Hereford heifers had 144 kg (the lowest), Aberdeen Angus 208 kg, Red Angus 214 kg, Hungarian Simmental 245 kg (the highest) 205-day weight. Very strong correlation (r=0,97) were found between the days of life and standardized weights. Similar growth dynamic were found at the four breeds. Hereford and Angus heifers reached the turn-over point 30 to 40 days sooner than the Hungarian Simmental.

ANIMAL GENETICS [G]

Poster G6.15

Effects of Crossbreeding on Fertility Traits in a Synthetic Dairy Cattle Population
M. Stenske[1], G. Seeland[1] and O. Distl[2]. [1]Institute of Animal Sciences, Humboldt-University, Invalidenstraße 42, 10115 Berlin, Germany, [2]Institute of Animal Breeding and Genetics, School of Veterinary Medicine, Bünteweg 17p, 30559 Hannover, Germany*

The population of the German Black Pied Dairy Cattle - a synthetic breed - bred by crossing Holstein-Friesian (HF) and Danish Jersey (DJ) with the local Black Pied Cattle, was used to estimate crossbreeding parameters for fertility traits (n=2.37 to 2.96 mio). First to third lactation records of cows that had calved for the first time between the years 1985 and 1989 were included using the traits calving interval (CI), breeding delay (BD), number of inseminations per service period (NI) and NonReturnRate (NR). Models used to estimate crossbreeding parameters were derived by DICKERSON (1969), KINGHORN (1987), GROSSHANS et al. (1994) and WOLF et al. (1995). They were integrated in a general mixed linear sire model with the fixed effects of calving year season, herd, and interval from calving to first breeding. The additive breed effects were relatively consistent between different models. Estimates varied for CI in HF from 23.8 to 85.3 and in DJ from -109 to 10 days, and NI in HF from .122 to .541 and in DJ from -.108 to .587, and BD in HF from 2.6 to 13.2 and in DJ from 6.1 to 15.3 days, and NR in HF from -.02 to .074 and in DJ from -.332 to -.086 in the first lactation. All traits were influenced by non additive effects, which differ between models for one breed combination also in sign. Missing pure breed informations may have caused the low reliability of the estimates.

Poster G6.16

Physico-chemical and sensorial evaluation of beef from Polish bulls
A. Węglarz, E. Gardzina, P. Zapletal, J. Szarek. Department of Cattle Breeding, Agricultural University of Cracow, al.Mickiewicza 24/28, 30-059 Kraków, Poland

The studies were carried out on 24 bulls of three breeds: Black-and-White (B), Red-and-White, Simmental (S) and crossbreds BxS. The bulls were fattened using grass silage and meadow hay supplemented with a concentrate. The experimental slaughters were done at the age of 20 months and average weight about 500 kg. The meat samples of musculus longissimus dorsi were taken between 13th thoracis and 1st lumbar vertebrae. No significant differences between genetic groups in the contents of basic chemical components were found. The meat from Red-and-White bulls had the lighter colour compared to other groups. Also, in this group the lower values of meat pH_{48} were found. The lowest level of cholesterol was in meat of Black-and-White bulls - 56,7 mg per 100g and the highest in crossbreds BxS - 62,4 mg per 100g of meat. Statistically significant differences ($P<.05$) between these groups were found.
In all studied groups the sensorial traits were highly evaluated and there were no differences among groups with reference to these traits. Moreover, the level of fatty acids was determined. The highest level of linolenic acid was observed in Black-and-White and the lowest in Red-and-White bulls. Statistically significant differences between these groups ($P<.05$) were found.

ANIMAL GENETICS   [G]

Poster G6.17

Relationship between type traits and production in Polish Black and White sire evaluation
A. Żarnecki*, W. Jagusiak, Department of Genetics and Animal Breeding, Agricultural University, 30-059 Kraków, al. Mickiewicza 24/28, Poland.

Three years ago the national program for linear evaluation of type traits was introduced in Poland, and recently the first official sire proofs were made available. For milk production traits multi-trait BLUP animal model based on the first three lactations was implemented in 1998.
The data were breeding values of 616 Black and White sires. Correlations were calculated between the breeding values for milk, fat and protein yields, fat and protein content, and standardized breeding values for 21 linearly scored type traits, and 2 body measurements.
Final score, kaliber, type and muscularity showed the highest correlations with yield traits (0.56 - 0.66), slightly lower udder support and width. Low correlations with yield traits were found for chest and rump width and leg set. The multiple regression equations calculated for prediction of the yield traits included up to ten traits and multiple determination coefficients ranged from 0.55 to 0.57.

Poster G6.18

Identification of genomic regions linked to atresia coli in cattle
A. Kempers*[1], U. Thieven[1], S. Neander[1], C. Drögemüller[1], J. Pohlenz[2], O. Distl[1] and B. Harlizius[1]. [1]Department for Animal Breeding and Genetics and [2]Institute for Pathology, School of Veterinary Medicine, Hannover, Germany

The congenital defect atresia coli is a lethal intestinal malformation with absence of a colonic segment. Planned matings in a US Holstein herd suggested an autosomal monogenic recessive inheritance. Blood or tissue samples of 188 affected animals (108 males, 80 females) have been collected most of them belonging to German Holsteins.
A whole genome scan was performed on a subset of 38 affected and closely related animals plus ancestors, using 180 bovine microsatellite markers with an average spacing of 17.1 cM. The data were analysed for two-point (Linkage 5.1) and multipoint (Genehunter) linkage assuming a recessive inheritance and modelfree with the nonparametric approach of Genehunter.
In four genomic regions with a two-point Lod score above 1.0, additional markers were genotyped on larger families with a total of 82 affected calves. Lower Lod scores were obtained but suggestive linkage (two-point Lod score of 2.5) was observed for one family in one of the regions. Although multipoint analysis across all families gave no evidence for linkage, the nonparametric analysis showed a p-value of 0.01 in the same region mentioned before. This is supported especially by one family with a NPL-Score of 4.9 (p-value 0.002).

ANIMAL GENETICS   [G]

Poster G6.19

Hereditary determination of cattle to mastitis
N.N. Kochnev, Institute of Veterinary Genetics and Selection, 160 Dobrolubov Str.,630039 Novosibirsk, Russia

Mastitis is one of the most common disease problems in milk production in West Siberia and may reduce milk yield, fertility, productive life. The average morbidity in dairy cows has been recorded to constitute 12-55 %. The role of mastitis in heritance is revealed in cattle. Estimated heritabilities of resistance to mastitis are typically low (0.08-0.16), but the genetic variation has been shown to be reasonably high. Differences in morbidity in cow families have been revealed (6.7-17.0 %). Differences in morbidity in lines have not been established. This may be due to the absence of artificial selection for this character that is why the lines are not differentiated. Much greater genetic variability in the population of Black-and-White cattle was identified in sires for their daughters resistance (2.7-23.6 %). The inheritance of the mastitis resistance is suggestive of the polygene character.

Poster G6.20

The influence of the genetic factors on immune reactivity and natural resistance in cattle
V.G. Marenkov, V.L. Petukhov, Y.A. Krinskiy. Institute of Veterinary Genetics and Selection of Novosibirsk Agrarian University, 160 Dobrolubov Str.,630039 Novosibirsk, Russia

Determination genetic diversity of immune system indexes in animal populations for the further selection is one of the way to increase fitness and the genetic resistance to diseases. In this connection the study of some traits: cell's reaction on intradermal phytohemagglutinin injection (PGA-test), lysozyme (LA) and bactericidal (BA) activity of blood serum, phagocytosis activity (PhA) and immunneattraction leukocyte ability (ILA) was carried out in Black-and-White cattle on genetically different groups of animals (sire offspring, sire line, type of stressresistance - I,II,III,IV by E.P. Kokorina). The maximal relative differences in volumes of traits between animal groups were for LA - 126.2%, PGA-test - 66.5%, Pha -49,4%, ILA - 46.8% and BA - 34,3%. The PGA-test was characterised by highest genetic variability. The influence of sire genotype, the type of stressresistance, sire line on this index was 37.9, 17.4, 9.1 and 5.8, respectively. Also, the influence of sire genotype on ILA (16.9%) and mother's origin on PhA (8.9%) was established. The humoral indexes of resistance were not authentic.

ANIMAL GENETICS   [G]   Poster G6.21

Significance of Lys-mic genotyping in the genetic improvement of mastitis resistance in dairy cattle
C.S. Pareek[1], M. Schwerin[2] and K. Walawski[1].[1]Department of Animal Genetics, University of Agriculture and Technology, A.R.T., Olsztyn-10-718, Poland. [2]Forschungsbereich Molekularbiologie, Institut fur die Biologie Landwirt-schaftlicher-Nutztiere, Wilhelm-Stahl-Allee-2, D-18196, Dummerstorf, Germany.*

Widely distributed in nature Lysozymes (EC 3.2.1.17) together with lactoferrin and lacto-paraoxidase system are well known for its non-specific anti-bacterial action in the body. In dairy cattle population, lysozyme has been poorly expressed in mammary gland tissue as compare to its expression in other body fluids and secretions. Recent achievements in the exploration of macrophage expressed lysozyme gene (*mLys*) revealed polymorphic microsatellite alleles within the *mLys* gene co-segregates with high and low serum lysozyme activity (Pareek et al. 1998). By Lys-mic genotyping, a total of 5 gene variants were observed in Polish Black-and-White cattle (Pareek et al. 1998a) with one gene variant (Lys-mic allele 7) linked to high serum lysozyme activity. The identification of such marker alleles by means of Lys-mic genotyping is essential to uncover the casual mutation responsible for high lysozyme expression in the mammary gland tissue (Seyfert et al. 1996). In this poster presentation, we are elaborating the lys-mic genotyping procedure and its significance in the context of udder health improvement of dairy cattle.

Poster G6.22

Vital study of Bovine granulosa cells during follicular development by differential staining with acridine orange for detection of apoptosis
S.N. Proshin, T.A. Smirnova, T.I. Kuzmina, A.F. Yakovlev, All-Russian Research Institute for Animal Genetics and Breeding, Moscowskoye shosse, 55a, St.-Petersburg-Pushkin, 189620, Russia*

Apoptosis is important factor of organ and tissue morphogenesis. Previous studies have shown that apoptosis is a mechanism underlying the process of bovine granulosa cell death during follicular atresia. The aim of this study is to test technique of vital staining for bovine granulosa cells by acridine orange. As known acridine orange staining is a conventional technique for detection of apoptosis in living cells. Follicles were classified by morphometric criteria as healthy (n=7) or atretic (n=13). Apoptosis was detected in granulosa cells from all atretic follicles as well as from all healthy follicles. For healthy and atretic follicles the frequency of apoptotic granulosa cells was varied from 0,1% to 11,3% (X=2,54±1,4%) and from 23% to 90% (X=67,11±5,6), respectively.
The advantages offered by the present technique is the possibility of detecting apoptosis in living granulosa cells, enabling identification and possible selection of apoptotic cells within a population of cells that are in the process of dying.

ANIMAL GENETICS   [G] Poster G6.23

Cytogenetic and its relation to fertility in bovine
Karima F. Mahrous[1]*, T. M. El-Sayed[2] and I. A. Abd El-Hamid[2]. [1]Department of Cell Biology, National Research Center, Giza, Egypt, [2]Department of Obstet,Gynaec.& A.I, Fac. of. Vet. Med., Moshtohor , Zagazig. Univ. (Benha branch.)

Seventy-one Holstein-Friesian divided into five groups: Three groups of females and two groups males were investigated cytogenetically to identify and score the chromosomal aberration occurred. The female groups were repeat breeder group (25 animals), and freemartin group (4 animals) in which the females are heifers aged from 3 to 4 years and the third groups was pregnant heifers (25 animals control) aged 2 to 2.5 years. The male groups were young males aged from 12 to 18 month (8 animal with unilateral testicular orchidy) and breeding bulls (8 control animals) aged 3 years. Blood samples were collected from these groups for cytogenetical analysis. The highest percentage was detected for the total structural aberrations (25.68%) which are significantly higher than that observed for total numerical aberrations (11.84%) in the repeat breeders group. The freemartins had equal numbers of metaphases with xx and xy, and were found to have numbered (20 and 5 %) of structural and numerical variations, the percentage of total chromosomal aberrations was 25%. The total structural aberrations in the affected male group (26.75%) were significantly higher than numerical variations (11.0%). The total numerical and structural aberrations, the percentage in the affected males group (37.75%) is significantly higher than that mentioned for the normal control group (16.3%).

Poster G6.24

The level of somatic genome mutations in farm animals
M.L. Kochneva*, V.L. Petukhov, S.G. Kulikova, G.N. Korotkova, T.B. Paramonova, Novosibirsk Agrarian University, Novosibirsk, Dobrolubov Str. 160, Russia.

Cytogenetic investigations were conducted in farm animals of different age groups: 132 A.I. bulls, 10 breeding boars (from 2 to 6 years old), 144 cows of 1-4th lactations, 29 sows of 1-2nd farrows, 29 calves and 21 piglets at the age of 2 months.
An the result of karyotypic analysis of the animals a spectrum of genome mutations presented by poliploidy and aneuploidy was determined. The highest percent of poliploid cells was marked in the cows that constituted 4.12 and was 2 and 5 times higher than in the sires and calves, respectively. The frequency of poliploidy in the boars, sows and calves was similar. The investigated species were found to increase the number of poliploid cells related to their age ($P<0.05; 0.001$).
The level of aneuploidy in all groups of the cattle as well as sows appeared to be much higher than that in boars and piglets. Aneuploid cells were presented mainly by hypoploid ones (with 1-2 chromosomes lacked). The lowest frequency of aneuploidy (2.2 %) was observed in the piglets that was 4.5 times less than the highest level (10.36 %) revealed in the sires ($P < 0.001$).
Thus, it was established that the cattle was characterized by the highest frequency of genome mutations in comparison with the pigs.

ANIMAL GENETICS [G]

Poster G6.25

Chromosome mutations in cattle: consequence of the Tomsk Siberian chemical plant (SCP) accident
S.R. Masun, M.L. Kochneva, P.L. Petukhov, S.G. Kulikova, B.L. Panov, N.N., Shipilin, Novosibirsk Agrarian University, 160 Dobrolubov Str.,630039 Novosibirsk, Russia.

Chromosome analysis was performed in blood lymphocytes of 12 Black and White cows from "Naumovka" farm. This region was heavily contaminated by nuclear expells resulted from the radioactive accident at Tomsk SCP that happened on April 6, 1993. On relatively pure farm "Kaltai" 9 cows were investigated. The contents of Pb, Cs-137, Sr-90 in the soil "Naumovka" farm several times higher than the norm. "Naumovka" cows showed a significantly higher ($P<0.05$) frequency of polyploidy with a high level of radionucleatide doses (1.70 ± 0.28) than those from the farm with the relatively low level of contamination (0.70 ± 0.28). The differences were not revealed in the level of chromosome breaks and total rate of chromosome aberrations.

Poster G6.26

Somatic chromosome instability of phenotypically healthy and abnormal calves in different ecological areas
S.G. Kulikova, V.L. Petukhov, M.L Kochneva, G.N. Korotkova. Novosibirsk Agrarian University, Novosibirsk, Dobrolubov Str. 160, Russia.

The chromosome instability of 29 healthy and 38 abnormal Black and White calves was investigated in chemically heavy polluted and pure areas. The frequency of chromosome aberrations in calves with congenital abnormalities was in 3.3 times higher in polluted area than that in pure area ($P<0.001$). In healthy animals the frequency of chromosome aberrations was only 2,2 times higher in the heavily polluted area than that in the calves from the pure area ($P<0.001$). Cells with breaks of chromosomes were revealed more often in animals. The frequency of cells with breaks in the calves from the polluted area was 14.61%, that was 2.5 times higher than in animals with congenital defects from the pure area (5.78%). The highest variability was determined for this type of aberrations (0-37%).
Thus, calves with congenital abnormalities are more sensitive to environment character as compared to healthy ones.

ANIMAL GENETICS   [G]

Comparative analysis of the immunological and molecular methods to identify the BLV - positive cows at the different stages of viral infection
N. Koutsenko*, D. Shayakhmetov, M. Smaragdov, N. Dementieva, T. Starozhilova. All - Russian Research Institute for Farm Animal Genetics and Breeding, Moskovskoe shosse 55a, 189620, St. Petersburg, Russia.

White and Black Breed cows of 2 - 7 years age from the collective farm "Krasnaya Slavyanka" were examined using a complex of haematological (the amount and proportion of leucocytes, lymphocytes), immunological (RID), molecular - biological (PCR, Sauthern Blot) methods. The primers used in PCR were complementary to the region of envelope gene. A total 59 cows were tested in this study. Among them 16 cows showed the positive reaction in RID testing and were haematology negative. 28 cows were positive both of the RID and haematology. It was not possible to reveal 3 cows in this group of animals by PCR. It is about 5% of all animals tested. 15 healthy cows in control group were negative in immunological tests but 5 animals showed the positive reaction in PCR testing. At the present time 2 of them are transported to the BLV(Bovine Leukemia Virus) infections group because of the positive reaction in the following RID testing. Thus, the molecular - biological methods may be used for the purposes under examinations. They are more sensitive and permit to reveal ill animals at the early stages of leucosis in control group.

ANIMAL NUTRITION [N]   Paper NC1.1

Economic evaluation of different dairy cow types under heavy roughage rations
D. Erdin, R. Schwager-Suter, D. von Euw and N. Künzi. Institute of Animal Science, Animal Breeding Group, Swiss Federal Institute of Technology Zürich, Clausiusstr. 50, CH-8092 Zürich, Switzerland.

Results from bookkeeping dairy farms show a steady proportion of milk produced with concentrates of 25 % in terms of net energy. This relation was stable over the last 10 years, whereas the level of production increased by 100 kgs of milk per cow and year.

A herd model was developed to estimate profits for different cattle types (Holstein, Jersey, Simmental). Milk and beef yields as well as feed intake in milk, beef and replacement animals were based on results from the research station Chamau. Costs for housing, labor and various were taken from the agron. research station standards. The best results were obtained with the high yielding dairy type, needing the lowest number of cows for a given milk quotum.

Four feeding strategies (2 forage qualities x 2 concentrate levels) were evaluated with data of 212 lactations (Jersey, Holstein, Je x Ho). The highest income above feeding costs per cow was obtained by Holstein cows, fed with the better quality forage and the higher level of concentrate, while no distinct difference was found in the income per kg energy corrected milk between the 3 types of cows.

Paper NC1.2

Breeding to optimise use of grass and forages for milk production
R.F. Veerkamp. Department of Animal Breeding and Genetics, ID-DLO, P.O. Box 65, 8200AB Lelystad, The Netherlands.

Many aspects of the feed utilisation complex are heritable, and hence there is scope for genetic improvement to optimise grass and forage utilisation. Although the obvious breeding goal appears to be selection for improved gross feed efficiency (i.e. more milk produced from less grass) and smaller animals (because lower maintenance costs), some care must be taken using this approach. This is because of the link between the energy complex and fertility. A more negative energy balance (i.e. perceived as a better gross feed efficiency), a lower live weight and more live weight loss during lactation all have an unfavourable genetic correlation with days till first heat (range -0.40 to -0.80). Selection on an index including protein yield, and energy balance or live weight (change) allows 0.71 to 0.80 of the response in yield, whilst there is no negative selection effect on days till first heat anymore. However, all these indices have a positive weight for energy balance, live weight or feed intake, which is opposite to selection for improved feed efficiency and lower maintenance costs. The probable solution is to select for the functional components of weight that are important for energy balance (i.e. feed intake and body condition) and select against the components of weight that just add maintenance costs (i.e. body size). In all cases, however, additional traits need to be recorded to select cows that have high yield and good fertility simultaneously.

ANIMAL NUTRITION [N]

Paper NC1.3

Feeding high genetic merit dairy cows in grass based systems
F.J. Gordon[1] *C.P. Ferris*[1] *and A. Cromie*[2] [1]*Agricultural Research Institute of Northern Ireland, Large Park, Hillsborough, BT26 6DR,* [2]*ICBF, Bandon, Co Cork*

The continued trend towards increasing genetic merit of dairy cows for milk production poses problems in grass and forage based systems. Recent research in Ireland has explored genotype x environment interactions and demonstrated considerable scaling effects between environments but provided no evidence that sires selected in a high concentrate environment were not equally appropriate in high forage regimes.
Calorimetric studies have shown that while increasing genetic merit improves overall gross feed efficiency it does not influence the partial efficiency with which energy is converted to milk production. These data have however raised concern about the present maintenance requirements adopted with high genetic merit cows and suggest that requirements are higher than those traditionally adopted. These studies support the view that the continued high performance of high genetic merit cows can only be sustained by increasing nutrient intake.
There are considerable constraints to increasing nutrient intake in grass and forage based systems and component studies in Northern Ireland have examined the influence of aspects of pasture grazing, conserving forage and concentrate supplementation on animal performance. These studies have been further developed into full lactation studies aimed at defining systems which will enable the potential of high genetic merit cows to be exploited in a grass based forage environment.

Paper NC1.4

Planning and monitoring herbage growth and utilization to feed grazing animals under uncertain conditions
M. Duru, Station d'Agronomie, BP 12, 31326 Castanet, France

The design of sustainable livestock systems which reconcile production, economic returns and environmental protection, is encouraged through regulations. This often necessitates reappraising the way plants and animals are managed at the farm level, particularly for grazing. Research could contribute to these adaptations by providing integrated models allowing comparison of scenarios including planning rules to establish feed budgets and indicators for short term grazing plans. We present models for plant resource planning and monitoring: (i) herbage growth as a function of nitrogen supply, herbage nitrogen status being assessed through herbage N content, (ii) net herbage production, a function of grazing pressure and monitored using sward height, (iii) herbage digestibility, a function mainly of grazing pressure and monitored using sheath height. Secondly, we show how these models could be used (i) to establish feed budgets over a grazing period, defining key periods at which adjustment rules could be used to deal with climatic uncertainty, (ii) to monitor plant resources at the whole grazing area level in order to fit the grazing area to the feed demand. To assess *a posteriori* the ability of grazing management to fit a defined production objective, a practical method usable at commercial farm level is presented. Finaly, we discuss how these models could be used to establish multiyear business plans.

ANIMAL NUTRITION   [N]

Paper NC1.5

Grazing managements with Belgian Blue growing bulls before an indoors finishing
I. Dufrasne[1], A. Clinquart[2], J.-L. Hornick[3], C. Van Eenaeme[3], L. Istasse[3] , [1]Experimental Station, [2]Technology, [3]Nutrition, Veterinary Faculty, University of Liège, Belgium*

A fattening system including a grazing period was tested over six consecutive years as an alternative to the usual indoors system. Thirty two young growing double muscled bulls were divided in 4 groups. The 3 first groups were firstly grazed in pastures managed either with a normal level of N fertilizer (3/3N) and at normal stocking rate (6 bulls/ha), with N fertilizer reduced to 0.66 of the previous group (2/3 N) or with no N fertilizer but with a reduced stocking rate of 4 bulls/ha (0N). After the grazing season, the bulls were finished indoor on a sugar beet pulp based concentrate diet. The 4th group was fattened indoor with the same diet. There were nearly no differences in terms of grass characteristics and grass composition, one on the largest differences being the Ca content (7.04 vs 8.45g/kg DM in 3/3 N and 0N). The average daily gain was 1.05, 1.03 and 1.09 kg/d ($P > 0.05$) in the 3/3 N, 2/3 N and 0N groups. The corresponding total gain per Ha was 903, 895 and 629kg. During the finishing period, the gain was 1.32, 1.28 and 1.34 kg/d ($P>0.05$) respectively. When the grazing period was pooled with the finishing period, the daily gain was 1.15, 1.13 and 1.17kg/d ($P>0.05$) and was significantly lower than 1.51kg/d ($P<0.001$) in the indoors system. There were no large differences in terms of carcass composition but the meat of the bulls previously grazed was significantly darker (lower L^* at 43.2, 42.0 and 42.3 vs 45.3%, $P<0.05$) and redder (higher a^* at 18.2, 18.3 and 18.5 vs 16.4 %, $P<0.05$).

Paper NC1.6

Comparison of Friesian and Charolais x Friesian steers in grass-based production systems
M.G. Keane and E.G. O'Riordan, Teagasc, Grange Research Centre, Dunsany, Co. Meath, Ireland

The objective was to quantify carcass output in different beef systems. Ninety-eight spring-born Friesian (FR) and Charolais x Friesian (CF) steers were reared from calfhood in two production cycles. In cycle 1, FR were slaughtered after 783 days (second winter) and CF were slaughtered after 902 days (third grazing season). In cycle 2, all were slaughtered after 712 days (second winter). Fertiliser N inputs were 232 and 255 kg/ha in cycles 1 and 2, respectively. Concentrate inputs per animal in cycle 1 were 1024 and 404 kg for FR and CF, respectively. Corresponding inputs in cycle 2 were 1116 kg for both. Slaughter weights (s.e.d. 7.7) were 632 and 727 kg (cycle 1) and 611 and 630 kg (cycle 2), for FR and CH, respectively. Corresponding carcass weights were 327, 392, 317 and 340 (s.e.d. 4.3) kg. In the same order, carcass conformation scores were 2.1, 3.1, 2.2 and 2.9 (s.e.d. 0.09), and carcass fat scores were 3.6, 3.6, 4.2 and 4.2 (s.e.d. 0.09). Carcass output per ha was 785 (FR) and 672 (CF) kg in cycle 1 and 793 (FR) and 850 (CF) kg in cycle 2. It is concluded that with pasture finishing a carcass output of 670 kg/ha can be achieved with CF on a concentrate input of 400 kg per animal.

ANIMAL NUTRITION [N]

Paper NC1.7

Use of extensive breeding system for beef production in Italy: the Maremmana case
S. Gigli*, M. Iacurto. *Meat Department - Animal Production Research Institute Via Salaria 31 - 00016 Monterotondo (Roma) - Italy*

In Italy the incidence of lowland areas is scarce: only 23% of the total. The land use and the incidence of productive and permanent forages differ with the orographic and climate difference in North, Centre and South of Country (35%, 9% and 18% of plains respectively). This situation affects also the period and the quantity of the forage production that is very low in summer and winter. So we can rear beef calves in the open air on pasture only for short time (maximum 6 months) in the first part of life (from birth to weaning) with their mothers. The Maremmana breed (and its crosses) was used in extensive and mixed breeding in Centre of Italy for beef production. Here we compare three rearing systems on MM crossbreed young bulls: intensive, extensive and mixed; the last appeared the best situation. In fact, the final weight (594 kg), the ADG (0.930 kg/d) and the slaughter (net dressing percentage 65.5%, conformation 3+, fatness 2-) and dissection (side 169 kg, meat 71.6%, fat 8.5%) performances were comparable with production reached in other countries in similar conditions and meat quality was profitable.

Paper NC1.8

Use of the automated gas production technique to determine the fermentation kinetics of carbohydrate fractions in maize silage
E. Deaville and D.I. Givens*. *ADAS Feed Evaluation and Nutritional Sciences, Alcester Road, Stratford-on-Avon, CV37 9RQ, United Kingdom.*

As forage maize matures, starch concentrations increase whilst neutral detergent fibre (NDF) declines. Such changes may affect the nutritive value (NV) of the resultant silages. Whilst current *in vitro* procedures aim to estimate NV in whole feeds (WF), assessment of the different carbohydrate (CHO) fractions will become increasingly important for rationing models designed to optimise rumen function and predict supply of nutrients.

This study used an automated *in vitro* gas production (GP) technique with samples of both WF and isolated NDF of four maize silages (cv. Hudson) of increasing maturity (DM contents 230, 280, 330 and 380 g kg^{-1}), in order to establish the fermentation kinetics of the NDF and neutral detergent solubles (NDS, as WF-NDF).

The conventional approach of incubating WF resulted in small differences in gas production values which were not related to the changes in the principal CHO fractions. However, incubating the NDF fractions separately and calculating the NDS values produced strong relationships between the changes in the starch and cell wall contents and the gas asymptote values for the NDS ($r = 0.74$) and NDF ($r = 1.0$) fractions respectively. The gas asymptote values for the NDF fraction were also highly correlated ($r = 0.99$) with NDF degradability. The use of this approach in nutrition models needs consideration.

ANIMAL NUTRITION [N]

Paper NC1.9

Long term effects of a large concentratre decrease for dairy cows.
V. Brocard*[1], J. Kerouanton[2], D. Le Meur[2]. [1]Institut de l'Elevage, BP 67, 35652 Le Rheu Cdx, [2]EDE-CA du Finistère, Ferme expérimentale de Trévarez, 29520 St Goazec.

This trial belonged to a program aiming to reduce the feeding costs of dairy herds. In July 1992 the Holstein herd of Trévarez experimental farm, whose production level reached 8000 kg/cow/year, was split into two feeding groups. The multiparous cows of the Control group were given 1600 kg of concentrates per year, with a protein level in early lactation of 110 g of PDI per kg of DM. The Low group multiparous cows were given 650 kg of concentrates per year, with a protein level lowered to 90 g of PDI/kg of DM.

After the first 3 years, the following results have been stated in the Low group: a forage to concentrate substitution rate of 0.4; a better fodder valorization ; a change in the shape of the lactation curves; a decrease of 0.9 kg of milk per kg of saved concentrate; the same protein content, but an increased fat content; no significant change in the body condition evolutions ; a relative degradation of the reproduction performances; less herd diseases, in particular around calving; a reduced culling rate; and a higher gross margin by 0.12 FF/liter milk.

The results of the 6 consecutive years (92-98, 300 lactations in each group) will be presented during the session.

Paper NC1.10

Energy supplementation of high altitude grazed dairy cows: effect on pasture intake, blood and milk parameters as well as nitrogen utilization
N.R. Berry*[1], F. Sutter[1], R.M. Bruckmaier[2], J.W. Blum[2] and M. Kreuzer[1]. [1]Institute of Animal Sciences, ETH Zurich, ETH centre/LFW, CH-8092 Zurich, and [2]Institute of Animal Breeding, University of Berne, Bremgartenstr. 109a, CH-3012 Berne, Switzerland.

At 2000 m above sea level 12 cows (29 kg milk/head/d) were supplemented 0, 2.2 or 4.6 kg DM/d of a high energy (8.6 MJ NEL), low protein concentrate. The concentrate ration provided 0, 50% and 100% of the calculated ME. Pasture intake was estimated in wk 3, 7 and 11 by alkane markers using controlled release capsules (CRC). The concentrate caused a gross depression ($p<0.01$) in pasture intake which was not compensated for by the supplement fed causing the same reduction of -3 kg/d DM intake in both groups. CF and N intake were consequently reduced. Crude fiber digestibility declined ($p<0.001$) in all periods with increasing supplement level ($0.68<0.63<0.55$). Blood levels of insulin and IGF-1 in comparison to pre-alp levels reduced more in cows receiving concentrate than controls ($p<0.05$ for IGF-1). Milk protein and ECM remained unchanged between groups but milk fat was higher in controls. Utilization of dietary N for milk N excretion numerically increased with concentrate application ($0.24<0.31<0.35$). This was confirmed by corresponding alterations in blood and milk urea as well as N excretory pattern. The data suggest that the energy supplement had no direct production advantage but increased production efficiency and reduced N emission potential.

ANIMAL NUTRITION   [N]

Poster NC1.11

Evaluation of nutritive value of different meadow communities in Lower Silesia
K. Wolski, P. Nowakowski*, A. Szyszkowska, Agricultural University of Wrocław, Skłodowskiej-Curie 42, 50-369 Wroc3aw, Poland

Evaluation of feeding value of herbage from four meadow communities supplied with different level of N (0, 90, 180 kg/ha) was performed during 1993 - 1995 period. Data were collected from meadows located on alluvial soil with dominance (> 25% share) of: (1) *Poa pratensis*, (2) *Dactylis glomerata* L., (3) *Festuca pratensis* Huds and (4) *Trifolium pratense* L. Meadows were cut 3 times/season. Standard chemical analysis of herbage were performed. Feeding value of herbage was expressed according to INRA88. The highest content of crude protein (CP) was observed for grass communities fertilized with 180 kg N/ha and for *Trifolium repens* community without N fertilization (15,09 and 14,69 g/kg DM, respectively), while the lowest CP content was observed in grass communities without N - 12,75 g/kg DM. Energy value of herbage for communities without N application was the most favourable for clover dominated sward - 0.92 UFL or 0.87 UFV/kg DM. Nitrogen application lowered energy value of this sward by 5 to 10 %. In climatic and soil conditions of Lower Silesia grass-clover swards without N fertilization are able to yield 4179 UFL or 3867UFV/ha/season and they are the base for extensive systems of production only. The highest yield (9319 UFL or 8582 UFV/ha/season) was obtained from *Dactylis glomerata* community supplied with 180 kg of N.

Poster NC1.12

Evaluation of milk performance of Red Angus cows in first three lactations
J. Makulska, A. Węglarz, J. Szarek. Department of Cattle Breeding, Agricultural University of Cracow, al. Mickiewicza 24/28, 30-059 Kraków, Poland

The studies were carried out on the group of 73 Red Angus cows calved in 1996-1998 in the area of Western Pomerania, Poland. In 1995 pregnant heifers of that breed were imported from Canada Then the cows were inseminated using the semen from Canadian and American bulls. Most calvings took place at the end of February and the beginning of March. The cows calved in open wooden shelters with straw bale windbreak in three sides. Winter feeding of cows based on withered grass silage, hay and mineral mixture (200 g/cow/day) supplemented with wheat bran (300g/cow/day). Calves had a free access to hay, concentrate and rolled oat. From May to the end of October cows with calves were grazed on pastures. A supplementary feed for calves was rolled oat in amount of about 1 kg/calf/day.
Milk performance standardized on 210 days of lactation was estimated basing on the weight of weaned calves. The estimated average milk performance in the first lactation was 1889 kg of milk, the average weaning weight of calves 264 kg and the average age of weaning 239 days. The relative values in the second lactation were: 2230 kg of milk, 289 kg and 221 days and in the third lactation 1847 kg of milk, 267 kg and 247 days. A high milk performance resulted in high daily body gains from calving to weaning (about 1000g) and high weaning weights of calves.

ANIMAL NUTRITION   [N]

Poster NC1.13

Milk yield and milk composition of buffalo and friesian cows as affected by supplementation of milk plus
F.M.R.El-Feel*, A.K.I.Abd El-Moty, AA.A.Abd-El-Hakeam, M.A.A.El-Barody and A.A.Baiomy. Dept.of anim.prod., fac. of agric., Minia Univ. El-Minia, Egypt.

An experiment was conducted to study the effect of incorporation of milk plus (milk plus is a mixture of fermented and dried extracts of several medicinal herb and edible plants) in the concentrate mixture on the yield of milk and milk composition of both buffalo and Friesian cows. The design of the experiment followed the swing over method. Ten lactating buffalo and Friesian cows (5/each) were chosen during their high lactation period, shortly after peak, to carry out the experiment.The experiment was divided into four periods (initial and final control periods; two tested periods) and lasting 76 days. During initial and final control periods animals were fed a concentrate mixture without milk plus supplementation. During second and third periods 50mg and 75mg of milk plus powder/head/day were used respectively. Data revealed that the addition of milk plus increased significantly ($p < 0.01$ or $p < 0.05$) the yield of milk,fat corrected milk, fat, protein, solid not fat, ,total solids of buffalo milk. Similar trend was observed in Friesian cows except the yield of milk and solid not fat were not affected significantly by addition of milk plus in diet. The differences between two levels (50 mg and 75 mg) milk plus used were not significant; therefore, it is beneficial to use 50 mg/head/day in diets of lactating buffalo and Friesian cows. Estimation saturated and unsaturated fatty acids of buffalo and Friesian milk were taken into consideration.

Poster NC1.14

Differential redox properties of meat after biocatalist as quality indexes
H.-D. Matthes,[1] V. Pastushenko,*[1] H. Heinrich[2]. [1]Department of Agricultural Science, Rostock University, Vorweden 1, D-18069 Rostock, Germany, [2]LABO TECH Labotechnik GmbH, Richard-Wagner Str. 31 D-18119 Rostock-Warnemünde, Germany

The redoxpotential is the qualitative and quantitative index of the total antioxidant status and plays a significant role as an indicator of oxidative damage. In the samples of beef meat were investigated the reduction-oxidation-potentials. Measurements were made in untreated meat homogenate and after adding of biocatalist to the samples. FAD^+, p- benzochinone, ATP, Coffein and GTP were used as an adding substances for the differential analysis. Treated meat samples showed various alterations of their redoxpotential. In this study the addition of p- benzochinone, ATP and FAD^+ to meat homogenate in vitro gave not a clear alteration of the redoxpotential. Coffein + ATP and GTP tests produced significant changes in the redoxpotential behaviours. These reactions give information concerning the state of the metabolism of rich-energy phosphate groups, reflect the power of the muscle tissue to regulate the proportions of cAMP and cGMP and hence to regulate the redox conditions. The behaviour of the resulting redoxpotential reflects the pathophysiological and biochemical relevant properties of meat in respecting the power of the redox buffering capacity and the radical scavenging properties.

ANIMAL NUTRITION   [N]   Poster NC1.15

The interrelation between economically useful characters and biological characters in ayrshire race cows and use their in the selection
O.V. Proshina*, Yu.V. Boykov, All-Russian Research Institute for Animal Genetics and Breeding, Moscowskoye shosse, 55a, St.-Petersburg-Pushkin, 189620, Russia

To predict an indirect effect of the selection during direct selection for various signs with the method of imitative modeling, we have studied changes in basic useful and biological values in ayrshire race cows. The animals of this culling were divided into 5 groups with the interval of 0,67 σ according to the following performances: first lactation milk yield; best lactation milk yield; lifelong productivity; average fat percentage; duration of the economic use. With increase of the milk yield first lactation an increase in productivity was observed during best lactation, in service period, in one-day milk yield. The lifelong productivity, the milk fat content and the duration of the economic use have diminished. In our study the correlation between milk yield during first lactation and lifelong milk yield was -0,12, between milk yield during best lactation and lifelong milk yield 0,40. The modeling of the selection for fat milk content has shown in this case a decrease in milk yield during the best lactation as well as during the best lactation takes place. Such marked correlation was not observed with respect to the lifelong productivity. It is particularly remarkable that the first animals that were culled out were those with a high milk yield during first lactation, at old age of the first calving and with a shorter service period.

Poster NC1.16

Potentionals for spent *Pleurotus ostreatus* compost ensiling
M. Adamović[1], R. Jovanović[2], G. Grubić[3], Ivanka Milenković[4], Ljiljana Sretenović[5], Lj. Stoićević[1], [1]Institute Agroekonomik, Belgrade-Padinska Skela,[2]Maize Research Institute, Belgrade- Zemun Polje, [3]Faculty of Agriculture, Belgrade - Zemun, [4]HK Agroekonomik, Belgrade, [5] Institute for Animal Husbandry, Belgrade - Zemun Polje, Yugoslavia

Results of investigation of potentials for spent *Pleurotus ostreatus* compost ensiling combined with ground maize plant and grain are given in this paper. Compost was made based on pasteurized wheat straw inoculated with *Pleurotus ostreatus* mycelium. Ensiling was done in laboratory conditions in metal boxes with volume of 5 kg. Spent compost (23% DM) immediately after the third mushroom collection was ensiled mixed with ground maize plant (27% DM) in following ratios: I - 0:100, II - 25-75, III - 50-50 and IV - 75:25%, and with ground maize (V) grain in ratio 90:10%. Silages were analyzed two months after material was closed in boxes. All five obtained silages were high quality, and were ranked as first class. Values of pH in silages was between 3.68 and 4.01, while butyric acid was detected only in silage I. Amount of total acetic acid was near optimal (22.73 - 31.20%). Lactic acid was dominant (68.80 - 77.27%) which enabled optimal conditions for successful and quality preservation of silages.

ANIMAL NUTRITION [N]

Poster NC1.17

Wirtschaftlichkeit der Bullen- und Färsenmast mit Kreuzungstieren aus sechs verschiedenen Fleischrassen und bayerischen Fleckviehkühen.
J. Kögel*, M. Pickl und A. Obermaier. Bayerische Landesanstalt für Tier-zucht, Grub, Postfach 1180, D-85580 Poing (bei München)

In Bayern lief ein Kreuzungsversuch, in dem in drei Versuchsteilen Charolais (Ch), Blond d'Aquitaine (BA), Limousin (Li), Piemonteser (Pi), Deutsch Angus (DA) und Weiß-blaue Belgier (10 Prüfbullen/Rasse) auf Fleckvieh (FV) - Kühe eingesetzt wurden. Die Kreuzungsgruppen (insgesamt 1100 Tiere) wurden intensiv gemästet und nach Reifegrad geschlachtet. Es erfolgten auch Ver- zehrsermittlungen und grobgewebliche Zerlegungen von Schlachthälften. Bei der Bewertung der Schlachtkörperqualität auf der Grundlage der Gewebean-
teile und eines abgeleiteten Wertes für die Konformation (Erlös nach EUROP-
Punkten minus Wert der Fleischausbeute-Differenz) ergab sich bei Bullen im Vergleich zu FV im Deckungsbeitrag (DB)/Tag die Folge (SF) Pi 0,718 > Ch 0,561 > BA 0,558 > Li 0,328 > DA -0,184 und in der Tierwert-Differenz (SF):
Pi 323 > Ch 260 > BA 258 > Li 141 > DA -79. Bei der Bewertung der Schlacht-körperqualität allein nach EUROP- und Fettgewebsklasse und amtlichen No-tierungspreisen resultierte im DB/Tag die Folge (SF) Ch 0,821 > BA 0,718
> Li 0,609 > Pi 0,418 > DA -0,173 und nach der Tierwert-Differenz: Ch +228
> Pi 188 > BA 180 > Li 120 > DA -74. Über die Weiß-bl. Belgier im Referat.

Poster NC1.18

The Hereford animals of Siberian selection.
N.G. Gamarnic*, A.G. Antropov and N.V. Konskich. Siberian Res. Designing & Techn. Institute for Animal Husbandry 633128 P.O. Krasnoobsk, Novosibirsk, Russia

In recent years there is a great resources potential for beef cattle breeding in Siberia. The young bulls and heifers much are in demand, especially beef and dual-purpose breeds. This demand increases constantly. We have been engaged the Hereford breed in Siberia since 1952 when the first pedigree bulls and heifers were imported from Canada. Itisrequired 9-10 feed units per 1kg of gain. At present we have bred the stable Hereford population of Siberian selection, that is perfectly adapted to the continental climate of our region. The animals of the population are of high tehnological qualities and beef productivity. The young bulls and heifers are characterized by high intensity of growth. A daily live weight gain is about 900-1000 g, live weight of 15-month bulls exceeds 500 kg. Later animals were imported from England too. Agreat polled animals (the line of Shalun D-50). It includes 3 male lines and several female line. They are excellent for erossing that is showed by numerous experiments. We have developed the maintaining technology for beef cattle (the "cow-calf" system). At present we are interested in animals and semen exchange.

ANIMAL NUTRITION [N]

Poster NC1.19

The vinasse utilization in dairy cattle nutrition
R. Valizadeh and A. Ziaei. *Department of Animal Science, College of Agriculture, Ferdowsi University of Mashhad, P.O. Box 91775-1163, Mashhad, Iran.*

The effect of vinasse utilization on dairy cattle performance was studied in a change-over design with four treatments and four experimental periods. Eight Freshen dairy cattle were fed with the total mixed rations containing 0, 2.5, 5 and 7.5 percent vinasse which was substituted with 0, 20, 40 and 60 percent cottonseed meal respectively. In a total period of 84 days there was no significant difference between the average daily dry matter intakes. The mean milk production of the cattle fed by the rations was 18.9, 18.6, 17.9 and 17.3 respectively. The milk production, milk-fat content and rumen liquor pH (6.68, 6.40, 6.21 and 6.26) also were not significantly affected by the treatments. A significant reduction in dry matter digestibility of the rations was observed in accordance with increasing the level of vinasse. The cost of each Kg milk produced was 438, 436, 428 and 419 Rails for the experimental rations respectively.

Poster NC1.20

Modelling a cow's energy balance using repeated measurements
R. Schwager-Suter*, C. Stricker, D. Erdin and N. Künzi. *Institute of Animal Sciences, Swiss Federal Institute of Technology (ETH), CH-8092 Zürich, Switzerland.*

The present study is part of a larger project concerning biological efficiency of dairy cows differing in body size. The aim of modelling individual energy balances (EB) was to know the energy content of one unit of the following covariates: live weight change corrected for gut fill (lwc), body condition score change (bcsc) and change in backfat thickness (bfc). Additionally, the impact of live weight (lw), body condition score (bcs) and backfat thickness (bf) on EB was investigated. These variables are needed for further use in efficiency calculations. A total of 6359 records from 213 lactating dairy cows, 71 Holstein-Friesians, 71 Jerseys and 71 Holstein-Jersey F1-crosses was analysed (1st, 2nd and >2nd parity). Data was collected during 210 days of lactation, from calving to week 30. Individual EB was calculated as follows: total energy intake (MJ NEL) minus energy necessary for maintenance and production.
The analyses were performed using a repeated measures design with an autoregressive error structure. Models contained type of dairy cow, parity and type of roughage as fixed effects, week of lactation and one out of lwc, bcsc, bfc, lw, bcs, bf, respectively as covariates. lwc, bcsc and bfc didn't show energy contents as described in the literature. Further investigations concerning the time lag between EB and changes in lwc, bcsc and bfc will be performed. lw, bcs and bf had a negative impact on EB. Coefficients of determination were around 0.8.

ANIMAL NUTRITION [N]

Poster NC1.21

Crossbreeding of dairy cattle in Kenya. An economic evaluation
A.K. Kahi[1], G. Nitter[2], J.A.M. van Arendonk[3]*, W. Thorpe[4] and C.F. Gall[1]. [1]Institute of Animal Production in the Tropics and Subtropics, Hohenheim University, 70593 Stuttgart, Germany, [2]Institute of Animal Breeding and Husbandry, Hohenheim University, 70593 Stuttgart, Germany, [3]Wageningen Institute of Animal Science, P.O. Box 338, 6700 AH Wageningen, The Netherlands, [4]International Livestock Research Institute, P.O. Box 30709, Nairobi, Kenya.

Data on accumulated life performance of crosses of Ayrshire (A), Brown Swiss (B), Friesian (F) and Sahiwal (S) cattle collected over a 21-year period from a dairy ranch in the lowland tropics of Kenya were analysed to estimate additive and non-additive genetic effects on economic performance and to predict performance of alternative crossbreeding strategies. For profit per day of herdlife, F was superior to the other breeds for additive breed effect. The B additive breed effect was not significantly different from that of A. Dominance effect for profit in the cross B x S was substantial (25 %) and significant. The additive x additive interaction was negative in all the crosses. Profit would be lowest in $(ABFS)_{Syn}$ and highest in $(3/4F\ 1/4S)_{Syn}$. F x S would be the second-best strategy and would attain 92 % of the expected $(3/4F\ 1/4S)_{Syn}$ profit while $(FS)_{Syn}$ would attain 83 %. In the tropics, potential economic benefits may be derived by use of the F breed for continuous crossbreeding in production systems with adequate management standards or in the other production systems (e.g., smallholder sector), from the development of an F-based two-breed synthetic breed.

Poster NC1.22

Adiposity and winter feeding of dairy heifers
M. Marie*[1], V. Thénard[2] and C. Bazard[2]. [1] Sciences Animales-INRA, ENSAIA-INPL, B.P. 172, 54505 Vandoeuvre-lès-Nancy Cedex, France, [2] INRA-SAD, B.P. 35, 88501 Mirecourt Cedex, France.

While adiposity has been correlated to body condition in adult cows, no such observations does exist for heifers. 27 heifers (Holstein, n=15, Montbéliarde, n=12), 10.5 to 14 months old, received from end of November to April one out of three regimen, based on straw and barley, or straw, grass silage and aftermath, equivalent respectively to 2.9, 3.7 and 4.5 UF/animal/day. Body condition score and weight were estimated at the beginning and at the end of the period, and a biopsy of subcutaneous adipose tissue and a blood sample were performed in April.
Winter mean growth rate was respectively -49, +257 and +485 g/day. Mean diameter of adipocytes was 110.6 ± 8.5 µm, and loss of one point of body condition score corresponds to a 50 µm reduction of adipocyte diameter. The size of adipocytes was highly correlated (r=0.606, P=0.001) to the blood level of Free Fatty Acids. For females with growth rate < 200 g/d, low adipocytes diameter (75.7 ± 11.3 vs 131.1 ± 8.6 µm with growth > 200 g/d) and low FFA levels (0.238 ± 0.035 vs 0.332 ± 0.027 mEq/l) are indicative of the depletion of lipid reserves associated with low dietary inputs.
A better estimation of body reserves status would help to optimize nutritional management of heifers in order to attain a target body condition favourable to subsequent performances.

ANIMAL NUTRITION   [N] Poster NC1.23

Testing stability of performance in young sires?
L. Panicke[1], R. Staufenbiel[2], O. Burkart[2], E. Fischer[3], F. Reinhardt[4]
Research Institute for the Biologie of Farm Animals (FBN), D-18196 Dummerstorf[1], Freie Universität Berlin, Klinik für Klauentiere[2], Universität Rostock[3], VIT Verden[4]

Insulin plays an outstanding role based on its central position in energetic metabolism. The function of insulin may be recorded by means of the intravenous glucose tolerance test. The reaction of insulin and glucose was investigated after infusion of 1 g Glucose/kg^{0,75} because of the probable genetic determination of the reactive ability. The coefficients of heritability range from $h^2 = 0.16 \pm 0.10$ to $h^2 = 0.28 \pm 0.16$. Investigating 45 sires the correlation coefficients amount to $r = 0.4 - 0.5$ for parameters of GGT and estimated breeding values (EBV) which is closer than those of pedigree breeding value (PBV) and GGT: for fat yield $r = 0.40^{xx}$(PBV) and $r = -0.46^{xx}$ (GGT), for protein yield $r = 0.29$ (PBV) and $r = -0.45^{xx}$ (GGT). The results suggest that the protein yield of cows is suited to be a parameter of stability. Cows show a higher protein yield when their metabolism is stabile. This gain on information is available before testing of young bulls. It is influenced by age and simultaneity of test bulls to be compared. In elder bulls we can not expect results with sufficient certainty. The based expectations to the investigations of bulls during the 2nd and 3rth half year of their life agree with the results of REINICKE *et al.* (1993 EAAP). The study mentioned was based on 44 cows and their calves ($r = 0.41$).

Poster NC1.24

An observation of the significance of the relationship between the dry matter intake and body weight and between the DMI and milk production (zield)
Z. Mudřík, Z. Němec, B. Hučko, J. Obadálek. *Department of Animal Nutrition, Czech University of Agriculture in Prague. 165 21 Praha 6 - Suchdol, Czech Republic.*

One hundred and two (102) cows were observed in the experiment. The condition of breeding and the nutrition of all the observed cows were identical. The cow's body weight, daily milk yield, and the DMI were closely monitored. Correlation and regression were established utilising regression analysis.
A significant correlation was discovered between daily milk yield and DMI.
  $y = -1,6325 + 0,8771x$, ($r = 0,775423$) - Czech pied cattle   and
  $y = -1,8597 + 0,7197x$, ($r = 0,884732$) - Holstein cattle.
A less significant correlation was found between the body weight and the DMI
  $y = -205,2277 + 0,3566x$, ($r = 0,34876$) - Czech pied cattle   and
  $y = 32,6 + 0,0922x$ ($r = 0,359131$) - Holstein cattle.
In actual breeding situations it is more advantageous to utilise the daily milk yield correlation for the prediction of the DMI.

ANIMAL NUTRITION   [N]   Poster NC1.25

Indicators influencing economy of beef cattle breeds breeding in Czech Republic
A. Ježková. Department of Cattle Breeding and Dairying, Czech University of Agriculture Prague, 165 21 Prague 6, Czech Republic

Twelve breedings with mean yearly number of cattle - 25 to 300 animal units farming in different production conditions were included into the evaluation. Height above sea level was from 386 up to 785 m. Six different beef cattle breeds (Charollais, Hereford, Simenthal, Aberdeen Angus, Scotch Highland, crossbreds of Czech Pied x Charollais) were observed. The observation lasted 3 years. Reproduction indicators, quality of calve rearing, utilization of permanent grass stands and realization of animals were evaluated. In the field of reproduction reserves in percentage of cows calved in the range of 43.9 % to 100 % were found. Death rate of calves varied in the range from 1.67 % to 11. 3 %. In the breedings seasonal variability was not observed. In a half of breedings extension of calving was found, 13.36 % of calves were born after finishing the period of calving. Expenditures on plant production increase the low mean yearly stocking rate of grazing land in the range from 0.32 AU/ha up to 2.58 AU/ha and relatively large area harvested per AU stocked in winter season (0.18 - 1.41 ha/AU).

Poster NC1.26

Substitution of protein supplement by urea in diet of fatted steers.
D. Vrzalová, A. Krása, J. Třináctý
Research Institute of Animal Nutrition, Ltd., Pohořelice, Czech Republic.

An experiment was designed to determine an influence of urea concentrate (with an urea content of 15%) substitution on average daily gain (ADG) of young steers. The trial was provided on 8 three-month-old steers, which were divided into two groups. The initial average live weight was 126,0 kg. Both of groups were fed barley flakes (ad lib), alfalfa hay (15% of fed ration) and protein concentrate (1 kg/day), twice a day for 36 weeks. The protein concentrate contained components as follows: control group (CG) - rap-seed pressing (30%), soya bean meal (30%), microbial protein (30%), minerals and vitamins (10%); group with urea (UG) - rap-seed pressings (20%), soya bean meal (20%), microbial protein (20%), urea concentrate (30%), minerals and vitamins (10%). Average daily consumption of feeds (by 1 steer) for CG and UG group were as follows: barley flakes 4,7 and 5,6 kg, alfalfa hay 0,87 and 0,87 kg, protein concentrate 1,0 and 1,0 kg, respectively. CG group tended to have a lower ADG 1160 g, compared to UG group 1360 g. Digestibility of nutrients (5th month of trial, live weight of 380 kg) for the CG and UG group provided the following values: dry matter 79,9 and 78,6 %, crude protein 78,8 and 80,0 %, crude fiber 61,2 and 61,5 %, fat 83,4 and 78,6 %, respectively.

ANIMAL NUTRITION [N]

Poster NC1.27

Feed intake and feed utilization of suckling cows during early lactation
S. Teichmann[*1], *R.-D. Fahr*[1], *G. v. Lengerken*[1], *F. Mörchen*[2], [1]*Institute of Animal Breeding and Husbandry with Veterinary Clinic, Martin-Luther-University Halle-Wittenberg, Adam-Kuckhoff-Str. 35, 06108 Halle, Germany,* [2]*Teaching and Experimental Institute of Husbandry and Technology Iden, Sachsen-Anhalt, Germany*

Within two years the daily feed intake of 40 suckling cows of German Angus (GA) and German Simmental (GS) was registered from the second week a.p. to the 14th week p.p.. A mixed ration of wilted grass silage and maize was given ad libitum. Body weight (BW) and backfat thickness (BT) of the cows were measured by weighing and by real-time ultrasound, respectively. The cows were milked on one day in the second, fourth, eighth, 12th and 14th week p.p. by milking machine immediately after the injection of 20 I.U. oxytocin into the Vena jugularis. Their calves have been separated 12 hours before. The average feed intake of GS and GA was 14.8 kg DM/d and 13.7 kg DM/d respectively, resulting in 90.5 MJ NEL/d and 83.7 MJ NEL/d. The daily milk yield in the first 100 days of lactation was 11.7 kg in GA and 19.5 kg milk in GS cows. From calving to the 14th week p.p. the BW of the GA increased from 550 kg to 577 kg while the weight of GA cows decreased slightly. BT was unchanged (26 mm vs. 27 mm) in GA and decreased up to 25 % in GS. The different energy requirement of suckling cows of several breeds before and after calving must be taken into account, when feed rations are formulated.

Poster NC1.28

Study on the relations between nutritive and environment factors for growing-fattening steers
Coculeana Bunghiuz, D. Georgescu, Laura Rădulescu, F.Beiu, Elena Bugaru. Research and Production institute for Bovine Breeding, 8113 Baloteşti, Romania

The purpose was to determine some nutritional and environmental factors influencing on liveweight gain and specific intake. The experiment was organised on two lots of steers Romanian Black spotted breed. One lot was introduced into a climatic chamber and the other inside a common shelter. There were 4 experimental periods. During the first period both lots were fed with same forages ad libitum. During the next 3 periods the diets consist of 3 different energy and protein levels. Considering individual data, the influence of the factors taken into account and interrelations between them was established, using linear and multiple regression equations. Body weight, daily dry matter (DM), energy and protein intake, air temperatures were considered as independent variables. As dependent variables, daily weight gain, DM and energy specific intakes were considered. The significance was established using Fisher test. Body weight, DM and DPI intake have a significant influence on daily weight gain. DM and energy intake influenced the specific intake of DM. Also, DM and energy intake have a significant influence on specific energy intake.

ANIMAL NUTRITION   [N]

Poster NC1.29

Growth ability of Limousin breed and crossbreds of Czech Pied cattle with beef breeds
V. Nová, F. Louda, L. Štolc. Department of Cattle Breeding and Dairying, Czech University of Agriculture Prague, 165 21 Prague 6, Czech Republic

Live weight, mean daily gain and 17 body measurements at the age of 120, 210 and 365 days in calves of following genotypes were followed: Limousin (Li - 47 head), crossbreds of Czech Pied x Simenthal (C x Si - 21 head) and crossbreds Czech Pied and Limousin (C x Li - 54 head). At the age of 120 days the highest live weight and mean daily gain was found in C x Si (132.95 kg, 1.11 kg) and the lowest values in Li (123.74 kg, 1.03 kg). At the age of 210 days the highest live weight and mean daily gain was again in crossbreds C x Si (234.99 kg, 1.12 kg) and the lowest value in Li (192.98 kg, 0.92 kg). At the age of 365 days both indicators were the highest in C x Si (335.37 kg, 0.92 kg) and the lowest value in Li (306.66 kg, 0.84 kg). Up to the age of 120 days the highest growth intensity of body measurements showed the crossbreds C x Li. The lowest growth intensity was found in Li breed. Since the age of 120 days up to the age of 210 days the crossbreds C x Si showed higher growth intensity than crossbreds C x Li. At the age of 210 days most of the body measurements were the highest in C x Si. In this period the lowest growth intensity was in Li breed. Since the age of 210 days up to the age of 365 days the highest growth intensity showed again the crossbreds C x Si and the lowest value was in Li breed. The musculature of the thigh was in all age categories higher in crossbreds C x Li and in Li breed than in C x Si.

Poster NC1.30

Slaughter value of bulls of Czech Pied breed, Limousin breed and their crossbreds
F. Louda, V. Nová, L. Štolc. Department of Cattle Breeding and Dairying, Czech University of Agriculture Prague, 165 21 Prague 6, Czech Republic

Slaughter bulls were divided by their genotype into three groups: Limousin breed (Li - 25 head), Czech Pied breed (C - 24 head) and crossbreds between the two breeds (Li x C - 26 head). Bulls of these genotypes were slaughtered in mean slaughter weight 550-580 kg. Weight of carcass was the highest in Li bulls - 320.8 kg, lower in C x Li - 312.3 kg and the lowest value was in bulls C - 307.8 kg. The highest dressing percentage reached the bulls of Li breed - 62.3 %, in bulls C x Li the value of 59.8 % was found and in bulls C - 58.6 %. The highest percentage share of the carcass was in bulls Li in following joints: shoulder (7.54 %), brisket with ribs and bones, short plate and flank with bones (14.50), chuck and neck without bones (11.07 %), rib without bones (5.01 %), sirloin (1.87 %), round (19.51 %). Percentage share of bones was the lowest in Li bulls - 16.36 %. Percentage share of total quantity of meat calculated from carcass was the highest in Li bulls - 82.01 %, in bulls C x Li it was 77.07 % and the lowest share of meat was in C breed - 76.28 %. The most favourable share meat : bones (kg meat /kg bones) was in Li breed - 5.01, in C breed it was 3.81 % and in crossbreds C x Li it was 3.60.

ANIMAL NUTRITION [N] Poster NC1.31

Variability of chemical composition and heavy metal content in commodity milk produced in industrial and agricultural regions
A. Litwińczuk, A. Drozd-Janczak, M. Florek, A. Filozof. Subdepartment of Animal Materials Estimation and Utilization, Lublin Agricultural University, ul. Akademicka 13, 20-950 Lublin, Poland.

Chemical composition (protein, fat, lactose and lactose and dry matter) was evaluated in 680 samples taken directly from suppliers and in 343 the lead and cadmium content was also assessed. Chemical content varied considerably during the evaluation of the commodity milk, which, in the case of protein, fluctuated in individual samples from 2.21-4.14%, fat from 2.66-5.33%, lactose from 3.12 to 4.89% and dry matter from 9.48- 13.68%. The results obtained in the survey indicate significant differences in lead and cadmium content in commodity milk between industrial and typically agricultural production regions. Milk samples taken at a collection centre in the same town where a cement factory was located contained twice as much Pb (41.7 µg/l) and Cd (12.9 µg/l) in comparison with the samples taken in towns which were located 8-12 km from the cement factory (24.4- 26.5 mg/l Pb and 4.9-5.3% mg/l Cd). The lowest content of both elements (7.2- 8.3 µg/l of Pb and 1.0- 1.1 µg/l Cd) was found in milk from agricultural regions. It should be noted, however, that even in industrial regions the heavy metals content (Pb, Cd) in commodity milk did not exceed accepted standards in this respect for any of the samples.

Poster NC1.32

Possibilities of increasing nutrition quality of milk
R. Toušová, F. Louda, L. Stádník. Department of Cattle Breeding and Dairying, Czech University of Agriculture Prague, 165 21 Prague 6, Czech Republic

Recently an interest in polymorphism of milk proteins increased. Milk protein systems in Jersey dams (89 head), their genetic control and possibilities of gene linkage on heritability and variability were observed. Also different genotypes of kapa-casein and their effect on increasing of milk utility and content of solid components (fat, proteins and lactose) were observed. It was found that from the point of view of genotypes, the genotype of the alelle AB predominates in 54 % of dams over the genotype of the alelle BB in 39 %. Genotype of the alelle AA was found in 7 % of dams only. Significant increment of milk production, proteins and fat in dams of genotype of allele BB was obtained.

ANIMAL NUTRITION [N]　　　　　　　　　　　Poster NC1.33

Lifetime milk production of purebred and crossbred Russian Black Pied cows
D.V. Karlikov*, A.P. Pyzhov and O.G. Tsvetkova.. Russian Academy of Livestock Breeding Management, 142023 Bykovo-Podol'sk, Moscow region, Russian Federation.

Records of all lactations of purebred Russian Black Pied (RBP) and Holstein (H) x RBP crossbreds F1 (50% H inheritance) and those obtained by upgrading (87% H inheritance), rotational crossing (37-62% H inheritance), "inter-se" breeding of crosses (60% H inheritance) and backcrosses (12-25% H inheritance) (initially 38, 290, 815, 260, 370 and 328 cows of the 6 breed types respectively) were analyzed. For the 6 breed types, the lifetime 4%-fat corrected milk yield averaged 21124, 18318, 163811, 16921, 14051 and 19065 kg, lifespan 2140, 1957, 1855, 1867, 1766 and 2030 days, productive life length 1054, 922, 834, 855, 778 and 977 days, percentage of survival to 6th lactation 26.3, 13.4, 9.4, 11.5, 7.0, 17.8, milk yield per day of productive life 19.3, 19.5, 18.9, 19.3, 17.8 and 18.8 kg with small or no differences in fat and protein percentages. Statistical significant differences on the lifetime fat corrected milk yield (2800 - 7000 kg) between RBP and F1 groups on one hand and other crossbreds on the other hand were estabilished. Holstein crossbreds calved earlier (up to 18-27 days), gave more milk in the 1st lactation (up to 366-648 kg) but tended to be culled from the herd ealier than RBP cows. It is concluded that prediction of herd life from 1st lactation yield traits estimated from a mixture of purebred and crossbred cows may not be accurate.

Poster NC1.34

Genetic relationship of Romanian Black and White with different Holstein breeds
Elena Florescu, I Granciu. Research and Production institute for Bovine Breeding, 8113 Balotesti, Romania

Based on herd book data, the active population of Romanian Black and White breed was analysed. The degree of genetic resemblance to different Holstein populations was studied. All official milk recorded cows in Romanian during 1987 - 1997 and official data from import origin countries were compared. There were some 2523 cows and 157 sires originating in USA, Canada, Denmark, The Netherlands, Germany and Swiss. Wright formula was used to determine genetic relationship coefficient. In previsions study (Draganescu C. 1986) a 90% resemblance of Romanian Black and White to European Friesian and 10% to Holstein breed (USA, Canada, Italy, Australia) was estimated. Ten years later accorded our study, genetic relationship with European type decreased to 60% and with Holstein increased to 40% but the reproductive isolation dropped from 0,62 (1987) to 0,49 (1997). It is desirable to increase the reproductive isolation in order to keep genetic potential this population.

ANIMAL NUTRITION   [N]　　　　　　　　　Poster NC1.35

Genetic variability in docility in German Angus and Simmental cattle
H. Mathiak, M. Gauly, K .Hoffmann, R. Beuing, M. Kraus and G. Erhardt*. Department of Animal Breeding and Genetics, Justus-Liebig-University, Ludwigstr. 21B, D-35390 Giessen, Germany

The economic situation in farms keeping suckler cows depends not only from production traits but also from the input of labor during handling periods. Whenever handling problems occur, both the stockperson´s safety and the animal´s welfare are at risk. To include behavioral traits in breeding programs, the accurancy of the test system has to be proved. In addition less docile animals cause more working hours and the costs will increase.

Therefore five progeny groups within each breed (Simmental n = 123, German Angus n = 109) were tested at weaning age of 8 months by a modification of the handling test of Le NEINDRE et al., 1995. The suckler herds were kept on pasture. Both sexes were subjected to the handling test. The test resulted in genetic differences within as well as between the breeds. Heritability estimates of different characteristics of the handling test ranged from 0.03 to 0.41 by the German Angus cattle and from 0.02 to 0.27 by the Simmental cattle.

Some of the docility characteristics could be used in breeding programs based on their heritability. Therefore the economical importance and the correlation with other traits will be studied.

Acknowledgement: This study was supported by the German Research Foundation (SFB 299).

Poster NC1.36

The negative influence of intensive farming on the antioxidants property of meat.
H.-D. Matthes,[1] V. Pastushenko,*[1] H. Heinrich[2] , Z. Holzer[3]. [1]Department of Agricultural Science, Rostock University, Vorweden 1, D-18069 Rostock, Germany, [2]LABO TECH Labotechnik GmbH, Richard-Wagner Str. 31 D-18119 Rostock-Warnemünde, Germany, [3]ARO, Institute of Animal Science, Newe Ya'ar Research Center, P.O.Box 1021, Ramat Yishay 30095, Israel.

The objective of this investigation was to determine the influence of farming type and hormone supplementation on reactive oxygen metabolites scavenging reactions. The samples of beef meat from the organic farmed animals and intensive managed calves, supplemented with the hormones, were investigated. The antioxidant reactions after adding of p- benzochinone and the decontamination capacity against free radicals, corresponding to Lambert-Beer's low, were measured calorimetrically. The results showed significant increase in the antioxidants titter and decontamination capacity again reactive oxygen metabolites in the ecological meat samples. These data indicate the negative influence of the intensive farming and hormone supplementation on the radical scavenging ability of meat.

ANIMAL NUTRITION [N]

Poster NC1.37

Effect of three contrasting calving patterns on the performance of dairy cows
G. Ryan[*1], S. Crosse[1] and M. Rath[2]. [1] Teagasc, Moorepark Research Centre, Fermoy, Co. Cork [2] Faculty of Agriculture, University College Dublin, Ireland*

A highly seasonal milk supply pattern is a feature of the Irish dairy industry. This results from the milk price system and the lactation period in Ireland is associated with the availability of grazed grass (low cost) in the production system. This results in a limited portfolio of products which can be manufactured. The objective of the study was to compare the inputs and outputs for systems of milk production with contrasting calving patterns. Twenty six cows were randomly assigned to the following treatments: 100% autumn calving (A); 100% spring calving (S) and 50% autumn:50% spring (AS). The study lasted for two years. The average yield was 6,532, 6,358 and 6,142 kg/cow for systems A, AS and S respectively. The difference in yield was not significant. There was no significant difference between treatments in milk fat, protein and lactose yield in year one and year two of the experiment. The fat and protein content was high for all treatments. Milk fat concentration was significantly lower ($p<0.05$) for Treatment S (39.6 g/kg) than for Treatment A (41.4 g/kg) or Treatment AS (42.2 g/kg) in year two with no significant difference in year one of the study. The feed budget was very different for the three treatments in terms of grazed grass, grass silage and concentrates.

Poster NC1.38

Production of quality milk for processing in a grass based system
B. O'Brien, Teagasc, Dairy Production Department, Moorepark Research Centre, Fermoy, Co. Cork, Ireland.

The manufacturing milk sector in Ireland has a highly seasonal supply. While this system maximises grass utilisation for conversion to milk, it has inhibited diversification of the product portfolio and can have adverse implications for product quality. Milk compositional changes occurring over the lactation are emphasised with this system and may result in deterioration of cheese texture and functionality, flavour impairment of butter and unstable cream liqueurs. Milk quality defects in early and mid lactation milk generally arise from low protein concentrations in milk. Protein may be increased through nutrition at this time by (i) inclusion of a grass component in the diet and (ii) concentrate supplementation of pasture in spring/early summer. During the main grazing season, studies have shown that grazing management systems offering either (i) a daily herbage allowance of 20 to 24 kg dry matter/cow/day or (ii) a mean stocking density of 3.8 c/ha in a 3-day rotational paddock system resulted in a processable milk and a quality cheese. Milk quality defects are more profound in late lactation. Suggested methods of overcoming these problems include (i) improved herd management practices in relation to cow nutrition, milking management, drying-off strategy and milk assembly procedures, (ii) segregation of milk supplies by the processor and (iii) blending good quality late lactation milk with early lactation autumn milk.

ANIMAL NUTRITION [N]

Poster NC1.39

Research regarding the intake capacity for hay and green forages in young buffaloes and steers

D. Georgescu, F. Beiu, Coculeana Bunghiuz, Gabriela Macarie, H. Macarie. Research and Production institute for Bovine Breeding, 8113 Balotești, Romania

The study presents the results of some tests which permitted to establish the intake capacity of young buffaloes. It was as a ratio between dry matter (DM) intake and 100 kg live weight. In the first trial using Glyceria aquatica as green forage (a plant with the highest production in flooded areas) the intake was 1.17 kg DM/100 kg live weight. Glyceria aquatica hay was ingested 60 % comparatively with alfalfa hay. The study also presents the results of comparative trials involving, on the one hand, 228 to 329 kg Black Spotted steers and on the other hand 230 - 324 kg young buffaloes whose ingestion capacity for a wide range was investigated. The trials lasted 110 days; ingestion capacity was assessed by monitoring the daily dry matter intake per 100 kg live weight. The first trial, using green forages revealed an ingestion capacity of 2.71 kg DM for buffaloes, as against 3.01 kg DM for steers. The second trial using hay and silage found an ingestion capacity of 2.75 kg DM for buffaloes and 3.06 kg DM for steers. Over the given body weight range, buffaloes were thus found to have 10 % lower ingestion capacity than steers.

Poster NC1.40

Effect of soyoil on the DM disappearance of high fiber diets from nylon bags
A.A. Nasserian Dept. Animal Science Ferdosi university of Mashhad P.O.Box: 91775-1163 Mashhad - IRAN

The use of fat in the rations of dairy cattle is underfocused in recent years. Five mature rumen cannulated sheep were used in a randomized block design in two periods: a 14 days adaptation period which was accompanied with four days DM disappearance of the samples. The diets of the sheep were: lucerne (60%) and oaten chaff (40%) mixed in the required proportions (diet A) , while diet B was diet A mixed with 20 g soyoil (5.5% curde fat in the diet B) before feeding. The diets were fed at maintenance level. DM disappearance was examined in the rumen of the sheep, when fed with diet A and diet B. It was concluded that DM disappearance of the incubated samples was not significantly affected by the diet supplementation with soyoil.

ANIMAL NUTRITION   [N]   Paper N5.1

The nutritive value of palm kernel meal measured in vivo and using rumen fluid and enzymatic techniques
F. P. O'Mara*, F. J. Mulligan, E. J. Cronin, M. Rath and P. J. Caffrey. Department of Animal Science and Production, University College Dublin, Belfield, Dublin 4, Ireland.

The objectives of this experiment were to determine the in vivo digestibility of solvent extracted (n = 4) and expeller (n = 8) palm kernel meal (PKM) and to establish how well this was predicted by laboratory techniques. The in vivo digestibility was measured by total faecal collection using four wether sheep per sample. The laboratory methods for predicting digestibility were the in vitro rumen fluid (RF), neutral detergent cellulase with gammanase (NCG), and pepsin cellulase with gammanase (PCG) methods. The solvent extracted samples had higher digestibility of organic matter (OM; 691 vs 653 g/kg, sed 15.7) and crude protein (727 vs 597 g/kg, sed 36.6). However, the expeller samples had higher (P= 0.12) contents of digestible energy (13.4 vs 12.5 MJ/kg of DM, sed 0.48) due to their higher gross energies (20.6 vs 19.1 MJ/kg of DM). None of the laboratory digestibility tests predicted digestibility satisfactorily when the laboratory OM digestibility was regressed on the in vivo OM digestibility. The residual standard deviation was 29, 32 and 31 g/kg for the RF, NCG and PCG methods, respectively. In particular, two expeller samples and two solvent extracted samples were grossly underpredicted by the laboratory methods. These results indicate that PKM is a medium energy feed for ruminants and that enzymatic procedures including gammanase or the in vitro rumen fluid method do not accurately predict its digestibility.

Paper N5.2

Effects of altering hourly nitrogen supply to the rumen on metabolism in growing lambs fed either barley or sugar beet based diets.
J.M.Richardson*, L.A.Sinclair and R.G.Wilkinson. ASRC, Harper Adams University College, Edgmond, Newport, Shropshire TF10 8NB, UK.

The *in situ* rumen degradability coefficients of nitrogen (N) and organic matter (OM) of 5 feed ingredients were determined and used to formulate two diets, based on either barley (B) or sugar beet pulp (SBP). Diets were fed at a restricted level in two equal meals at 09.00 and 16.00h. Within each diet the hourly rumen N supply was altered by changing the sequence of allocation of the protein components of the diet, providing either: equal amounts of protein in each feed, slowly degradable protein in the morning and quickly degradable protein in the afternoon, or the majority of the protein in the morning. Blood and rumen fluid samples were taken at intervals during the day. Lambs on B diets showed patterns of rumen and plasma ammonia and plasma urea which generally followed the pattern of dietary N intake. Lambs on SBP diets showed a pattern of rumen ammonia which followed that of N intake, but no clear pattern of plasma N metabolites. Other blood metabolites showed no significant differences between treatments with mean concentrations of b-hydroxybutyrate (mmol/l) 0.547, 0.475 (sed 0.043), glucose (mmol/l) 4.807, 4.769 (sed 0.076) and insulin (iµ/l) 19.97, 17.58 (sed 1.198) for lambs on B and SBP diets respectively. In conclusion, energy source affected metabolic response to synchronisation of N and OM supply to the rumen.

ANIMAL NUTRITION [N]

Paper N5.3

Comparison of wheat and corn for dairy cows
R. Daenicke*, D. Gädeken and P. Lebzien. *Institute of Animal Nutrition of the Federal Research Centre (FAL), Bundesallee 50, D-38116 Braunschweig, Germany*

It is well established that starch from corn is degraded in the rumen to a remarkably lower extent as compared to wheat, barley or rye. Due to this fact forage intake and efficiency of energy utilisation might be influenced positively. In continuation of experiments performed previously a feeding trial involving two groups of 29 cows each (duration: 90 days, individual feeding conditions) was carried out to study the effect of wheat replacement in the concentrate mixture by corn on silage intake and performance. Both groups were fed 9.5 kg concentrates per day on average, given in 8 portions. The concentrate mixture for group A was based on wheat (incorporation rate 60 %) and that for group B was based on corn (incorporation rate 58 %). Both mixtures were formulated to have nearly the same protein and energy contents. Wilted grass silage (122 g uCP, 5.7 MJ NEL/kg DM) was offered ad libitum. Replacing wheat by corn in the concentrate mixtures did not influence silage intake, amounting to 9.1 kg or 9.2 kg per day. Milk yield was registered to be 25.8 kg/d in group A and 26.8 kg in group B (p > 0.05). Feeding corn based concentrates amounts of milk fat and milk protein tended to increase. Related to FCM a not significant effect of 1.2 kg/d in favour of the corn fed group was detected. Daily live weight gains were low in both groups, amounting to 0.13 kg vs. 0.11 kg.

Paper N5.4

Prediction of nutrient rumen turnover from animal and dietary information
A. Cannas*[1,2], P. J. Van Soest[2]. [1]*Dipartimento Scienze Zootecniche, Università di Sassari, via de Nicola, 07100, Sassari, Italy,* [2]*Department of Animal Science, Cornell University, 14853, Ithaca, NY, USA.*

A database was developed from the literature (9 publications) to explore mechanisms controlling rumen turnover and to predict it. The database included the treatment means of 40 experiments (23 on cows and 19 on sheep) in which rumen contents were measured by manual emptying or by slaughter. Predictors were intake, body weight, metabolic weight, and dietary components (CP, NDF, neutral detergent solubles (NDS), lignin, ash). Models obtained to predict rumen turnover for NDF, NDS and DM are in table 1.

Table 1 - Regressions to predict rumen turnover (in parentheses are the standard errors of the coefficients; all coefficients are statistically significant, P<0.01).

Equations	R^2	SEE [a]
$T^b_{NDF} = 25.1\,(3.87) + 8.64\,(1.65) * (1/NDFI^c) - 0.718\,(0.20)\,CP^d$	0.532	0.82
$T_{NDS} = 18.1\,(0.71) - 33.43\,(3.05) * \log_{10} NDSI^e$	0.760	0.69
$T_{DM} = 21.1\,(0.65) - 12.6\,(1.22) * \ln NDSI$	0.739	0.63

a = standard error of the mean of Y estimated from the mean of X; b = rumen turnover (h); c = NDF intake (% of BW); d = diet CP (% of DM); e = NDS intake (% of BW).

ANIMAL NUTRITION   [N]   Paper N5.5

Effect of casein supply to the abomasum on post-ruminally starch digestibility and glucose metabolism of dairy cows
S.M. Abramson [1,2], A. Arieli [1], I. Bruckental [2], S. Zamwel l[1] and Z. Shabi [1]. [1] Department of Animal Science, The Hebrew University of Jerusalem. [2]Institute of Animal Science, ARO, The Volcani Center, Bet Dagan, Israel.

The effect of protein level in abomasum on post-ruminal digestibility of starch was determined on six multiparous Israeli Holstein dairy cows equipped with ruminal and abomasal cannula and with catheters in subabdominal vein and abdominal artery. The cows were arranged in a 3X3 Latin square design. Cornstarch in a water slurry was infused continuously into the abomasum to provide 1600g/d. Sodium caseinate in aqueous solution was infused to the different treatment as follows: 1. No casein. 2. 400g/d and 3. 800g/d. Post-ruminally casein infusion increased digestibility of starch (%) and arterial and venal levels (mg/dl) of glucose and insulin.

Treatment	1	2	3	p<	L<	Contrast
PRSD	90.00	94.00	94.70	.04	.07	.001
Venal glucose	33.49	38.46	34.33	.09	.04	.14
Arterial glucose	47.79	53.48	51.31	.12	.09	.06
Venal insulin	17.52	24.40	23.12	.02	.04	.007
Arterial insulin	21.22	28.22	29.80	.02	.21	.007

PRSD, Post ruminally starch digestibility; P< significancy; L< Linear Trend, Contrast: 2+3 vs. 1.

Paper N5.6

A comparison of the feed value of triticale, winter wheat and winter rye for growing pigs including the effect of variety, soil and nitrogen fertilisation
J.A. Fernández, H. Jørgensen and A. Just[*]. Danish Institute of Agricultural Sciences, Research Centre Foulum, P.O. Box 50, 8830 Tjele, Denmark.

In the past years barley accounted for approx. 70% of the cereal crops in Denmark and for 70-80% of the cereals in diets for pigs. Due to genetic improvements of frost resistance and yield and thereby more reliable yields of the crops triticale, winter wheat and winter rye have become more important to the pig producers. Thus the aim of the investigation is to throw light on sources of variations in the yield and production value of some species and varieties of winter cereals. The investigation comprises three varieties of triticale (Dagon, Lokal, and Uno), two varieties of winter wheat (Kraka, Sleipner) and one winter rye (Petkus II). All six samples were grown on two soil types (heavy clay and sand) each at two levels of N-fertilisation i.e. the normal level for each soil type and normal plus 40 kg N/ha. The chemical composition of the 24 samples, the digestibility of nutrients and the content of metabolizable energy were determined. The digestibility and N-balance experiment were conducted with female pigs weighing 40-60 kg. The crude protein contents of the cereals were significantly influenced by cereal species, by type of soil and by N-fertilisation. Type of soil had an important effect on the content of selenium in the grains as cereals grown on clay soils contained three times more selenium than those grown on sandy soils. Furthermore type of soil had effect on digestibility of crude protein and on the metabolizability. Application of 40 kg N-fertiliser per ha had only influence on the digestibility of crude fat.

ANIMAL NUTRITION   [N]

Paper N5.7

The effect of feed mixtures on histopathological changes of mucous membrane of digestive tract
Z. Němec,[1] M. Doskočil,[2] Z. Mudřík[1], A. Kodeš[1], B. Hučko[1], J. Obadálek[1]. [1]Department of Animal Nutrition, Czech University of Agriculture in Prague, Prague 6 - Suchdol, 165 21. [2] Department of Anatomy, Medical Faculty of Charles University in Prague

This contribution is engaged in the study of the effect of fattening pig technology on the status of digestive tract mucous membrane. Immediately after slaughter samples of tissues from stomach (fundus, body and pylorus) were taken and microscopically the status of epithelium was evaluated.

In the first experiment determination of cereals, which were broken to mixtures on screen of different size of mesh, was performed. These diets were fed to pigs of live weight 30 kg for the period of 30 days. In another experiment status of mucous membrane in piglets before weaning and after feeding of starter mixture was observed. In another experiment starter mixture dry and starter mixture moisten was fed to the piglets. Results of the experiments proved that the size of broken particles of cereals have no effect on the degree of injury of mucous membrane. Further it was found out that the mucous membrane is injured already after feeding starter mixture. Entirely different results were after feeding of moisten mixture.

Our results univocally proved that feeding of moisten mixtures to pigs is expressively better and from the health point of view more convenient.

Paper N5.8

Effect of protein concentration in the diets balanced according to ileal digestible amino acids content on protein utilization in lean pigs
H. Fandrejewski H., St.Raj, G. Skiba and D.Weremko. The Kielanowski Institute of Physiology and Animal Nutrition, 05-110 Jablonna, Poland.

Two diets containing 18 or 15% of CP were identical in the concentration of ileal digestible amino acids and ME. The P pigs (Pietrain breed, n=15) were fed *ad libitum*. The H pigs (synthetic line 990, n=23) were fed *ad libitum* either received rations equivalently to Pietrain breed. Protein and energy balance was measured by using a comparative slaughter method from 25 to 70kg body weight.

The P pigs daily eaten 1.9kg of feed, gained 692g and deposited 124g of protein and 105g of fat. Their freely fed analogies from the H group consumed daily 0.3kg more feed, grew 186g faster and deposited 16g more protein and 68g more fat. However, at the same rations, the daily protein deposition in the body was similar in both P and H genotypes, but the P pigs grew still slower (by 50g/d), presumably due to the lower rate of fat deposition. The P pigs produced more heat in the body and less efficiently utilized feed for body gain. Protein utilization was better in pigs receiving diets with 15 than 18% of CP (43.4 vs 35.4%). However, if a concentration of lysine in protein gain was assumed as 7.2%, the gross efficiency of ileal digestible lysine did not differ between treatments. It was 61%.

ANIMAL NUTRITION   [N]   Paper N5.9

Substitution of yellow corn by sorghum in layer ration and the effect of metionine and kemzyme supplementaion on hen performance and egg quality
Samia M. Hashish and A.A. El-Ghamry, Department of Animal and poultry production and nutrition, National Research Center, Dokki, Cairo, Egypt

A study was conducted to invistigate the effect of substituting 50% of yellow corn in layer rations with sorghum grains and the effect of supplementing methionine and kemzyme (an enzyme preparation) on laying hen performance. In a 12 weeks experiment 60 brown layers (Bovan strain) were used and divided into 4 treatments (5replicates each),the first was used as a control based on yellow corn soybean meal, the second included sorghum (50% of yellow corn)+0.15% methionine, the third contained sorghum + 0.15% methionine +0.05% kemzyme and the fourth consisting of sorghum + 0. 1% kemzyme. All egg production parameters were recorded on daily bases.Egg shell thickness (in mm), yolk index,Haugh unit, yolk color, and meat and blood spots were determined . The results indicated that sorghum could successfully substitute 50% of yellow corn in layer rations especially when supplemented with 0.15% methionine and 0.05% kemzyme.

Paper N5.10

Influence of fusariotoxins on the growing performance of broilers and turkeys
Leitgeb[*1], R., H. Lew[2], R. Khidr [3], W. Wetscherek[1], J. Böhm[4], W. Zollitsch[1]. [1]Inst. für NUWI, Univ. für Bodenkultur, A-1180 Wien, [2]BA für Agrarbiologie, A-4020 Linz, [3]Des.Res.Centre, Mataria, Egypt, [4]Inst. für Ernährung, Vet.Med.Univ., A-1210 Wien.*

In a feeding trail with 180 broilers the impact of non (FG1), low (FG2), medium (FG3) and high contaminated diets (FG4) on growing performance was investigated. The proportion of contaminated maize was 0, 18.2, 36.4 and 54.6 %. In the contaminated maize were 9,8 mg DON, 1,04 mg MON, 1,43 mg BEA and 0,105 mg FB1. At the end of growing period on day 37 the weights of FG 1, 2, 3 and 4 were 1896, 1942, 1904 and 1943 g and the feed conversion rates per kg LW-gain 1.81, 1.77, 1.83 und 1.82 kg.
In the feeding trail with 60 turkeys (Big 6) were used phasefeeds with 27 % protein in the first 4 weeks, with 23 % protein from 5th to 8th and with 21 % protein from 9th to the end of the 11th week. Phasefeeds of FG1 consists of 36.8, 48.9 and 59.3 % of uncontaminated maize and in FG2, FG3 and FG4 uncontaminated maize was substituted with 33, 66 and 100 % of contaminated maize (5 mg MON and 2 mg DON). At the begin of the trail the weight of the turkey chickens was 52 g and at the end on day 77 of FG1, FG2, FG3 and FG4 6.69, 6.27, 6.35 and 6.26 kg and the feed conversion rate 2.07, 2.16, 2.23 and 2.19 kg, respectively. There were no significant differences in LW-gains and feed conversion rates between the feeding groups.

ANIMAL NUTRITION [N]

Paper N5.11

A comprehensive approach of the rabbit digestion : consequences of a reduction in dietary fibre supply

T. Gidenne*[1], V. Pinheiro[1,2] and L. Falcão e Cunha[3]. [1]INRA. Station de Recherches Cunicoles, BP 27, 31326 Castanet, France, [2]UTAD, Secção de Zootecnia, 5000 Vila Real, Portugal, [3]ISA, Secção de Produção Animal, Tapada de Ajuda, 1399 Lisboa, Portugal.

The impact of a linear dietary fibre level reduction (20-16-12% ADF), without variations in fibre quality, was studied using a comprehensive study of rabbit digestion (ileal digestibility + transit + microbial activity) combined with a zootechnical study. During the two weeks postweaning (28-42d), the reduction of dietary ADF led to a lower feed intake (-25%) and weight gain (-9%). From 42 to 70d, we registered a higher (P<0.01) morbidity with the lowest ADF level. A linear reduction of OM faecal digestibility was observed according to ADF level (-1.6 point per % ADF), without changes in fibre digestibility. From 20 to 12% ADF, ileal digestibility of OM rose from 38 to 61%, while caecocolic digestion decreased from 27 to 17%. In parallel, starch ileal digestibility was lowered by 6 points, although it remained very high (> 93%). A sharp increase in ileo-rectal retention time (+38%) was observed, reaching 19h for the lowest ADF level. The bacterial fibrolytic activity was reduced with the lowest ADF level, affecting mainly the pectinolytic activity (50%). Two weeks after weaning the VFA level decreased linearly (P<0.01) with reduction of ADF level, while 4 weeks later no significant effect was observed.

Paper N5.12

The effect of amino acid supplement and feeding system in gestation on reproductive performance of sows

L. Schmeiserová, K. Šimeček, L. Klapil, J. Třinácty. Research Institute of Animal Nutrition, Ltd., Pohořelice, Czech Republic.

One hundred and eighty nine sows (1. - 4. parity) were used in 2 x 2 factorial experiment to determine the effect of amino acids supplement and the feeding system during gestation on reproductive performance. The experiment was conducted from mating to 110 days of gestation. Sows were housed in groups of 25 and fed individually by an electronic feeder with a basal wheat-barley diet (12.6 MJ ME, 4.94 g/kg of lysine, 2.17 g/kg of methionine) - control group (C) which was for experimental group (E) supplemented with lysine and methionine (12.6 MJ ME, 6.58 g/kg of lysine, 2.23 g/kg of methionine). Sows fed according to simple feeding system (S) consumed on average 2.4 kg of mixture a day. Average amount of daily feed intake of sows fed with difficult feeding system (D) was regulated according to changes in energy requirements during gestation as follows: 1. - 21. day - 29.1 MJ (2.4 kg/day), 22. - 84. day - 23.3 MJ (1.9 kg/day) and 85. - 110. day - 44.3 MJ (3.6 kg/day). Differences in average gain of sows during gestation were significant (P<0.05) between groups S and D regardless of amino acid supplement (41.02 kg vs. 51.42 kg for C or 40.54 kg vs. 47.08 kg for E respectively). Reproductive performance was not significantly influenced by amino acid supplement.

ANIMAL NUTRITION   [N]

Poster N5.13

Application of an in vitro incubation technique for estimating utilisable amino acids in feedstuffs for ruminants

P. Lebzien, H. Valenta, R. Daenicke* and H. Böhme. *Institute of Animal Nutrition of the Federal Research Centre (FAL), Bundesallee 50, D-38116 Braunschweig, Germany*

The amount of utilisable amino acids available at the small intestine of ruminants depends mainly on the complex nitrogen metabolism in the forestomachs. Up to now, there are still methodical problems to estimate amount and the amino acids composition of rumen bypass and microbial protein. An in vitro incubation technique was developed to estimate the utilisable amino acids (uAA) and was applied to 21 feedstuffs. The method is based on the first stage of the in vitro incubation technique published by TILLEY and TERRY (1963). The feedstuffs were incubated for 24 hours in a buffer-rumen fluid-mixture at 38° C. $(NH_4)HCO_3$ served as nitrogen source for the microbes and replaced partly the $NaHCO_3$ in the originate buffer solution. After incubation the residues were freeze dried and analysed for their amino acids contents. Results were corrected for blanks without feedstuffs. Results show a significant regression relationship between uAA values measured (y, g uAA/kg DM) by the in vitro incubation technique and the amount of utilisable crude protein (uCP) calculated on the base of the regression which is derived from in vivo trials and used in the German protein evaluation system (x, g CP/kg DM): $y = 0.97 x - 9.05$, $r^2 = 0.89$. Amino acids compositions of the residues after incubation of a concentrate mixture were found to be nearly the same as the mean composition of duodenal protein. It is concluded that in the future uAA determination of feedstuffs might be performed applying this in vitro incubation technique. We acknowledge DEGUSSA, Hanau (Germany) for analyses of amino acids.

Poster N5.14

Nylon bag degradability and mobile nylon bag digestibility of crude protein and amino acids in hydrolysed feather meal

J. Harazim*[1], L. Pavelek[1], L. Pavelková[1], J. Třinácty[2], P. Homolka[3]. *[1]Central Institute for supervising and testing in Agriculture, Opava, [2]Research Institute of Animal Nutrition, Ltd., Pohořelice, [3]Research Institute of Animal Production, Uhříněves, Czech Republic.*

Feather meal samples were incubated in situ to evaluate rumen effective degradability of crude protein (CP) and amino acids. Nylon bags containing incubated samples were suspended in the rumen of two cannulated steers for 0, 2, 4, 8, 16, 24 and 48 h. In each steer, three repetitions were in each time interval. Intestinal digestibility of CP values were determined by mobile nylon bag method provided on two cows with duodenal cannula. The nylon bags and mobile nylon bags were made of nylon with pore size 42 um. The rumen effective degradability was calculated at a rumen outflow rate 5 %.
The content of CP and amino acids in original sample was at CP, Cys, Met, Asp, Thr, Ser, Glu Pro, Gly, Ala, Val, Ile, Leu, Tyr, Phe, His, Lys and Arg 887.0, 30.1, 5.0, 57.5, 35.2, 72.0, 97.3, 88.7, 67.6, 40.4, 75.6, 42.3, 68.5, 12.0, 44.0, 9.3, 16.2 and 63.4 g per kg DM, effective degradability values were 40.2, 24.5, 32.3, 36.2, 28.2, 27.9, 33.8, 34.0, 39.1, 34.2, 37.0, 36.1, 38.2, 25.4, 35.0, 44.5, 35.3 and 28.7 %. Intestinal digestibility of CP was found 89.1 %.

ANIMAL NUTRITION [N]

Poster N5.15

Effect of rape lecithins on growth, carcass quality and body fat composition of bulls
H.-R. Wettstein*, F. Sutter, M.R.L. Scheeder and M. Kreuzer, Institute of Animal Sciences. Animal Nutrition, ETH Zurich, ETH centre/LFW, CH-8092 Zurich, Switzerland.

Plant lecithins could be a preferred feedstuff over plant oils as they are dispersible in water and so may not reduce ruminal fiber degradation to the same extent. In a fattening trial with 6 x 6 bulls rations containing three different rape lecithins, raw (RL1), deoiled (RL2) and deoiled partially hydrolysed lecithin (RL3), were opposed to rations containing either rape oil (RO), whole crushed rapeseed (RS) or deoiled soya lecithin (SL). Dietary lipid addition was kept equal with 3 % of DM on a fatty acid base. No significant differences were found in live weight gain, feed conversion efficiency, killing-out percentage, grading for fatness and proportion of kidney fat. Kidney fat had a lower proportion of C16:0 with RL1 (20.6 %), RL2 (20.7%) and RO (20.8%) than with SL (22.5%). The proportion of the C18:1 trans $\omega 7$-$\omega 9$ relative to total C18:1 was significantly higher with SL (13.2%) than in all other treatments (~ 11%). This was less pronounced when related to total fatty acids because of the lower content of total C18:1 with SL. With rape and soya lecithin kidney fat contained less C18:3 and C20:1 than with RO and RS. The results indicate that 3 % rape lecithins in fattening rations do not affect growth performance and carcass grades when replacing rape oil or rapeseed. Compared with the more common soya lecithin rape lecithin addition resulted in a slightly different body fat composition. A potential advantage of rape lecithins over oils at higher dietary percentages remains to be confirmed.

Poster N5.16

The balance of microelements (Fe and Mn) in green mass and hay of natural pastures in the nutrition of sheep
S. Muratović. Faculty of Agriculture, University of Sarajevo, Zmaja od Bosne 8, 71000 Sarajevo, Bosnia and Herzegovina

On the Kupres tableland (Central Bosnia), there were done four experiments (I and II experiment - the green mass and III and IV experiment - the hay), with aim to investigate the nutritive value of forage in the natural pastures. As a part of that, there were also held the balance experiments of microelements (Fe and Mn), on the basic of Winterberg breed. The balance experiments have shown that:
– the balance of Fe is positive in all experiments: I (3503,94), II (7836,27), III (2518,2) and IV (118,51) mg;
– the balance of Mn is positive in I (202,36), II (473,48) and IV (1089,91) but negative in the III experiment (1237,81) mg;
– the differences between the average rate merits of the examined microelements are statisticalll highly important (P<0,05).

ANIMAL NUTRITION [N]

Poster N5.17

The effect of a mineral drink on a dry metter intake, a milk yield and a physiological state of dairy cows

V.Kudrna, P.Lang, P.Mlázovská, Research Institute of Animal Production, Prague,104 01, P. O. Box 1, Praha 10 - Uhříněves, The Czech Republic

The effect of the after-calving drink *"Drenč - minerální směs G" (Drenč - mineral mixture G)* on a milk yield and a physiological state was pursued in a group experiment with 30 high-yielding dairy cows. Animals were divided in two well-balanced groups - the experimental (Ps) and the control (Ks) ones. The drink was given in two ways according to its amount: in the variant A 5 litres and in the variant B 25 litres, given to 8 hrs after calving, through a sound direct to the rumen. The application of the drink presented oneself in an ingrease of the average daily dry matter intake by 365,51 g per head ($P > 0,05$), and in statistically conclusive increasing ($P < 0,01$) of the average daily milk yield (33,17 kg) by 1,54 kg in cows of the Ps. In the same time was found out higher contents of a milk fat by 0,45 % (to 4,84 %) in the Ps. This fact was presented also in the FCM production (Ps = 37,35 kg, Ks = 33,48 kg), because the difference of 3,87 kg was statistically significant ($P < 0,01$). Levels of a milk protein and a lactose were approximately same in both groups. In 1., 10. and 30. day after calving were pursued both basic parameters of a rumen liquid and a blood. Repeated filling of three cows with some digestive problems renewed an activity of their digestive systems.

Poster N5.18

The effects of different ration energy levels on fattening results of lambs

D. Ruzic[*][1], D. Negovanovic[1], A. Pavlicevic[2], M. P. Petrovic[1]. [1]Institute for Animal Husbandry, Belgrade-Zemun Po Box 23, [2]Faculty of Agriculture, Belgrade, Yugoslavia.

Ration energy level is very important factor which influence the lambs results in intensive fattening. The aim of our examination was to evaluate the production effects of energy levels, changed by adding tallow, in isoprotein diets for lamb fattening. The experiment was conducted on 60 suckled Wirtemberg x Pirot Pramenka lambs, divided in 3 groups. Rations were of alfalfa hay and concentrate mixtures, fed ad libitum, which differ in groups A, B and C, in % of included tallow in mixtures - 0, 4 and 6 %, respectively. Daily consumed energy, in diets A, B, and C, was 7.29, 7.70 and 7.94 MJ NEM and rations contained 160 g of crude protein in kg DM. The lambs were fattened from 15 - 30 kg L.W. At the and of experiment 8 lambs from each group were slaugh -tered. The average daily weight gain (0.28, 0.30 and 0,28 kg/h) was highest in animals on treatment B ($P>0.05$). Average daily dry matter (DM) consumption was 0.75, 0.79 and 0.81 kg day and conversion 2.68 , 2.58 and 2.93 kg/h/d. Energy conversion was 19.08, 18.37 and 20.71 MJ NEM and crude protein 481, 451, and 514 g for kg of body weight gain. Increasing the energy levels by adding 4 % tallow has lead to slight increase of more valuable meat cuts ($P>0.05$). Dressing % (cold carcass with head, liver and lung) was best in B group (58.0:58.6:58.5%). On the basis of results, optimal energy level of ration was, in our experimental conditions, 7.70 MJ NEM.

ANIMAL NUTRITION [N]

Poster N5.19

Effect of length of vegetation season of some maize hybrids on their nutritional characteristics
Loučka, R.., Machačová, E. Research Institute of Animal Production, Přátelství 815, Prague 10 - Uhříněves, 104 01, Czech Republic.

The FAO number characterises an estimated length of vegetation season of maize, in The Czech Republic one hybrid gets it after 3 years of testing by ÚKZÚZ (Central control and testing institute of agriculture). One day of difference of the length of vegetation season or one percent of dry matter of cob harvested in the optimum of maturity phase of the choice hybrid in comparison with control hybrids, equals the difference of 10 FAO numbers. In our experiments we tested 6 hybrids with the different FAO number for 4 years. Every year maize plants were harvested in the optimum of maturity phase (2/3 milk line of corn).
In addiction to the increasing of the FAO number, the dry matter content was increased, but the proportion between dry matter of cobs and dry matter of green parts of plants was decreased. The annual differences were higher than the differences among hybrids in one year, the global differences of the FAO number among hybrids were not statistically significant. For that reason the FAO number is only supplementary signpost for the choice of maize hybrids for growing.

Poster N5.20

Estimation of usability of intermediate method AIA in research of feeding value green forage *amaranths* for nutrition of ruminants
Z. Łukaszewski, K. Petkov, K. Antczak. Department of Animal Nutrition and Feed Management, University of Agriculture, Judyma Street 2, 71-466 Szczecin, Poland.

At the present time, the studies of intestinal digestibility of feed on ruminants are very expensive. However, the results of studies *in vitro* often are understated. The aim of this paper was determination of usability the intermediate method AIA (Acid insoluble ash) estimation digestibility nutrients of green forage *amaranths* in half flowering, in comparison with balance immediate method. The experimental part was carried out among 8 sheeps. The main advantage of AIA method is simplicity and short time of realisation (after preliminary period, one day's analysis of feeds and faeces for crude ash and nutrients components). The experiment was divided on period: preliminary - 14 days and proper - 10 days. On the basis of the results obtained digestion coefficients (%) in the immediate method, the intermediate method, together with coefficients of variability (%), properly: organic matter 65.7; 65.9; 10.6, crude protein 75.6; 73.4; 9.8, crude fibre 54.6; 50.2; 12.6, NFE (Nitrogen-fee extractives) 69.3; 66.8; 16.3, cellulose 68.6; 61.4; 12.9, pentosans 71.9; 66.2; 12.2.There weren't differences of statistical among the methods. The purpose of studies showed possibility usage of intermediate method AIA for ruminants to determinate digestibility value organic matter and remaining components.

ANIMAL NUTRITION [N]

Poster N5.21

Critical size of plastic particles passing digestive tract of cows.
J. Třináctý[1], P. Homolka[2]*, M. Šustala[1]. [1]Research Institute of Animal Nutrition, Ltd., Pohořelice, [2]Research Institute of Animal Production, Uhřiněves, Praha, Czech Republic.

Two cows in lactation and two dry cows were used in two trials. The diet of dairy cows consisted of 20 kg maize silage, 4 kg alfalfa hay and 6 kg mixture. The diet of dry cows consisted of 6 kg alfalfa hay and 1.4 kg mixture. In trials plastic particles of a cylindrical shape and specific gravity 1.25 ± 0.05 g/cm3 were used. Their sizes were 8/8, 11/11, 14/14 and 17/17 mm. The total count of particles simultaneously applied one cow was 4 x 20 pieces. The plastic particles were applied orally as a paper bolus by a balling gun device after feeding in the morning.
The recovery of 8/8, 11/11, 14/14 and 17/17 mm particles during 384 hours was 92.5, 92.5, 88.8, 5.0 % at dairy cows and 90.0, 77.5, 35.0, 5.0 % at dry cows, respectively. The total mean retention time (TMRT) of 8/8, 11/11, 14/14 particles was 47.0, 53.8, 95.6 hours at dairy cows and 115.8, 132.7, 170.7 hours at dry cows, respectively. The regurgitation of 11/11, 14/14 and 17/17 mm particles was 1.3, 7.5 and 27.5 % at dairy cows and 5.0, 40.0 and 80.0 % at dry cows. At dairy cows was found even 90 % recovery at particles bigger than 11/11 mm with satisfactory TMRT values.

Poster N5.22

Improved equations for predicting the metabolizable energy (ME) content of feeds for ruminants
D. Palic* and M. Brits. ARC-Animal Nutrition and Animal Products Institute, Private Bag X2, Irene 0062, South Africa.

In a preliminary study, Palic et al. (1998) validated original equations (Palic et al., 1997) for predicting the metabolizable energy (ME) content of feeds for ruminants. The aim of this study was to improve the original prediction equations.
The best prediction of the ME in original study was achieved by the Tilley and Terry (1963) and the Pepsin/Enzyme-Mix (PEM) (Weisbjerg and Hvelplund, 1993) methods. Fifty new feeds in total (21 feedstuffs and 29 complete diets for ruminants), with pre-determined in vivo organic matter digestibility (OMD), were used in the study. The original equations were firstly validated. The in vitro Tilley and Terry and PEM values of fifty new feeds were substituted into appropriate original prediction equations, thus obtaining estimated in vivo value for each laboratory result. The estimated in vivo ME values were then regressed against the determined in vivo ME values and the corresponding R-square and SEP were calculated.
The statistical analysis showed that the original and new feeds belong to the same population, (as there was no significant difference in slopes and intersepts of their regression lines) thus pointing out that the validation of the original equations was successful. The general regression equations for combined original and new feeds (49 feedstuffs and 35 complete diets in total), derived through validation, are regarded as improved prediction equations.

ANIMAL NUTRITION [N]

Poster N5.23

Effect of donor feeding in the gas production capacity
A.E.S. Borba*[1], P.M.M.O. Gonçalves[1], C.F.M. Vouzela[1] and A.F.R.S. Borba[2]. [1]University of the Azores, Department of Agrarian Sciences, 9700 Angra do Heroísmo, [2] SRAPA, SDAT, 9700 Angra do Heroísmo Azores.

In the present work we tested the donor feeding capacity of gas production by two sources of inocula. Forage samples of 24 grammineae (oat, Italian ryegrass, perennial ryegrass and maize), with known chemical composition and in vivo digestibility were taken. Gas production was measured at 24 h of incubation. The inocula sources utilised were liquor from rumen of sheep and suspension of sheep faeces (obtained from animals fed with hay and concentrate, and alfalfa and concentrate).

Values for the total gas production after 24h were submitted to an analysis of variance on the basis of two sources of feeding, two methods of incubation, four species, three growth stages and two methods of conservation. Nine individual measurements were taken.

The analysis of variance for gas production showed that the difference was statistically different ($p \leq 0.05$). The lower value was obtained with the inocula obtained from animals feeding with alfalfa. Gas production was significantly ($p \leq 0.05$) greater with liquor from rumen of sheep than with suspension of sheep faeces. Was significantly ($p \leq 0.05$) greater from material at the intermediate growth stage than at earlier or later stages, and greater from fresh material.

Poster N5.24

Determination of maintenance requirement of limiting amino acids for pigs
A. Kodeš,[1] D. Vodehnal,[1] J. Heger,[2] B. Hučko,[1] [1]Department of Animal Nutrition, Czech University of Agriculture in Prague, Prague 6 - Suchdol, 165 21, [2]Biofaktory Prague s.r.o., Prague 9 - Horní Počernice.

Purpose of biological experiments performed was the determination of maintenance requirement of limiting amino acids as the basic presumption for making more aaccurate the standards of lysine, threonine, methionine + cystine and tryptophan requirement for growing fattening pigs. We organised 4 series of balance experiments (four amino acids observed) in each series five half-synthetic diets were verifies (different dosing of amino acids observed). Response of each diet was observed on 5 gilts (together 100 balances.

When determining the basic requirements following premises were respected:
Nitrogen balance is a function ($Y = a + bx$) of intake of the first limiting amino acid
Maintenance requirement of the given amino acid is covered if $y = a + bx = 0$, i.e. that maintenance requirement follows from the relation $x = -a / b$).

On the basis of the results obtained daily maintenance requirement of amino acids :
lysine- 37.6 mg per $W^{0.75}$ and day, threonine 46,0 mg per $W^{0.75}$ and day, methionine + cystine 49,2 mg per $W^{0.75}$ and day, tryptophan 16,4 mg per $W^{0.75}$ and day.

ANIMAL NUTRITION [N] Poster N5.25

N and P reduction from livestock waste
F.Yano*[1], D.Kondo[1], H.Osako[2] and S.Kato[2]. [1]Department of Biotechnological Science. Kinki University,649-6493 Wakayama,Japan, [2]Oishi Experimental Farm, Kinki University, 643-0531, Wakayama,Japan.

Sixteen crossbred growing barrows (28.6kg B.W.) were used to evaluate the effect of feeding 19% CP diet, 18% CP diet, 17%CP-lowP diet supplemented with phytase and 15% CP-lowP diet supplemented with phytase, lysine and threonine on N and P balance trial. There are four replications of each treatment in a randamized block design. All pigs were fed 2% of their body weight level of diet per meal twice a day. After 7 days adaptation period, Total urine and feces were collected for 5 days. Pigs recieving 15%CP-lowP diet(+ phytase and amino acids) excrete less N and P than pigs fed other diets. The biological value of pigs fed 15%CP-lowP diet showed 71.6%, which was similar level to those of 18%CP diet group. The P digestibility of 15%CP-lowP diet(+ phytase and amino acids) was improved from 49% to 62%. Simultaneous reduction of N and P from animal waste might be available by supplementation of amino acids and phytase.

Poster N5.26

Ileal digestible amino acid requirements of Belgian lean meat-type pigs
R. De Schrijver* and J. Vande Ginste. Department of Animal Science,
Catholic University of Leuven, Kardinaal Mercierlaan 92, B-3001 Leuven, Belgium.

The amino acid requirements of starting (20-40 kg), growing (40-70 kg) and finishing (70-100 kg) high-lean-growth barrows (Belgian Landrace x Piétrain) were determined using the nitrogen balance technique. Animals from each body weight class received a low-protein basal diet consisting of barley, wheat, soybean meal and cassava, which was fortified with various amounts of lysine, methionine, threonine and tryptophan, respectively. The optimum dietary amino acid supplementation for each weight class was verified in feeding experiments involving barrows as well as gilts which were maintained in practical conditions. The ideal protein composition regarding lysine, methionine + cystine, threonine and tryptophan as well as the optimal dietary contents of ideal protein for starting, growing and finishing pigs were derived from the combined results of the nitrogen balance and feeding assays. The apparent ileal digestible amino acid requirements were determined for each body weight class using barrows fitted with a T-cannula replacing the cecum which received the optimally amino acid fortified diets. In order to find the amino acid requirements based on their true digestibility, endogenous ileal nitrogen secretion was measured in separate experiments using diets with different protein levels and extrapolating to zero protein intake.

ANIMAL NUTRITION [N]

Poster N5.27

Effect of ensiled corn grain on the hens laying
J.M.A. Stekar[1], E. Tkalčič[2], A. Holcman[1] and M. Kovač[1]. [1]Zootechnical Dept., Biotechnical Fac., Univ. of Ljubljana, Groblje 3, SI-1230 Domžale, [2]Kmetijski zavod Maribor, Vinarska 14, SI-2000 Maribor, Slovenia.*

In the feeding trial the incorporation of ensiled corn grain on the laying of 600 layers isa brown proveniences was tested. At the 32nd week of age animals from the flock of 1400 animals were random selected and ordered into six groups. Experimental diets were given to the animals from the 40th week of age when they already reached the peak of laying. The experiment lasted for 150 days. Two hybrids, raissa and lotus, were compared. In one diet 40% of DM was ensiled grain and in the second one 60%. In both cases the supplementary mixture was the same one. The animals in the control group were given the complete mixture ad lib. Other groups, that were fed with silage and also one group fed with complete mixture, were fed restrictively. The results of laying in the experimental period were processed by statistical package SAS by two models. With the first one the laying among groups was compared. With the second model hybrids and rations were compared. Between the hybrids were no significant differences. The groups on the 40% DM from silage had significantly worse laying as the control group, which was explained by smaller energy content. The groups on the 60% DM from silage had significantly better laying.

Poster N5.28

Nutritional evaluation of corn-barley diets supplemented with different levels of enzymes preparation for layers and broiler chicks
Samia M. A. Hashish, G. M. El-Mallah and A. A. El-Ghamry. National Research Center,12311 Dokki Cairo, Egypt.

Two experiments were conducted to study the effect of dietary supplementation of Kemzyme (an enzyme preparation comprises of alpha-amylase, beta- glucanase, protease, lipase and cellulase) at an inclusion rates of 0, 0.5 and 1 gm/kg of corn-barely-soybean diet (barley sunstituted 50% of yellow corn) on laying hens and broiler chicks performance. A total of 60 commercial egg-type brown Bovans hens of 34 weeks old were used in the first experiment, and 216 one week-old-Arbor-Acres chicks were used in the second experiment. The result of the first experiment showed that the control group had the highest egg production, while the least value was recorded for the group fed diet containing 1gm Kemzyme /kg diet, without signigicant difference. No significant differences were observed among treatments in egg mass or feed conversion. However, the data of egg quality revealed no significant differences among treatments in all parameters except for yolk colour score which was declined upon feeding barely-diets. The results of the second experiment indicated that the enzyme addition to the corn-barley-soybean diets had no significant effect on live body weight and weight gain. The results of apparent digestion coefficients of the nutrients of the experimental diets revealed that no significant differences were observed among treatments in digestibility coefficients for organic matter, crude protein, crude fiber or NFE. While , EE digestibility improved significantly by dietary Kemzyme supplementation at inclusion rate of 0.5/kg.

ANIMAL NUTRITION   [N]

Poster N5.29

Effect of extrusion of rapeseeds on performance of hens

L.Zeman[2], D. Klecker[1], M. Lichovníková[2], *Faculty of Agronomy, Mendel University of Agriculture and Forestry Brno. [1]Department of Breeding of Farm Animals, Branch of Poutlry Breeding, [2] Department of Nutrition and Feeding of Farm Animals, Zemědělska 1, 613 00 Brno, Czech Republic*

The effect of extrusion of rape seeds was investigated in a factorial experiment with 1624 hens of hybrid combination Hisex Brown. As compared with control (C), one experimental variant received feed containing 9% of extruded seeds of rape (EX) while the other 9 % of untreated seeds (NOEX). The highest egg production was observed in the EX variant (+ 5 eggs) while in the variant NOEX the egg performance was lower than in control (- 1.5 eggs) The highest production of egg mass were recorded in the EX variant (15.26 kg); in C and NOEX variant the corresponding values were 15.09 and 15.07 kg, respectively. The lowest and the highest proportions of non-standard eggs were recorded in EX and NOEX variants (1.3 % and 1.5 %, resp.). The highest mortality was recorded in NOEX variant (6.28 %) while in EX and C these values were 2.42 %). As compared with NOEX variant, use of extruded feed reduced the occurrence of blood stains in eggs and limited depigmentation of yolk and egg shell.

Poster N5.30

Effect of sodium and calcium phosphates on egg production and quality

D. Klecker[1], M. Lichovníková[2], L. Zeman[2], *Faculty of Agronomy, Mendel University of Agriculture and Forestry Brno. [1]Department of Breeding of Farm Animals, Branch of Poultry Breeding, [2] Department of Nutrition and Feeding of Farm Animals, Zemědělska 1, 613 00 Brno, Czech Republic*

The effect of phosphate source, i.e. either from dinatrium phosphate (DNP) or dicalcium phosphate (DCP) or their combination (50 : 50) (MIX) on egg production and quality was investigated in a factorial experiment with 72 hens of hybrid combination ISA Brown placed individually in cages during the last phase of egg production (up to 72 weeks of age).
As compared with DCP, the DNP variant with showed the highest intensity of egg production as well as the highest egg weight (66.82 g *vs.* 65.36 g), thicker egg shell (0.40 mm *vs.* 0.39 mm), the highest weight of egg shell (6.28 g *vs.* 6.11 g) and the highest value of Haugh's units (79.94 *vs.* 75.94 g). Markedly lower values of all parameters were obtained in variant with a combination of DNP and DCP (MIX).

ANIMAL NUTRITION   [N]   Poster N5.31

Productive efficiency and safety of broiler's diets containing wheat treated with fungistatic drugs
J.Harazim*[1], E.Mareèek[1,] L.Pavelek[1], P.Suchý[2], I.Herzig[2], P.Suchý[3], E.Straková[3]. [1]*Central Institute for Supervising and Testing in Agriculture, Jaselská 16, Opava, CZ.,* [2]*Veterinary Research Institute, Hudcova 70, Brno, CZ.,* [3]*University of Veterinary and Pharmaceutical Sciences, Palackého 1/3, Brno, CZ.*

For an experimental purpose, various poultry feed mixtures (BR) were prepared to feed broilers in the I and II phase. About 50% of the feedstuffs was composed of wheat. In each diet, the moisture and drug treatment of the incorporated wheat was different. In BRK and BRA, the wheat was intact and a moisture of 11.9% and 16.4%; in BRB and BRC, it was treated with drug X and Y, and had a moisture of 16.8% and 16.2%, respectively. Next task was to follow the appearance of different fungal species, colony forming units (CFU) and the level of mycotoxins in the feedstuffs. Eight groups of broilers, each with 100 birds, were given these diets for 32 days. Subsequently, flock performance, haematological and biochemical parameters were investigated. Besides, pathological examinations and physical assessment of meat quality were conducted. Fungal contamination of the feeds was: BRK-I 7700 or BRK-II 1700, BRA 26800 or 28500, BRB 20000 or 48000, BRC 47300 or 44000 CFU/g. This indicated a close association with the occurrence of different fungal species. However, the negative impact of feed contamination on broiler's performance, haematological and biochemical parameters, and pathological findings, was not confirmed. Also, physical meat assessment did not reveal any difference among the broilers.

Poster N5.32

Amino acid pattern and biological protein value of spelt and winter wheats
M. Chrenková[1], Z. Čerešňáková[1], A. Sommer[1], Z. Gálová[2], L. Dahlstedt[3].
[1]*Research Institute of Animal Production, Hlohovská 2, 949 92 Nitra, Slovak Republic,* [2]*Slovak University of Agriculture, Tr. A. Hlinku 2, 949 76 Nitra, Slovak Republic,* [3]*Swedish University of Agricultural Sciences, S-750 07 Uppsala, Sweden.*

While the world wheat crop arises from production of common (*Triticum aestivum* L.) and durum (*Tritituc durum* Desf.) cultivars, there is increasing interest in ancient wheat species, especially spelt (*Triticum spelta* L.). The aim of this study was to evaluate the nutritional composition and to compare the quality of two spelt samples and two common winter wheats grown in Slovakia and in Sweden. The chemical analyses indicate that T. spelta has higher content crude protein, protein and most essential amino acids except lysine. The higher content of crude protein in spelt wheat deteriorated the already unfavourable amino acid composition, mainly the surplus of proline in proportion to arginine and lysine decreases the total utilization of proteins. We found lower values of this proportion in spelt wheat compared with values in winter wheat. Low proportion of albumins and globulins conditions also the low concentration of lysine, threonine, methionine and arginine as well as high concentration of glutamic acid and proline. The results achieved in biological experiment by rats are in good agreement with the content of essential amino acids. The protein digestibility varied from 77 to 85 % for winter wheat and spelt, and biological protein value varied from 72 to 75 % for spelt and 73 to 81 % for winter wheat.

ANIMAL NUTRITION   [N]

Poster N5.33

Antinutritional substances and their effect on the quality of the rape protein
A. Kodeš, Z. Němec, B. Hučko. Department of Animal Nutrition, Czech University of Agriculture in Prague, Prague 6 - Suchdol, 165 21

This contribution is engaged in evaluation of the effect of select antinutritional substances (erucic acid, tannins, phenolic substances, glucosinolats total and out of it progoitrine, gluconapine, glucobrasicanapine) occurring in rapeseed on the quality of rape protein diets.
To reach this aim 2 series of balance experiments were organised, in each series diets with different content of antinutritional substances were verified. Response of each diet was observed on 12 male laboratory rats strain Wistar originating from SPF rearing (together 48 balances). Balance observation took place in individual cages in the stable with controlled light regime and microclimate according to usually accepted usage, on animals of average live weight 60 g.
Results obtained proved different intensity of negative effect of individual antinutritional substances on coefficients of apparent and true digestibility of rape protein, its biological value, net protein usage (NPU) and faeces digestibility of several amino acids (lysine, threonine). The highest mutual relation was found between the quantity of feed intake and content of erucic acid ($r = 0.990$), resp. of glucosinolates content ($r = - 0.999$)

Poster N5.34

Fermentation of soybean meal improves zinc and iron availabilities in rats
M. Hirabayashi, T. Matsui and H. Yano*. Division of Applied Biosciences, Graduate School of Agriculture, Kyoto University, Kyoto, 606-8502, Japan

Soybean meal contains considerable amounts of zinc and iron. However the availabilities of these elements are not high because of a large amount of phytate which forms insoluble complexes with these elements in soybean meal. Additionally, phytate also makes insoluble complexes with extrinsic iron and zinc in the digestive tract. We showed that the fermentation of soybean meal with Aspergillus usamii almost completely degraded phytate without adversely affecting crude protein content and amino acid composition. The present experiment examined the effects of fermented soybean meal on availabilities of iron and zinc. Rats were given a diet containing 40% soybean meal (SBM) or 40% fermented soybean meal (FSBM) for 4 weeks. Iron concentrations in plasma and liver, and zinc concentrations in plasma and femur were measured. Iron and zinc solubilities were also determined in the digesta of small intestine. Femoral and plasma concentrations of zinc were higher in the FSBM group than in the SBM group. Liver and plasma concentrations of iron were higher in the FSBM group than in the SBM group. The FSBM group showed higher solubilities of iron and zinc in the small-intestinal digesta than the SBM group. These results suggested that dephytinized soybean meal by fermentation improved iron and zinc availabilities through increasing solubilities of these elements in the small intestine.

ANIMAL NUTRITION   [N]

Poster N5.35

The use of *N*-alkanes as markers for intake and digestibility determination of hay fed to Islandic toelter horses

L. Mølbak*[1], O. Gudmundson[2], P.D. Møller[1]. [1] Institute of Anatomy and Physiology, Section of Animal Physiology, Royal Veterinary and Agricultural University, Copenhagen, Denmark, [2] Agricultural Research Institute, Reykjavik, Island.

The present experiment was established to investigate the possible use of *n*-alkanes as markers to determine intake and DM digestibility of 4, 6 and 8 kg hay DM fed to 9 Icelandic toelter horses. When the horses were eating 4 and 6 kg of hay daily the marker technique generally overestimated the intake. When the horses were eating 8 kg the marker method appears to be more accurate. In spite of this deviation the differences estimated when using the n-alkane pairs $C_{31/32}$ and $C_{32/33}$ is quite similar. The average DM digestibility coefficient of hay measured by using the *in vivo* method gave a lower digestibility (43-60%) than the *in vitro* method (52-70%) but higher than the marker method (36-54%) for low and high hay quality respectively. The results showed that the estimates using *n*-alkane C_{31} were within the confidence intervals for all hay qualities. This was not the case using *n*-alkane C_{33}. Tests for each factor in ANOVA for digestibility showed a highly significant influence of hay quality at the 0.001% level. Horse nested within hay quality showed a significant interaction at the 0.05% level. It can be concluded, with some caution, that *n*-alkanes can be used to estimate intake and digestibility in grass eating horses when fed at or close to ad libitum.

Poster N5.36

Determination of fat-soluble vitamins in animal feeds

L. Wágner*[1], J. Csapó[2], L. Vincze[1], G. Szüts[1], G. Kovács[1] and K. Dublecz[1]. [1]Department of Animal Nutrition, Pannon University of Agriculture Georgikon Faculty, H- 8360 Keszthely, Deák Street 16, Hungary, [2]Faculty of Animal Science, Pannon Agriculture University, P.O. Box 16., H-7401 Kaposvár, Hungary

Vitamins are organic compounds necessary for maintaining normal body processes (growth, health, fertility, performance). Generally, the animal cannot itself synthesise these natural biologically active agents, which is why they must be supplied with the feed. Some of these vitamins are sensitive to the effects of moisture, heat, light, heavy metal ions, pH value and oxidation and reduction processes.
HPLC has developed during the past decade as a separating technique with many advantages relevant to vitamin analysis. It can be performed at room temperature. Light and oxygen are easily excluded during chromatography.
A simple high-pressure liquid chromatographic method for the simultaneous determination of vitamins A and E in animal feeds has been developed and evaluated. After saponification and extraction the sample was run on a reversed-phase column with water-methanol as the mobile phase. The method has been tested on different types of feeds. The recovery of added vitamins was 92 % for retinyl acetate and 90 % for α-tocopheryl acetate with a standard deviation of 5 % for both.

ANIMAL NUTRITION   [N]

Poster N5.37

Relation between ochratoxin A content in cereal grain and mixed meals determined by the Elisa and HPLC methods and attempt to evaluate their usability for monitoring studies

Antoni Jarczyk[1], Lucjan Jędrychowski[2], Barbara Wróblewska[2], Roman Jędryczko[3]
[1] Chair of Swine Breeding, Agricultural-Technological University, Olsztyn; [2] Institute of Animal Reproduction and Food Studies of Polish Academy of Sciences, Olsztyn; [3] Dept of Veterinary Hygiene, Olsztyn, Poland

Cereal grain and mixed meal samples were collected for analyses in two periods:
1st - from harvest to the end of 1996 and 2nd- from June to August 1997. The content of mould, toxicogenic fungi and ochratoxin A (OA) was determined by immunoassay (ELISA) and high performance liquid chromatography (HPLC). In the 1st period, when OA level ranged from 0.05 to 2.02μg/kg, the correlations of results obtained by the two methods were low (r=.047 for cereals and r=.170 for mixed meals). In the 2nd period, after 9-11-month storage, when OA level ranged from 0,13 to 28,12 μg/kg the correlation was high (r=.946 for cereals and r=.682 for mixed meals at $P<0.001$). HPLC was evaluated as more useful for both analysing single samples and monitoring studies due to the possibility of simultaneous determination of OA, fumonisine and aflatoxin in the extract as well as its lower costs.

Poster N5.38

The effect of anthropogenic activity on fodder crop quality in an area of CR with long-term pollutant load

Z. Wittlingerová, A. Kodeš, B. Hučko. Department of Animal Nutrition, Czech University of Agriculture in Prague, Prague 6 - Suchdol, 165 21

The effect of anthropogenic activity on fodder crop quality was studied in a several-year period in an area of East Bohemia with intensive farming, exposed to long-term pollutant load; it is one of the most productive areas. Main nutrients and some hazardous substances were studied: nitrogenous matters, P, K, Ca, Mg, S, NO_3, Cd, Pb, Hg, Cu, Zn, as well as coefficients of soil:plant transfer. Standard techniques in agreement with the relevant methods were used for chemical analyses of fodder crops and soil. Data acquired by measurements were evaluated with respect to effective feed limits. It was confirmed that fodder crops complied with the effective limits in spite of the long-term pollutant load of farm land. Evaluation of fodder crops pursuant to feed limits was satisfactory and the limits were not demonstrated to be exceeded.

ANIMAL NUTRITION [N] Poster N5.39

"Flushing method" to improve the reproductive efficiency in rabbit does

F. Luzi[1], S. Barbieri[1], C. Lazzaroni*[2], C. Cavani[3], M. Zecchini[1], C. Crimella[1]. [1]Istituto di Zootecnica, Facoltà Medicina Veterinaria, via Celoria, 10, 20133 Milano, Italy. [2]Dipartimento Scienze Zootecniche, Facoltà Agraria, via L. da Vinci, 44, 10095 Grugliasco, Italy. [3]Istituto di Zootecnica, Facoltà Agraria, via S. Giacomo, 9, 40126 Bologna, Italy.

Over eight months period (March-October 1998) the effect of "flushing method" to improve the reproductive parameters of rabbit does reared in intensive system were studied. Five consecutive reproductive cycles were investigated on nulliparous and multiparous (lactating or not) does, submitted to artificial insemination (AI). Three experimental treatments were performed: PMSG group (20 IU injected 48 hrs. before AI); PG group (2% of propilenic glycol, as energetic source (flushing), administered into drinking water during the 4 days before AI) and control group. Furthermore, the physiological status of does and month of insemination were considered. Fertility rate, number of total and alive born, number of weaned kids, mortality rate at birth and weaning did not show any significant differences among the treatments and the physiological status of does. During the summertime, all the reproductive parameters showed the worst values ($P<0.05$). These results evidenced it could be avoided hormonal treatments and energetic dietary surplus on the receptive and good healthy does.

ANIMAL MANAGEMENT AND HEALTH [M] Paper M1.1

Impact of animal behaviour research on new developments in housing design for farm animals
E. von Borell*. Institute of Animal Breeding and Husbandry, Martin-Luther-University Halle-Wittenberg, Adam-Kuckhoff-Str. 35, 06108 Halle, Germany.

Animal behaviour research has contributed to standards of housing systems, but its influence on the development of new housing designs is actually less than expected. Behavioural science has, however, made a major contribution to define animal needs and to assess the impact of housing and management procedures on the welfare of farm animals. Current behavioural research is mainly focussed on the assessment of existing housing designs with emphasis to modify them in order to improve welfare of farm animals. This emphasis has to do with the funding situation, as most of the resources are invested in problem areas where public concern is most evident (i.e., individual confinement of sows or cage housing of laying hens). Most of the standards that we have today were mainly developed under technical and economical considerations aimed to maximise animal performance (including health) and minimise labour input. There are few examples, where housing systems were initially developed according to the behavioural needs of farm animals. Those systems (i.e., family housing for pigs or aviaries for hens) are only viable, when economic output and labour input are considered. It is therefore necessary that animal behaviour scientists collaborate with engineers and economists in order to improve housing and welfare of farm animals without compromising competitiveness within the markets in which the farmer has to produce.

Paper M1.2

Relationships between farmers' attitudes, retailers' actions, legislation and welfare research
D.M. Broom*, Department of Clinical Veterinary Medicine, Cambridge University, Madingley Road, Cambridge CB3 0ES, UK.

Farm animal welfare is much affected by the systems and management procedures used by farmers. The attitudes and actions of farmers depend on their knowledge and beliefs, the pressures of purchasers, legislation and the attitudes of other people to what the farmer is doing. Government ministries and advisory committees within Europe now take account of animal welfare research. Retailers take notice of consumers and have had increasing effects on farm practice, especially in the last three years. However, the effects of legislators and retailers on breeders, have been very slight. Many breeders operate internationally and their responsibility for poor welfare in farm animals is not appreciated by most of the public. Future improvements in farm animal welfare depend to some extent on World Trade Organisation agreements. Poor welfare in animals must be regarded as a moral issue which can be a criterion for refusing imports. Retailers have much influence because they can require good welfare on farms, irrespective of the country of production. Current pressures are likely to encourage a more uniform high level of concern for animal welfare throughout the world.

ANIMAL MANAGEMENT AND HEALTH [M] Paper M1.3

Dutch farm animal welfare research in interaction with society, legislation, consumer demand and industry
H.J. Blokhuis*[1] and G.F.V. van der Peet[2]. [1]Institute for Animal Science and Health ID-DLO, P.O.Box 65, 8200 AB Lelystad, The Netherlands, [2]Ministry of Agriculture, Nature Management and Fisheries, Department of National Reference Centre, P.O.Box 482, 6710 BL Ede, The Netherlands.

In The Netherlands the issue of farm animal welfare developed over the last thirty years.
Worries in society about the welfare of production animals focussed on negative effects of housing and management systems, risks of high production levels and the application of modern production technologies.
Following the public concerns research in this field was stimulated both by government as well as the industry. This research did contribute to the improvement of different aspects of animal production. Conclusions were drawn about effects of housing conditions and regulations in this field were formulated. Also alternative systems were developed, tested and introduced.
Society as a whole on a more general level as well as more specifically consumers of animal products will require high welfare standards for production animals. Good welfare is more and more considered as an important aspect of the total quality of the product. This will encourage a continuing effort of scientists to analyse the welfare status of animals and to come up with innovative solutions. The animal industry as well as retailers should aim at the further implementation of this knowledge and to specify welfare standards to improve consumer acceptance of animal production.

Paper M1.4

The impact of welfare research on legislation and animal housing in Sweden
L.J.Keeling, Department of Animal Environment and Health, Swedish University of Agricultural Sciences, P.O. Box 234, SE-532 23 Skara, Sweden

The generally high standard of living in Sweden, the fact that it is self-sufficient in most animal products and, until recently, was outside the European Union have all contributed towards Sweden having probably the most comprehensive animal welfare legislation in the world. Legislation was originally based mainly on practical experience, but recent revisions are increasingly based on scientific research. However, in the absence of other type of knowledge, the tendency in Sweden has been to promote what is natural. Thus, for example, Sweden has never allowed some management practices, such as beak trimming of laying hens, and the clause in the Animal Welfare Act, stating that animals should be able to behave naturally, has been enforced to ban some housing systems. For example, it is no longer allowed to confine sows or keep cows inside all year round. Of note is the fact that the most recent legislative changes have been based on research identifying features in the environment that are important for animal welfare, such as straw for pigs and perch, nest box and dustbath for laying hens. Another example of this positive approach to animal welfare is the increasing emphasis on research to promote health, rather than treat symptoms, and there are several examples where the industry has been able to turn this to advantage in marketing the end product e.g. chicken meat and ecological milk.

ANIMAL MANAGEMENT AND HEALTH [M] Paper M1.5

Direct payments to promote housing systems that are especially adapted to the behavioural needs of farm animals
W. Meier. Swiss Federal Research Station for Agricultural Economics and Engineering, FAT, 8356 Tänikon, Switzerland.

In Switzerland, an increasing number of consumers ask for animal products that are in accordance with high standards with regard to ecology and animal welfare. In 1993, Swiss government introduced direct payments to promote housing systems for farm animals that are especially adapted to their behavioural needs. Two programs were launched. The first program focuses on indoor housing systems and puts special emphasis on the quality of the lying area. The second program aims at promoting regular access to outside yards and pasture. Participation in these programs is voluntary.

In 1997, 27 and 6.3 million Euro, respectively, were invested in the two programs. A total of 17'000 farms housing 350'000 cattle equivalents participated in the programs. This accounts to 27% of the Swiss livestock population. The aim of the government is to have 50% of the livestock in these programs by 2005. According to calculations made by the Swiss Federal Research Station for Agricultural Economics and Engineering (FAT), this goal can be achieved.

Paper M1.6

French people's willingness to pay for farm animal welfare
K. Latouche*. INRA, Unité d'Economie et Sociologie Rurales, Equipe Politique Environnementale et Risques, rue Adolphe Bobierre, CS 61103, 35011 Rennes cedex, France. Email: latouche@roazhon.inra.fr

This paper deals with French's concerns about Farm Animal Welfare (FAW) and their Willingness To Pay (WTP) to support FAW improvement. To elicit respondents' WTP, Contingent Valuation Method (CVM) was used, whereas this method is still criticised. The WTP to support FAW improvement in different production systems is expressed as an increase of the market price of four different products: one kilo escalope of veal, one kilo white ham, a dozen eggs and one kilo beef (steak cut from the ribs). This survey, interviewing 1009 persons, shows that French people state a great concern about FAW. Some production systems have clearly been classified as unacceptable. French producers and public decision makers will have to take into account these results. They are in fact a good indicator of French public opinion and assess the perceived benefits of farm animal welfare policies. Four econometric models aiming at determining the main variables which explain the WTP elicited to support improvement of animal welfare are proposed. They show that people who feel concerned with farm animal welfare express a greater WTP than the others for veal welfare but express the same WTP for the welfare of hens, pigs and cattle as the others.

ANIMAL MANAGEMENT AND HEALTH [M] Paper M1.7

Legislation of farm animal welfare requirements in Slovakia
S. Mihina[1], B. Lovas[2], L. Hetenyi[1], J. Sokol[2]. [1] Research Institute for Animal Production, Hlohovska 2, 94992 Nitra, [2] State Veterinary Office of MoA, Botanicka 17, 84213 Bratislava, Slovakia

During the early years following the political changes of 1989, Slovakia began to formulate regulatons for animal welfare. At first, they were directed only at protecting animal for maltreatment. Later, regulations were added in connection with housing, facilities and husbandry of farm animals. In 1995 an Animal Protection Law No 115 Z.z. was approved by Parliament. Following that, a working group began to work on a detail code of practice. Members of the working group were specialists from The State Veterinary Office, the Research Institute for Animal Production, the Veterinary and Agricultural Universities and some other planning and breeding organisations. The code of practice was developed on the basis of Slovak and foreign research knowledge of animal behaviour and scouting of relevant foreign codes as well. It was introduced in the summer of 1998. It contains general rules on animal welfare and detailed requirements for individual categories of animal covering housing and husbandry factors. The nature of these requirements, and early experiences of using them, will be presented in the paper.

Paper M1.8

Implications of changes to legislative space allowance for performance, aggression and immune competence of growing pigs at different group sizes
S.P. Turner[1]*, S.A. Edwards[2], M. Ewen[1] and J.A. Rooke[1]. [1]Scottish Agricultural College, Craibstone Estate, Bucksburn, Aberdeen, AB21 9YA, UK. [2]Department of Agriculture, University of Aberdeen, 581 King Street, Aberdeen, AB24 5UA, UK.

Floor space requirements of pigs on deep straw bedding are poorly understood. The possibility of an interaction between group size and space requirement has not been investigated under large group conditions. Growing pigs, allocated to one of two space allowances representative of commercial practice (low; 50 kg/m^2 or high; 32 kg/m^2), were housed on deep straw in groups of 20 or 80 in a 2x2 factorial design with 4 replicates. Low space allowance did not affect average daily gain (ADG), but tended to reduce feed intake while improving feed efficiency ($p<0.1$). Groups of 80 pigs had depressed ADG ($p<0.01$), but no greater variation in ADG. Aggression, measured by number of skin lesions, was worse in pens with low space allowance over the six weeks on trial ($p<0.05$), but small groups experienced the most aggression immediately after mixing ($p<0.1$). Response to antigen challenge (Newcastle disease virus) suggested poorer humoral immunity in pigs at the smaller space allowance ($p<0.01$). No interactions were apparent between space or group size main effects, but the poorer ADG of large groups and greater aggression and poorer immune response of pigs at the lower space allowance suggest that both conditions were sufficient, in isolation, to compromise welfare and profitability.

ANIMAL MANAGEMENT AND HEALTH [M] Poster M1.9

Schallanalyse in Ferkelerzeugerbetrieben
V. Marquardt *[1], D. Schäffer [1], G. Marx [2] und H. Prange [1] [1] Institut für Tierzucht und Tierhaltung mit Tierklinik, Martin-Luther-Universität Halle-Wittenberg, Emil-Abderhalden-Str. 28, 06108 Halle/ Saale, Deutschland, [2] Institut für Tierzucht und Tierverhalten Mariensee, Bundesforschungsanstalt für Landwirtschaft (FAL), Höltystr. 10, 31535 Neustadt, Deutschland

Die Einbeziehung der Messung des Schalldruckpegels und die Analyse der Schall-ereignisse in der tierhygienischen Bewertung von Schweinehaltungssystemen ist eine neue Herangehensweise zur Beurteilung von Haltungssystemen. Hierbei hat nicht nur das technisch bedingte Grundgeräusch, welches in erster Linie von der Lüftungstechnik hervorgerufen wird, sondern auch die Schallpegel tierischen Ursprungs einen ent-scheidenden Einfluß auf die Schallsituation im Stall. Zur Differenzierung dieser Schall-ereignisse wird ihr Frequenzspektrum analysiert. In Abferkelställen zeigt das lüftungs-bedingte Grundgeräusch ein typisches Frequenzspektrum, wobei die Maximalpegel in den Terzbändern von 40 Hz bis 1.000 Hz liegen. In den mit Tieren belegten Ställen ist eine generelle Erhöhung des äquivalenten Dauerschallpegels (L_{eq}) zu beobachten. Die Maximalschallpegel (L_{max}) treten insbesondere im Frequenzbereich von 1.000 Hz bis 10.000 Hz auf. Diese L_{max} sind auf Lautäußerungen der Ferkel, die hauptsächlich während des Saugaktes auftreten, zurückzuführen. Unterschiede zwischen verschiedenen Haltungs-systemen sind vor allem im Grundschallpegel während der Ruhephasen der Tiere und im leeren Stall zu beobachten.

Poster M1.10

The prises of the Polish Polar blue foxes skins on international auction
S. Socha *[1], D. Pomykała[1], J.Sławon [2] and G. Jeżewska[3]. [1]Department of Animal Breeding, Agriculture Faculty, Agricultural and Pedagogic University, 08-110 Siedlce, ul. B. Prusa 12. Poland, [2] National Research Institute of Animal Production, 05-510 Konstancin -Jeziorna, ul. Kolobrzeska 54 . Poland, [3]Department of Biological Basis of Animal Production University of Agriculturale, 20-950 Lublin, ul. Akademicka 13, Poland.

The aim of the work was to analyse and determine of the cause of variation in the prices of polish skins of blue foxes reached on Helsinki auctions during 1991-1995 years. Skin size and quality, colour type and intensity were taken into consideration.
 Higly significant effects of auction and skin size, quality and colour was found on the price of the skins. The interactions of auction effects with skin size, fur quality and colour type were highly significant. The skin prices were more related to their size than to the quality Standard deviation and coefficient of variability were differentiated in both skin size and colour type, which shows a considerable variability of the skins within these traits. It was observed, that the skin prices increase by 15% for each sucsessive size of skin. Lower prices reached the darker coloured skins. The reasons of lower price of skins from Poland than those from Finland were: smaller size, lower quality and improper preparation of the skins for auction.

ANIMAL MANAGEMENT AND HEALTH [M]   Paper MC2.1

Prevention of disease transmission by semen in cattle
G.H. Wentink*, J.C. Bosch, J.E.D. Vandehoek, Th. van den Berg, Holland Genetics, P.O.Box 5073, 6802 EB, The Netherlands

The main goal for artificial insemination in cattle next to achieve genetic improvement is to avoid transmission of infectious diseases. Semen used for AI therefore must be free of infectious agents. For safety testing of semen two approaches can be applied: checking the end product, or continuous surveillance of the bulls before and after semen production.The first method is investigation of semen for the presence of infectious agents. This method completely depends on one single investigation and therefore relies on the sensitivity of the test method only. The second method is testing the bulls for diseases before and after semen collection, based on sequential investigations of bulls for the absence of either antibodies against infectious agents or pathogens.

The EU-directive 88/407 prescribes that the bulls in AI stations must be monitored for the absence of diseases, but with the severe disadvantage that an interval of 12 months between successive investigations is allowed.. Furthermore, the directive is neither specific in the tests to be carried out nor in the specification of some pathogens (e.g. Campylobacter fetus).

In the presentation a programm will be presented based on monthly testing of a limited number of bulls for the absence of diseases, on the basis of Hazard Analysis of Critical Control Points (HACCP). This method only applies to diseases with high transmission rates. These highly contagious diseases (e.g. IBR) can be monitored by testing some 20% of the animals on a monthly basis. However, diseases with slow transmission rates in bulls (e.g. Campylobacter fetus) do not fit in such a system and can only be monitored on an individual basis.

Paper MC2.2

Prevention of diseases transmission by the use of semen in the porcine AI industry
C. Leiding, Besamungsverein Neustadt/Aisch, P.O.Box 1220, D-91402 Neustadt/Aisch, Germany

For more than 40 years the AI in pigs is used in piglet producing farms. Only in the last 20 years it has become the dominating method in comparison to natural service. Boar ejaculates can be extended in average 20 times or more. One boar can produce more than 2000 dosis of semen annually. AI studs with sometimes more than 300 boars are therefore required to fulfill all regulations to prevent the spreading of diseases. OIE regulations, an EU-directive are enforced to have a common standard in semen production all over Europe. In addition, certain disease prevention programmes exist, e.g. against PRRS or Parvo virosis.

The methods to prevent spreading of diseases by AI centres are documented and valued. Both the animal hygiene and the personal hygiene of people working with the boars are required. To a high extent the success of a prevention programme is due to the regional situation of the diseases. Only if the semen producing organisations value the disease prevention methods highest in their work, infections of boars and semen dosis can be prevented.

ANIMAL MANAGEMENT AND HEALTH [M]   Paper MC2.3

Prevention of disease transmission through the transfer in-vivo-derived bovine embryos
D. A. Stringfellow* and M. D. Givens. *Department of Pathobiology, College of Veterinary Medicine, Auburn University, Alabama 36849-5519 U. S. A.*

Investigation and experience have demonstrated that movement of in-vivo-derived bovine embryos can be accomplished while effectively limiting spread of infectious diseases between populations of cattle. Experimental and theoretical justifications of current strategies for production of specific-pathogen-free, in-vivo-derived embryos are reviewed. Hazards of spreading bovine viral diarrhea virus via in-vivo-derived embryos are dealt with specifically. It is concluded that established sanitary procedures for producing pathogen-free, in-vivo-derived embryos are efficacious if the ethical and technical excellence of those performing the procedures can be assured.

Paper MC2.4

Embryo transfer in small ruminants: the method of choice for health control in germplasm exchanges
M. Thibier*[1] and B. Guérin[2]. *[1]Centre National d'Etudes Vétérinaires et Alimentaires, BP 19, 94701 Maisons Alfort, France, [2]Union Nationale des Coopératives d'Elevage et d'Insémination Artificielle, BP 65, 94703 Maisons Alfort, France.*

Several thousands of in vivo derived sheep and goat embryos, as well as significant numbers from deer, are transferred annually within countries and internationally. The technologies used for superovulation, collection and transfer of such embryos are now well developed, and pregnancy rates of over 60% can be achieved.
Such transfers offer a unique opportunity to safeguard the health status of the flocks and herds even when the embryos originate from countries with a radically different health status. Viral diseases which have been investigated with regard to their risks of transfer by in vivo derived small ruminant embryos include blue tongue, border disease, pulmonary adenomatosis, maedi/visna and caprine arthritis-encephalitis. Bacterial diseases investigated include brucellosis, campylobacteriosis, mycoplasmosis and chlamydial abortion. Scrapie, a prion disease, has also been studied and will be discussed in another paper. Provided that the sanitary procedures recommended in the IETS Manual (1998) and here presented, are strictly followed, the risks of transmitting diseases are minimal.

ANIMAL MANAGEMENT AND HEALTH [M]   Paper MC2.5

The effective control of sanitary risks associated with the production of *in vitro* produced bovine embryos.
B. Guérin*[1], Brigitte Le Guienne*[1], M. Thibier*[2]. [1] *Union nationale des Coopératives d'Elevage et d'Insémination Artificielle, BP 65, 94703 Maisons-Alfort, France,* [2] *Centre National d'Etudes Vétérinaires et Alimentaires, BP 19, 94701 Maisons-Alfort, France.*

The commercial transfer of in *vitro* produced bovine embryos is now widely practiced in some areas of the world (Thibier, 1998). The majority of such transfers are domestic and are associated with integrated breeding programs. The international movement of such embryos has been limited due to cryo-preservation related technical difficulties. However, recent advancements in this area provide the potential for a dramatic increase in such movements. A reliable control of the sanitary risks associated with the production of such embryos is therefore extremely critical. There are specific concerns associated with *in vitro* produced embryos which provide the justification for certification standards and production controls which differ from those established for *in vivo* produced embryos. The risk factors involved will be reviewed in the full paper. Particular references to the following sources of disease causing agents will be made: pathogens found in the bull semen (*Brucella sp., Haemophilus somnus, Campylobacter fetus, Leptospira sp etc*), pathogens associated with the oocytes (*Campylobacter fetus*) or more frequently adhering to the zona pellucida (BHV-1, BVDV, *Leptospira hardjo*). New recommendations from the OIE for handling such embryos are under discussion and when finalized should be strictly adhered to in order to effectively manage the risk of disease dissemination through the international movement of in *vitro* produced embryos.

Paper MC2.6

Risks of transmission of spongiform encephalopathies by reproductive technologies in domesticated ruminants
A.E. Wrathall*, *Veterinary Laboratories Agency (Weybridge), New Haw, Addlestone, Surrey KT15 3NB, United Kingdom*

This paper considers whether transmissible spongiform encephalopathies (TSEs or prion diseases) could be spread by artificial insemination, embryo transfer and some other more advanced reproductive technologies which are used for genetic improvement and also for production of recombinant drugs for medical use. Although the technologies are most used in cattle, they are increasingly used in sheep, goats and deer as well, all of which can be naturally affected by TSEs. In general, provided appropriate precautions are taken, the risks of TSE carriage specifically by the gametes (spermatozoa and oocytes) or by in vivo derived embryos *per se* appear to be negligible, but further research, some of which is already in progress, will be helpful to give assurance on this point. Greater concerns relate to the many biological products that are used in the technologies, e.g. pituitary hormones used for the superovulation of donors, and various tissues and blood products used in semen and embryo culture/transport media, some of which have the potential to carry TSE infectivity if derived from infected animals. The myriad instruments and items of technical equipment that are used also give cause for concern because if they become contaminated with TSEs they may, due to their construction, be impossible to sterilise properly.

ANIMAL MANAGEMENT AND HEALTH [M]   Paper MC2.7

Application of survival analysis to identify management factors related to the rate of BHV1 seroconversions at Dutch dairy farms
G. van Schaik[1]*, Y.H. Schukken*[2]*, M. Nielen*[1]*, A.A. Dijkhuizen*[1]*, R.B.M. Huirne*[1]*.* [1]*Department of Economics and Management, Wageningen University and Research centre, Hollandseweg 1, 6706 KN Wageningen, The Netherlands,* [2] *College of Veterinary Medicine, Cornell University, Ithaca, USA.*

The prevalence of BHV1 at dairy farms is dependent on several factors. First, the prevalence is influenced by introduction of BHV1 at the farms, which is dependent on the risk factors for introduction. Second, the BHV1 prevalence might also be influenced by reactivation of BHV1 within the farm, which might be affected by the management of the farmer. In this study the relations between risk factors, management factors and the estimated time since latest BHV1 outbreak were investigated by means of Cox regression analysis. The results showed that direct animal contacts (i.e. purchase of cattle and returning export cattle) and occasional visitors increased the rate of BHV1 outbreaks on dairy farms. Management factors related to reactivation of BHV1 at dairy farms were all related to a loose housing system which incurred an increased risk of reactivation of BHV1 at the farm. The reactivation was facilitated when the barn was overcrowded (i.e. more cows than cubicles in the barn).

Poster MC2.8

Results of a MAS study in dairy cattle with respect to longevity
Freyer, L. Panicke: Research Institute for the Biology of Farm Animals D-18196 Dummerstorf, Germany,

The increasing possibilities of biotechnology which can make use of marker assisted selection raise the question of the methods efficiency with respect to the breeding aim and possible influences on other traits. A marker assisted selection is used to improve important traits in the next generation f.e. increasing milk protein yield. But it has to be ensured that all negative influences are excluded which arise by direct increase of homozygosity at several loci. Especially traits describing health, fertility and the longevity are of importance. Sustainability in every sense has to be considered. Selection to increase milk yield strongly affects longevity besides of health and functional traits.

The study presented is based on simulation on real data involving information of nearly 3000 Black and White cows genotyped for several loci. The period of birth years ranges from 1984 to 1993. The milk protein loci κ-Casein, β-Casein, α_{S1}-Casein and β-Lactoglobulin were used as markers for yield and fat content in different scenarios. First results show that during the investigated period no negative effects caused by selection using the above described markers occurred in the population. The highest longevity was attained by animals being heterozygous for at least two of the markers.

ANIMAL MANAGEMENT AND HEALTH [M]   Poster MC2.9

The evaluation of mastitis pathogenic agents and their possible influence on consumers health
A.Jemeljanovs*, J.Bluzmanis, V.Mozgis, V.Jonins, A.Reine. Research Centre "Sigra", Latvia University of Agriculture, 1 Instituta Street, Sigulda, LV-2150, Latvia

The bacteriological tests of udder secretion taken up from 439 cows diseased with mastitis showed, that in 55.17% cases mastitis was caused by *Staphylococcus aureus*, in 22.99% by *Steptococcus agalactiae*, in 17.24% cases by *Coccus* and in 4,60% by *Escherichia coli*, whereas before thirty years the main pathogenic agents were *Streptococcus agalactiae* 44.74%, *Staphylococcus aureus* 7.32%, *Coccus* in 47.94% cases. From all isolated Staphylococcus aureus cultivations 10.76% were pathogenic and all those were isolated from cows diseased with acute mastitis. The ability of *Staphylococcus aureus* cultivation to produce α and β toxins was observed in 53.74% cases. During the determination of the sensitivity of pathogenic microorganisms against 34 different medicines, including 22 which contain antibiotics, it was ascertained that pathogenous were more sensitive against *Brulamycinum, Ampiocum, Lincomycinum hydrohloridum* and *Niphthiolum* because due to their use recovered 94.6% animals. To prevent cows from the infection special teats disinfection means are elaborated, giving effect up to 90% of cases. Because mastitis is caused by *Staphylococcus aureus* the elaboration of vaccine against pathogenic agent to prevent cows' disease with mastitis will be completed soon.

Poster MC2.10

Microbiological and mycological investigations on reproductive organs of A.I. bulls and cows
A.Jemeljanovs. Research Centre "Sigra", Latvia University of Agriculture, 1 Instituta Street, Sigulda, LV-2150, Latvia

The microbiological and mycological investigations of reproductive organs of A.I.bulls show pollution by *Bac.proteus vulgaris, Ps.aeruginosa, Staph.aureus, Strept.pyogenes, E.coli, Bac.myxoides, Bac.urobacter etc.* We have isolated from semen 420 different kinds of mycological agents including pathogenic - *Actinomicetes, Mucor, Aspergillus, Fusarium, Penicillium, Alternaria* and others. Infection was not ascertained in testes and epididymis of 6 slaughtered tested bull sires. Mentioned before microorganisms were ascertained in additional sexual glands as well as in full length of *urogenital canal* and *praeputium*. Semen infected by *Ps.aeruginosa* was used for the insemination of 9 heifers which had sterile rinses from sexual organs, high levels of blood biochemical indices without negative deviations from clinical indices. Four days after insemination these animals were slaughtered and bacteriological analysis of sexual organs was carried out. The analysis showed that for all heifers in *corpus* and *cervix uteri* and in the *vagina* were ascertained only few *Streptococcus* were ascertained, but for one which during this time had organism resistance decrease, *Ps.aeruginosa* was isolated from cornu, *corpus et cervix uteri* and *vagina*. By applying different antimicrobial measures and using effective semen sanitation means a complex semen antimicrobial system was elaborated.

ANIMAL MANAGEMENT AND HEALTH [M] Poster MC2.11

Comparison of two staining methods used in morphometry of bull spermatozoa heads by computer-assisted image analysis.
J. Yániz[*1], F. López-Gatius[1] and P. Santolaria[2]. [1]Departament de Producció Animal, Universitat de Lleida, Rovira Roure 177, 25198 Lleida, Spain, [2]E.U. Politécnica de Huesca, Universidad de Zaragoza, 22071 Huesca, Spain.*

The present study was carried out to compare the relative merits of toluidine blue and haematoxylin as staining method procedures to use in the computer automated morphometric study of bull spermatozoa. Sperm samples from a single ejaculate were air dried, fixed for 40 min in glutaraldehide 2.5 % and stained with toluidine blue or with haematoxylin for 30 min. Videoimages were captured, and sperm heads where detected via image segmentation and particle analysis. At least 200 properly digitised sperm heads were analysed for each group by computer-assisted sperm morphometry analysis (ASMA) technology. The mean measurements of length, width, area and perimeter were recorded and group means compared by a two-sample t-test. A noticeable increase in contrast and a decrease in artefacts is observed when spermatozoa were stained with toluidine blue. The sperm head parameters measured were significantly larger ($p < 0.001$) when toluidine blue stain was used. Differences are associated with the frequency of digitalization errors that resulted in misidentification of spermatozoa.

Poster MC2.12

Relationship between peripartum climatic conditions and retained placenta in dairy cows
J. Yániz, F. López-Gatius. Departament de Producció Animal, Universitat de Lleida, Rovira Roure 177, 25198 Lleida, Spain

The objective of this study was to evaluate the influence of temperature and relative humidity on the incidence of retained placenta in dairy cows. Data from 1487 calving of Holstein-Friesian cows at a commercial dairy herd in the north-eastern of Spain were used. The cows were kept in open stalls and were maintained on a weekly reproductive health program. The risk of retained placenta increase in summer, only when the minimum relative humidity exceeds 40% ($P<0.05$). When mean relative humidity the day before calving was analysed, a tendency to increase the incidence of retained placenta was observed in spring ($P=0.07$) and summer ($P=0.06$) with values over 70% and 65%, respectively. Any direct influence of maximum, minimum and mean daily temperature of the day before calving on the incidence of retained foetal membranes were found. In summary, our results suggest that high values of humidity increase the effect of heat stress on the retained placenta in dairy cows.

ANIMAL MANAGEMENT AND HEALTH [M] Poster MC2.13

Ewe fertility following cervical AI of fresh or frozen semen
A. Donovan[1], J.P. Hanrahan[1], P. Duffy[2] and M.P. Boland[2]. [1]Teagasc Research Centre, Athenry, Co. Galway, [2]Faculty of Agriculture, UCD, Ireland*

Conception rates of ~60% have been reported for Norwegian ewes following cervical insemination with frozen semen. Objectives were to compare frozen semen from Norwegian (NOR) and Irish (IRL) rams and evaluate the effects of ewe breed, synchronisation and operator. Exp.1: Parous ewes (various breeds, n=297), were inseminated to a natural (N) or synchronised (S) oestrus with fresh or frozen (IRL or NOR) semen. Ewes were assigned, within breed, to: i) N fresh: n=28, ii) S fresh: n=30, iii) N-IRL: n=62, iv) S-IRL: n=50, v) N-NOR: n=68, vi) S-NOR: n=59. Two operators performed inseminations. Pregnancy rate was not significantly affected by synchronisation or by source of frozen semen but was influenced by operator and ewe breed ($P<0.05$). Fresh semen yielded pregnancy rates of 0.82 and 0.70 (treatments i and ii) compared with 0.40, 0.52, 0.34 and 0.37 (treatments iii to vi) for frozen semen ($P<0.001$). Exp.2 was designed to test ewe breed effects. Purebred (Suffolk (n=84), Texel (n=110), Finn (n=43), Belclare (n=25)) and crossbred (Scottish Blackface, cross, n=45; Suffolk cross, n=56) ewes were synchronised and inseminated with fresh (n=29) or frozen (IRL or NOR; n=334) semen. Pregnancy rates, using frozen semen, were 0.17, 0.30, 0.77, 0.44, 0.43 and 0.19 for above breeds, respectively ($P<0.001$), compared with 0.76 for fresh semen ($P<0.001$).

Poster MC2.14

Time scheduled insemination of lactating dairy cows
Gy. Gábor, Research Institute for Animal Breeding and Nutrition, Herceghalom, HU*

Lactating dairy cows have poor reproductive efficiency because of low fertility and low rates of estrus detection. The aim of this study was to examine, whether synchronization of ovulation without estrus detection could be an effective method for managing of the reproductive problems in dairy farms. Holstein-Friesian cows (n=231) were examined routinely by rectal palpation and/or ultrasound scanner, if estrus was not detected between 40-60 d postpartum. Several reproductive disorders were found (metritis, different ovarian cysts, etc.). Those cows, which were apparently found healthy (n=109) received a treatment following the synchronization protocol of Pursley et al. (1997). The treated cows received an initial injection of 100 µg GnRH (Fertagyl, Intervet) 10 d prior to scheduled AI. Seven days later they received 35 mg $PGF_{2\alpha}$ (Dinolytic, Upjohn). Forty-eight hours after the $PGF_{2\alpha}$ cows were treated again by 100 µg GnRH. AI was performed 20-24 h after the second injection of GnRH. Twenty-one cows (19.2%) after the first GnRH injection, and 13 cows (11.9%) after the $PGF_{2\alpha}$ treatment showed estrus symptoms and were inseminated. The pregnancy rates (checked on days 30th and 60th) were 42.8 and 53.8% respectively. Seventy-five cows received the original protocol, and 32 were found pregnant (44%). On the basis of these findings we concluded that synchronization of ovulation without estrus detection is an effective method for managing of reproductive problems in a dairy farm.

ANIMAL MANAGEMENT AND HEALTH [M] Poster MC2.15

Factors influencing conception of cows
L. Stádník, F. Louda, R. Toušová. Department of Cattle Breeding and Dairying, Czech University of Agriculture Prague, 165 21 Prague 6, Czech Republic

In a herd of 400 head of dams of Holstein-Friesian breed with the performance on the level of 8 741 kg of milk, 376 kg of fat and 287 kg of protein, observation of factors effecting conception of cows was performed. Among these important factors the effect of milk production by the dam in the day of insemination, influence of preparedness of sexual organs of the dam for insemination, effect of season of the year, effect of body condition of the dam, effect of environmental teperature at the day of insemination and effect of body temperature of the dam at insemination were included. Quantity of the daily milk output in inseminated dams was observed. Preparedness of sexual organs to insemination was checked by technician and was classified in the range 1 - 3 points. Body conditions varied in the range 1.5 - 5 and mean body condition was 2.9. Environmental temperature varied in the range from - 15°C in the winter season up to 28°C in summer season. Body temperature of dams measured the day of insemination varied in the range 37.6°C up to 40.1°C. Results of conception were confirmed by the help of sonographic examination at 30th to 45th day after insemination.

Poster MC2.16

Cryopreservation in long-term selected lines of laboratory mice
Langhammer, U. Renne, H. Alm. Research Institute for the Biology of Farm Animals, Wilhelm-Stahl-Allee 2, D-18196 Dummerstorf, Germany

Mice of the Fzt:DU strain were long-term selected in genetic different lines for high fertility (DU-FL1), high body weight (DU-6, DU-6P) and fitness (DU-6+TP), high locomotory activity (DU-hof) and endurance fitness (DU-hTP, DU-lTP). The genetic status in the lines were conserved by cryopreservation of 4102 embryos after superovulation in about 1100 females. The embryos were gathered 2 and 4 days p.c., respectively.
In dependence on selection line the results of embryo production, embryo quality and survival rate after freezing and thawing differed. Fertilisation rate ranged from 55 % in Fzt:DU to 20-30 % in growth lines. Fertilised females in the growth lines (DU-6, DU-6P, DU-6+TP) produced smaller numbers of embryos in the blastocyst stage versus fitness and fertility lines.
Embryo quality varied between the lines but not in dependence on the selection direction. After thawing no line specific differences in the survival rate were noticed. But if the thawed embryos were cultivated in vitro distinct effects in survival rate and developmental competence were shown between the different lines.
In conclusion the specific long-term selected mouse lines responded in a different manner to superovulation treatment and cryopreservation procedure. If the genetic status of those long-term selected lines should be conserved with a high genetic diversity, the varying results in embryo production and survival rate has to be taken into account.

ANIMAL MANAGEMENT AND HEALTH [M]   Poster MC2.17

Comparison of viability and acrosome status of boar spermatozoa frozen in mini-or maxi - straws
A. Bali Papp[1], SZ. Nagy[1], J. Ivancsics[1], A. Kovacs[2], T. Pecsi[3] and J. Dohy[1]. [1] Institute of Animal Breeding, PANNON Agricultural University, Var 2., H-9200 Mosonmagyarovar, Hungary, [2] Institute for Animal Breeding and Nutrition, Gesztenyes ut 1., H-2053 Herceghalom, Hungary, [3] Hungarian AI Corp., Mezogazdasz u. 2., H-4014, Debrecen-Pallag, Hungary.

The aim of present study was to compare the difference of membrane integrity and acrosome status of boar spermatozoa frozen in mini- (0.25 ml) and maxi (5 ml) straws.

Boar semen was collected once from six fertile Hungarian White boars. Ejaculates with >70 % motility were packed in mini- and maxi straws and frozen by the Beltsville method. Smears from thawed semen were stained by the Kovacs-Foote (trypan blue-Giemsa staining) method and 200 spermatozoa were examined by light microscopy (400x magnification) in each sample for their viability and membrane -acrosome status.

The percentage of membrane- ("live") and acrosome-intact spermatozoa were significantly higher in the smears made from the semen packed in maxi straws: 58.75 % and 27.75 % in maxi- and mini straws, respectively ($P<0.005$).

Our results showed that the trypan blue-Giemsa staining method was a reliable tecnique to evaluate the membrane integrity and acrosome status of spermatozoa in processed boar semen.

Poster MC2.18

Laparoscopic insemination in sheep
G. Catillo[1], G.M.Terzano*[1], L. Taibi[2], G. Ficco[1], G. Noia[3]. [1]Istituto Sperimentale per la Zootecnia, Roma, Italy. [2]Istituto Sperimentale per la Zootecnia, Foggia, Italy. [3]Istituto di Clinica Ostetrica e Ginecologica dell'Università Cattolica del S. Cuore - Policlinico "A. Gemelli", Roma, Italy.

The aim of the present study was to evaluate the effect of laparoscopic insemination (LAP) and natural mating (NM) on fertility rate in Gentile di Puglia ewes during the month of June. For the experiment, 97 ewes were used. The LAP was performed with frozen semen of 3 different Romanov rams: LAP_1 (n=24), LAP_2 (n=26), LAP_3 (n=28) and NM was performed with 2 different Gentile di Puglia rams with proved fertility: NM_1 (n=10), NM_2 (n=9). Estrus was synchronized with fluorogestone acetate (FGA) impregnated intravaginal sponges (45 mg,14 d). The PMSG (Folligon, Intervet International) at a dose of 400 UI was given intramuscularly at sponge removal. AI was carried out at 60 h after the removal of the progestagen sponges in LAP groups. The mean pregnancy rate at ecographic diagnosis performed at about 36 days from sponge removal for LAP and NM groups were 62.8% and 78.9% with no significant difference ($P \leq 0.18$). The mean fertility rates for LAP and NM groups were 59.0 and 89.5 respectively with significant difference ($P \leq 0.05$).

ANIMAL MANAGEMENT AND HEALTH [M] Poster MC2.19

Bovine oocytes recovering by the method of slicking and scaryfication of ovarian cortex
R. Toborowicz[1], J. Żychlińska[1], J. Szarek[1], J. Buleca[2] . [2]Department of Cattle Breeding, Agricultural University of Cracow, Al Mickiewicza 24/28, 30-059 Kraków Poland [2]University of Veterinary Medicine, Kosice, Komenskeho 73, Slovakia

Although the bovine ovary contains thousands of immature egg cells, relatively few can be recovered by current procedures. Over the years, various workers have attempted to recover large numbers of oocytes from ovaries. The aim of our work was to estimate the possibilities of recovering oocytes by slicking the ovary's cortical layer and attempt at their in vitro maturation and fertilization. A total of 104 ovaries obtained at slaughter from heifers at a local abattoir were used in the experiment. Oocytes were recovered by cutting the ovarian cortex and its scarification. Retrieval and morphological estimation of oocytes was performed according to a generally accepted methodology. The conducted experiments resulted in recovering 2445 oocytes out of 104 ovaries, averaging 23 oocytes from one ovary. Over 53% of oocytes obtained by this method were morphologically normal and suitable for culture. Morphological evaluation of preovulatory oocytes obtained showed that the highest percent of oocytes was classified as class II oocytes 23% (569), further 17,4 % (426) oocytes represented class III and 12,4% (297) class I. There is an association between the oocyte class and the in vitro fertilization rate, which is connected with the number of the cumulus cells. The rate of fertilization obtained in the present study was 22%.

Poster MC2.20

In vitro embryos production in cattle
I. Vintilă, A. Grozea, A. Guler, Gh. Ghişe. Banat University of Agricultural Sciences Timişoara, Calea Aradului 119, 1900 Timişoara, Romania.*

The aim of this study was to find the best ways of oocytes retrieval, maturation (IVM), fertilisation (IVF) and embryos development in our laboratory conditions. Oocytes retrieval was performed on 98 ovaries through ovary punction (88 ovaries) and dissection (10 ovaries). Punction retrieval was faster, but only 3.375±0.082 oocytes resulted, in comparison with 8.2±0.554 oocytes obtained through ovary dissection. TCM 199 medium supplemented with 10% (V_1) and 20% (V_2) FCS (fetal calf serum) or 10% (V_3) and 20% (V_4) BES (bovine estrus serum) was used for IVM of 367 oocytes. The nuclear and cytoplasmatic maturation were estimated after IVF and IVC (in vitro cultivation). After 8 days of IVC with cumulus cells monolayer, in the same maturation medium, the obtained morulae (M) and blastocyst (B) rates were: V_1-14.46% M; V_2-13.82% M, 3.19% B; V_3-10.77% M; V_4-12.5% M. IVF was performed in TALP (Tyrod, Albumine, Lactat, Pyruvat) medium and the sperm was processed in two ways: swim-up and Percoll gradient. Using swim-up and Percoll gradient processing, and IVF and 8-days IVC, resulted 4.35% M and 10.71% M, respectively. Co-culture for 8 days of 33 zygots with cumulus cells monolayer in TCM 199 supplemented with 20% FCS resulted in 4 M (12.12%) and 3 B (9.09%), meaning 7 (21.21%) transferable embryos.

ANIMAL MANAGEMENT AND HEALTH [M] Poster MC2.21

Effect of selenium supplementation prepartum on reproductive function of dairy cows at pasture
A.S. Erokhin, All-Russian Research Institute of Animal Breeeding, Box Lesnye Polyany, Pushkin Disrtict, Moscow, 141212, Russia.

The goal of this study was to investigate the effect of new organic selenium compound DAFS-25 (diacetophenonilselenide) supplementation on the postpartum reproductive dysfunction and glutathione peroxidase activity in erytrocytes of Kholmogore cows. During 8 wk prepartum 1 and 2 group cows each were given daily per os 6 or 9 mg DAFS-25. Control animals received no suplementation. Concentration of Se in whole blood before experiments in cows 1 and 2 gr. were 0,13 and 0,12 mg/l vs. 0,12 mg/l in control cows, indicating adequate Se content. Treatment with DAFS-25 increased by 2 d postpartum concentration of Se in blood cows 1 and 2 group: 0,19 and 0,22 mg/l respectively, vs. 0,13 mg/l in control, and increased activity of Se-depended glutathion peroxidase in erytrocytes: 5,7 and 6,5 U/g Hb respectively, vs. 4,2 U/g Hb in control ($P<0,01$). Supplementation DAFS-25 reduced the rise of metritis in 1 and 2 group: 19 and 12% respectively, vs. 26% in control ($P<0,05$), but the incidence of retained placenta was inaffected by supplementation: 13 and 12% vs. 11% in control. No statistical significance occured in the reproductive interval and conception rate from difference groups. This results indicate that supplementation with DAFS-25 may increase antioxidant status and decrease insidence of metritis in dairy cows grazed on pasture.

Poster MC2.22

MHC and milk production
Kostomakhin N.M. Omskgosplem, 30 let VLKSM, 45, Omsk, 644070, Russia.

Many scientists are looking for genetics markers of commercial traits in dairy cattle now. MHC is very important because its genes are related with some animal diseases and their milk production. In our experiment we used lymphocyte antigens (BoLA)- A locus, wich are class I locus of the bovine lymphocyte antigens and economically important traits was examined.
The results show that only 11 antigens were present in Holstein cows. Antigenes W19 and W11 were closely and positively associated with milk production and fat yield and protein yield. Antigen W8.2 had negative correlation with these traits. Antigens W6.1 and W10 were closely and positively related with fat and protein percentage.
These results suggest that genes of the bovine MHC are associated with economically important traits, but more research is necessary to confirm these findings and to determine the biological mechanisms underlying these associations.

ANIMAL MANAGEMENT AND HEALTH [M] Poster MC2.23

The immunologic factors in pig production
E.G. Akulich, Siberian Res. Designing & Techn. Institute for Animal Hasbanry 633128 P.O. Krasnoobsk, Novosibirsk Russia.

The immunotechnique for forecasting the reproductive traits in the sires, as well as the combinability of the sires with the sows on the basis of the identity of the sperm and blood antigens, was studied.
A number of antigens identical to those of the red blood cells were found on the spermatozoa and the white blood cells/Ac; Eb, d, c; Fa; Bb; Kb/. The natural autosperm antibodies were registered in 66.6 % examined sires. The autosperm antibodies significantly reduced the sperm quality in the sires. The ejaculate volume and ejaculate sperm concentration decreased by 14.8 ml and by 10.6 billion, respectively. The immune reactivity of the sows to the blood and sperm antigens of the sires affected their fertility. The sows negatively responsive to the sperm and positively responsive to the blood of the sires used for impregnation revealed significantly low litter size/1.6 per litter/ and the greater number of 'failure' farrows /by 18.8 %/. The difference between these data and the average data in the herd was $P<0.01$.

Poster MC2.24

Research of incident factors and minimization of embryonic mortality of animal at early stages embryogenesis
E.P. Karmanova, I.A.Hakana, M.E. Huobonen, A.E.Bolgov. Department of Animal Breeding, University of Petrozavodsk, Karelia, 185640 Russia

The researches carried out within 4 years on the cows (1300) Finnish Ayrshire of breed. Studied variability of a parameter of early embryonic mortality (EEM) at the cows under the influence of genetic and paratypical factors. The dependence of frequency EEM from a level of security is revealed the cows by forages. Defective nutrition in the greater degree influences on young animals. Decrease of a feeding level causes in the first-cow growth EEM with 10,6 up to 20,7%. In adverse conditions nutrition the influence of the productivity factor and the growth EEM was observed at achievement yield 4000 kgs. At sufficient power security increase of cases embryonic mortality it is marked at the cows with yield 6000 kgs. or more. The frequency EEM depend on a calving season. In the autumn-winter period the frequency EEM in 2 times is higher, than in a spring-year's season. Most is sharp seasonal features are shown at the cows with indifference-period less than 40 days the after calving (EEM-29,4-35,4 %). The typical parameters EEM are observed at duration indifference-period 2,5-3,5 months. Some influence of a genotype of the bulls is established on the daughters frequency EEM. The frequency EEM at the daughters of the different sires varied from 5,5 up to 44 % ($h^2 = 0,137$). Interrelation of attributes " yield-EEM " at the cows in the certain measure is supervised by a genotype of the bulls.

ANIMAL MANAGEMENT AND HEALTH [M] Poster MC2.25

The programmable freezer for use on farm
A.M. Malinovskiy, T.A. Moroz. All-Russian Institute of Animal Breeding, Box Lesnye Polyany, Pushkino district, Moscow region, 141212, Russia.

The optimal regime of cryoconservation of bovine embryos was determined. Bovine embryos were cryoconservated by the programmable freezing device EMBI-K (Russia) with six regimes. The following regime among all checked was best: the temperature of seeding -7°C, the rate of cool -0,5°C/min, the temperature of plunging into liquid nitrogen -35°C. The quality of thawed embryos in optimal regime by device EMBI-K (gr.1) was compared with the thawed embryos after freezing by device Minicool A-25 (gr.2). The quality characteristics of thawed embryos were decreased by 5% (gr.1) versus by 20% (gr.2). The survival of embryos after transfer (gr.1) was 51% versus 49% embryos (gr.2). Although these results were same we concluded that the freezer EMBI-K had some advantageous characteristics compared the freezer Minicool A-25: high economical (the consumption of liquid nitrogen - 150-200 g per one cycle of freezing), little dimensions of freezing chamber and processor; high precision and by exploitation on farm. We recommended the EMBI-K for use on farm.

Paper MH4.1

The ISO standards for animal Radio Frequency Identification, its present status and future extensions.
M.B. Jansen[1]*.* [1]*Institute of Agricultural and Environmental Engineering (IMAG-DLO), P.O. box 43. 6700 AA, Wageningen, The Netherlands.*

Animal Radio Frequency Identification (RFID) is being introduced in many countries all over the world now. Some years ago, the ISO standards 11784 and 11785, which describe the bit structure and the air interface of animal RFID transponders (tags), were accepted. This means that a world wide standardisation of animal RFID systems will be established. After successfully passing an ICAR (International Committee for Animal Recording) acceptance test, seventeen RFID transponder manufacturers have obtained a manufacturer's number from ICAR for as much as 32 different types of transponders. This means that their products for RFID are allowed for use in ICAR member countries. The RFID transponders are available as injects, ear tags and bolus transponders. This paper describes briefly the RFID technology, and gives an overview of the so called advanced transponders for which an extended standard is being defined now. These advanced transponders will contain a number of additional features such as authentication, multi page read/write possibilities and sensors. Furthermore, the outlines of the extended ISO standard for these advanced transponders will be discussed and how the upward and downward compatibility with the existing technology will be guaranteed.

ANIMAL MANAGEMENT AND HEALTH [M] Paper MH4.2

IDEA project (IDentification Electronique des Animaux): evaluation of the feasibility of a community-wide electronic animal identification system
O. Ribó*[2], M. Cropper[1], C. Korn[2], A. Poucet[2], U. Meloni[2], M. Cuypers[2], P. De Winne[1], [1]General Directorate of Agriculture, Rue de la Loi 200, B-1049 Brussels, Belgium, [2]Safeguards and Verification Techniques Unit, ISIS Institute, Joint Research Centre, Ispra, Italy.

The aim of the IDEA Project is collect information on the feasibility of the EU introduction of an animal electronic identification system as referred in Directive 92/102 and Regulation 820/97. The project will run for a 3 years period (1998-2001) with 1 million animals in 6 countries (France, Germany, Italy, Netherlands, Portugal and Spain). The Joint Research Centre (EC) provides the technical support in terms of performance testing and certification of electronic identification devices, establishment of the central experimental database, data transmission and recording and evaluation of the results.

Three species (cattle, sheep and goats) will be identified applying three types (ruminal bolus, eartag and injectable transponders) of electronic tags. The performance of these will be checked periodically until their recovery in the slaughterhouse. The evaluation of the results in terms of performance of the electronic identification devices and the organisational structure needed to perform the electronic identification activities, will validate the system to trace the animals individually from birth to slaughterhouse, and advise on the appropriate technology to be applied in any eventual generalisation of the system in the EU livestock. The utilisation of the electronic identification system could also make farm management more effective and contribute to productivity.

Paper MH4.3

Electronic identification with passive transponders in veal calves
E. Lambooij[1], C.E. van't Klooster[2], A.C. Smits[2] and C. Pieterse[1]. [1]DLO - Institute for Animal Science and Health. P.O. Box 65, 8200 AB Lelystad. The Netherlands, [2]DLO - Institute of Agricultural and Environmental Engineering. P.O. Box 43, 6700 AA Wageningen, The Netherlands.

The readability at different subcutaneous positions of transponders and newly developed ruminal boluses were evaluated in veal calves. In 3 experiments 89, 421 and 199 veal calves, respectively were identified subcutaneously in the nose, the armpit, the ear base or in the rumen by different operators and at ages of 1 or 4 weeks.

The results of our study show that readability decreased during fattening up to 3%, 7%, 10% and 19% when placed in the armpit, the rumen, the nose and the ear base, respectively. Recovery from the armpit and nose took too much time and succeed only in 26 out 86 and 21 out of 80 carcasses, respectively. Recovery of the remaining 30 transponders in the armpit damaged the carcass. When placed in the ear base the readability decreased significantly during fattening and is affected by the operator and age of the animal. A bolus placed in the rumen was lost in 14 out of 221 animals, and was easy to recover in the slaughter line

Transponder positions in the nose and armpit cannot be recommended due to problems with recovery in the slaughter line. The ear base position may not optimal due to losses resulting in a lower readability. The use of a bolus positioned in the rumen may be promising

ANIMAL MANAGEMENT AND HEALTH [M] Paper MH4.4

First results about electronic cattle identification in the german part of the european project IDEA (Identification élèctronique des animaux)
M. Klindtworth[1]; *G. Wendl*[1]; *H. Pirkelmann*[2] *and W. Reimann*[2]. [1] *Bavarian State Institute of Agricultural Engineering, Technical University of Munich, D- 85350 Freising;* [2]*Bavarian Animal Husbandry Centre Grub; D-85586 Grub; Germany*

The introduction of an electronic system for individual cattle identification is intensively discussed in Europe. The technology will be useful at any place, where automatic animal identification is required. Both, administration and farmers will have benefits. It is expected, that the new transponder technology is a cheap replacement of old neck collars which are frequently used in calf rearing and dairy farming. Other aspects are i.e. improvements of quality management and regional health control.

The IDEA-project is carried out in five different european countries under various production conditions. In Germany three different types of electronic tags (electronic eartags; injectable and ruminal (bolus) transponders) will be compared and evaluated.

The announcement of the project lead to an increased development of ISO-compatible readers and tags, which had to be certified at the Joint Research Center of the EC in Ispra (Italy) before any use in the project. The investigations will show if reading range, reading efficiency and finally the handling of tag application and recovery is comparable to earlier field results and if all components have reached a level for official practical use in cattle husbandry.

Paper MH4.5

Comparison of injectable and bolus transponders for the electronic identification of sheep on farm conditions
Caja G.[1], *Nehring R.*[1], *Conill C.*[1] *and Ribó O.*[2]. [1]*Producció Animal, Universitat Autònoma Barcelona, E-08193 Bellaterra.* [2]*Institute Systems Informatics & Safety, Joint Research Centre, I-21020 Ispra.*

A total of 6,789 sheep (18 flocks) were identified during 4 years using injectable (IT; n=5,907) and bolus (BT; n=882) transponders (32mm HDX). Right armpit (A; n=4,854) and ear-base (E; n=1,053) were used for injections. Readability and reading efficiency were recorded twice a year in static (SRE) and dynamic (DRE) conditions. Two hand-held (simple and intelligent) and two stationary (field strengths: 137 and 140dbμV/m at 3m; antennas of 94x52cm placed according to body site) transceivers were used.

Readability was higher in BT (100%) than IT (A: 97.8%; E: 92.8%), due to losses (0, 1.7 and 4.7%) and breakage (0, 0.3 and 2.5%), respectively. Total electronic failures were 0.13%. SRE was 100% in all cases but DRE varied with experimental factors, the 140dbμV/m (BT and A-IT, 100%; E-IT, 97.5%) showing higher values than 137dbμV/m (BT and A-IT, 99.9%; EB, 94.2%).

Transponders were recovered by farmers from slaughtered and dead ewes (34.4%). Recovery of BT (99%, n=276) was easy and quick. Most IT were retrieved after palpation (A: >95%, n=1,577; E: >80%, n=482) but migration was reported (A: 1.3%, to 3rd and 5th ribs and foreleg; E: 6%, to atlas and mandible). Moreover, the IT in 150 sheep (A: n=137; E: n=13) were lost (7.3%). In conclusion BT are recommended for electronic identification of sheep on farm conditions.

ANIMAL MANAGEMENT AND HEALTH [M]   Paper MH4.6

The use of electronic identification with horses
Heinrich Pirkelmann. BLT Grub. Bavarian Institute for Animal Production Grub, P.O. Box 1180, D-85580 Poing, Germany

The actual methods of horse marking as description and the numbering brand are difficult to handle, insufficiently safe and relatively disputed to animal welfare. For these reasons electronic identification shows an raising acceptance and is already used by some breeding organizations especially trotters or Frisian horses. A basic prerequisite for the approval as an official marking method is the use of ISO compatible systems according to the standards ISO 11784 and ISO 11785. From the wide range of the available transponder types only injectables are suitable for horse marking. The recommended injection site is the left side of the neck below the forth vertebra and a one hand's distance below the mane. Both reading distance and practical use are influenced by transponder size. Small transponders of 12 mm only can be read by handheld readers, whereas stationary readers require at least transponders with a length of 20 - 25 mm. This dimension is necessary for automated processing like computer controlled feeding. If sufficient hygienic conditions during the injection are guaranteed, the transponders neither provoke irritations in tissue nor infections. The necessity of local sedation depends on the size of the used transponders. An authorized data base system containing allocation, registration and administration of all animal numbers with links to all national and international organizations in horse husbandry has to be installed. As a result the security of marking and handling in all fields of horse husbandry can be improved by using electronic identification especially using fixed programmed transponders.

Paper MH4.7

Vergleichende untersuchungen zur kennzeichnung von pferden mit transponder und mit Heißbrand
U. Pollmann. Fachbereich Ethologie und Tierschutz, Tierhygienisches Institut Freiburg, P.O. Box 5140, 79108 Freiburg, Germany

In der BRD haben die Zuchtorganisationen sicherzustellen, daß Zuchttiere dauerhaft so gekennzeichnet oder bei Pferden so genau beschrieben werden, daß ihre Identität festgestellt werden kann. Übliche Verfahren zur Kennzeichnung von Pferden sind der Heißbrand und neuerdings die Kennzeichnung mittels Transponder.
An insgesamt 68 Warmblutfohlen wurden Untersuchungen zur Belastung der Tiere und zur Effizienz der Kennzeichnung durch Schenkelbrand bzw. mit Transponder durchgeführt. Aufgrund des Verhaltens der Fohlen sowie der Herzfrequenzwerte infolge der Kennzeichnung war zu schließen, daß der Heißbrand für die Tiere belastender ist als die Transponderimplantation. Eine über den Kennzeichnungsvorgang hinausgehende Beinträchtigung der Fohlen konnte jedoch weder für den Heißbrand noch für die Transponderimplantation festgestellt werden. Die Identifizierung der implantierten Transponder erwies sich in allen Fällen (n=12) als zweifelsfrei möglich. Voraussetzung hierfür ist je-doch der Einsatz eines kompatiblen Lesegerätes. Dagegen war die Qualität der Brand-zeichen sehr unterschiedlich ausgeprägt. Als Ursache hierfür wurde die beim Brennen offensichtlich nicht berücksichtigte Hautdicke der Fohlen in Beziehung zu Andruckdauer und Andruckstärke des Brenneisens ermittelt.

ANIMAL MANAGEMENT AND HEALTH [M] Paper MH4.8

Electronic identification in horses
E. Søndergaard*[1], L.L. Hansen[2], H. Staun[3] and H. Schougaard[4]. [1]*Department of Animal Health and Welfare and* [2]*Department of Animal Product Quality, Danish Institute of Agricultural Sciences, P.O.Box 50, 8830 Tjele, Denmark,* [3]*Department of Animal Science and Animal Health, Royal Veterinary and Agricultural University, Grønnegårdsvej 3,1870 Frederiksberg, Denmark,* [4]*Nørlund Equine Hospital, Rodelund, 8653 Them, Denmark.*

The aim of this project was to look at the function and position of chips placed in the neck area of horses for a long period. In 1991 each of 17 experimental horses at Research Centre Foulum (Report no.38, Danish Institute of Agricultural Sciences, Denmark) were injected with two chips - one on each side of the neck - one from Trovan Electronic Syst-ems and another from Rhone Merieux. Additionally 21 horses at *Nørlund Equine Hospital* and in 1992 12 horses at Research Centre Foulum, were injected with a chip from Rhone Merieux. These chips were placed on the left side of the neck in the area between *lig. nuchae* and the muscle, in the middle of the neck. The first two years after injection the horses were followed thoroughly and in 1999 all horses still in Denmark were controlled. One chip was rejected right after injection due to infection of the injection place. In the first two years all chips were found in the injection place and there was no signs indicating that the chips had moved under the skin. In 1999 approx. 50% of the horses were still alive in Denmark and were controlled for functioning and position of the chip.

Poster MH4.9

An exponential smoothing model in time series analysis of milk electrical conductivity data for the clinical mastitis detection
P. Secchiari[1,2], M. Mele[1,2]*, R. Leotta[3]. [1]*C.I.R.A.A., Pisa University,* [2]*D.A.G.A., Sector of Animal Husbandry, Pisa University, Via del Borghetto 80, 56124 Pisa, Italy,* [3]*Dpt. of Animal Production, Pisa University, Viale delle Piagge 2, 56124 Pisa, Italy.*

Moving average model is the standard method to analyse milk electrical conductivity (EC) data in commercial informatic system for the detection of clinical mastitis in dairy cows. In a recent study, sensitivity and specificity of a commercial system which uses a moving average model of lag 20 (Afimilk, S.A.E. Afikim, Israel) was just evaluated. The aim of the present study is to compare sensitivity and specificity of this commercial system with the same parameters estimated using the same EC data and an exponential smoothing model ($\hat{Y}_{N+1} = \alpha Y_N + \alpha(1-\alpha)Y_{N-1} + \ldots + \alpha(1-\alpha)^{N-1}Y_1$; for $0 < \alpha < 1$). EC data was elaborated using several smoothing constants ($\alpha = 0.095; 0.2; 0.4; 0.6; 0.8$) and 2 levels of significativity (2σ and 3σ) to value the prediction errors. Using the exponential smoothing model the results show that sensitivity increased with the increase of the smoothing constant values and reached 100% for $\alpha = 0.6$ and 2σ. Therefore, exponential smoothing model is able to detect correctly more clinical mastitis than moving average one (64.71%). Specificity showed a trend similar to sensitivity and reached a very high level (98.1% for $\alpha = 0.8$ and 3σ) but never reached the specificity of moving average model (99.4%). Further research on time series model is needed.

ANIMAL MANAGEMENT AND HEALTH [M]   Paper M5.1

An emerging problem in the Pig: PMWS (Postweaning MultisystemicWasting Syndrome)
*F. Madec, E. Eveno, P. Morvan, L. Hamon, D. Mahé, C. Truong, R. Cariolet, E. Albina, A. Jestin
CNEVA (National Centre for Vet. Res. And Food Safety) BP53 22440 Ploufragan France*

PMWS is taking a growing importance, several countries being now concerned. In France the first cases were found in 1996. The present paper is aimed at describing the condition through a series of cohort studies undertaken in affected farrow-to-finish farms. The pig when 2 to 3 months of age is at its critical period. Unthriftiness, emaciation, muscle wasting, pallor are the most typical clinical signs. In addition, hyperthermia, dyspnoea, cough or diarrhoea can also be observed. Wasting can clearly establish in 3 to 4 days and part of the concerned pigs will die rapidly while others will survive several days in a miserable body state. Drugs are not efficient. In a pen the problem has a strong individual expression and in the cohorts the pigs of some litters were much more susceptible to PMWS expression. Mortality in severely affected farms was as high as 15%. Surprisingly no other clinical sign was perceivable either in the sows or in the piglets before weaning. Necropsy showed a variety of gross lesions affecting the lymph nodes, the spleen, the lung and often other organs. Histopathology showed severe alteration in the lymphoïd tissues. A novel strain of porcine circovirus (PCVII) was found associated to PMWS.

Paper M5.2

Economic solution of heating the piglets nests
I. Štuhec[1], *Marija Vogrin-Bračič*[1], *Milena Kovač*[1] *and Špela Malovrh*[1]. [1]*University of Ljubljana, Biotechnical Faculty, Department of Animal Science, SI-1230 Domžale, Slovenia.*

The aim of this study was development of a system with a hover (closed nest) in the farrowing pen with infra-red heater, sensor and automatically controlled heating regulation. Hovers reduce losses of heat energy and consequently provide warm environment for piglets and nevertheless, the farrowing unit can be thermal comfortable for sows too. During two-year period, electricity usage was measured from farrowing to weaning during 60 lactations standardised to 28 days. On average, the electricity consumption was 42.03 kWh per lactation (52.26 kWh in winter and 34.39 kWh during summer season). Major part of the energy was used in the first week of lactation (41.47 %). After that, usage of energy per week decreased (29.47 %, 19.33 %, 9.46 % for second to fourth week, respectively) because of increase in heat production in piglets which resulted in smaller needs for supplemental heating. Calculation of total costs for heating had shown that nearly 195 DM per farrowing pen per year were saved up comparing to the system without hover and without regulation of warming which is common practice in raising up the piglets.

ANIMAL MANAGEMENT AND HEALTH [M] Paper M5.3

Pathological changes in the periparturient period induced by fumonisin B1 fed to pregnant sows
M. Zomborszky-Kovács*[1], F. Vetési[2], Á. Bata[2], Á. Tóth[1], G. Tornyos[1]. [1]*Department of Animal Physiology and Hygiene, Pannon Agricultural University, Faculty of Animal Science, P.O.Box 16, 7401 Kaposvár, Hungary,* [2]*University of Veterinary Science, P.O.Box 2. 1400 Budapest, Hungary.*

Fumonisin B1 (FB1) toxin (in *Fusarium moniliforme* fungal culture) was administered to three pregnant sows in a daily dose of 300 mg. FB1 resulted in damage to the foetuses *in utero*. Of the changes *intraalveolar*, *subpleural* and *interstitial* pulmonary oedema could be detected in the piglets examined immediately subsequent to parturition, prior to the first suckling. Pathological change was demonstrated in the histopathological finding for the liver, and in the activity of the plasma AST, GGT and ALKP. The serum sphinganine/sphingosine ratio, regarded as a bioindicator of fumonisin B1 toxicosis, varied in accordance with the degree of severity of the changes which occurred.
In the piglets of two sows fed the toxin for a further 7 days subsequent to parturition mild pulmonary oedema could be detected after colostrum suckling, 24 hours after parturition and also on the 7th day.
In the milk samples taken from sows FB1 was detected in quantities of 18.0-27.5 ppb.
No change was observed on the 7th day in the piglets of the 3rd sow, the diet of which contained no toxin after parturition.

Paper M5.4

The effects of feeder space allowance and group size on finishing pig welfare
H.A.M. Spoolder[1], S.A. Edwards*[2] and S. Corning[1]. [1]*ADAS Terrington, King's Lynn, PE34 4PW, UK,* [2]*SAC Aberdeen, Craibstone, Aberdeen, AB2 9YA, UK.*

Housing finishing pigs in large groups offers certain advantages to the producer, compared to small groups. However, there may be disadvantages for low ranking pigs who may need to compete for access to resources. This study investigated the interactive effects of food availability (L: 1 single space hopper per 20 pigs; H: 1 per 10 pigs) and group size (20, 40 or 80 pigs per pen) on welfare, at constant stocking density in pens without straw. Feeder related aggression per pig was not affected by group size, but higher in L compared to H (12.5 and 8.3 interactions per 12 hours, $P<0.01$). The number of skin lesions increased with group size (7.8, 9.1 and 10.2 lesions for 20, 40 and 80; $P<0.01$). Average daily gain in the first half of the finishing period was negatively influenced by group size (771, 737 and 735 g/day for 20, 40 and 80; $P<0.05$), and higher in H compared with L (760 and 721 g/day for H and L; $P<0.001$). No effect on weight gain was found subsequently. No differences between treatments were found on within group variation in growth, or the number of pigs removed for health reasons. It is concluded that the number of pigs per feeder space should be lower than 20 for welfare reasons, although performance appears acceptable at this level. Further research will have to identify whether the effects found on aggression are common to all finishing systems, or whether straw bedding can serve as a mitigating factor.

ANIMAL MANAGEMENT AND HEALTH [M]   Paper M5.5

Behaviour of slaughter pigs during separation
D. Schäffer* and E. von Borell. *Institute of Animal Breeding and Husbandry with Veterinary Clinic, Martin-Luther-University Halle-Wittenberg, Adam-Kuckhoff-Str. 35, 06108 Halle, Germany.*

Electrical prods are generally used for the separation of pigs in slaughter houses. As a result, noise levels are often exceeding 110 dB(A) and animals are slapped while balking or turning back.

In this experiment, different separation strategies in conventional slaughter houses were compared to a patented method for a stress reduced separation of individual animals within a passage equipped with triangles on the inner side of the side walls. Locomotory behaviour of a total of 200 slaughter pigs was recorded by a video system. Groups of 4 or 6 animals were driven through a passage between elements of different widths (1.00 m or 1.22 m). On the basis of the highest separation success with 86 % in each group of 6 animals and the least usage of driving aids (0.02 and 0.05 flap blows / animal, respectively), a passage width of 1.00 m between elements is recommended.

The results demonstrate that the usage of a passage with appropriately spaced elements reduces stress in the process of individual separation of slaughter pigs.

Paper M5.6

Production results from sows and their offspring kept outdoors or indoors with respect to their maternal abilities
M. Wülbers-Mindermann*, B. Algers, *Department of Animal Environment and Health, Faculty of Veterinary Medicine, SLU, P.O.Box, 234, SE - 532 23 Skara, Sweden.*

The aim of this study was to investigate how physical maternal characteristics were associated with the housing system and the offspring performance. The study included 42 indoor and 57 outdoor sows (multiparous sows and gilts). Sows were kept indoors in conventional single pens or outdoors in farrowing huts. Sows' weight and backfat thickness was recorded 5 days prior to parturition and at weaning. Piglets individual weights were recorded 4 days p.p. and at weaning. The piglet mortality was recorded on day 1, 2, 3, 4 p.p. and on the day of weaning. Outdoor litters grew significantly faster (b-value: 0.01109 vs 0.00885, $p<0.001$), had significantly less within litter weight variation at weaning (16.15% vs 20.72%, $p<0.01$) and a lower mortality rate from day 4 p.p. until weaning (0.58 vs 1.06, $p<0.05$) than indoor litters. The production results were not determined by the maternal characteristics, instead smaller litter sizes at day 4 p.p., rearing piglets during the cold season of the year and keeping them outdoors was the most favourable combination of factors. Sows' weight and backfat loss were more a consequence of the production results. Both were lowest when the sows had a smaller litter and were reared indoors during the cold season of the year.

ANIMAL MANAGEMENT AND HEALTH [M]   Paper M5.7

Health-control costs and dairy farming systems in western France
H. Seegers*, C. Fourichon, F. Beaudeau and N. Bareille. Unit of Animal Health Management, Veterinary School & INRA, BP 40706, 44307 Nantes Cedex, France

Current dairy systems are moving towards more differentiation. The study aimed to explore the relationships between system and health control costs, as a summary of health features. Data were issued from 248 farms involved in a 2-year survey. Farms were classified into groups according to specialisation, size of the dairy unit (measured by the quota) and intensification pattern (by use of upward hierarchic classification). Health control costs averaged 85.63€ per cow-year and 1.14€ per 100kg milk. Calf- and heifer-related costs accounted for 13.2% and 12.6%, respectively. Udder disorders, nutritional disorders and infertility accounted for 43.7%, 18.4% and 16.5% in cow-related costs, respectively. Differences in costs within groups were very large. Costs were higher in diversified farms (especially in farms holding also beef cattle). Costs did not vary with size of the quota. Non intensive Holstein farms had lower costs per cow-year (62.03€) and per 100kg milk (9.09€) than intensive farms. Among the latter ones, those feeding large amounts of concentrates had very high costs (103.3 and 12.70€). Dual-purpose-breed (or mixed-breed) farms also had very high costs per 100kg milk (13.5€), but not per cow (88,97€). Further investigation is needed to determine if the differences result from differences in disease occurrence or in heath management practices.

Paper M5.8

The meat's thermodynamic parameters correlate with the farming type
H.-D. Matthes,[1] V. Pastushenko,*[1] H. Heinrich[2], Z. Holzer[3]. [1]Department of Agricultural Science, Rostock University, Vorweden 1, D-18069 Rostock, Germany, [2]LABO TECH Labotechnik GmbH, Richard-Wagner Str. 31 D-18119 Rostock-Warnemünde, Germany, [3]ARO, Institute of Animal Science, Newe Ya'ar Research Center, P.O.Box 1021, Ramat Yishay 30095, Israel

The samples of beef meat from the organic farming animals and intensive managed calves, treated with the hormones, were investigated. In the meat homogenate the following parameters were determined: the zero level of redoxpotential and the level after adding ATP + Coffein and GTP. We found that the redox parameters were associated with the type of animal managing. The ecological samples of meat characterised by steady thermodynamic reactions to the catalyst, compared to samples of the hormone treated animals. The influence of the second messengers on the meat redox balance and the reaction of the redox dependent enzymatic system significant correlated with the animal management. Thus complex meat redox measuring can be used for the distinction of animal management, indication of meat antioxidant properties and estimation of redox dependent enzymatic balance of muscle tissues.

ANIMAL MANAGEMENT AND HEALTH [M]   Paper M5.9

Growth and Survival of Holstein and Brown calves reared outdoors inindividual hutches
O. Alpan*[,1], O. Ertugrul[2], N. Unal[2], F. Azeroglu[3] and O. Kaya[3]. [1]Faculty of Veterinary Medicine, J. University of Science and technology, Irbid, Jordan, [2]Veterinary Faculty, Ankara University, Ankara, Turkey. [3]General Directorate of State Farm Enterprises, Ankara, Turkey.

Indoor raising of dairy calves is common practice in Turkey and in Jordan. The purpose of this project was to study the effects of outdoor raising of Brown and Holstein replacement calves in individual hutches. The study was carried on a farm having Northern Mediterranean climatic conditions. A total of 240 female calves were used with 30 calves born in each breed and in each of the four seasons. An animal kept in hutches for two months and then in groups of 10 calves until 6 months. The birth weight of Brown and Holstein calves were 37.2 and 37.9kg, respectively ($p<0.01$). The average birth weights ranged from 37kg to 38.7kg according to seasons ($P<0.01$), being highest in spring and lowest in winter. The breed and season effects were present at 6 months, the live weights in breeds were 114.9 and 125.7kg ($P<0.01$). Number of health disorders was significantly higher ($P<0.01$) in the first month of life. Mortalities came in later months, 2.5% in Holsteins and 5% in Browns. The results indicated that calves could be reared outdoors successfully.

Poster M5.10

Information network - the base for finding economical decisions in the animal production
L. Döring[1], H. Saage[1], R.-D. Fahr[2], G. v. Lengerken[2]. [1)] MRA of Saxony-Anhalt; [2)] Martin-Luther-University Halle-Wittenberg, Institute of Animal Breeding and Husbandry with Veterinary Clinic

The increase in economic efficiency in the professional animal production of agricultural companies is subject to a permanent pressure. Like in all considerable lines of production, the agriculture business also needs objective and assured information for finding economical decisions to the right extent and at the right time.
The MRA for cattle breeding system and milk quality testing in Saxony-Anhalt offers his member companies a number of precise management tools for controlling the economic process in the animal production.
The information system for the „Agrargenossenschaft Cobbelsdorf eG" shows as an example the model of offered management tools from the MRA. The base for all tools are the results of milk recording from each milk cow and the calculated statistic data of the herd as well as the results of milk quality testing. All this data comes from a work process which is quality assured in the norms DIN EN ISO 9002 and EN 45001 and guarantee a highest level in precision.
Special milk analysis, fodder testing and the recording of a number of physical, technical and biological values as well as the imparting of science knowledge make the information network usefuly.

ANIMAL MANAGEMENT AND HEALTH [M]   Poster M5.11

Heat production Cherbage intake, and heart rate of grazing heifers
Satoshi Ando*, Kazuo Otsuki, National Grassland Research Institute, Senbonmatsu 768 Nishinasuno Tochigi 329-2793 Japan

The relationships among heart rate (HR), heat production (HP), and herbage intake on grazing were studied. Two Holstein heifers with a mean body weight of 406 kg were used. Each heifer was placed on a treadmill with a head cage. There were five levels of walking speed ,from 6.7 m/min. to 30 m/min. Heart rate (HR) and heat production (HP) were measured while the heifers were walking and resting. The regression equations between HP and HR were calculated. Then,the two heifers were grazed under two different grazing conditions,namely , with a large quantity of grass and a small quantity. HR and HP were measured continyously for 24 hours,and compared between the two conditions. HP was calculated from HR and the regression equation described above. When dry matter intake was high , HR and HP were high , and when dry matter intake was low, HR and HP were low. HR and HP were influenced by dry matter intake in grazing. HR correlated well with HP. It was considered that mean HR during , 24 hour period could be used as an index of total HP per day.

Poster M5.12

Dominance relationships and mating efficiency in boars kept with gilts in a dynamic service system
D.F. Grigoriadis[1], S.A. Edwards[1]*, P.R. English[1] and F.M. Davidson[2]. *[1]Department of Agriculture, 581 King Street, University of Aberdeen, Aberdeen AB24 5UA, UK, [2]A. Simmers Ltd., Mains of Bogfechel, Whiterashes, Aberdeenshire AB2 0QU, UK*

The reproductive behaviour of three teams of 4-5 boars was recorded for 54 days in a dynamic service system for gilts. The female population of each service pen (20 gilts) was changed weekly (4 in-4 out) while boars were resident. Altogether 933 mating attempts (MAs) and 98 sodomies were videotaped and recorded in detail. Differences in mating quality (defined using quantitative behavioural criteria) between and within boar teams were significant (all $p<0.05$). The hierarchies (as determined through competition for food) of all teams were linear. Male dominance rank was not associated with boar age and weight. High-ranking boars did not achieve better quality MAs than their low-ranking counterparts, did not have more frequent or longer duration MAs, or a higher mating replacement index (calculated from interactions between boars during MAs). Boar social rank also failed to influence the adoption of passive and/or active roles in homosexual activities. Sodomies did not seem to distract boars from their mating duties since teams and individuals with high homosexual activities had also high number of MAs. It was concluded that no simple relationship exists between feeding and mating dominance in a dynamic service system, and that subordinate males are not deprived of sexual partners.

ANIMAL MANAGEMENT AND HEALTH [M] Poster M5.13

Sexual courtship of Andalusian stallions in a directed mount
E. Rodero[1]*, B. Alcaide[2], M.Anglada[1], J.R.B. Sereno[3,4]. [1]Departamento de Producción Animal. Universidad de Córdoba. Avda. Medina Azahara s.n. 14005 Córdoba, Spain. [2]Centro de Reproducción Equina y Remonta n⁰ 3. Ecija, Sevilla, Spain. [3]Embrapa Pantanal, 79320-900, Corumbá, MS, Brasil. [4]Becario AECI.

The object of this work was to objectively study the conduct of the sexual courtship and the exploration of the Andalusian stallions in a natural directed mount. 69 services were controlled during the reproductive season of 1998. All the stallions submitted to study belong to the CRE and R n°3, located in Écija, Seville. Through direct observations and real times the records correspondent to quantitative as much as qualitative characteristics of behavior and referring to various parts of the mare which were smelled and licked (84% are in the perineal area or escutcheon) or bit with the lips (from 1 to 4) and teeth (1 to 5). The stallions showed Flehmen close to a 52% and the number of times this reflex repeats itself during the mount oscillates between 1 and 14 times with a mean of 1.71 (sd =2.73 and se =0.32). This concludes that the stallions in a natural directed mount perform a discreet sexual courtship, probably due to the condition of the mount, since the stallion approaches the mare in a direct form until reaching the mount.

Poster M5.14

Sexual behavior of Andalusian stallions in a directed mount
E. Rodero[1]*, B. Alcaide[2], M. Anglada[1], J.R.B. Sereno[3,4]. [1]Dpto. Producción Animal. Universidad de Córdoba. Medina Azahara. 14005 Córdoba, Spain. [2]Centro de Reproducción Equina y Remonta 3. [3]Embrapa Pantanal, Brasil. [4]Becario AECI.

The data was obtained by direct observation of 69 services realized by 25 Andalusian stallions in the season of 1998. The environmental conditions were controlled in each service, showing: firmness of hind limbs while leaning on them (44% of the horses kicked); grade of erection or depletion of the penis in the first service (87% totally erected); incidence of bite on the mane (44%). It was controlled in relation to the previous sexual experience: services without erection (18%); elevations or missed impulses (10%); mounts through the side (14%). In 97% of the observations, the ejaculation was produced in the first service. It was observed during the ejaculation a mean of 6.6 shakes of the tail (sd=2.6, se=0.04) and these signs were manifested greatly in the 25% of the cases. The time observed from the beginning of the exploration of the mare until the end of the ejaculation oscillated between 1 and 65 minutes with a mean of 10m (sd=16, se=0.26). In the post-service behavior it was observed that 76% of the stallions showed to be relaxed after coitus. Within them, 73% showed a hanging flaccid penis as opposed to a 12% that maintained it erected and a 15% that had it in normal conditions. These data show the great capacity for mounting of the Andalusian stallion.

ANIMAL MANAGEMENT AND HEALTH [M]   Poster M5.15

Sexual behavior of Andalusian stallions during their conduction towards a direct mount

E. Rodero[1]*, B. Alcaide[2], C.Fernández[1], J.R.B. Sereno[3,4]. [1]Departamento de Producción Animal. Universidad de Córdoba. Avda. Medina Azahara s.n. 14005 Córdoba, Spain. [2]Centro de Reproducción Equina y Remonta n[0] 3. Ecija, Sevilla, Spain. [3]Embrapa Pantanal, 79320-900, Corumbá, MS, Brasil. [4]Becario AECI.

A descriptive study of behavior during the conduction towards a direct mount has been realized in 69 services corresponding to 25 Andalusian horses. These stallions are owned by the Center of Equine Reproduction, located in Ecija, Sevilla. During the reproductive seasons of 1998 and 1999, keeping in mind the climatic and special conditions of each service, the conduction times (time from box until the stallion reaches the mare) were obtained (x = 2,3±1,1 minutes) and the behavior and resistance while being conducted were controlled in the following critical moments of the process: when leaving the box (88% were easily conducted); when reaching the training wheel (only 37% allow an easy conduction); during the run (50% have a stringhalt gait). The moment in which the erection was produced and lost in relation to the interaction with the mare was registered (93% of the erections begin while the stallions explore the mare and 25% while looking at the mare). In 57% of the cases, there is not a loss of erection during the whole process of courtship. It was concluded that these stallions have a lot of sexual experience and did not show dangerous for mare or manager.

Poster M5.16

Sexual behavior of the mares directed mount

E. Rodero[1]*, J. Cantatrero[2], C..Fernández[1], J.R.B. Sereno[3,4]. [1]Departamento de Producción Animal. Universidad de Córdoba. Avda. Medina Azahara s.n. 14005 Córdoba, Spain. [2]Centro de Reproducción Equina 3. [3]Embrapa Pantanal Brasil. [4]Becario AECI.

In this work a study of the symptoms of acceptation or rejection by the stallion, in relation to the follicular size (mm) (μ =47, sd =5.2 and se=0.7) was realized. Close to 58% of the mares were of the Andalusian breed. All of the mares were serviced by Andalusian stallions during the season of 1998. It was taken into consideration the presence of the foal in the 55% of the cases and the waiting time (minutes) until the arrival of the stallion (μ =7, sd =6, se =0.8). The following manifestations have been considered as signs of acceptation: fluttering of the vulva (73% of the cases); flux and quantity of mucosity (38% lacked of it); turned and offered the croup (63%); it was observed that 51% of the mares urinate while in heat with a mean of 3 times (sd =2.3, se =0.4); 58% incline offering the croup and a 66% vocalize. As signs of reactivity or rejection the following manifestations have been reported: threat of biting the stallion in 7% of the cases; gestures of threat by turning the ears (13%); charges (4%) and kicking, reaching or not the stallion (3 to 9%, respectively). This concludes that the mares in heat show themselves to be relaxed and very receptive to be mounted.

ANIMAL MANAGEMENT AND HEALTH [M]   Poster M5.17

Mechanical damage to the body of turkeys raised on different types of floors.
A. Wójcik, J. Sowinska, K. Iwanczuk-Czernik. Department of Animal Hygiene, Olsztyn University of Agriculture and Technology, 10-718 Olsztyn, Oczapowski str. 5/107, Poland

The aim of the study was to evaluate mechanical damages to the body of meat turkeys which occurred during rearing and can cause disqualification of part or the entire carcass. Turkeys (female - 16 weeks old and males - 22 weeks old): heavy (h), medium (m) and light (l) were been reared on different types of floors: slotted (s) and bedding (b). Most of the birds with damaged body and weight loss were characterised by:
1/ males (hs - 95.2%; 5.03%, ms - 94.4%; 4.01%, ls - 88.9%; 1.59%, hb - 53.6%; 2.64%, mb - 25.0%; 0.54%, lb - 35.7%; 0.2%) which were reared 7 weeks longer than females (hs - 44.8%; 3.53%, ms - 16.7%; 0.07%, ls - 13.3%; 0.06%, hb - 16.7%; 0.36%, mb - 6.9%; 0.0%, lb - 3.7%; 0.1%);
2/ all of the turkeys which were reared on slotted floors;
3/ hevy turkeys which were reared on all types of different floors.
It was mostly noted the following damages to the body of turkeys: oedema of the talocrural joint, deformity and keratosis of talus and digiti, contusion and inflammation in areas of the wings, bedsore and cystic swellings on the breast.

Poster M5.18

Growth curves of intensively reared ostriches (*Struthio camelus*) in Northern Italy
A. Sabbioni*, P. Superchi, A.Bonomi, A.Summer and G. Boidi. Istituto di Zootecnica, Alimentazione e Nutrizione, Università di Parma, 43100 Parma, Italy.

Growth curves parameters, obtained by the Gompertz equation, were calculated for 151 ostriches (63 African Blacks, 88 Blue Necks ; 78 males, 73 females), reared under typical intensive farm conditions of Northern Italy with *ad libitum* feeding. Ostriches were weighted at monthly intervals from birth to adults (last weight recording at 676 days of age). Mean ($\pm$ s.d.) birth weight was kg 0.937 $\pm$ 0.082. The estimated mature weights ($\pm$ s.e.) were 109.1 ($\pm$ 4.3); 111 ($\pm$ 3.7); 110.4 ($\pm$ 5.7) and 109.4 ($\pm$ 4.2) kg, respectively for African Blacks, Blue Necks, males and females. Maximum daily weight gain occurred at 256.6 ($\pm$ 5.7); 246.3 ($\pm$ 9.6); 256.9 ($\pm$ 13.4) and 231 ($\pm$ 8.7) days of age and rates of growth (x 10^3) were 5.551 ($\pm$ 0.395); 5.621 ($\pm$ 0.301); 5.590 ($\pm$ 0.447) and 6.062 ($\pm$ 0.321), respectively. Mature weight and age of maximum weight gain were significatively influenced by breed ($P<0.01$) and sex (respectively $P<0.05$ and $P<0.01$). Data suggest that environmental conditions of Northern Italy strongly affect the growth parameters of ostriches, compared to those calculated for birds reared in hotter regions.

ANIMAL MANAGEMENT AND HEALTH [M]   Poster M5.19

Vergleich der verkaufsform chinchillapelzen in den jahren 1997/98
M. Sulik[1], *L. Felska*[1], *G. Mieleńczuk*[2]. [1]*Department of Cattle Breeding, Laboratory of Fur Animals, Agricultural University in Szczecin, ul. Dra Judyma 12, 71-460 Szczecin, Poland,* [2]*Alex Chinchilla Farm, ul. Gen. Bema 14A, 72-200 Nowogard, Poland.*

Der Anfang 90er Jahre charakterisierte sich durch eine intensive Entwicklung der Chinchillazucht in Polen. Sie wurde durch eine große Nachfrage nach Chinchillapelzen auf den Weltbörsen verursacht. Die in dieser Zeit auf den Versteigerungen erreichten Pelzpreise waren sehr hoch. Polnische Züchter verkaufen die Chinchillapelze haupsächlich mit der Vermittlung der kanadischen Firma Canchilla Ass. Ltd. und des dänischen Auktionshauses Copenhagen Fur Center (CFC). Die vorliegende Arbeit setzt sich zum Ziel eine Analyse des Verkaufs der Chinchillapelze in Jahren 1997-98 durch die obengenannten Vermittler. Die Zusammenstellung umfaßt: Pelzklasse, Menge der angebotenen Pelze, Verkaufsgröße und ökonomische Verkaufsergebnisse. Analysiert wurden auch die Form der Pelzannahme und die Zahlungsbedingungen. Höhere Preise erreichten Pelze, die durch Vermittlung von CFC (Durchschnittspreis auf der Börse für reguläre Pelze 61-75$) als diejenigen, die durch Canchilla Ass. Ltd. (Durchschnittspreis auf der B(rse f(r reguläre Pelze 20-30$) verkauft wurden. Der Verkaufsvorgang verläuft in der Firma Canchilla Ass. Ltd. schneller, worauf den entscheidenden Einfluß die Lieferungen von bereits gegerbten Pelzen haben.

Poster M5.20

Nutzung des internet für effektive forschung in der tierproduktion
M. Andres. Zentralstelle für Agrardokumentation und -information (ZADI), Villichgasse 17, 53177 Bonn, Deutschland

Die Nutzung der im Internet verfügbaren Informationsangebote ist in zunehmendem Maße Voraussetzung für eine effektive Forschungsarbeit in der Tierhaltung und Tierzucht.
Informationssysteme, Forschungsnetzwerke und Datenkataloge bieten im Internet einen schnellen und zielgerichteten Zugang zu den verfügbaren Informationen. Forschungsrelevante Literatur- und Projektinformationen stehen in Datenbanken, Verzeichnissen und Bibliothekskatalogen zur Verfügung, die in zunehmendem Maße online recherchiert werden können. Telekooperation ist bereits Grundlage für institutionen- und länderübergreifende Forschungsaufgaben in der Tierproduktion. Sie ermöglicht modernes Datenmanagement, nutzt das Internet für schnellen Informationsaustausch und für weltweite Diskussion. Repräsentative Beispiele für Informationsangebote im Internet und für Forschungskooperation auf dem Gebiet Tierproduktion werden dargestellt. Dabei wird sowohl auf Netzwerke und Datenangebote mit internationaler und nationaler Orientierung eingegangen als auch auf spezielle Lösungen im Deutschen Agrarinformationsnetz (DAINet).

ANIMAL MANAGEMENT AND HEALTH [M]   Poster M5.21

Emission burdening of an area in connection with putting an incinerator into operation
E.Ziková* and T.Adamec. Research Institute of Animal Production, Praha 10 - Uhříněves, 104 01 Czech Republic

The study concerned a recently constructed Prague municipal waste incinerator and its effect on forage produced in surrounding areas and in direction of predominant winds. It concentrated on emission burdened forage in the period of one year before putting the incinerator into operation and one year after it. Concentrations of Cu, Zn and Hg were measured.
Average concentrations in mg.kg-1 of absolute dry matter:

plant	before putting into operation			after putting into operation		
	Cu	Zn	Hg	Cu	Zn	Hg
lucerne	8.35	69.26	0.014	8.79	124.88	0.011
clover	8.67	82.56	0.016	10.56	112.19	0.016
maize	4.05	113.6	0.005	3.45	66.68	0.002
grass	8.75	71.38	0.017	7.28	131.60	0.011
beet tops	3.79	84.57	0.056	6.50	74.69	0.037

Significant differences ($P<0.05$) were found among metals in maize and in beet tops between Hg only. All found values do not exceed norms valid in the Czech Republic for the highest admittable amount of metals in feed.

Poster M5.22

Effects of selenium on health and productivity of farm animals in Western Pomerania (Poland)
A. Ramisz[*1], J. Malecki[1] and A. Balicka-Ramisz[1]. Dep. of Animal Hygiene and Prophylaxis, Szczecin Agricultural University, Dr. Judym str. 6, 71-466 Szczecin, Poland.

Western Pomerania belongs to a selenium deficient regions of Poland. The aim of our studies was to established the influence of selenium on health and productivity of animals in this region. The monitoring studies were carried out on 266 animals: 81 horses, 32 cattle, 76 sheep, 25 gouts and 52 piglets. The influence of selenium deficiency on health, production and reproduction was realised on 394 pigs and 204 ewes. The experi mental groups (Se supplemented animals) consisted of 235 pigs and 104 ewes. The Se level in serum was estimated by using the Grzebuala et al. (!977) modification of the Watkinson's (1966) methods. In 71,1 % of animals the Se level in blood serum was low (24-40 µg/l) or very low (<10 µg/l). The reproduction performance of non-treated sows and ewes was reduced. From one sow 1,6 piglets and from one ewe 0,41 lambs less in the control group compared with the Se supplemented groups has been obtained. The piglets from the experimental group were more resistant against alimentary canal diseases compared with control piglets. The piglets mortality in the control group was two time higher (10,4 %) as in the experimental group (5,6 %). Selenium as feed additive effected the weight gain of porkers. The body weight of the experimental porkers was 4,6 kg higher, per animals, compared with control porkers.

ANIMAL MANAGEMENT AND HEALTH [M]   Poster M5.23

Einfluß der sommerhitze auf ethophysiologische parameter bei Holstein kühen im Holstein Deutschland

R. Schahidi[*1], *M. Steinhardt* [2] *und H.H. Thielscher* [2] . [1]*Tierärztliche Fakultät derUniversität Teheran , P.O. Box 19395-3789 Teheran Iran* [2] *Institut für Tierzucht und Tierverhalten der (FAL),Trenthorst! Wulmenau, 23847 Wusterau*

Rinderrassen passen sich klimatischen Umgebungsbedingungen gut an, weisen jedoch in einigen Entwicklungs- und Leistungsstadien eine größere Empfindlichkeit besonders gegenüber tieferen oder höheren Umgebungstemperaturen auf. Thermoregulation während erhöhter Umwelttempera-tur findet bei Tieren auf verschiedenen Wegen wie Respiration, Evaporation und Durchblutungs- steigerung statt. Es standen vierzig Milchkühe der Rasse Holstein für dieses Experiment zu Verfügung. Die Tiere sind auf der Weide gehalten worden. Täglich wurden drei verschiedene Merkmale: Rektaltemperatur, Herzfrequenz und Atemfrequenz bestimmt. Die Messungen fanden täglich in der Sommerzeit vom 9. bis 27. Juli um 7.00 und 16.00 Uhr statt.

Bei gleicher Umgebungstemperatur (16-24 °C) lagen drei gemessenen Parametern bei allen Tieren am 07,00 Uhr niedriger als am 16,00 Uhr. Wehrend die Körpertemperatur und Herzfrequenz der Tiere bei den Temperaturen zwischen 24 und 28 °C fast gleich blieben, erhöhte sich die Atemfrequenz vom 30 auf 41 Schläge pro Minute an. Bei einer Umgebungstemperatur großer als 25°C stiegen alle drei Parameter mit steigender Alte an. Bei einer Umgebungstemperatur großer als 25 °C wiesen die gleichalterige Tiere höhere *Rektaltemperatur, Herzfrequenz* und *Atemfre-quenz* auf, wen die Tiere eine tägliche Milchmenge großer als 25 kg produzierten.

Poster M5.24

Effect of type of castration on meat quality in rabbits
M.N. Kenawy. Dept. of Food Sci. and Nutrition, Faculty of Agric., Minia University, Egypt.

The effect of type of castration (hemicastration, castration and control) on meat quality in New Zealand white rabbits was studied. The results showed that except the significant effect of type of castration on percentage of cooking loss, no significant differences were observed in the carcass wt., dressing percentage, carcass boneless meat, percentage of yield of 4 cut up parts (shoulder, breast, back and leg), ratio of meat to bone and chemical composition. Although the effect of castration was not significant on previous traits, but we can observe that fat % was higher for complete castration rabbits meat in different cut up parts compared to hemicastration and control, back cut up and meat to bone ratio of breast tend to be higher in hemicastration than other type of castration. On the other hand, the value of the ratio of meat to bone of back cut up in control was lower compared to other type of castration. The results of panel tests revealed that hemicastration rabbits meat more tenderness and control was more juiciness than other type of castration

ANIMAL PHYSIOLOGY   [Ph]

Paper Ph1.1

The importance of muscle cell division in myogenesis and postnatal muscle growth
C. Rehfeldt. Division of Muscle Biology and Growth, Research Institute for the Biology of Farm Animals, D-18196 Dummerstorf, Germany.

During myogenesis myoblasts develop from mesenchymal precursor cells by proliferation and myogenic commitment. Myoblasts subsequently fuse to form multinucleated myofibres accompanied by an increased synthesis of muscle specific proteins. Recent research suggests that these processes are not exclusively under genetic control, but are also influenced by the maternal environment. For example, exogenous administration of porcine somatotropin to gilts or increased feeding during early pregnancy has been shown to cause an increase in muscle fibre number and/or myonuclear number.

During postnatal life the increase in skeletal muscle mass is mainly attributed to an increase in muscle fibre size (hypertrophy). The number of muscle fibres remains almost unchanged, apart from changes induced by extreme exercise, disuse, or starvation. However, myonuclear number and DNA content increase with postnatal growth. Mitotically active satellite cells which are situated between the basal lamina and the sarcolemma of the muscle fibre are the source of new myonuclei. Myonuclei themselves are considered to be postmitotic. Muscular DNA and protein accumulation are under the influence of both genetic and environmental factors. The ratio of these cellular components is significantly changed by selection for lean meat, and seems to be important in guaranteeing normal muscle function.

Paper Ph1.2

The importance of cell division in udder development and lactation
C.H. Knight, Hannah Research Institute, Ayr KA6 5HL, UK

The mammary secretory cell population increases in an exponential fashion during pregnancy in all species studied, as a consequence of very high rates of cell division. After parturition the mitotic index drops dramatically, but a limited amount of proliferation does continue, at least until the time of maximum milk yield. This is particularly true of rodents but does also occur in dairy species. During declining lactation cell apoptosis exceeds cell division, so the size of the cell population falls and it is this decrease which is responsible for the reduction in milk yield. Many factors influence cell division. In addition to well known hormones such as ovarian steroids, prolactin and GH and growth factors such as EGF and IGF1 there are also effects of milking frequency and of nutrition. Some of these same factors are now also known to regulate apoptosis. The challenge for the future is to understand more about the relationships between apoptosis and cell division in the mammary gland; are the two mutually exclusive and independent or is apoptosis important in preparing the gland for renewed cell division, for instance? To this end, we have developed a lactation rescue model which will allow us to study interactions between apoptosis and cell division in lactating mouse and cow mammary glands. Initial results indicate considerable coordination of the two processes.

ANIMAL PHYSIOLOGY   [Ph]

Paper Ph1.3

The components of protein synthesis (intra and intercellular regulation)
B. Riis*. Department of Animal Product Quality, Danish Institute of Agricultural Sciences, Research Centre Foulum, P.O.Box 50, DK-8830 Tjele, Denmark

Animal production is heavily dependent on proteins, because these molecules perform many different physiological tasks. Protein synthesis is the final outcome of gene expression, where the "DNA language" of four different nucleotides is transformed into the "protein language" of 21 basic amino acids.

Protein synthesis is divided into three separate steps: Initiation, elongation and termination. The initiation step is facilitated by at least 24 different initiation factor subunits and is not fully characterized. Two soluble enzymes, the elongation factors, $eEF-1_H$ and $eEF-2$ facilitate the elongation step in mammals. The covalent bond between the growing peptide chain and the activated amino acid is formed during the elongation cycle. The termination step is relatively simple. The release factor assures that the nascent polypeptide leaves the ribosome. All newly synthesized polypeptide chains are post-translationally modified and folded into a specific three-dimensional structure, before they can perform their tasks as proteins.

Regulation of the protein synthesis is performed at many levels, including transcription and mRNA processing. Direct control of the protein synthesis is exerted at multiple points during the initiation- elongation- and termination steps. During initiation and elongation two points are known, where reversible control is performed.

Paper Ph1.4

Quantitative aspects of protein synthesis in muscle growth
B. Sève, INRA Station de Recherches Porcines 35590 Saint-Gilles, France

In young rats, re-feeding after fasting increases muscle protein synthesis and intravenous infusion of insulin alone can simulate this effect. Oxidative muscles are much less sensitive to this effect of insulin than glycolytic muscles. There is stimulation of synthesis per unit of RNA (K_{RNA}), rather than an increase in the RNA : protein ratio. By contrast, in adult animals, protein synthesis is not sensitive to insulin and, after a fasting period, repletion of muscle occurs through the anti-proteolytic effect of insulin. In the muscle of young pigs, the decrease in K_{RNA} associated with fasting differs greatly between genotypes, parallel to differential variations of plasma insulin, IGF-1 and cortisol. There is still uncertainty as to whether the activation of muscle growth involves an increase in fractional protein synthesis rate (K_S) or a decrease in fractional protein degradation rate (K_D). We have shown in growing pigs that exogenous growth hormone stimulated both in similar proportion, while the increase in dietary protein increased less K_D than K_S. Protein seems to stimulate muscle protein synthesis through enhanced sensitivity of the muscle tissue to insulin and the effect is simulated by intravenous branched-chain amino acids. Dietary tryptophan was also shown to stimulate muscle protein synthesis rate. By contrast, improving the balance of the most frequently limiting dietary amino acids, lysine and threonine, at constant protein supply, decreases muscle protein synthesis and (or) turnover, in the pig as well as in other species.

ANIMAL PHYSIOLOGY [Ph]

Paper Ph1.5

Myogenesis and expression pattern of myogenin and myf-3 in fetal pigs
M. Christensen*[1], N. Oksbjerg[2], P. Henckel[2] and P.F. Jørgensen[1]. [1]The Royal Veterinary and Agricultural University, 1958 Frederiksberg C, Denmark, [2]Danish Institute of Agricultural Sciences, P.O. Box 50, 8830 Tjele, Denmark.

In the fetal pig muscle fibres develop as two distinct populations. At d 60 the primary fibres have been formed. Formation of secondary fibres occur from d 50 to d 90. Myogenin and myf-3, two muscle specific transcription factors, are important for myogenesis. An experiment were conducted to investigate myogenesis and the expression pattern of myf-3 and myogenin during formation of secondary fibres in pigs. Two muscles (*M. longissimus dorsi*, LD and *M. trapezius*, TRP) from 24 fetuses (12 male and 12 female) at d 60, 80 and 100 of gestation were sampled and analysed. Muscle fibres were histochemically characterised by the myosin ATPase stain, and nuclei expressing myf-3 and myogenin were characterised immunohistochemically. The frequency of primary fibres decreased and the frequency of secondary fibres increased from d 60 to d 80, with no further changes until d 100 of gestation. From d 60 to d 100 primary fibres atrophied, whereas secondary fibres hypertrophied. The frequency of nuclei expressing myf-3 did not change, while that of nuclei expressing myogenin decreased from d 60 to d 100 of gestation. Our data suggest, that myogenesis is concluded at app. d 80 in fetal pigs. Myf-3 and myogenin appears to exhibit different expression patterns during myogenesis. The significance of decline in the frequency of nuclei expressing myogenin during secondary muscle fibre formation is not clear.

Paper Ph1.6

Growth and mammary development in gilts provided different energy levels from weaning to puberty
M.T. Sørensen*, N. Oksbjerg and M. Vestergaard, Danish Institute of Agricultural Sciences, Research Centre Foulum, 8830 Tjele, Denmark

We investigated the effect of feeding level from weaning (d 28) to puberty (d 170) on growth and mammary development in gilts. From d 28 to d 90 (period 1) and from d 90 to d 170 (period 2), pigs were fed either ad lib (A) or restrictively (R) in a 2x2 factorial design with treatments named AA, AR, RA and RR. In period 1, growth rate of A-gilts was 674 g/d vs. 497 g/d for R-gilts ($p<0.001$). Blood samples taken at the end of period 1 showed that A-gilts compared with R-gilts had higher plasma IGF-I (237 vs. 168 ng/ml, $p<0.01$), lower plasma IGFBP-2 (78 vs. 163 arbitrary units, $p<0.001$), while IGFBP-3 was unaffected. In period 2, growth rates of RA- and AA-gilts were 25% higher than for RR- and AR-gilts and they tended to have the lowest plasma IGFBP-2 at slaughter ($p<0.10$). Furthermore, body weight at slaughter ($p<0.001$), weight of M. semitendinosus ($p<0.02$) and amount of mammary tissue ($p<0.001$) were highest in AA- and RA-gilts. However, since mammary DNA concentration was highest in RR- and AR-gilts ($p<0.04$), there was no significant difference in total mammary DNA content between the 4 treatment groups, although RA-gilts tended to have the highest. We conclude that gilts fed restrictively from weaning to d 90 and ad lib after that deposit as much muscle tissue as do continuously ad lib fed gilts and at the same time seem to have the best mammary gland development.

ANIMAL PHYSIOLOGY [Ph]

Paper PhN2.1

Postnatal adaptation of the gastrointestinal tract in neonatal animals: a possible role of milk-borne growth factors
R.J. Xu. *Department of Zoology, The University of Hong Kong, Pokfulam Road, Hong Kong*

During the postnatal period, the gastrointestinal (GI) tract in neonatal animals encounters numerous challenges and server physiological stress, particularly at times of birth and weaning. At the time of birth, the GI tract encounters the challenges of nutrient recruitment and of exposure to micro-organisms. At the time of weaning, the GI tract switches from handling liquid nutritious diet, milk, to solid feed of often plant origin. Failure of the GI tract to adapt to these challenges often results in retarded body growth, diarrhea or even death.

Maternal milk apparently plays an important role in regulating postnatal adaptation of the GI tract in neonates. It has been know for a long time that milk not only contains easily digestible nutrients but also bioactive compounds such as immunoglobulins and digestive enzymes. More recently, a number of growth factors has been detected in maternal milk of various species. There is an increasing evidence showing that milk-borne growth factors can survive in the GI lumen of the suckling young and exogenous growth factors administered orally stimulate GI maturation in neonates. The biological significance of milk-borne growth factors, however, remains to be fully illustrated. Understanding the role of milk-borne growth factors may help us to prevent many GI disorders in postnatal growth animals.

Paper PhN2.2

Intestinal maturation induced by spermine in young animals
G. Dandrifosse, C. Grandfils, O. Peulen, P. Deloyer and S. Loret. *Department of Biochemistry and general Physiology, Institute of Chemistry (B6c), Liege University, Sart Tilman, 4000, Liege, Belgium.*

Appropriate doses of spermine orally administered to suckling rats induces all modifications occurring naturally at weaning in the digestive tract, e.g. -1. an variation of the specific activity of disaccharidases and peptidase in intestine; an induction in the expression of genes, related to intestinal maturation, accompanied by the acquisition of mature phenotype of intestinal mucosa, a rise in the level of the receptor to the polymeric immunoglobulin (RPI), and an diminution of intestinal permeability to macromolecules; -2. an increase of the specific activity of enzymes contained in the pancreas; -3. a change in the growth fraction and biochemical properties of the liver, e.g., the synthesis of RPI; -4. a maturation of the gut immune system. The mechanism of spermine action is actually partly understood. In the intestine, it implicates a desquamation of the epithelium resulting from an activation of apoptosis, probably induced by an accumulation of spermine in the enterocytes. This phenomenon is accompanied by a hormonal cascade comprising IL-1β, IL-6, TNF alpha, bombesin, ACTH and corticosterone. Maturation is induced at least by increase in plasma corticosterone level and in enterocyte spermine concentration. A hypothesis taking into account all the results obtained will be presented as well as other observations recorded in the polyamine field in the case of Mammals.

ANIMAL PHYSIOLOGY   [Ph]   Paper PhN2.3

Nutritional and hormonal regulation of development in the neonatal pig
I. Louveau[1], *M. J. Dauncey*[2] *and J. Le Dividich*[1]. [1]*INRA, Station de Recherches Porcines, 35590 Saint Gilles, France, and* [2]*The Babraham Institute, Cambridge CB2 4AT, United Kingdom.*

The neonatal period is a particularly critical stage during which long-term development of the individual can be affected. Rapid somatic growth is accompanied by significant changes at the anatomical, physiological, cellular and molecular levels. This period is also associated with marked changes in the endocrine system, which in turn plays a central rôle in the control of metabolism and growth. After birth, the ability of the young to express its growth potential is related to both the quantity and composition of the food intake. Moreover, food intake can exert profound indirect effects on development *via* hormones acting as nutritional signals. This review focuses primarily on the somatotropic and thyroid axes in relation to plasma levels of hormones and their binding proteins, and to receptor expression in target tissues. In addition to nutrients, colostrum and milk contain a large group of biologically active components including enzymes, hormones, growth factors and immunological agents that may be involved in neonatal development. The complex interactions between nutrition, hormones and growth are discussed in relation to the effects of food intake and milk-borne growth factors on development at the whole-body and cellular levels of the neonatal pig, with particular emphasis on the gastrointestinal tract and on protein synthesis.

Paper PhN2.4

Aspects of GIT motility in relation to the development of digestive functions in neonates
H.N. Lærke[1], *V. Lesniewska*[2], *M.S. Hedemann*[1], *B.B. Jensen*[1] *and S.G. Pierzynowski*[3]
[1]*Dept. of Animal Nutrition and Physiology, Danish Institute of Agricultural Sciences, P.O. Box 50, 8830 Tjele, Denmark,* [2]*Warsaw Agricultural University, Dept. of Animal Physiology, Nowoursynowska 166, 02-787 Warsaw, Poland,* [3]*Dept. Animal Physiol. Helgonavagen 3b, S-223 62 Lund, Sweden and Gramineer Int. AB, Ideon, Scheleevägen 17, S-22370 Lund, Sweden.*

GIT motility is responsible for mixing and transport of digesta transport and elimination of undigested residues. The basis for the motility is the electrical activity of the GIT smooth muscle. The electrical activity as well as motility has a cyclically recurring pattern, which is associated with chronologically periodic fluctuations of mesenteric blood flow, GIT secretion (gastric, pancreatic and biliary), and absorption (e.g. glucose). Mostly in mammals feeding abolish the cyclic pattern, replacing it with a continuous postfeeding pattern. The duration of the postfeeding pattern is dependent on animal species, composition of the diet, and feeding regime. The perinatal and weaning period are periods of drastic changes in digestive function. Weaning induces enzymatic and structural changes in the digestive system. However, due to difficulties in performing studies in perinatal and neonatal animals only few data on the development of GIT motility, and its synchronisation with other digestive functions are available. Certainly, changes in GIT motility in response to weaning could be of paramount importance for a proper digestive function.

ANIMAL PHYSIOLOGY [Ph]

Paper PhN2.5

The prenatal gastrointestinal tract: hormonal and luminal influences on its functional maturation
P.T. Sangild. Division of Animal Nutrition, Royal Veterinary and Agricultural University, Groennegaardsvej 3, DK-1870 Frederiksberg C, Denmark.

At birth, the gastrointestinal tract (GIT) must be able to support a shift from parenteral nutrition before birth (via the placenta) to enteral nutrition after birth (oral intake of milk). In preparation for this event, the GIT grows and matures rapidly in the weeks before birth (the prenatal period). Both hormonal and luminal factors influence this rapid phase of GIT development in domestic animals. Among the hormonal factors, cortisol plays a pivotal role. In the fetal pig and sheep, the normal development of stomach acid and gastrin secretion and of intestinal enzyme activities (chymosin, pepsin, amylase, lactase, aminopeptidases) is stimulated by cortisol. Cortisol has limited effects after birth. The luminal ingestion of amniotic fluid by the fetus also modulates GIT growth, GIT hormones and GIT enzymes. These effects are mediated via growth factors, hormones and nutrients present in amniotic fluid. However, luminal influences on the developing GIT are less pronounced in the fetus than they are in the neonate. In conclusion, both hormonal and dietary (luminal) factors affect prenatal GIT development to ensure that the fetal gastrointestinal tract is sufficiently mature to support the dramatic nutritional transitions at the time of birth.

Paper PhN2.6

Nutrition, metabolism and endocrine changes in neonatal calves
J.W. Blum* and H. Hammon, Division of Nutrition Pathology, Institute of Animal Breeding, University of Berne, 3012 Berne, Switzerland

Newborn calves must adapt to changes from a primarily carbohydrate-based energy supply during the fetal period to a high fat/relatively low carbohydrate energy supply with colostrum (C). C contains essential nutrients and non-nutrient components. Early intake of C in high amounts is important for the development and function of the gastrointestinal (GI) tract and for the absorption of nutrients and metabolic, endocrine and nutritional status. C feeding increases the absorptive capacity of the mucosal surface of the gut and changes of metabolic and endocrine traits express enhanced anabolic metabolism. The somatotropic axis, which is important for the regulation of postnatal growth, is basically functional in neonatal calves and greatly affected by the amount and by the time-point of C intake. Although colostral insulin and insulin-like growth factor-I (IGF-I) are barely if at all absorbed by neonatal calves, the status of insulin, IGF-I and IGF binding protein are markedly influenced by C feeding. Various other endocrine systems, such as glucocorticoids, prolactin, glucagon, and thyroid hormones are involved in metabolic control, change after birth, but are variably influenced by feeding. There are marked differences of feeding C relative to feeding milk replacer or mature milk with respect to effects on the GI tract, metabolism and endocrine systems.

ANIMAL PHYSIOLOGY   [Ph]

Paper PhN2.7

Development of digestive and immunological function in neonates. Role of early nutrition
D. Kelly*, Rowett Research Institute, Greenburn Road, Bucksburn, Aberdeen, AB21 9SB.

The developing intestine undergoes morphogenesis and cytodifferentiation, events which are exquisitely regulated and highly organised, both spatially and temporally. The outermost epithelial layer of the intestine, the site of digestion, absorption and secretion, is organised into two morphological and functionally distinct compartments. Within the crypt, stems cells undergo several cycles of cell division, and migrate onto the villus where they exhibit changes in phenotype and functional characteristics. Enzyme, transporter and glycosylation patterns change dramatically during the transition from birth to adulthood. These transitions permit dietary change/adaptation and also are important determinants of disease susceptibility. In addition to its role in nutrient uptake, the intestine provides a key innate component of intestinal defence by providing a physical barrier to the external environment. Maintenance of tight junctions and the ability to secrete mucin are critical components of its protective function. Innate immunity is particularly important to the neonate, as the active or specific arm of the immune system is immature. The expansion of intestinal epithelial and lymphoid tissues is driven by the presence/processing of luminal antigen, particularly from bacterial sources. Early nutrition is also recognised to play a very significant role in modulating intestinal differentiation and function and in regulating beneficial and damaging immune responses.

Paper PhN2.8

Effect of cation-anion difference and EDTA on performance, ruminal fermentation, blood acid-base status and Fe availability in grain-fed calves
A. Ghodratnama, S. L. Scott, J. R. Seoane and G. St-Laurent*. Département des sciences animales, Université Laval, Ste-Foy, Québec, Canada. G1K 7P4

Effects of dietary cation-anion difference (CAD) and EDTA on performance, ruminal fermentation and blood acid-base status of grain-fed veal calves were investigated. Forty Holstein calves (85.6 ± 9.5 kg) were used in a randomized complete block design. Calves were allocated to five dietary treatments comprising 5 mg EDTA mg^{-1} Fe and a CAD of either 150, 300, 450, 600, or 750 mEq kg^{-1} DM. The trial included a 6-week grower phase (71% whole corn and 29% protein supplement) and a finisher phase (83% whole corn and 17% protein supplement) lasting until slaughter at 260 kg. Neither feed conversion nor digestibility were affected by treatments ($P>0.05$). Ruminal fluid pH increased and soluble Fe in ruminal fluid decreased with increasing CAD ($P<0.05$). While total VFA concentration (mg ml^{-1}) was unaffected by treatments ($P>0.05$), acetate concentration increased ($P<0.05$) and propionate concentration tended to decrease ($P<0.1$) with increasing CAD, resulting in a greater acetate:propionate ratio. At slaughter, treatments had no significant effect on blood pH, pCO$_2$, pO$_2$, or [HCO$_3^-$], although base excess tended to increase ($P<0.1$) with increasing CAD. While blood hematocrit and haemoglobin concentration did not vary among treatments, serum Fe decreased with increasing CAD ($P<0.05$). Therefore, EDTA and a CAD up to 600 mEq kg^{-1} DM reduces Fe bioavailability.

ANIMAL PHYSIOLOGY [Ph]

Paper PhN2.9

Contribution to the study of gut hypersensitivity reactions to soyabean proteins in preruminant calves and early-weaned piglets
D. Dréau* and J.-P. Lallès. *Laboratoire du Jeune Ruminant, Institut National de la Recherche Agronomique, 65, rue de Saint-Brieuc, 35042 Rennes cedex, FRANCE. lalles@roazhon.inra.fr*

Dietary antigens may lead to immune-mediated gut hypersensitivity reactions in preruminant calves and in piglets post-weaning. Recent advances in the understanding of these disorders are reviewed mostly based on studies conducted in preruminant calves and early-weaned piglets fed regimens containing soyabean proteins treated using various ways. Our results and others indicate that immune-mediated gut hypersensitivity reactions to soyabean proteins were associated with alterations in the morphology and function of the small intestine, with changes in the distribution of various sub-populations of T and B lymphocytes in both species. Detailed analysis of the local and systemic dietary antigen/allergen repertoires obtained in preruminant calves emphasized the critical importance of the storage globulin β-conglycinin and well as of more conventional antinutritional factors in immune-mediated gut hypersensitivity reactions to soyabean. The results also indicated that a quantitative assessment of the nutritional and health value of soyabean products is possible and can be used in a predictive way. It should take into account the antigenicity of key soybean proteins notably β-conglycinin, but also residual anti-tryptic activity. Future investigations should be aimed at measuring the impact of particular dietary components on the gut mucosa, especially in terms of cytokine profiles, and understand the interactions between luminal stimuli including diet, digestion products and bacteria, the enterocyte, and the underlying immune and nervous systems. This should provide new alternatives to the use of additives in animal feeds for optimizing digestive function.

Paper PhN2.10

Effect of milk intake level on the development of digestive function in piglets during the first postnatal week
I. Le Huërou-Luron[1]*, M.J. Lafuente*[2]*, F. Thomas*[2]*, V. Romé*[1] *and J. Le Dividich*[2]. [1]*Laboratoire du Jeune Ruminant, INRA, 35042 Rennes cedex, France,* [2]*Station de Recherches Porcines, INRA, 35590 St Gilles, France.*

The present trial was designed to evaluate the effect of milk intake (MI) on the development of digestive function in 7 d-old piglets. Eighteen newborn unsuckled piglets were alloted to 3 treatments consisting in 3 levels of sow colostrum and milk intake -low (L), intermediate (I) and high (H)-. L and H treatments correspond to maintenance requirement and *ad libitum* intake, respectively. Total body weight (BW) gain of H, I and L piglets averaged 1184g, 677g and 176g, respectively. Increasing MI has no significant effect on pancreas weight (g/kg BW) and enzyme activities (U/g tissue and U/kg BW). SI lenght was 25% higher ($P<0.01$) in L than in H and I piglets, whereas both mucosa (g/kg BW) and segmental weights (g/cm) increased linearly ($P<0.01$) with MI level. Lactase activity (U/g mucosa and U/kg BW) was 35% lower ($P<0.05$) in L compared to H and I piglets. Expressed as segmental activity (U/cm), there was a linear effect of MI on lactase, maltase and sucrase activities. Aminopeptidase A and N and dipeptidyl peptidase IV activities decreased linearly ($P<0.01$) with the level of MI whatever the mode of expression of the enzymes. Overall, a low level of MI does not impair pancreatic function during early life but has important effects on the SI function and may delay its maturation.

ANIMAL PHYSIOLOGY   [Ph]   Poster PhN2.11

Mid term effect of formula milk on the development of digestive function in piglets
I. Le Huërou-Luron*[1], B. Codjo[2], F. Thomas[2], V. Romé[1] and J. Le Dividich[2]. [1]Laboratoire du Jeune Ruminant, INRA, 35042 Rennes cedex, France, [2]Station de Recherches Porcines, INRA, 35590 St Gilles, France.

The present trial was designed to evaluate the mid term effect of feeding formula on the development of digestive function in 7 d-old piglets. Twelve unsuckled newborn piglets were alloted to 2 treatments consisting in formula (F) or sow colostrum then milk (SM). Piglets were bottle fed for 7 d to achieve similar weight gain. F was supplemented with immunoglobulins during the first 24 h. Levels of crude protein, fat, lactose were 12.9%, 5.1% and 3.3% in colostrum, 6.3%, 8.2% and 4.9 % in sow milk, and 5.3%, 4.3% and 6.6% in F, respectively. Corresponding levels of insulin and IGF-I were 825, 194 and 33µU/ml and 856, 19 and 11ng/ml, respectively. Total body weight gain of SM and F piglets were 950g and 930g, respectively. Diet had no effect on pancreas, SI and mucosa weights, SI lenght and SI differentiation parameters (protein, RNA, DNA). Feeding F resulted in a decrease (40-66%; $p<0.01$) in most of pancreatic enzyme activities (U/g tissue) as well as in lipase mRNA. In F piglets, expressed per g mucosa, aminopeptidase A activity was increased ($P<0.01$) whereas that of dipeptidylpeptidase IV was decreased in intermediate and distal SI. Lactase was 56% ($P<0.01$) lower in distal SI whereas sucrase and maltase were 74% and 51% ($p<0.01$) higher in proximal SI. Results suggest that both pancreas and SI function can be altered in the neonatal period by diet composition.

Poster PhN2.12

Sow metabolism and piglet performance
A. Lindqvist[1], H. Saloniemi*[1], M. Rundgren[2], B. Algers[3], [1]Animal Hygiene, P.O. Box 57, 00014 University of Helsinki, Finland, [2] Dept. Animal Nutrition and Management, Swedish University of Agricultural Sciences, P.O. Box 7024, 75007 Uppsala, Sweden, [3] Dept. Animal Environment and Health, Swedish University of Agricultural Sciences, P.O. Box 234, 53223 Skara, Sweden.

The reasons for variation between sows in piglet growth and survival are not fully understood. Therefor, sow metabolism and piglet performance were studied during a five-week lactation on eleven Yorkshire sows and litters kept in straw-bedded pens, fed twice daily and individually weighed on days 1 and 22 postpartum. On days 7 and 21 postpartum, blood was sampled from a permanent catheter in *vena jugularis*, once every hour for sixteen hours, starting before morning feeding, and every fifteen minutes the first hour after each feeding. Results were analyzed using linear regression models.
Mean glucose, urea and insulin levels did neither correlate with piglet growth and survival nor with sow weight loss ($p>0.05$). Mean NEFA-level day 7, but not day 21, was correlated to piglet growth ($r=0.55$, $p=0.02$), and to death rate of piglets ($r=-0.85$, $p=0.001$). Mean NEFA-level day 21 was correlated to sow weight loss ($r=-0.64$, $p=0.03$). Number of piglets correlated with piglet growth ($r=-0.57$, $p=0.02$), but neither with death rate nor with sow weight loss.
This indicates that sows, which invest much of their own body reserves into milk production, and thus are catabolic, produce litters with high growth rate and survival.

ANIMAL PHYSIOLOGY   [Ph]

Poster PhN2.13

Effects of colostrum intake on diarrhoea incidence and growth of neonatal calves
A. Gutzwiller, Federal Research Station for Animal Production, 1725 Posieux, Switzerland

In the first of two studies, we measured immunoglobulin G (IgG) in the first colostrum of 51 cows and of 22 heifers of the research herd and the IgG concentration in the serum of their calves, which had been receiving fixed amounts of colostrum of their dams. Heifers and their calves had lower colostrum IgG and serum IgG concentrations than cows and their calves. Although there were no seasonal differences in colostrum and calf serum IgG concentration, growth rate was significantly lower, and diarrhoea incidence was significantly higher in calves born in winter than in calves born in summer. Serum IgG concentration was not correlated with diarrhoea incidence.
In the second study, the principal calves were fed colostrum and later on fresh cows' milk plus 0.5 l of colostrum conserved with propionic acid twice daily for 10 days. The control calves were fed identically except that the conserved colostrum was replaced by fresh cow's milk. 2 of the 28 principal and 11 of the 32 control calves contracted diarrhoea ($p < 0.05$) . Weight gain of the principal and of the control calves during the first 10 days was 7.2 ± 2.0 kg and 6.2 ± 1.9 kg respectively ($p < 0.05$). The results suggest that continuous ingestion of small amounts of colostrum protects neonatal calves from diarrhoea.

Poster PhN2.14

The dependence between AVP and electrolytes concentration in the blood of calves in neonatal period.
M. Ożgo, W.F. Skrzypczak* . Department of Animal Physiology, Agricultural University, Str. Doktora Judyma 6, 71-466 Szczecin, Poland.

The experiment was carried out on 11 black-white breed calves. Blood was taken from the external jugular vein, just after delivery, after 24 and 72 hours and on the 7th and 30 day of life.
The concentration of vasopresin (RIA), sodium, potassium (flame photometry method) and chlorides (potentiometric method) was determined in blood plasma. During first 24 hours of life the AVP concentration had decreased ($p<0,05$), from 0,86 to 0,62 pg/ml. During next days, till the end of the first week of life the increase of AVP concentration ($p<0,01$) was obserwed, to the value of 1,35 pg/ml. Sodium concentration in plasma blood, in the first month of life was from 130,7 to 136,8 mmol/l, and potassium concentration was from 3,74 to 4,12 mmol/l. The direction of concentration changes of these electrolytes was other to the changes of AVP (especially in the first week of life). Chlorides concentration in the neonatal period was steady. The changes of AVP and electrolytes concentration observed in the experiment are probably the result of changes in calves kidneys¢ functions during neonatal period, and the loss of extracellular fluid.

ANIMAL PHYSIOLOGY [Ph]

Poster PhN2.15

Cereals as partial replacements of the milk ration in veal calves
I. Morel, Swiss Federal Research Station for Animal Production, 1725 Posieux, Switzerland

The effects of either rolled barley or whole grains of corn offered *ad libitum* in addition to whole milk on feed intake, fattening performance and carcass quality were studied in two trials with respectively 4 x 20 and 4 x 18 veal calves which were kept in loose boxes on straw bedding. In each trial, three experimental groups were compared with a control group which received a standard liquid ration containing whole milk plus a milk replacer (70 % and 30 % respectively on a dry matter (DM) basis).

In both trials, neither barley nor corn (both containing little iron) had a negative influence on meat colour. The intake of barley and corn increased during the fattening period; the average daily intake varied between 210 and 390 g DM. The lack of milk replacer had a negative effect on fleshiness, but this effect occurred only as a tendency when high amounts of milk were consumed. Here, the results were different in trials 1 and 2: in one trial (winter season, automatic milk feeder), milk intake and growth rate decreased, whereas in the other trial (summer period, bucket feeding), milk intake and growth rate did not decrease compared to the control group.

The economic advantage of this feeding practice is less evident than expected. Further trials are planned in order to complete the present results.

Poster PhN2.16

Immune response and performance of chickens under different factors
A.M. El-Kaiaty and F.A. Ragab. Department of Animal Production, Faculty of Agriculture, Cairo University, Giza, Orman, Cairo, Egypt.*

Five hundred one day old chicks from each of Hubbard and Fayoumi (as a local breed) were used to study the immune response to Sheep Red Blood Cells (SRBC's) and performance according to different diet ingredients. Birds from each breed were divided into ten equal groups according to type of feeding as followed. Groups 1 and 2 fed a diet supplemented with animal fat 3% and 8%, respectively. Groups 3 and 4 fed a diet supplemented with cotton seed oil 3% and 8%, respectively., Groups 5 to 8 fed dietary zinc "as an oxide" at levels 50, 100, 150 and 200 ppm/kg diet. Group 9 was fed a diet containing 100 ppm zinc plus 4g methionine/kg diet, while group 10 served as a control. The control diet was a commercial broiler diet containing 21% C.P. and 2960 k cal ME/kg. All diets and water were provided ad libitum. The light was 24 hrs/day. Birds were kept on floor and raised under the same environmental conditions. The experiment lasted until 9 weeks of age. Fayoumi breed showed greater immune response at both primary and secondary injections in all treatments than Hubbard. Animal fat increased total white blood cells (WBC) in both Hubbard and Fayoumi at 7 days after the first injection, but it decreased at 7 days after the second injection. Addition of vegetable oils caused reduction in total W.B.C. in both breeds at both times. Increasing zinc levels caused reduction in W.B.C. in Hubbard at 7 days after the first and second injections but that did not occur in Fayoumi. Fayoumi breed has a greater number of lymphocytes than Hubbard at both times and both treatments.

ANIMAL PHYSIOLOGY [Ph]

Paper PhM3.1

Genetically caused retarded growth
P. Sellier, INRA, SGQA, 78352 Jouy en Josas cedex, France.

Growth process of animals is regulated by a multitude of physiological pathways among which components of the somatotropic axis play a key role. A number of severe, simply inherited growth disturbances have been identified in humans, laboratory and farm animals. These disorders are controlled by defective alleles at major loci referring to hormones or hormone receptors, e. g. growth hormone receptor for the recessive sex-linked dwarfism (*dw*) in chickens and the recessive autosomal Laron-type dwarfism in man, and growth hormone releasing factor receptor for the recessive « little » mutation (*lit*) in mice. Apart from these particular cases, growth rate is a quantitative polygenic trait which has a moderate heritability (close to 0.30) and is influenced by prenatal and postnatal maternal effects. Divergent selection experiments have shown that upward and downward selection on growth are effective, sometimes with asymmetrical responses, but patterns of changes in underlying physiological traits somewhat differed among experiments. Also, increase of inbreeding is known to result in lower growth rate.

Paper PhM3.2

Cytokine-related free radical modulation of reduced growth performance
T.H. Elsasser*[1], S. Kahl[1], T.N. Rumsey[1] and J.W. Blum[2]. [1]U.S. Dept. of Agriculture, Agricultural Research Service, Growth Biology Laboratory, Beltsville, MD, USA and [2]Div. of Nutrition Pathology, School of Veterinary Med., University, Berne, Switzerland

A significant part of the host response to invading pathogens is the destruction of microorganisms by oxidative chemical reactions within immune cells. The formation of peroxides, superoxide, nitric oxide, and peroxynitrite is mostly directed by cytokine factors such as tumor necrosis factor-α through specific enzyme reactions or the degradation of these compounds Often there results an overproduction of these reactants which can result in frank additive pathology. One of the components of the immune response initiated and maintained by the pattern of cytokines elaborated during infection stress is the upregulation of the both constitutive as well as inducible isoforms of nitric oxide synthase (NOS) and the generation of nitric oxide (NO) from substrate arginine. Once formed, NO decays to other forms of reactive nitrogen intermediates (RNIs) and the path of decay dictates whether NO is beneficial or harmful to cells adjacent to those in which NO is produced. The path of decay depends in part on the internal oxidation/reduction atmosphere inside the cell. Evidence presented here suggests that the formation of peroxynitrite from NO and superoxide radical contributes to perturbed cell function and impacts growth through several mechanisms. One of these mechanisms postulated here is the nitrosylation of liver proteins that impairs the regulation of IGF-I.

ANIMAL PHYSIOLOGY [Ph]

Paper PhM3.3

Reduced growth of calves and its reversal by use of anabolic agents
J.L. Sartin*, M.A. Shores. Department of Anatomy, Physiology and Pharmacology, College of Veterinary Medicine, Auburn University, Auburn, AL 36849-5518, USA.

Catabolic disease processes cause, among other changes, tissue wasting, reduced growth, and reduced plasma growth hormone (GH) and insulin-like growth factor (IGF-1) concentrations. Thus, a hypothesis has been advanced that administration of anabolic agents may assist in recovery from catabolic events such as disease, surgery, trauma, etc. Early studies in humans and rats suggested that GH might improve recovery time, enhance protein synthesis, improve wound healing, etc. However, not all studies have found positive effects of GH. Other studies in the human have suggested that insulin or alpha-receptor agonists might be used similar to GH. GH administered to cattle was effective in blocking TNF release in response to endotoxin and was shown to have positive effects on the immune system. In contrast, GH treatments were ineffective in reversing the decline in IGF and in reversing catabolic events seen in Sarcocystis cruzi or Eimeria bovis infections in cattle. Other anabolic agents, estrogen/progesterone (EP) and insulin, were also shown to have positive effects on immune function. Moreover, EP can reverse many of the detrimental effects of Eimeria infection on growth and food intake as well other physiological responses to the infection. Similar protective effects of EP were observed in endotoxin treated cattle. Although individual anabolic hormones may not be effective in all diseases or catabolic circumstances, these studies provide intriguing evidence that anabolic hormones may have beneficial effects on the responses to disease as well as recovery from disease.

Paper PhM3.4

Mechanisms involved in the process of reduced and compensatory growth
J.L. Hornick*, C. Van Eenaeme, O. Gerard, I. Dufrasne, L. Istasse. Nutrition B43, Faculty of Veterinary Medicine, University of Liège, 4000 Liège, Belgium.

Growth is an integrated process resulting from the dynamic response of cells to endocrine status and nutrient availability. During restricted feeding, digestibility of the diet increases but plasma nutrients levels tend to be lowered, leading to hypersecretion of GH by the pituitary gland. Owing to low synthesis of GH receptors and GH binding proteins, the clearance of GH is low and the concentration is high with, as result, inhibition of the secretion of IGF-I by the target cells. On the other hand, high levels of GH exert positive effects on lipolysis, thus allowing use of fatty acids for energetic purpose. So, when feed restriction is moderate, dry matter weight gain is mainly composed of protein. By contrast, a severe feed restriction leads to secretion of catabolic hormones, such as cortisol, which stimulate protein catabolism and the liberation of amino acids by muscle cells. These amino acids are metabolised by hepatocytes for neoglucogenesis. During refeeding and compensatory growth, the secretion of GH and insulin is initially stimulated, promoting compensatory intake. There is, as result, an hypersecretion of insulin and thyroid hormones and settlement of conditions for enhanced synthesis of GH and IGF-I hormone receptors, the inhibition of the somatotropic axis being removed. The stimulation of protein synthesis allows the deposition of a tissue similar to that deposited during feed restriction. The phenomenon lasts some weeks. Then, protein synthesis decreases while compensatory intake is maintained, leading to increased fat deposition.

ANIMAL PHYSIOLOGY [Ph]

Paper PhM3.5

Effects of underfeeding during the weaning period on the growth and metabolism of the piglet
J Le Dividich* and B. Sève. INRA,Station de Recherches Porcines,35590 Saint Gilles,France

Weaning between 3 and 4 weeks of age implies that piglets are abruptly switched to a complex solid diet at an age when most of the nutrients are still obtained from milk. This is usually associated with a critical period of 10 to 14 days of underfeeding. This review aims at examining the growth and metabolism and its hormonal control in the piglet during this period. Whenever possible, effects of diet composition and environmental temperature are taken into account. The extent and the duration of the underfeeding period are examined in relation to pre-weaning solid feed intake , diet composition and age at weaning. Growth is analysed in terms of whole body and organ weight gain. Effects of underfeeding on protein metabolism are assessed from nitrogen balance, organ and tissue protein synthesis and its hormonal control (GH, IGF-1, T3, T4). Metabolism of fat including lipolysis and lipogenesis and of carbohydrate (gluconeogenesis)) is examined first, in the immediate period following weaning when underfeeding is the most pronounced and second, in relation to change in the type of diet and its effects on pancreatic hormones (insulin and glucagon). Finally, emphasis is put on the effects of underfeeding on the thermal requirement of the piglet and on the effects of the growth lag and on the post-weaning diet composition on the subsequent performance and body composition.

Paper PhM3.6

Body weight gain and reduced bovine mammary growth and development: implications for potential milk yields
K. Sejrsen, S. Purup & M. Vestergaard, Danish Institute of Agricultural Sciences, Research Centre Foulum, 8830 Tjele, Denmark

The available evidence concerning the relationship between body weight gain, mammary growth and milk yield potential in heifers leads to the following conclusions: 1) Increased growth rate due to high feeding level before onset of puberty may lead to reduced pubertal mammary growth and reduced milk yield potential. 2) Increased growth rate due to high feeding level after onset of puberty and during pregnancy have no influence on mammary growth and milk yield. 3) Higher body weight gain due to higher genetic potential for growth is positively related to yield. The negative effect of high feeding level before puberty seems to be present in all breeds, but the level of feeding causing reduced milk yield varies between breeds. The magnitude of response varies between experiments suggesting that it may be possible to develop high growth rate feeding regimes without negative effect on yield. A breakthrough most likely will originate from increased knowledge of the physiological relationship between nutrition and mammary development. Our investigations suggest that growth hormone (GH) is involved in mediating the effect of feeding level, but the glands do not bind GH. Evidence suggests that GH acts on the mammary glands via IGF-I, but IGF-I is increased by high feeding level - not decreased as GH. This paradoxical relationship cannot be explained by changes in circulating IGF binding proteins, but the sensitivity of mammary tissue to IGF-I is reduced by high feeding level, probably due to the action of locally produced binding proteins and/or growth factors.

ANIMAL PHYSIOLOGY [Ph]

Poster PhM3.7

Physiological mechanisms involved in the effects of concurrent pregnancy and lactation on foetal growth and mortality in the rabbit
L. Fortun-Lamothe[1]*, A. Prunier[2], G. Bolet[3], F. Lebas[1]. [1]Station de Recherches Cunicoles and [3]Station d'Amélioration Génétique des Animaux, INRA, BP 27, 31326 Castanet Tolosan, France. [2]Station de Recherches Porcines, INRA, 35590 Saint Gilles, France.

We studied the effects of lactation on foetal growth and survival in primiparous rabbits, and the origin of these effects. We showed that the foetal survival (-9.6%) and weight (-16%) at day 28 of pregnancy, were lower in females lactating (10 young) than in non lactating females. The harmful effects of lactation on foetal survival and growth were related to the size of suckled litter (4 vs 10 suckling youngs). Simultaneously pregnant and lactating does increased their feed intake (+56%). However, this increase was not sufficient to meet energy requirements for both milk production and foetal development, and the energy balance of these females was negative (-11.8 MJ in does lactating ten young). That led to a mobilisation of protein and lipid body reserves. The nutritionnal deficit that occured in lactating does induced a competition between mamary glands and pregnant uterus for the nutrient supply which prevailed foetal growth. The hyperprolactinemic status associated with lactation seemed to be responsible, at least partly, for the lower foetal survival and the reduced concentration of progesterone (-21%) in lactating does. However we could not establish any relationships between these latter parameters.

Poster PhM3.8

Comparison of intramuscular fat development in rabbits born from either simultaneously pregnant and lactating does or only pregnant does
F. Gondret*[1], L. Lamothe[2] and M. Bonneau[1]. I.N.R.A., [1]Station de Recherches Porcines, 35590 St-Gilles, [2]Station de Recherches Cunicoles, 31326 Castanet Tolosan, France.

The female rabbit can be fertilized shortly after parturition and be simultaneously pregnant and lactating. However, the competition between lactation and gestation induces a nutritional deficit that can impair foetal growth and subsequent development. The aim of this study was to compare the intramuscular fat development in youngs born from either concurrently pregnant and lactating does (PL group) or pregnant non-lactating females (P group). On day 29 (weaning, n = 5 in each group) or day 70 (commercial slaughter, n = 4 in each group), lipid content was measured on muscle homogenates, whereas adipocyte characteristics were investigated in cryostat-cut serial sections of *semitendinosus* muscles. On day 29, there were no significant differences between the two groups in the different lipid traits. On day 70, total lipid content tended to be depressed in PL rabbits compared with P animals (1.41 vs 1.65 g/100 g, P = 0.08). Adipocytes clustered along myofiber fasciculi were significantly fewer (84 vs 224 adipocytes per cm^2) and smaller (diameter: 25 vs 31 µm) in PL group compared with P group. Characteristics of adipocytes isolated among myofibers did not differ between the two groups. This study suggests that inadequate feeding *in utero* could impair intramuscular lipid deposition during rabbit postnatal growth.

ANIMAL PHYSIOLOGY   [Ph]   Poster PhM3.9

Effect of energy intake on protein deposition in the body of pigs during compensatory growth
H. Fandrejewski H. and St.Raj. *The Kielanowski Institute of Physiology and Animal Nutrition, 05-110 Jablonna, Poland.*

The experiment was arranged by 2 x 2 x 2 factorial method on 16 barrows and 16 Landrace gilts with the two crude protein diets (12 vs 17%) between 15 and 30kg live weight, and four daily intakes of metabolizable energy (26.2, 27.1, 28.3 and 29.9MJ) during subsequent stage of growth (30-60kg). A comparative slaughter technique was used to measure of protein deposition in the body of pigs.
The protein restriction reduced live weight gain (395 vs 550g/d) and rate of protein deposition (42 vs 65g/d) between 15 and 30kg. During realimentation, previously underfed pigs grew on average 7% faster (852 vs 791, $P<0.01$) and deposited 8% protein more than the control ones (132 vs 122, $P<0.01$). Increasing of energy intake from 26.2 to 29.9MJ linearly increased the rate of protein deposition from 114 to 141g/d ($P<0.01$) with somewhat higher extension in re-alimented pigs (7.5g) than in control ones (5.7g/MJ). At 60kg, the re-alimented and control pigs don't differ in respect to their chemical body composition. Gilts grew faster and deposited more protein in the body than barrows.
The study shown, that re-alimented pigs temporarily have a higher capacity to protein accretion, and the rate of compensatory response significantly depends for amount of consumed energy. High energy content in the diet improve of protein utilization in young pigs.

Paper Ph6.1

Interactions between growth of skeletal muscle and IGF-I in pigs of different sex
S. Biereder*[1], M. Wicke[1], G. v. Lengerken[1], F. Schneider[2], W. Kanitz[2]. *[1]Institute of Animal Breeding and Husbandry with Veterinary Clinic, Martin-Luther-University Halle-Wittenberg, A.-Kuckhoff-Str. 35, 06108 Halle and [2]Research Institute for Biology of Farm Animals Dummerstorf, Dep. Biology of Reproduction, Germany*

IGF-I is a pluripotent factor synthesized by hepatocyts and extrahepatic tissues (e.g. skeletal muscle). It is necessary to investigate interactions between plasma levels of IGF-I and the increase of muscle fibre diameter during the course of growth. The examinations were made on 42 crossbred pigs. Seven blood samples and four biopsy samples of two muscles (M.longissimus dorsi and M.triceps brachii) were taken for the deterinination of IGF-I blood plasma concentration and muscle fibre diameter as well as for further muscle structural and biochemical traits. IGF-I plasma concentrations show an increase during fattening with significantly highest results for boars. There are no significant differences in mean muscle area and fibre diameter of both muscles except for sows (mean muscle area of M.triceps, 180th day of life, significantly smaller area than boars and barrows). A group of animals with high mean diameter in muscle fibres (day 200) of M.triceps has significantly higher IGF-I concentrations in blood plasma than a group of animals with low muscle fibre diameter in the same muscle.

ANIMAL PHYSIOLOGY [Ph]

Paper Ph6.2

Effect of divergent selection for body weight on plasma growth hormone concentration in 90 generation mice
E.Wirth-Dzięciołowska*, B.Reklewska, A. Karaszewska. *Animal Science Faculty, Warsaw Agricultural University, 05-840 Brwinów, SGGW, ul. Przejazd 4, Poland*

The purpose of the study was to estimate a direct and correlated response to selection, for high or low body weight, conducted on many generations of mice originated from an outbred population developed from four inbred strains (A/ST, BALB/c, BN/a, C57B1/6JN). The animals of both lines (H and L) were kept in conventional environment (CvII). Highly significant differences ($p<0.01$) in body weight were found between lines from 12th to 500th days of life. Plasma growth hormone (GH) concentration was determined at the age of 42, 90, and 330 days using commercial heterogenic RIA hGH. Preliminary results indicate that the line affected highly significantly plasma GH concentration ($p<0.01$). At the pubertal age and in adult mice plasma GH levels in H line exceeded those in L and differences persisted up to 330 days of life. Males had higher GH levels in blood compared with females, reflecting sexual dimorphism in growth rate. The most distinct differences in plasma GH concentration between males and females occurred at pubertal age (in H and L line at 42 and 90 days respectively), but the effect was insignificant.

Paper Ph6.3

Plasma growth hormone levels in ewe lambs of two genotypes and association with weight gain and milk yield
G.M. Peclaris[1], K. Koutsotolis[2], G. Kann[3], E. Nikolaou[2], and A. Mantzios[2]. *[1]Department of Animal Production, Aristotle University, Thessaloniki, Macedonia Greece, [2]Agricultural Research Station, Ioannina, Greece, [3]INRA, Jouy - en - Josas, France.*

Plasma GH was determined in ewe lambs of two genotypes Boutsiko (B) (n=35) and Karagouniko x Boutsiko (KB) (n=35) at a mean age of 151, 200 and 256 d.
GH concentrations were 3.24, 2.59 and 1.24 in B and 3.13, 2.10 and 1.30 ng/mL (SEM 0.10) at the 3 sampling periods, respectively. Mean concentrations at the 2nd sampling period was greater in B than KB. Correlation coefficients of GH between sampling periods were significant in B but not in KB. Measures of growth were lower in B compared to KB. Correlation coefficients between plasma GH and measures of growth were not signifigant in both genotypes except that of the 2nd sampling period and daily gain in the next period in B. Correlation coefficients between GH (3rd sampling period) and commercial milk yield at the first lactation (2 year old) were 0.038 and - 0.416 ($P< 0.02$) in B and KB, respectively. Positive correlation coefficients between measures of growth and milk yield were significant in B and not in KB. These results suggest that differences in the association of plasma GH and milk yield may be related to developmental differences between genotypes.

ANIMAL PHYSIOLOGY [Ph]

Paper Ph6.4

Effects of Mg^{2+} on prolactin and somatotropin binding to bovine granulosa cells
I.Yu. Lebedeva*, V.A. Lebedev and T.I. Kuzmina. Research Institute for Farm Animal Genetics and Breeding, St. Petersburg-Pushkin, 189620 Russia

We have previously shown the specific binding of ^{125}I-labelled bovine PRL (bPRL) and ST (bST) to bovine granulosa cells (bGC), suggesting a direct effect of these hormones on the cells (Lebedeva et al., Biotech Agron Soc Envir 1998, 2 (Sp. issue): 67). The aim of the present study was to assess Mg^{2+} effects on bPRL and bST binding to bGC, since the constitution of extracellular medium may affect the interaction of the hormones with their receptors. Levels of bPRL and bST binding to bGC from follicles of 1-5 mm in diameter were determined by radioligand binding assay. Experimental data were tested by analysis of variance. The cation effect caused a significant rise in specific binding of ^{125}I-labelled bPRL to bGC, being dose-dependent between 1mM and 70 mM. The addition of only 1 mM Mg^{2+} to an assay buffer enhanced the binding from 7.5±1.3 to 29.2±5.7 fmol per 10^6 cells ($P<0.05$). In contrast, ^{125}I-labelled bST binding to the cells was not affected by Mg^{2+} concentrations of up to 5 mM, while 70 mM Mg^{2+} reduced 1.5-fold the level of the binding (from 38.0±2.0 to 25.8±3.7 fmol per 10^6 cells, $P<0.05$). Competition experiments with respective unlabelled hormones have shown the lack of Mg^{2+} action on the receptor affinities. The findings of the present study demonstrate that changes in the follicular content of the cation may influence the level of PRL binding to bGC, whereas somatogenic binding to the cells is independent of physiological concentrations of Mg^{2+}.

Paper Ph6.5

Effect of somatotropin on nuclear status and morphology of bovine cumulus oocytes complexes
T.I.Kuzmina,T.V.Shelouchina and N.N.Neckrasova *. Research Institute for Farm Animal Genetics and Breeding, St. Petersburg-Pushkin, 189620 Russia

The study was undertaken to assess the role of granulosa cells (GC) in the realisation of the effect of recombinant bovine somatotropin (rbGH) on nuclear status and morphology of bovine cumulus oocytes complexes (COCs). Three systems (S) were used for culture of COCs: S1. Control (C)-TCM 199 (Sigma) with 20% bovine serum (BS); S2.C with 5 ng rbGH ; S3. Ñ with 5 ng rbGH and $1\cdot10^6$ GC . COCs were cultured during 24 h. Nuclear status of the oocytes and cumulus cells(CC) was controlled by the Tarkowski's cytogenetic method. The data from 3 replicate experiments were compared between systems on the basis of Chi-square analysis. The rates of maturation, percentage of oocytes with chromatin degeneration, percentage of COCs with a high degree of cumulus expansion (CE) and a high level pycnotic index (PI) of CC(more than 20%) were respectively: S1 -71,1% (38/53), 28.3% (15/53), 50.9 % (27/53), 35.8% (19/53); S2 - 77.1% (37/48), 20.8% (10/48), 68.8% (33/48), 22.9 % (11/48); S3- 87.5% (56/64), 14.1% (9/64), 82.82% (53/64), 17.2% (11/64). There was no difference in the chromatin degeneration of oocytes between the groups. RbGH in the presence of GC (S3) enhanced percentage of COCs with a high degree of CE ($P<0.001$) and decreased percentage of COCs with a high level of PI of CC($P<0.05$). The positive effect of rbGH on CE and chromatin status of CC (PI) is dependent on the granulosa cells.

ANIMAL PHYSIOLOGY [Ph]   Paper Ph6.6

Role of prostaglandins in ovulation in the swine
Y. Yamada*[1], H. Kadokawa[1] and Y. Kawai[2]. [1]Hokkaido National Agricultural Experiment Station, Hitsujigaoka-1, Sapporo, 062-8555 Japan, [2]Tokyo University, Hongo-1, Bunkyou-ku, Tokyo, 113-0033 Japan.

Prepubertal gilts were injected intramuscularly with hCG 72 hours after receiving PMSG. Twenty-four hours after hCG injection, they were injected intramuscularly either phisiological saline solution (Control group) or 7.6 mg/kg.b.w. of indomethacin (Indomethacin group). In Control group, ovulation occurred 32 to 48 hrs after hCG injection and progesterone levels were low at the time of hCG injection, but markedly increased after ovulation. Estradiol-17 β and testosteron levels which reached maximum values 72 hrs after PMSG injection, and decreased after hCG injection and remained low until about 36 hrs after ovulation. Levels of $PGF_{2\alpha}$ in ovarian vein blood plasma reached maximum values near the time of ovulation. Concentrations of PGE_1, $PGF_{2\alpha}$ and 6-keto-$PGF_{1\alpha}$ in follicular fluid remained relatively low until 18 hrs after hCG injection. And then they showed an increase 36 to 42 hrs after hCG injection and reached maximum values as expected time of ovulation approached. In Indomethain group, ovulation was inhibited in a dose-dependent manner. But patterns of steroid secretion were similar to those of Control group. Levels of PGE_1 and $PGF_{2\alpha}$ in the follicular fluid were low 48 and 72 hrs after hCG injection.

Paper Ph6.7

Relationships between follicular fluid composition and follicular/oocyte quality in the mare
N. Gérard*, G. Duchamp, M. Magistrini. Reproduction Equine, INRA-Haras Nationaux, Physiologie de la Reproduction des Mammifères Domestiques, 37380 Nouzilly, France.

In the mare, no preovulatory LH surge occurs but LH rises gradually with a maximun 1 days after ovulation. Moreover, *in vitro* maturation rate of oocytes remains low, and *in vitro* fertilization methods give only limited success. Thus, follicular development and maturation could involve some species-specific mechanisms and studies on follicular fluid composition may contribute to understanding of these events. We report here studies on the evolution of intrafollicular content of steroids, protein patterns and IGFBP profiles during follicular growth, maturation and atresia in the mare, in relation to follicular and oocyte quality. We demonstrated that 1) follicular growth of the dominant follicle is associated with an increase in intrafollicular estradiol-17β and progesterone levels and a decrease in IGFBP-2 and IGFBP-5, 2) follicular maturation is associated with a decrease in intrafollicular estradiol-17β level, a further increase in progesterone level, a slight increase in IGFBP-2 and finally, the induction of the follicular expression of a 200kDa protein, which identity is still unknown, and 3) follicular preovulatory regression and atresia are characterized by alterations in steroidogenesis, as well as an increase in intrafollicular levels of IGFBP-2, IGFBP-4 and IGFBP-5. Interestingly we obtained data which suggest that the expression of the 200kDa protein relates to oocyte resumption of meiosis and cumulus expansion.

ANIMAL PHYSIOLOGY [Ph]

Paper Ph6.8

Time of LH peak in oestrus synchronized buffaloes in two different seasons
A. Borghese*[1], V.L. Barile[1], G.M. Terzano[1], A. Galasso[1], A. Malfatti[2], O. Barbato[2], A. Debenedetti[2]. [1]Istituto Sperimentale per la Zootecnia, 00016 Tormancina Monterotondo, Roma, [2]Istituto di Fisiologia Veterinaria, Università di Perugia, Italy.

The trial was carried out on 20 water buffaloes, bred in southern Italy. Ten animals were synchronized in November, 10 in March with PRID (1.55 g progesterone intravaginal device + 10 mg oestradiol benzoate) kept for 10 days. On the 7th day after PRID application, 1000 UI PMSG and 0.15 mg cloprostenol (prostaglandine $F_2\alpha$ analogue) were injected. Blood samples were taken every 4 hours starting from 30 h to 110 h, after PRID removal. A qualitative detection of prevulatory LH surge was done by ELISA (Reprokit-Sanofi). In 16 animals out of 20 was possible to detect LH preovulatory peak. This phenomenon appeared 46.87 hours ± 21.53 after PRID removal in buffaloes synchronized in November, 61.00 ± 12.05 hours in other ones synchronized in March without significant difference. The ovulation (checked daily by rectal palpation) occured within 72 hours from PRID removal in 8 buffaloes in November and in 4 in March, while occured within 96 hours in 3 buffaloes in March. On the basis of our results we can suggest that 72 and 96 h after PRID removal are proper time for A.I. in March synchronized buffalo cows, while 48 and 72 hours could be better for November synchronized buffaloes.

Paper Ph6.9

Effects of prenatal stress on endocrine and immune responsiveness in neonatal pigs
E. Kanitz, M. Tuchscherer and W. Otten. Research Institute for the Biology of Farm Animals, Wilhelm-Stahl-Allee 2, D-18196 Dummerstorf, Germany.

Stressful events during pregnancy are known to alter offspring behavior, morphology and physiology. The aim of the present study was to determine the effect of prenatal stress on endocrine responsiveness and immune function in neonatal pigs. Sows were daily restrained for five minutes during the last five weeks of pregnancy. Piglets of stressed sows showed higher catecholamine levels under emotional stress conditions on the first days of life, indicating an enhanced activation of the sympathetic nervous system. However, there were no differences in resting levels of cortisol and in stress-induced cortisol release after ACTH in prenatally stressed and control piglets. Futhermore, changes in prenatally stressed piglets include a reduction of steroid receptors in the hypothalamus, enhanced concentrations of plasma corticosteroid-binding globulin, an increased size of adrenal gland and a decreased thymic weight. Prenatal stress suppressed immune function as shown by decreased lymphocytes responses to T-cell and B-cell mitogens and by reduction of serum IgG levels on the first postnatal days. These findings and the higher mortality of prenatal stressed piglets support the hypothesis that stress during pregnancy can induce alterations in immune function of the offspring, thereby making them more susceptible to infection. The mechanisms underlying this effect are yet not clear but could result from an action of maternal stress hormones on the fetal neuroendocrine system.

ANIMAL PHYSIOLOGY [Ph]

Paper Ph6.10

Effects of milking frequency on milk proteins
A. Sorensen, D. Muir and C. Knight, Hannah Research Institute, Ayr KA6 5HL, UK.

Long-term increased milking frequency improves casein as a proportion of total milk protein, purportedly due to the shorter storage time and hence reduced plasmin proteolysis. To test this, 11 cows were milked three times daily (3x) on one udder-half commencing in lactation week 9 while the other half continued on twice daily milking (2x). During lactation week 52 milking frequency was reversed in the two halves for 2 days. Milk samples were collected immediately before (3x/3x, 2x/2x) and during the reversal (3x/2x, 2x/3x). Samples were analysed for content of crude protein and casein protein and the integrity of the mammary epithelium was measured by the milk concentrations of sodium and potassium. Casein as a proportion of total protein (cn) was significantly higher in 3x3x (77.36±0.22) compared to 2x2x. (75.67±0.20). A short term increase in milking frequency increased cn (2x/3x: 76.96±0.22) while a reduction decreased it (3x/2x: 76.62±0.20), however, significant differences remained (Anovar), so the recovery or loss was incomplete. α-casein and β-casein measured in 2 cows were higher in 3x3x while γ-casein was lower, as were plasmin and the ratio of sodium to potassium. Reversal of milking frequency only partially eliminated these differences. In conclusion, two phenomena contribute to the milking frequency-related differences in milk proteins. One is proteolysis during storage, a short-term phenomenon, the other is a change in the integrity of the mammary epithelium, which has long-term importance.

Poster Ph6.11

The changes of electrolytes concentration and plasma osmolality in the blood of cows in perinatal period.
W.F. Skrzypczak*, M. Ożgo. Department of Animal Physiology, Agricultural University, Str. Doktora Judyma 6, 71-466 Szczecin, Poland.

The experiment was carried out on 11 black-white breed cows. Blood was taken from the external jugular vein -30, 14, 7 and 3 days before delivery; on the day before delivery; just after delivery; 24 and 72 hours after and on the 7th and 30th day after delivery. Sodium concentration in plasma blood was steady, from 139,2 to 147,1 mmol/l. Statistically significant decreasing of Na concentration was obserwed 24 hours before delivery. Potassium concentration in plasma blood was from 3,48 to 3,79 mmol/l. In the last week of pregnancy and during 30 days after delivery the concentration of this elektrolyt in plasma blood was steady. The changes of chlorides concentration in plasma blood were statistically significant, from 107,9 to 99,5 mmol/l. The highest concentration was observed in the last week of pregnancy. The plasma osmolality in the research period was steady, from 255,1 to 258,6 mmol/kg H_2O. The highest value was observed a month before delivery, and the lowest value - on the 30th day after delivery. The results of the experiment show that the mechanisms regulating of water-electrolyte balance are efficient.

ANIMAL PHYSIOLOGY [Ph]

Poster Ph6.12

On immunoprotective effect of cerebral neuropeptides
J.Kumar[1], A.Karus*[2], T.Schattschneider[1], M.-A.Kumar[1], A.Kaljo[3] and Ü.Pavel[1] [1]Estonian Agribiological Centre, Rõõmu tee 10, Tartu 51013, Estonia; [2]Estonian Agricultural University, Institute of Animal Science, Kreutzwaldi 1, Tartu 51014, Estonia, [3]Tartu University, Faculty of Physics and Chemistry, Jakobi 2, Tartu 51014, Estonia

In our preliminary investigations on the porcine brain neuropeptide fractions (cortex cerebri, CC, frontal lobe) and hypophysis plus hypothalamus (HH) extraction with basic pH 7.4 (CCb and HHb, correspondingly) and acidic pH 5.4 buffer (CCa and HHa, correspondingly)(A.Karus et al., EAAP 1998) it was demonstrated, that the fraction CCb had the most active immunoprotective power (CCb>HHb=HHa>CCa). The immunoprotective effect of brain neuropeptide fractions was determined by subcutaneous administration of different doses and five days later the experimental infection with *Salmonella typhimurium* (about LD_{100}, i.e. 1.4×10^4 bacterial cells intraperitonally) of mice. The mortality of mice was registered on the 10th day after the infection. Present experiments with the fraction of CCb confirm our previous results on immunoprotective effect of CCb (dose ~70 ng per animal), although, the 10-fold dose (747 ng) of CCb did not affect the immunoprotection of mice, and had high toxic effect. Also, the 10-fold lower doses (6.75 ng) had neither a significant immunoprotective, nor toxic effect. That shows that the CCb dose of approximately 68-70 ng is close to the optimum.

Poster Ph6.13

The role of protein kinase C in realization of effect of prolactin on pig granulosa cells
V.Yu.Denisenko and T.I.Kuzmina*. Research Institute for Farm Animal Genetics and Breeding, St. Petersburg-Pushkin, 189620 Russia

The purpose of these studies was to determine the source(s) of the increase in the intracellular free calcium ($[Ca^{2+}]_i$) in response to bovine prolactin (bPRL) in pig granulosa cells (pGC) from ovarian follicles (diameter of 3-6 mm). Fluorescence spectrophotometer (Hitachi) and fluorescent dye fluo-3AM (4.4 mM) and chlortetracycline were used for the measurement of the $[Ca^{2+}]_i$ and intracellular store calcium ($[Ca^{2+}]_{is}$) levels in pGC. Levels of $[Ca^{2+}]_{is}$ were expressed in the conventional units (C.U.) of the intensity of cell fluorescence. Influence of bPRL (5 ng/ml) and effects of protein kinase C (PKC) inhibitor, Ro-31-8220 (Ro-31) (10 ng/ml) on mobilization of Ca^{2+} were studied. BPRL increased $[Ca^{2+}]_i$ levels in pGC. Ro-31 increased level of $[Ca^{2+}]_i$ in pGC. The addition of BPRL to the incubation medium containing Ro-31, further enhanced the basal level by 195 ± 10 nM (P<0.01). Level of $[Ca^{2+}]_{is}$ decreased after the influence of bPRL (0.48 ± 0.01 C.U.). Ro-31 decreased concentration of $[Ca^{2+}]_{is}$ to 0.36 ± 0.02 C.U. (P<0.001). Influence of bPRL on the pGC treatment with Ro-31 have induced the decreasing the level of $[Ca^{2+}]_{is}$ to 0.38 ± 0.02 C.U. These data suggest the involvement of PKC in the realization of the effect of bPRL on pGC and have demonstrated that the PKC is mobilized Ca^{2+} from intracellular stores and increased level of the $[Ca^{2+}]_i$.

ANIMAL PHYSIOLOGY [Ph]

Poster Ph6.14

Relationship between mitochondrial activity and intracellular stored Ca^{2+} levels in parthenogenetically activated bovine oocytes
E.D. Fedoskov*, L.D. Galieva, T.I. Kuzmina, and Ju.S. Kirukova. Research Institute for Farm Animal Genetics and Breeding, St. Petersburg-Pushkin, 189620 Russia

The aim of the present work was to study a relationship between mitochondrial activity (MA) and intracellular stored Ca^{2+} ($[Ca^{2+}]_{is}$) levels in parthenogenetically activated (PA) bovine oocytes. In vitro matured oocytes were treated with 8.0 % ethanol in PBS for 5 min at room temperature and then cultured for 0.5 - 24 h. Levels of $[Ca^{2+}]_{is}$ and MA in PA oocytes were expressed in conventional units (CU) of a fluorescence intensity (FI) of the oocytes after their incubation with the fluorophore chlortetracycline (40 µM) or the lazer dye rhodamine 123 (Rh123, 25 µg/ml), respectively. The FI of the oocytes was measured immediately before and after the activation. The data obtained were analyzed by analysis of variance. Prior to activation, the mean level of $[Ca^{2+}]_{is}$ was 5.66±0.64 CU, and that of FI of Rh123 was 1.53±0.23 CU. After 1, 8, 10, and 24 h of activation, the average levels of $[Ca^{2+}]_{is}$ in the oocytes were 5.83±0.80, 3.55±0.56, 3.21±0.44, and 5.85±0.33 CU, whereas those of Rh123 FI were 1.47±0.19, 2.94±0.38, 2.35±0.43, and 1.49±0.19 CU, respectively. Thus, $[Ca^{2+}]_{is}$ levels in PA oocytes decreased first with time and subsequently increased up to their initial values, while levels of Rh123 FI in the oocytes did conversely, with r=-0.677±0.212 (P<0.01). These results show the inverse relationship between mitochondrial activity and $[Ca^{2+}]_{is}$ levels in PA bovine oocytes.

Poster Ph6.15

Embryo cloning by nuclear transfer: experiences in sheep
P. Loi[1]*, Sonia Boyazoglu[2], J. Fulka Jr[3]., S. Naitana[4], and P. Cappai[1]. [1]Istituto Zootecnico e Caseario, 07040 Olmedo, Italy, [2]Istituto de Produzioni Animali, Facoltà di Medicina Veterianaria, Perugia, Italy, [3]Research Institute of Animal Production, Prague, Czech Republic, [4]Dipartimento di Biologia Animale, Università di Sassari, Italy.

This article reviews a series of interrelated experiments carried out from 1992 to 1997 regarding the co-ordination of nuclear and cytoplasmic events after embryo reconstruction. The experiments were undertaken at the Istituto Zootecnico e Caseario per la Sardegna (IZCS) and the sheep used belonged to the Sardinian dairy breed. The use of enucleated, pre-activated recipient cytoplasts resulted in an increased frequency of development to blastocyst of embryos reconstructed with unsynchronized blastomere nuclei. This increase in development was due to the absence of chromatin damage and unbalanced ploidy in nuclei transferred after the decay of MPF (Maturation Promoting Factor). The combination of synchronous S-phase cytoplast-karyoplast nuclear transfer together with the reduction of embryo losses from the oviduct of temporary recipients allowed for the first time the production of large number of clones of genetically identical lambs thus bringing nuclear transfer closer to practical application. Finally, the last series of experiments deals with the establishment of an alternative protocol for embryonic nuclear transfer where the combination of a chemical activating agent, ionomycin, with a protein kinase inhibitor, 6-Dymethylaminopurine, resulted in the highest development rates to blastocyst stage of metaphase II enucleated oocytes reconstructed with embryonic nuclei described so far.

ANIMAL PHYSIOLOGY   [Ph]   Poster Ph6.16

Some physiological responses related to dietary magnesium oxide in laying Japanese quail diets
A.M.M. Hamdy. Department of Animal Production, Faculty of Agriculture, Minia University, Egypt.

Two hundred layers and one hundred cocks of Japanese quail of 10 weeks old were used to study the physiological responses related to dietary magnesium oxide (MgO). All birds were housed in battery cages, each cage included two females and one male. The hens were divided into five groups each of fourty hens. Birds of group 1 were served as control those of group 2, 3, 4 and 5 were received dietary (MgO) by levels of 300, 400, 500 and 600 p.p.m. respectively. The experiment was extended till 24 weeks of age (14 week period), all birds were reared under the standard recommended management.

Results reveal that levels of 400 and 500 p.p.m. of MgO significantly ($P<0.05$) improved egg production (egg number) and increased ($P<0.05$) fertility and hatchability. While hatchability was tended to reduce at the level of 600 p.p.m.

Hematocrite values and hemoglobin; and total protein, globulin and vitamin C in blood plasma were significantly ($P<0.05$) increased by increasing the dietary MgO level upto 500 p.p.m.

It could be concluded that the level of 500 p.p.m. of MgO is more proper for the laying Japanese quail.

CATTLE PRODUCTION   [C]   Paper C3.1

Does the milk redoxpotential correlate with the variability of microbes?
H.-D. Matthes,[1] V. Pastushenko,[1] H. Heinrich[2]. [1]Department of Agricultural Science, Rostock University, Vorweden 1, D-18069 Rostock, Germany, [2]LABO TECH Labotechnik GmbH, Richard-Wagner Str. 31 D-18119 Rostock-Warnemünde, Germany*

The objective of this study was to determine the response of thermodynamic milk property to a bacterium metabolism. The initial level of redoxpotential and the level after adding 10, 20, and 30 mg of p-benzochinone (pbc) in 92 milk samples with different bacterium composition were investigated. The application of the pro-oxidant pbc make it possible to determine the reactions of the antioxidant system on the changed metabolism. The analyses of the antioxidant status showed advantage of thermodynamic stability of the milk samples free of bacterium, compared with the micro-organisms contained milk. There was no significant influence of not pathogenic staphylococci on the thermodynamic of milk. Our findings demonstrate, that metabolic alterations caused by corynebacterium destroyed the antioxidant quality of milk and increased thermodynamic disorders of the biocolloid. The observed reactions correlated with the variability of the enzymatic influences of microbes and common reactions of the immune system.

Paper C3.2

The effect of feeding intensity and protein supplementation during the last 6 or 12 weeks before parturition on performance of primiparous cows
P.E. Mäntysaari, Animal Nutrition, Animal Production Research, Agricultural Research Centre MTT, FIN-31600 Jokioinen, Finland

The effect of duration of high feeding level before parturition with or without protein supplement was studied. A group of 24 Finnish Ayrshire heifers were divided in 12- (12 weeks high feeding no protein supplement), 12+ (12 weeks high feeding with protein supplement), 6- (6 weeks high feeding no protein supplement) and 6+ (6 weeks high feeding with protein supplement) treatments. Rapeseed meal was used as protein supplement. Three weeks before parturition and after parturition all heifers were fed the same diet. The daily gain of the heifers during 12 to 6 and 6 to 3 weeks before parturition were 1130/1159, 1409/1190, 330/1399 and 480/1500 g/d on 12-, 12+, 6- and 6+ diets respectively. The increase in duration of high feeding level before parturition increased the calving body weight and mobilization of body reserves in early lactation ($P<0.05$), but had no effect on milk yields or milk protein content. Twelve weeks with high feeding level without protein supplement increased milk fat content during the first weeks of lactation. The rapeseed meal supplementation 12 or 6 weeks before parturition had no effect on milk yields or milk protein content, but tended ($P<0.10$) to increase the body weight of the calves at birth and body weight gain of the cows during the 120 days of their first lactation.

CATTLE PRODUCTION   [C]

Paper C3.3

Urea milk content in response to different supplementation on grazing dairy cows
A. González-Rodríguez* and O. P. Vázquez Yáñez.. Centro de Investigaciones Agrarias de Mabegondo. Apartado 10, 15080 A Coruña. Spain.

Protein nutrition of grazing dairy cows was estimated through the Milk urea nitrogen test. From spring to summer of two years, 32 dairy cows, calving from February to March, were divide in four groups, grazing ryegrass-clover pastures and supplemented with different levels of crude protein (CP). The treatments were: No supplementation, 140 g kg^{-1} CP ration, 170 g kg^{-1} CP ration and 200 g kg^{-1} CP ration.
Supplemented groups had a significantly higher milk production compared to no supplementation group. Milk protein was significantly higher for 170 g kg^{-1} CP treatment.
Milk urea response was directly proportional to CP in the ration. Values were significantly different ranging from 244, to 364 mg kg^{-1}, urea in late spring of the first year, or from 143 to 255 mg kg^{-1} urea, in early spring of the second year, for the first treatment, no supplementation, to the fourth, 200 g kg^{-1} CP treatment, respectively. Results shows the influence of the total CP content of the ration on milk urea content variation in grazing cows.

Paper C3.4

Changing of milk composition in first ten days of lactation in Black Pied Cattle
Ö. Sekerden*[1], İ. Tapkı[1], M. Sahin[2]. [1]Mustafa Kemal University, Faculty of Agriculture, Department of Animal Science, 31034, Hatay, Turkey, [2]Hatay State Farm, Hatay, Turkey

This research was conducted to determine changing of milk composition in the period from 1st to 10th days of lactation in Black Pied Cattle raised at Reyhanlı State Farm.
The material of the research was formed by the data belong to milk samples taken from 1st to 10th days of their lactation from the 17 cows calved 24.10.1997-29.03.1998 period in the morning milkings. All the trial animals were in their second lactation order.
Composition of the samples belong to 1st day is more different than other ones. The changing in protein rate was completed in first 8-day of lactation, the changing in dry matter (DM) was completed in first 9-day of lactation; The changing in fat, solids non fat (SNF) and ash rates continued until 10th day of lactation.
The averages of milk composition belong to colostrum period become stabile in the first 7 days of lactation for fat and lactose rates, but in 10 days of lactation for DM, SNF and ash rates.

CATTLE PRODUCTION   [C] Paper C3.5

The claw horn parameters in Holstein-Friesians in West Siberia
E.V. Telezhenko*, V.L. Petukhov, N.N. Kochnev, A.V. Kosolapikov. Institute of Veterinary Genetics and Selection of Novosibirsk Agrarian University, 160 Dobrolubov Str., 630039 Novosibirsk, Russia.

The imported dairy cows were characterised by high frequency of diseases including leg and foot ones. However, the offspring had no significant differences in the disease frequency with the local cattle. Nevertheless, the incidence of foot and leg problems remains rather high. In order to identify traits for genetic improvement of claw quality, the parameters of claw horn must be evaluated. The data of the claw horn black-and-white cattle with various degrees of Holstein-Friesians inheritance were used. The growth velocity, wearing velocity and growth degree were assessed. In addition, such physical indices as wearing resistance and water absorption capacity were estimated. The rear claws horn was characterised by low wearing velocity (P<0.05), but there were not significant differences between the rear and front claws on the growth velocity and growth degree. The water absorption capacity of the front claw horn was more significant than in the rear claws horn (P<0.001). The genetic variation of the claw horn resistance to wear was not found. The heritabilities for the parameters were generally low (h^2=0.02 - 0.18) which might be a consequence of the small data set.

Paper C3.6

Comparison of leg injuries and behaviour in dairy cows kept in cubicle systems with straw bedding or soft lying mats
B. Wechsler, J. Schaub, K. Friedli and R. Hauser. Swiss Federal Veterinary Office, FAT, 8356 Tänikon, Switzerland.

In order to reduce straw input, soft lying mats are increasingly used instead of straw bedding in cubicle systems for dairy cows. In the present study, data on leg injuries and cow behaviour was collected on 5 farms with straw bedded cubicles and on 13 farms using 4 different types of soft lying mats. The study was realised in winter when the cows did not have access to pasture. On each farm, all cows were checked once for leg injuries and lying behaviour of 10 cows was recorded automatically during 3 days by means of a girth that had been especially developed for this study.
Cows kept in cubicle systems with soft lying mats had a significantly higher incidence of both hairless patches more than 2 cm in diameter (P < 0.001) and scabs or wounds less than 2 cm in diameter (P < 0.001) located in the tarsal joints than cows in cubicle systems with straw bedding. With the carpal joints, the incidence of leg injuries did not differ significantly between the two housing conditions. There were also no significant differences with regard to the total time the cows spent lying per day and the number of lying bouts per day. In conclusion, soft lying mats were found to be equivalent to straw bedding in terms of lying behaviour but less favourable with respect to leg injuries located in the tarsal joints.

CATTLE PRODUCTION   [C]   Paper C3.7

Comparison of free stall cattle mattresses in a preference test
B. Sonck*, V. Vervaeke, J. Daelemans. Department Mechanisation, Labour, Buildings, Animal Welfare and Environmental Protection, Agricultural Research Centre, B. Van Gansberghelaan 115, B-9820 Merelbeke, Belgium.

The paper describes the results of a comparative cow preference test with 11 different types of mats or mattresses. The purpose of the study is to determine suitability and cow comfort (cow preference) for various materials. The study was initiated in December 1997 in a three-row cubicle house for 94 cows. About 30 cows were observed for twice 24 h before and after the installation of the mattresses to determine cow behaviour in the cubicle house. Significant differences have been found in their lying and standing behaviour. Video-recordings have been made before and after the installation of the mattresses. Significant differences have been found in 'lying rate' (% of 24 h that a mattress is occupied by a lying cow) and in 'occupancy rate' (% of 24 h that a mattress is occupied by a lying or standing cow). The soft lying floors are significantly more used than the hard ones. The degree of occupation of the cubicles with a relatively hard floor (rubber mat) can positively be influenced by using a soft supporting layer (25 mm) under a thin rubber mat. We found also remarkable differences in lying and occupancy rate between mattresses based on the same concept. Differences in lying rate of nearly 15% were observed.

Paper C3.8

Types estimated by using of linear scoring and measurements of dairy cows
J. Püski[1], I. Györkös[2], A. Gáspárdy*[3], S. Bozó[2] and E. Szűcs[1]. [1]Dept. of Anim. Husb. of the Gödöllö Univ. of Agr. Sci., H-2103 Gödöllö, Páter 1., [2]Res. Inst. for Anim. Breed. and Nutr., H-2053 Herceghalom, Gesztenyés 1., [3]Dept. of Anim. Husb. of the Univ. of Vet. Sci., H-1400 Budapest, István 2., Hungary.

The goal was to evaluate the similarity of the body scoring system and the body measurement kept generally for the best objective method. The target-traits (withers height, body depth, rump width at the gluteal tuberosity and body capacity) of Holstein Friesian heifers (n=159) were taken down simultaneously after the first calving in 1998. Three-three groups (-extreme, medium, +extreme) based on the standard deviation of the body measurements were created and compared in each traits, as well as phenotipical relationships were calculated. In interest of the type evaluation six type groups (small-narrow/wide, medium-narrow/wide and high-narrow/wide) were made.

The coefficient of variation of scored traits excepting the body capacity was high comparing the measured traits (cv%=18.9-27.7 vs. 1.4-8.1) in all groups. Therefore, significant differences ($p<0.01$) were found only between the extreme groups of scored body depth and rump width traits. The correlation coefficients between the values of the same traits taken down by scoring and measuring systems can be seen as low figures (r=0.71, 0.50, 0.43 and 0.59, respectively) since we have the same traits. According to this the differences between the type groups were smaller in each traits estimated by scoring.

CATTLE PRODUCTION [C] Paper C3.9

Survey of Albanian Dairy Feed Resources
R. O. Kellems*[1] and R. Balogh[2]. [1]Animal Science Department, 353 WIDB, Brigham Young University, Provo, Utah, 84602, USA., [2]Land O'Lakes, International, Minneapolis, MN, 55440, USA.

More than 300 feed samples of various feedstuffs used to feed dairy cattle were collected during July and August of 1995. Samples were collected in six regions (Tirana, Shkoder, Korce, Lusnja, Durres and Kavaja) of Albania. These samples were primarily, alfalfa, wheat by-products, grass hay, and straws. The samples collected were subjected to standard laboratory analysis to assess their mineral (Ca, Mg, K, Na, P, Zn, Fe, Mn and Cu) and nutrient contents (CP, ADF, NDF, RFV). Based on the analysis and information relating to feeding levels, the nutritional adequacy of the diets being fed were evaluated for 5, 10, 15, and 20 l/d of milk production. Dietary TDN and CP levels seemed to be adequate, except when high amounts of straw was being fed. Dietary ADF and NDF were higher than required in all rations. Dietary intake of Ca, K, and Mg were all found to be adequate. Dietary Na was adequate up to 15 l/d of milk, but inadequate for 20 l/d of milk. Dietary P was inadequate at all production levels. Dietary intake of Fe was marginal at all production levels. Dietary intake of Cu, Mn and Zn were inadequate for all production levels. These results indicate a need for additional mineral supplementation that would enhance performance of dairy cattle in Albania.

Paper C3.10

The adaptation of brown Swiss and Simmental cattle originated Switzerland to Kazova State Farm in Turkey
Ö.Sekerden*[1]. [1]Mustafa Kemal University, Faculty of Agriculture, Department of Animal Science, 31034, Hatay, Turkey.

The material of the research was formed by records of 761 lactation of Brown Swiss cows calved in 1980-1988 period and 510 lactation of Simmental cows calved 1990-1993 period in Kazova State Farm, in Turkey.
The standardized averages of 305-day milk yield, calving interval, lactation length, first calving age determined as 3499 ± 718.1 kg, 398.9 ± 71.76 days, 308.4 ± 52.24 days and 986.4 ± 159.95 days respectively for Brown Swiss Breed. The standardized averages of 305-day milk yield, calving interval and first calving age were determined as 3426.9 ± 618.3 kg, 405.9 ± 48.72 days, 839.7 ± 60.90 days respectively for Simmental Breed.

CATTLE PRODUCTION   [C]   Paper C3.11

Efficiency of imported Holstein heifers
Urban F.*, Bouška J., Štípková M. Research Institute of Animal Production, Přátelství 815, Prague 10 - Uhříněves, 104 00, Czech Republic.

Production records of 15374 Holstein animals including heifers imported from several countries totally from 65 herds were analysed. The performance of imported heifers (n = 5738) and their contemporaries (n = 7967) was compared. It was found that the production of milk and protein of imported first-calvers was significantly higher by +593,1 and +22,9 kg resp. The production of imported animals` daughters (n = 1 669) was lower by -232,7 kg milk and -11,3 kg protein when compared with imported animals. This tendency was confirmed by the analysis of the data set classified according to the herd performance level; differences in milk production of imported cows ranged from +643 to +821 kg in comparison with national cows. The comparison of animals born in the Czech Republic (daughters of imported heifers x daughters of national cows) resulted in significantly higher performance of daughters of imported heifers (by +369,0 kg milk and +11,4 kg protein for the first lactation). The relationship between daughters` performance and their sires` breeding values (BV) for protein production was analysed. Groups of daughters after sires with BV > x + s (x = 10,8 kg; s = 8,0) reached significantly higher production by +527 kg milk and +15,0 kg protein. The results correspond with the fact that the highest efficiency of first-calvers according to the country of origin was found in animals imported from France (+651 kg milk, +23,4 kg protein); BV for protein production of their sires was +13,9 kg.

Paper C3.12

Evaluation of lifetime production of top cows in different dairy breeds
M. Horvai, J. Dohy, G. Holló. Institute of Animal Husbandry, University of Agricultural Sciences, H-2103 Gödöllõ, Hungary

Cows of different breeds with high lifetime production have been analysed in order to choose the best producers for embryodonors. The breeds have been ranked based on their milk production (Jersey milk fat kg) and a rank correlation was calculated between milk and protein production, and between milk and fat production. The probability of appearance of a correlation breaker was also investigated. In a Swiss Simmental population (n=241), it was found that lower the protein production, lower the correlation. In Swiss Brown (n=52), the correlation was r=0-0.47. In the Osnabrück Holstein-Friesian population (n=47), a close correlation (0.74) was found in the best producing subpopulation, but not in the others out of the three (p<0,01). 27% of the population were correlation breakers. In populations of German Black and White Holstein (n=15) and of Red Holstein (n=9), close correlations (0.90 and 0.84, respectively) with no correlation breaker were found. In a Hungarian Holstein-Friesian population (n=22) the correlation was weak (0.29), while the proportion of correlation breakers was 27%. A weak correlation and a 18% rate of correlation breakers was found in a Jersey population (n=22). It can be concluded that low rank correlation coefficient makes the appearance of correlation breakers more probable.

CATTLE PRODUCTION   [C]

Paper C3.13

Effect of growth rate on tenderness of meat from heifer calves
M. Therkildsen[1,2], M. Vestergaard*[1] and L. Melchior Larsen[2], [1]Danish Institute of Agricultural Sciences, Foulum 8830 Tjele, Denmark, [2]Royal Veterinary and Agricultural University, 1871 Frederiksberg C, Denmark

The objective of this study was to determine the relationship between growth rate of heifer calves and tenderness of the meat. The 36 Friesian calves were allocated to three different feeding regimes, MM, HM or HH representing either a high (H) or a moderate (M) energy level from 4 days to 90 kg live weight (period I) and from 90 kg to slaughter at 240 kg live weight (period II). The high and moderate feeding levels allowed for a daily gain of approximately 900 g and 650 g in period I and 1200 g and 750 g in period II, respectively. *M. longissimus* (LD) and *M. supraspinatus* (SS) were removed from the carcass 24 hours after slaughter and part of the muscles were either frozen immediately or aged for an additional 6 days and frozen. The muscle samples were used for shear force measurement on cooked meat aged for 1 day and 7 days and for tenderness evaluation by sensory panel using LD aged for 7 days. Meat from calves with a high growth rate before slaughter (HH) had the lowest shear force both after 1 day and 7 days of ageing (p<0.05) and the highest tenderness score (p<0.01). The correlation between growth rate in period II and shear force at day 7 in LD or SS or tenderness of LD was B0.56 (p<0.001), B0.45 (p<0.01) and 0.67 (p<0.001), respectively. The results show that tenderness of meat is affected by previous growth rate of the animal. Thus, a high growth rate resulted in the most tender meat.

Paper C3.14

Investigation of a polymorphism in the growth hormone (GH) gene, plasma concentrations of GH and IGF-1 and carcass traits in Friesian bulls
R. Grochowska*[1], Lunden A.[2], Zwierzchowski L.[1], Snochowski M.[3], Oprzdek J.[1], Dymnicki E.[1], [1]Institute of Genetics and Animal Breeding, Jastrzębiec, 05-551 Mroków, Poland, [2]Swedish University of Agricultural Sciences, 750 07 Uppsala, Sweden, [3]Institute of Animal Physiology and Nutrition, 05-110 Jablonna, Poland

GH genotypes of 109 Polish Friesian bulls were identified by PCR-RFLP technique. The GH release in bulls aged 11 months was analysed after thyrotropin releasing hormone (TRH) stimulation in blood samples collected during a period of 2.5 hours. Animals were slaughtered and dissected at the age of 15 months. Associations of the Leu/Val polymorphism (substitution of leucine to valine) and the regression of basal GH concentration, GH peak amplitude and IGF-1 concentration with carcass traits were estimated using a sire model. GH genotypes significantly influenced meat and fat carcass traits. Bulls with the Leu/Leu genotype tended to have the highest carcass gain (P<0.05), weight of fat at valuable cuts and weight of fat at carcass (P<0.01) compared to others. The Val/Val homozygotes were superior to the rest of GH genotypes (P<0.05) in weight of valuable cuts and weight of meat at valuable cuts. There was no significant effect of IGF-1 concentration on carcass traits. Among GH parameters GH baseline approached significance (P<0.05) for ratios of meat to bones and fat to bones at the carcass. Further investigations are needed to determine suitability of GH gene for use in marker-assisted selection programmes.

CATTLE PRODUCTION [C]

Paper C3.15

Dynamic developement of the deposition of energy and nutrients in the carcass during growth and meat quality of different cattle breeds
B. Ender[*1], H.-J. Papstein[2], M. Gabel[1] and K. Ender[2]. [1]University of Rostock, Agricultural Faculty, D-18051 Rostock; [2]Research Institute for the Biology of Farm Animals, D-18196 Dummerstorf, Germany

It was the objective of this study to determine the effects of selection on dynamic development of tissue and nutrients during the course of growth and on meat quality. For this purpose bulls from the following breeds were selected: German Angus (GA), Galloway (Ga), German Holstein (GH), and White-blue Belgian (WBB) with muscular hypertrophy. The recording of the substantial components of the value of the carcass was carried out during the growth period from birth to the age of 24 months, under comparable conditions and using the method of serial slaughterings of different age groups which was then followed by an analysis of the carcass. In order to indicate the dynamic development of tissue and nutrients during the course of growth, a time-dependent growth function was applied. White-blue Belgians show very high protein contents and very low fat contents. Selection did not have an impact on the ratio of protein to water which is 1 : 3,4-3,7. This ratio remained constant during growth.

Selection did have an impact on the size of the rib-eye area and the intramuscular content of fat. The meat quality is causally related to the different chemical components of the carcass due to selection; it also depends on a multitude of other features, however.

Paper C3.16

Reducing cross-suckling in calves by the use of a gated feeding stall
R. Weber[*1], B. Wechsler[2]. [1] Swiss Federal Research Station for Agricultural Economics and Engineering, Center for proper housing of ruminants and pigs, 8356 Taenikon, Switzerland, [2] Swiss Federal Veterinary Office, Center for proper housing of ruminants and pigs, 8356 Taenikon, Switzerland.

The aim of the present study was to reduce cross-suckling in group-housed calves. A new type of a gated feeding stall was tested in comparison with an open stall. With the gated stall the calves fastened themselves when entering the stall and could not be forced out by other calves after milk ingestion. As a consequence, they stayed longer in the feeding stall than calves in a pen with an open feeding stall and spent more time suckling the artificial teat after milk ingestion (3.3 minutes versus 2.1 minutes per visit to the stall involving milk ingestion). After milk ingestion, cross-suckling was much reduced in the pen with the gated stall compared to the pen with the open stall (0.2 times versus 1.3 times). In the pen with the gated stall no subsequent cross-suckling was observed in 92% of stall visits involving milk ingestion. With the open stall, the corresponding figure was only 62%. Neither breed, gender nor age had a significant effect on the frequency of cross-suckling after milk ingestion. Cross-suckling which occured without temporal association to milk ingestion was not significantly reduced in the pen with a gated feeding stall.

CATTLE PRODUCTION [C]

Paper C3.17

Meat production in pure and cross-bred Piemontese cattle
C. Lazzaroni*[1], D. Biagini[1], G. Toscano Pagano[1], M. Iacurto[2]. [1]Dipartimento di Scienze Zootecniche, Università di Torino, via L. da Vinci 44, 10095 Grugliasco (Torino), Italy. [2]Istituto Sperimentale per la Zootecnia, via Salaria 31, 00016 Monterotondo Scalo (Roma), Italy.

Use of pure and crossbred double muscled animals, to improve quality of meat production, has become quite common, not only in the European countries, but all over the world. In 1997, only in the Italian Herd Book of Piemontese cattle, 104,585 subjects were registered and 282,393 semen straws were produced in the A.I. station. To better characterise the breed, and the crossbred obtained using Piemontese sire (on dairy cows, e.g. Friesian), data from different trials on fattening young bulls are reviewed. The studied parameters consist of live performance (feeding level, average daily gain, feed conversion ratio), slaughtering performance (slaughtering age and weight - between 450 and 550 kg in pure-bred -, dressing percentage - up to 70% of live weight in pure-bred -, carcass weight, carcass conformation and fatness), and amount of saleable meat obtained (more than 80% of carcass weight in pure-bred), so as some parameters on meat quality.

Paper C3.18

Effects of a restricted breeding season versus annual repartition of calvings on Belgian Blue suckler beef heifers reproduction performance
V. de Behr*, O. Gérard, M. Diez, J.L. Hornick, L. Istasse. Nutrition, Faculty of Veterinary Medicine, University of Liège, Belgium.

Belgian Blue -double muscled type- suckler beef production's income is based on fecundity. Management policies may interact with the herd reproduction performances. Some breeders group most calvings in February and March while others do not. Reproduction and weight data were collected from 134 heifers belonging to 'grouped calving' herds and 136 to 'no calving policy' herds. The mean weight at fecundation was similar for both groups, respectively 420,7 ± 44,7 kg and 416,9 ± 57,7 kg. Heifers from 'grouped calving' herds were significantly older at first fecundation than heifers belonging to herds with no calving policy, with respective means of 19,5 ± 4,4 months and 17,7 ± 3,1 months. The frequency distribution of the 'no calving policy' heifers age at first fecundation tended to normality, as opposed to a binomial curve for the 'grouped calving' heifers, wherein the first maximum was related to 15 months fecundated heifers (calving around their 2 years birthday) and the second maximum to 23 months fecundated heifers, voluntarily delayed (heifers born late in the calving season) or not (heifers very light at breeding). The management policy of grouping the calvings had a negative effect on Belgian Blue heifers fecundity expressed as age at first fecundation, which led to the lost of a sixth of the yearly potential calf production.

CATTLE PRODUCTION   [C]

Poster C3.19

Some aspects regarding water intake of milking buffalo cows
F. Beiu, Laura Rădulescu, Elena Bolocan. Research and Production institute for Bovine Breeding, 8113 Balotești, Romania

The experiment was carried on six tied milking buffalo cows introduced into a climatic chamber, during three periods, fifteen days each. The diets of the cows consist of alfalfa hay, concentrates and maize silage. The values of air temperature during each period were: $T_1 = +25°C$, $T_2 = +16°C$, $T_1 = +9°C$. The factors taking into account in this experiment were: dry matter intake, milk yield, body weight and air temperature. The influence of these factors (as independent variables) in a singular or combined action on water intake was described using regression equations. The significance of the influences was established using Fisher test. The results obtained until now reveals that only milk production has a significant influence on water consumed by cows from the drinking bowl. A multiple regression equation taking into account all factors revealed the same result. The mean value of water intake during the experiment was 47.1 l per head per day. The hours by hour mean measurement during 24 hours shows that water intake was higher in the morning and in the afternoon, after concentrate portions were consummated. This quantity represents about 1/3 from the water intake per head per day. The fluctuation of hourly water intake was statistically insignificant.

Poster C3.20

Effect of propylene glycol administration on blood metabolites on periparturient Holstein Dairy cows
Z. Yi[1], R.. O. Kellems*[2] and B. L. Roeder[2]. [1]Cargill, P.O. Box 56214, Minneapolis, MN, 55440-5614, USA; [2]Animal Science Department, 353 WIDB, Brigham Young University, Provo, Utah, 84602, USA.

Thirty periparturient Holstein (20 cows and 10 heifers) were dosed with five levels of propylene glycol (PG): 0, 72, 114, 227, and 454 ml (4 cows and 2 heifers each). Blood samples were collected via the coccygeal vein before and 60 minutes after orally drenching with PG. Blood samples were collected at day -1, 1, 3, 5, 10, 15, and 20 of calving. The mean concentrations of glucose, NEFA, and total Ca were: 49.3, 51.4, 54.9, 54.5, and 57.5 mg/dl; 0.821, 0.779, 0.705, 0.669, and 0.556 mEq/l; and 2.07, 2.20, 2.20, 2,15, 2,19 mg/dl at 0, 72, 114, 227, and 454 ml of PG, respectively. Dosing graded levels of PG linearly increased ($P < 0.01$) whole blood glucose concentrations (mg/dl), with the equation $Y = 50.7 + 0.016X$ ($r^2 = 0.80$) and decreased ($P < 0.01$) serum NEFA concentrations (mEq/l), with the equation $Y = 0.8046 - 0.00057X$ ($r^2 = 0.96$). Dosing PG increased serum total Ca ($P < 0.05$) but there were no differences among the administrated doses ($P > 0.10$). The ionized Ca, normalized Ca and pH in serum were not affected by dosing PG ($P > 0.10$). Under the conditions of dosing PG in this study, the normal levels of glucose, NEFA, and total Ca were at 55 mg/dl or higher, 0.700 mEq/l or lower and 2.15 mg/dl or higher, respectively. Results indicate that 114 to 227 ml of PG are the required doses to be used in dairy cattle for preventing ketosis.

CATTLE PRODUCTION   [C]   Poster C3.21

Cattle SOD and GPD Isoenzymes, GH, PRL and IGF-1
A.Karus* and V.Karus. Institute of Animal Science, Estonian Agricultural University, Kreutzwaldi 1, Tartu, 51014 Estonia

386 cattle (151 Estonian Black-and-White and 235 Estonian Red Cattle) blood samples were analysed. Blood cells were separated by differential centrifuging and destroyed by repeated (twice) freezing and defrosting in the presence of 3-fold volume of bidistilled water. Isoenzyme spectra of superoxide dismutase and glycerol-3-phosphate dehydrogenase were separated by horizontal IEF using the tetrazolium method for the isoenzyme spectra determining. Serum growth hormone and prolactin were studied using Immulite Kits. Serum insulin-like growth factor-1 was measured using the DSL-5600 IGF-I IRMA Kit. In leukocytes both SOD_A and SOD_B were found. SOD_B (a tetrameric enzyme) did not have allelic forms. Most frequent GPD phenotype was $GPD_1 1/GPD_2 1$ (0.857). $GPD_1 2-1/GPD_2 1$ frequency was 0.107 and frequency of rare phenotype $GPD_1 1/GPD_2 2-1$ was 0.036. Correlations analyse was done in three cattle age groups: calves up to six months, heifers 10-15 months and cows more than two years. The results of investigations about the relationships between physiological parameters, GPD phenotypes, weight gains, cattle age and body weights will be discussed.
This work was supported by Estonian Government Grants 1837 and 3600.

Poster C3.22

Prediction of the main productive and reproductive traits of calves
V.P. Shabla, Institute of Animal Science, P.O. Kulinichi, Kharkiv Region, 312120, Ukraine.

The models for prediction of future fat and protein yields, milk yield, contents of fat and protein, age at first calving, lactation length, calving interval and service period, milk flow rate, body weight at first lactation were developed on the basis of research of the exterior or hystological structure of ear tips of calves at age of 1-7 and 8-18 months. The models were received using step-by-step multiple regression procedure. The models were significant (P<0.001) with R^2_{adj}=0.999, standard prediction errors were low.
The test of the models using sliding control should that coefficients of correlation between predicted and actual value of the traits were 0.70-0.99 (P<0.01), the standard prediction error was within the limits of 3-25 % of the standard deviation.
The test using a parallel sample showed concurrence of predicted and actual negative estimations in 70-100 %.

CATTLE PRODUCTION   [C]   Poster C3.23

Method of early prognostication of cattle milk productivity
Ya. Z. Lebengartz. All-Russian Research Institute of Animal Breeding, Box Lesnye Polyany, Pushkin district, Moscow Region, 141212, Russia.

Method of early prognostication of cattle milk productivity has been developed which includes determinations of glucose concentrations in blood of 6 month heifers and its change under adrenalin action. Breeds used were: Holstein (Moscow region), Holmogory(Moscow region, Kaluga region) and their hybryds with Holstein bulls. Milk producrivity is evaluated by digital scale corresponding to these changes: animals having received number 16-20 are consedered to be high productive, those having received less than 15 are consedered low productive. The predict precision is 72-78%.

Poster C3.24

Development of behaviour in Holstein-Friesian cows and calves
I. Györkös[1], *K. Kovács*[1], *E. Szücs*[2], *G. Gábor*[1], *G. Borka*[1], *K. Bölcskey*[1] *and J. Völgyi Csík*[1]. *Research Institute for Animal Breeding and Nutrition, 2053 Herceghalom, Hungary*[1], *Institute for Animal Husbandry, University of Agricultural Sciences, P.O. Box 303, 2103 Gödöllö, Hungary*[2]

The behavioural development of Holstein-Friesian cows and calves in different housing systems was examined. There were no significant differences in calving ease between cows housed with different areas per cow and in groups or individually. 32-45% of the cows housed in groups separated themselves from the group just before calving. In the cow-calf relationship, the sensitive period in the cow was expressed for 6 hours after the first calving and increased following the next calving. This behaviour of the cow could help the new-born calf to obtain colostrum in time.

The early sensitive period of the calves developed in the first 10 days after birth. Social relations are relatively weak until 2-3 days of age and become stronger on the 4th-7th days. Fear reactions to unknown herdmates, objects or persons intensify during the same period. Surplus human care applied after weaning can improve the subsequent manageability of the calves.

CATTLE PRODUCTION   [C]   Poster C3.25

The problem of Multifetal ability in dairy cattle husbandry
T.V. Makeeva, V.I. Ustinova, N.S. Ufimtseva. . Dept. of Animal Breeding and Genetics, Novosibirsk Agrarian University, 160 Dobrolubov Str.,630039 Novosibirsk, Russia

Investigations vere carried out in cattle studs of West Siberia. The maximum quantity of twins(80.1%) relies upon the animals that calved 2-4 times. The highest and lowest quantities of twinings is observed in March and Desember, that constitute 15.4 and 4.3%, respectively. The frequency of them constitutes 30.8, 27.4, 21.3, and 20.5% in spring, autumn, summer and winter, respestively.
After twinings 35.7% of cows were culled due complications, 64% of cows stopped lactation, 43.1% of them calved in the years followed. Gynaecological diseases, infertility (23%), extremety diseases and osteomalacia (28.6%) were the main cause to cull multifetal cows. The period of multifetal pregnancy is 8.5 days reduced (276,5), heterjsexual twin pregnancy being 4 days shorter than that of homosexual twins.
Multifetal cows highly exceed average indexes of herd in yeilds and milk fat. After twinings the true decrease in cow milk productivity was not revealed but 30% of cows had the highest lactations.
Hence, it appears that in dairy cattle husbahdry multifetal ability does not restrict cow longevity, milking and reproductivity if the cows are optimaly fed and the level of breeding and veternary is high.

Poster C3.26

Twin calvings in Black-and-White cows of West Siberia
N.G. Khimich*, N.N. Nesterenko., Chair of Animal Breeding and Genetics, Novosibirsk Agrarian University, 160 Dobrolubov Str.,630039 Novosibirsk, Russia

The reproductive ability in twining calving Black-and-White cows of West Siberia was studied. The sex distribution in 260 twin calvings was following: 111 bisexual twins and 149 monosexual ones, in this number 72 female and 77 male twins. The highest frequency of twin calvings was observed in secondary calving cows - 25,3%, as compared to primary calving cows - 5,1%. The frequency of twinnings in cows of 2-5th calving was 80,5%. The frequency of twinnings, is found, to decrease with cow ageing. The distribution of twinnings for seasons was the following: in winter - 26,5%, in spring - 26,8%, in summer - 22,5% and in autumn - 24,2%.
In polyfetal cows the specific weight of reproductive system and udder diseases was 27,3 and 19,3%, respectively, that is the highest one in the total morbidity.

CATTLE PRODUCTION   [C]   Poster C3.27

The relationship between visually scored foot and leg traits and some foot disorders in West Siberian dairy cows
V.V. Eryomenko*, E.V. Telezhenko, N.N. Kochnev. *Novosibirsk State Agrarian University, 160 Dobrolubov Str., 630039 Novosibirsk, Russia*

The limb estimation remains uneffective for the linear scoring system so far has not been widely spread on the dairy farms of West Siberia. The objective of the study was to assess the relationships between scoring the primary and some secondary foot and leg traits and foot disorders of cows. The rear foot angle, front foot angle, front and rear pasterns, rear leg side view, color intensity of hooves and the curled and overgrown claws were estimated. The foot angle of the rear limbs was more related to weak pasterns than the front limbs. In addition, the correlation between the rear foot angle and the rear leg side view was +0.50. The relationship between rear leg side view and weak rear pasterns was negative (-0.45), indicating that a steep pastern was related to a straight rear leg assessed from the side. The foot angle was rather high negatively associated both with curled front and overgrown rear claws (-0.37 and -0.53, respectively). Also, the overgrown rear claws related to the rear leg side view (+0.42). The relationship between the curled claws and the intensity of hoof color, scored on the scale from 1 (light) to 4 (dark), has been found (-0.19).

Poster C3.28

Effect of housing and climate on the intensity of gorwth in heifers
I. Györkös*[1], E. Szücs[2], K. Kovács[1], G. Borka[1], G. Gábor[1], K. Bölcskey[1]. *Research Institute for Animal Breeding and Nutrition, 2053 Herceghalom, Hungary[1], Institute for Animal Husbandry, University of Agricultural Sciences, P.O. Box 303, 2103 Gödöllö, Hungary[2]*

The growth of Holstein-Friesian heifers up to 6 months of age was examined. The calves were placed individually in barns or outdoor pens and weaned 50 days after birth. The food consumption and body weight of the calves were measured in winter and summer. The body weight gain of the calves increased after weaning from milk replacer. The growth rate decreased more in summer than in winter.
The calculated limits of the growth phases indicate only approximately that the weight gain changed its intensity according to the early level of nutrition.
The reason for this early intensive growth was probably the artificially limited weight gain at the beginning. This is proved by the greater growth coefficients in this stage, indicating great growing ability which decreased only slightly in the next phases of growth.
The weight gain and growth rate were generally most intensive between 2-4 months of age.
The relatively high growth intensity biologically possible at this age can only be utilised if the calves possess the necessary dry matter consuming ability.

CATTLE PRODUCTION   [C]   Poster C3.29

Growth curve estimation in the Romanian Spotted heifers and Red Holstein crossbreds and some factors of influence
G. Stanciu, L.T. Cziszter*, S. Acatincăi. Banat University of Agricultural Sciences Timișoara, Calea Aradului 119, 1900 Timișoara, Romania.

The body weigths at 1, 3, 6, 12, 18 and 24 months of age of 1000 heifers, born between 1984 and 1992, were used as database. Four types of growth models were tested: biological (Richards, Brody, Gompertz and von Bertalanffy), logistics, exponential and polynomial (up to the third degree). The best fit to our data was obtained by using the biological models, out of which Brody equation had the best goodness-of-fit charac-teristics (determination coefficient for adjustment to the actual data was 0.97 and the residual mean square was 714.2 kg). Good fit was obtained also with polynomial equations, and less suitable were the exponentials. The estimated Brody equation was: $BW_t = 695.01(1 - 0.985 \exp(-0.001594t))$. Effects of year and season of birth and the degree of infusion with Red Holstein genes, as well as their interactions, were studied using Brody equation. Year of birth had a significant effect ($p<\!\!<\!.001$) on parameters b and k, but not on the parameter A. Only b parameter was significantly influenced ($p<\!\!<\!.05$) by the season of birth. Crossbreeding affected ($p<\!\!<\!.05$) all the three parameters, heifers with more than 31.5% of Red Holstein genes in their genotypes having larger values for maturity weight (A), rate of maturation (k) and b. Significant interactions ($p<\!\!<\!.01$) were observed between factors, mainly on the rate of maturation (k).

Poster C3.30

Milking routine has an effect on udder health and milk quality of cows
H. Kiiman, O. Saveli. Institute of Animal Science, Estonian Agricultural University, Kreutzwaldi 1, Tartu 51014, Estonia

Milk is one of the most important products for human consumption. The milk which is synthesized is stored in the alveolis, milk ducts, udder and teat cistern between milkings. Good milking technique is of utmost importance and a milker develops a routine that is followed during each milking. An optimum milking routine includes different working operations: cleaning udder and teats, manual pre-stimulation, fore-milking, attaching the milking unit to the cow, removing the milking unit, teat dipping. Careless milking technique can cause udder infection, mammary tissue damage, decreased production, poor milk quality. 32 milkers were in our trials and their working time observations were carried out after control-milking. We fixed somatic cell count in milk, milk yield, milk fat and protein content. Daughters of 21 bulls were studied in our trials. Correlation analysis was made to find out by continual factors the cow's preparing for milking, delay in applying the milking unit to cow, machine stripping, over-milking and how these factors affect somatic cell count in milk. Sufficient udder preparation was important concerning the average milk yield per cow as well as the cows' udder health ($P<0.001$). When milkers did not watch the machine very carefully, there occurred over-milking. In this case the somatic cell count in milk was high ($P<0.001$). Uniform work routine must be established, which guarantees good udder health.

CATTLE PRODUCTION   [C]

Poster C3.31

Thermographic study of milker's thermal comfort during milking in milking parlour
I.Knížková*[1], P.Kunc[1], M.Koubková[2]. [1]Research Institute of Animal Production Uhříněves, 104 01 Prague 10, ,[2]Czech Agriculture University, 165 21 Prague. Czech Republic

In December the hand temperature states of 4 milkers were measured by means of thermography in milking parlour during two workshifts (morning from 3.30 to 7.00 a.m., evening from 4.00 to 6.00 p.m.) and thermal comfort was evaluated during milking. The air temperature was from z 13,4 °C to 16,7 °C in milking parlour. The influence of working conditions got worse hand temperature of milkers, hand temperature decreased resp. The hand temperature was changed during workshift from z 27,98 °C ± 3,1 °C (state before milking) to z 25,71 °C ± 3,9 °C (state after milking). These changes were significant $p<0,05$. The negative influence on the hand temperature states had a vet cleaning of udders, using during milking. Hands of milkers are very often in contact with water, which increases cooling effect of cool working environment.

Poster C3.32

The changes of teat surface temperature during milking with 45 and 40 kpa
P.Kunc*[1], I.Knížková[1], M.Koubková[2]. [1]Research Institute of Animal Production Uhříněves, 104 01 Prague 10, ,[2]Czech Agriculture University, 165 21 Prague, Czech Republic.

The objective of this experiment was to investigate the influence of the reduced vacuum on teats in dairy cows evaluated by means of the changes of teat surface temperature. Nine dairy cows resp. their teats were masured by thermovision set in autotandem milking parlour. The measurements were carried out in two cycles: I.cycle - vacuum 45 kPa, II.cycle - vacuum 40 kPa. The teat thermal profiles were recorded immediately before milking [IB],immediately after milking [IA] and 2 minutes after milking [2A]. During milking with vacuum 40 kPa the teat surface temperatures were recorded significantly lower than during milking with vacuum 45 kPa particulary immediately after milking [IB:IA]. The differences of temperatures were from 1.63 to 3. 71ºC. The changes of teat surface temperature were not so emphatic between IA:2A (from -0.39 to 1.07ºC), but were more emphatic between IB:2A (from 0.56 to 3.01ºC). Milking with vacuum 40 kPa decreased the traumatization of teats But the teat surface temperatures did not reinstate to starting values before milking until 2 minutes after milking. The regress time to starting temperatures of teats will be longer than 2 minutes after milking.

CATTLE PRODUCTION [C]

Poster C3.33

Sanitary conditions of udders and the yield, composition and physicochemical properties of milk from cows imported from France and Germany

Zbigniew Puchajda, Maria Czaplicka, Elżbieta Radzka, Rajmund Szatkowski, Department of Cattle Breeding, University of Agriculture and Technology, 10-718 Olsztyn, ul. Oczapowskiego 5/142, Poland

The objective of this study was to determine the effect of the sanitary condition of udders of young BW cows imported from France and Germany on certain milk characters. The investigation was performed in 1996 and 1997 on the farm Parcz. It included 100 Holstein - Frisian primiparas: 50 purchased in France and 50 in Germany. The animals imported from France are pure - bred Holstein - Frisians of the newest generation (Prim Holstein). The animals purchased in Germany have in their genotype from 87,5 to 97,5% hf. The cows were managed in boxes and pasture. Both in boxes and on pasture the animals had a constant access to water. Once a month, the udder sanitary state was analysed using Mastirapid (Mastitis California Test). Yield and composition of milk as well as its density and active acidity were determined. The fat, protein, lactose and dry matter content was determined using a Milcoscan apparatus. Milk density was measured using a thermolactodensymeter and pH with a pH - meter V-628. It was found that the cows imported from France were slightly more susceptible to mastitis than their equals from Germany. The intensity of morbid symptoms resulted in a decrease in yield of milk, fat, protein and dry matter as well as in the content of lactose and dry matter. On the other hand, with worsening of the sanitary state of udder, the content of fat and protein increased. No greater differences in the density and active acidity of milk were observed.

Poster C3.34

Changes in the level of saturated and non-saturated fatty acids in cow milk as a effect of various forms of subclinical mastitis

E. Gardzina, A. Węglarz, J. Makulska, J. Szarek. Department of Cattle Breeding, Agricultural University of Cracow, al. Mickiewicza 24/28, 30-059 Kraków, Poland

Experiments were carried out on quarter milk (n=600) from the Black-and-White cows. The aim of this study was to observe changes in the level of basic fatty acids in milk fat, caused by asymptomatic forms of mastitis (disturbances in secretion, latent infection, subclinical mastitis, according to FIL/IDF). It was found that among volatile fatty acids: $C_{4:0}$, $C_{6:0}$, $C_{8:0}$, $C_{10:0}$ a pathogen had highly significant or significant effect on the levels of butric, capronic and capric acids. As regards long - chained saturated acids a significant effect of asymptomatic mastitis on the levels of acid $C_{12:0}$, $C_{14:0}$, $C_{18:0}$ was found.

The highest changes in the levels of the above-mentioned acids were found in secretion disturbances. It seems to be interesting that the pathogen did not affected the level of palmitic acid.

As regards non-saturated acids $C_{18:1}$, $C_{18:2}$, $C_{18:3}$ a „mastitis" factor had a highly significant effect on the level of linoleic acid.

Also, it was found, that the effect of pathogen was highest in long chained non-saturated acids. The disturbance in secretion was the form of mastitis responsible for the most significant changes in the level of fatty acids.

CATTLE PRODUCTION [C]   Poster C3.35

Interaction of daily milk records, parity and stage of lactation with SCC in Holstein-Friesian cows
Cs. Dorner[1], A. Gáspárdy[2], L. T. Cziszter[3], and E. Szücs*[1]. [1]Gödöllö University of Agricultural Science, Páter K. u. 1, Gödöllö, [2]University of Veterinary Science, István u. 2, Budapest, Hungary, [3]Banat's University of Agricultural Sciences, Timişoara, Romania

Attempts have been made to establish interactions among daily milk records (DMY), parity, stage of lactation and somatic cell count (SCC) in Holstein-Friesian cows. Data covered 73776 milk records of 6148 cows and were assigned into categories according to month (12), parity (3), peak milk yield (4) and SCC (4). Total SCC in DMYs was also calculated (TSCC). For SCC and TSCC logarithmic transformation was made (LSCC and LTSCC). Data were processed using ANOVA in three separate studies. Dependent variables in the 1st, 2nd and 3rd models were DMY, LSCC and LTSCC. Analysis 1 reveal lowest DMY in the 1st parity in spite of favorable persistency. Evaluating DMY according to SCC categories, low SCC is impaired with high DMY and vice versa. In analysis 2 lowest values for LSCC were present in the 1st parity whereas highest ones in the 3rd one. Low DMY is impaired with high LSCC and high DMY with low LSCC. In analysis 3 highest LTSCC as dependent variable DMY seems to be proportional to the milk produced with mid-lactation minimum at low production and maximum at high production level. Low values for LTSCC were present in first parity with increasing values in subsequent ones.

                                                              Poster C3.36

Estimation of relations between fat and protein content in Black and White cows at the first lactation
J. Gnyp, J. Trautman, T. Małyska, P. Kowalski. Department of Cattle Breeding. Agricultural University, ul. Akademicka 13, 20-950 Lublin, Poland

Content of fat and protein and relations between these components (difference between fat and protein-RTB and protein to fat ratio-SBT) in 884 cows at the first lactation in 4 BW cattle herds improved in dairy direction using HF breed, was analyzed. Those parameters were estimated in progeny of 6 BW and 10 HF bulls that had at least 15 daughters. Calculations were made applying the least square method. In statistical model, influence of the following items was taking into account: herd, age of the first lactation beginning, age and season of the first calving. It was found that cows resulted from crossbreeding after HF breed application (the 3rd and the 4th generation- 75% and more HF genes) were characterized by 0.22% lower protein content than BW animals. Difference between fat and protein content (RTB) increased from 0.75% up to 1.00%, and the ratio of protein to fat (SBT) decreased from 0.81% to 0.75%. Moreover, it was proven that differences between fat and protein (RTB) and protein to fat ratio (SBT) in daughters of particular BW bulls ranged: RTB-from 0.51% to 0.86% and SBT-from 0.79% to 0.87%, respectively. For progeny of particular HF bulls those values were as follows: RTB-from 0.60% to 1.17% and SBT-from 0.72% to 0.84%. Obtained results point to the possibility of breeding work tending towards increasing the protein percentage in milk, but efficiency of that work would be lower in animals improved with HF breed.

CATTLE PRODUCTION   [C]

Poster C3.37

Length of intercalving period and effectiveness of milk production in high milk yielding herds
A.Zachwieja*, A. Hibner, J. Juszczak, R. Ziemiński. *Department of Cattle Breeding and Milk Production Wrocław Agricultural University, 51-631 Wrocław, Kożuchowska str. 5b*

Analysis of milk performance of 1510 cows of Black-and-White cows kept in high milk yielding herd was performed. Cows with minimum 3 calving records and with finished life-time production had been taken into account only. Three groups of cows with: short (< 359 days), normal (360 - 390 days) and long intercalving period (> 390 days) were distinguished. Each group was then divided into subgroups due to the lenght of life-time performance. Signifficant influence of the lenght of intercalving period on effectiveness of milk production was stated in analysed herd. Based on obtained results it seems that through the control of the lenght of intercalving period it may be possible to optimize milk production, both the scale and effectiveness, in herds with high milk production capacity. The best reproduction scheme and level of reproduction traits should be chosen on individual base of single herd taking into account genetic merit of dairy stock as well as environmental circumstances responsible for genotype expression.

Poster C3.38

Effects of calving interval and extended lactation length on milk production of Holstein cows raised under Mediterranean conditions
A.Koç, A.E.Okan, H.Akçay, M.Ilaslan*. *ADU, Faculty of Agriculture, Department of Animal Science, 09100 Aydýn, TURKEY.*

The purpose of this study is to investigate the effects of calving interval and extended lactation length on the milk production of Holstein cows raised at Dalaman State Farm located in the southwest Mediterranean coast of Turkey. Data of 720 lactations and 469 calving intervals collected from 234 cows were analysed. Lactation milk yields were adjusted for 305-d 2X, year, season and lactation number.
In order to find out the effect of calving interval on milk production, lactation yields were classified in < 400 days, 400-450 days and >450 days calving interval groups. The average calving intervals and milk production of the groups were 362.4 ± 22.5 d and 8501.0 ± 1476.0 kg; 424.1 ± 14.2 d and 8658.0 ± 1492.0 kg; and 527.0 ± 71.4 d and 8807.0 ± 1549.0 kg, respectively. The effects of calving interval on milk production were not statistically significant. But, on the other hand, the effect of extended lactation length on the milk production was found to be statistically significant.

CATTLE PRODUCTION   [C]

Poster C3.39

Beef meat production with Belgian Blue bulls: comparison of production systems
V. de Behr*, M. Kerrour, J.L. Hornick, M. Evrard, A. Clinquart, C. VanEenaeme, L. Istasse. Nutrition, Technologie, Faculté de Médecine Vétérinaire, Université de Liège, Belgium.

Six feeding strategies were tested in different experiments with 160 Belgian Blue fattening bulls. A control group was offered a sugar beet pulp diet in all the experiments. Compensatory growth was induced after either a reduced growth period of 255 days (group RG) or a two months interrupted growth (group IG). Bulls were grazed and finished indoors (group PI) or grazed and slaughtered directly from pasture (group P). Whole milk was used at a rate of 8 and 11 l as a supplement of the concentrate diet (group M). Vitamin E was used as supplement in one group (group E). The overall average daily gain was lower in animals submitted to a period of low growth either at pasture or in restricted feeding conditions (1.1 kg/d *vs* 1.4 kg/d). A grazing period (groups PI and P) reduced the fat content of the meat (3.5 and 3.3 *vs* 4.6%). Compensatory growth after a period of low growth did not influence the fat content of meat while there was an increase in the fat content of the carcass (4.3 *vs* 4.6% and 14.3 *vs* 13.3%). Compensatory growth after both a period of low growth or an interruption of growth increased meat color as indicated by larger a* and b* values. The water retention was improved when the animals were firstly grazed before an indoor finishing (22.3 *vs* 24.9% for cooking loss and 4.69 *vs* 5.13% for drip loss). Tenderness was largely reduced when growth was interrupted (44.8 *vs* 36.1 N).

Poster C3.40

Genetic comparison of top Holstein bulls and their progeny in Hungary
A. Janosa and J. Dohy*, Institute of Animal Husbandry, University of Agricultural Sciences, H-2103 Gödöllő

The Hungarian Total Production Index (TPI) of 44 Hungarian Holstein-friesian sires were compared to their 282 sons' TPI and to their 16,435 granddaughters' milk fat and protein production performance. The goal of the present work was to determine if there were correlations between (1) the sires' TPI and their sons' TPI and also (2) granddaughters' fat and protein production performance. The Hungarian Total Performance Index was counted as follows: TPI=6*PTA Fat(kg)+14*PTA Protein (kg)+80*PTA Final Score+70*PTA Udder Score. Rank correlation between the indices was calculated by the method of Spearman. After the analysis for rank-correlation (r), a weak correlation was found between the sires' TPI and the sons' average TPI (r=.424); the sires' TPI and granddaughters' fat (r=.332) and protein (r=.341) production performance. It can be concluded that however modern the selection of the best sires is (BLUP), the weak correlations between TPI of sires and the fat and protein production performance of their granddaughters show that progeny-testing of sires' sons is indispensable. The application of TPI does not improve production directly, but prevents the constitutional degradation of the animals. Our other results of correlation between sires' TPI and granddaughters' type final score, udder score and feet, also reinforce this conclusion.

CATTLE PRODUCTION   [C]

Poster C3.41

The determination of the general and specific breeding values of Holstein-Friesien and Hungarian Simmental species concerning the changes of weight and figure
Polgár, J.P. - Szabó, F. Department of Animal Husbandry, Pannon Agricultural University, Georgikon Faculty, H-8361 Keszthely, Deák F. u. 16., Hungary.

The increase of the weight of young HF and HS bulls, which were tested in Szarkavár Testing Centre between 1977 and 1990., is linear until they are 365 days old. The sire influence on growth is evident, even when they are 60 days old. The influence of the breeding place is tendentions in the case of HF species but the order of performance of HS young bulls has changed during of PT. The sire influence is stronger in the case of the body measurements of HF than that of HS. As on influence of genetic trend, the chest girth has decreased and the hindquarter has be come narrower and shorter. The change trend towards the milking character but concerning the position of the udder and the calving the transformation of the hindquarters can have a negative effect. During the PT testing the HF species the performance of the seqent generations differs definitely. The chest and hindquarters of the progeny generation are narrower and their chest girth is smaller than of the sire generation. During the testing of HS species i could not find enough data to prove the generational influence. According the tests on the level of breeding i estabilished that in the case of both species the sire TPI is in a negative correlation wth the progeny's weight at the age of 365 days. Testing the specific breeding value i estabilished that the hogher performance of sire TPI and maternal milk production result the progeny's smaller weight at the age of 365 days. This influence is more proved and significant in the case of HF species.

Poster C3.42

Phenotypic correlations between levels of fatty acids in cow milk
E. Gardzina, A. Węglarz, J. Makulska, J. Szarek. Department of Cattle Breeding, Agricultural University of Cracow, al. Mickiewicza 24/28, 30-059 Kraków, Poland

When investigating changes in lactic composition certain regularities can be observed. They consist in the occurrence of some interdependencies between individual fatty acids. The obtained results indicate that the highest correlation, both in winter and summer seasons, as well as throughout the year occurs between the contents of higher saturated and higher unsaturated fatty acids (r>-0.9). In all studied periods a correlation coefficient between volatile fatty acids and the higher unsaturated ones ranged from r = -0.5 to r = -0.55 while between volatile fatty acids and the saturated ones from r = 0.16 to
r = 26.
The highest correlation coefficients among volatile fatty acids were estimated in the caprilic acid ($C_{6:0}$) and capric acid ($C_{8:0}$). As regards higher saturated acids, the acids $C_{12:0}$ and $C_{14:0}$ were highly significantly positively correlated with volatile acids and negatively correlated with acids $C_{18:0}$ $C_{18:1}$ $C_{18:3}$. The acid $C_{16:0}$ was negatively correlated with acids $C_{18:0}$ $C_{18:1}$ $C_{18:2}$ and $C_{18:3}$. The correlations of the stearic acid with other fatty acids were negative and statistically highly significant (an exception was a positive correlation with acid $C_{18:1}$). The higher unsaturated acids were generally negatively correlated with all studied acids and the highest correlation coefficients were found between acid $C_{18:1}$ and capric, lauric and myristic acids.

CATTLE PRODUCTION   [C]   Poster C3.43

Protein polymorphism in milk from Black-and-White cows kept in Poland and Lithuania
A. Litwińczuk[1], Z. Litwińczuk[1], M. Tumienie[2], J. Barłowska[1]. [1]Lublin Agricultural University, ul. Akademicka 13, 20-950 Lublin, Poland, [2] Vilnius Agricultural Institute, Biała Waka 4003, Vilnius region.

Polymorphism of 4 milk protein arrangements was compared i.e. β-lactoglobuline, $α_{s1}$, β- and κ casein from 475 BW cows bred in East- central Poland and 199 cows from the Vilnius region in Lithuania. The results indicate a considerably greater genetic diversification (except β-casein) in Lithuanian cattle. Considering the share of genotypes determining polymorphism of milk protein in BW cows in Poland and Lithuania, a great diversification in particular fractions was stated depending on the identified animal groups. The gene frequency showed 2 or 3 times greater frequency of "minority" genes of the C $α_{s1}$ type, gene B of β- casein and gene B of λ-casein, in the population of cows from individual farms in Poland and local Lithuanian BW cows in relation to the remaining genetic groups.

Comparing the results of current own research with the previous research on milk protein polymorphism in BW cattle carried out in Poland at the end of 70s and mid-80s, a significant decrease of the κ - frequency was observed.

Poster C3.44

Genetic trends of the ukrainian Black-and-White dairy breed for productive and reproductive traits
V. Danshin, Institute of Animal Science, P.O. Kulinichi, Kharkiv Region, 312120, Ukraine.

The genetic and phenotypic trends of the ukrainian Black-and-White dairy breed for the main productive reproductive and reproductive traits were obtained using an animal model. It was shown that despite of extreme decreasing actual production during the last 10 years (average milk yield fluctuated from -1400 kg relatively to the start point) genetic merits of the animals increased for milk yield (+210 kg) and fat percentage (+0.13%). This was result of crossbreeding local cattle with Holstein breed. In the same time genetic merits of the ukrainian Black-and-White dairy cows increased for calving interval (+40 d.) and almost did not change for age of first calving and productive life.

CATTLE PRODUCTION   [C]

Poster C3.45

Effect of genetic and environmental factors on milk production of crossbred cows
S. Ruban, Institute of Animal Science, P.O. Kulinichi, Kharkiv Region, 312120, Ukraine.

Records of fat yield from 5600 crossbred first lactation cows of the ukrainian Red-and-White dairy breed were utilized to evaluate the effects of genetic (crossbreed group) and environmental (level of feeding) on milk production. The crossbreed groups included four breeds: Simmental, Ayrshire, Monbelliard and Red Holstein.
Analysis of variance showed that influence of level of feeding was 23.57% (by Snedecor), twice more than influence of a crossbreed group (10.26%). Also, the interaction of these factors was large (26.28%). In all eases the influence of the factors was significant ($P>0.999$).

Poster C3.46

Adaptation of young cows imported from France to the conditions of north-eastern Poland
Zbigniew Puchajda, Maria Czaplicka, Anna-Maria Szymańska, Witold Czudy, Department of Cattle Breeding, University of Agriculture and Technology, 10-718 Olsztyn, ul Oczapowskiego 5/142, Poland

The investigation was carried out in 1996-1998 on the farm belonging to Agricultural Technical School at Karolewo. Fifty Prim Holstein primiparas born in 1994 and purchased in France were examined. They were imported to Poland in September 1996 as heifers in calves. Calvings began on December 17, 1996 and lasted till March 6, 1997. Management were boxes and pasture. During winter the animals stayed in a byre in 50 single boxes with a controlly situated feeding hallway. Boxes were mediun long and littered. Both in boxes and on pasture, the cows had sufficient water. Milking was twice a day with a Alfa Laval milker and a Milk Master milking apparatus. Once a month, the yield of milk, fat, protein, lactose and dry weight as well as the percentage content of these components in milk were determined. These parameters were established on the basis of control milkings and analyses of milk samples. Determinatios were performed using a Milkoscan apparatus in OSHZ in Olsztyn. The sanitary state of mammary glands was determined basedon MCT. Milk for analyses was sampled from each udder quarter. Differences in yields between imported cows (6435 kg of milk, 278 kg of fat, and 193 kg of protein) and their mothers in France (6685 kg of milk, 286 kg of fat and 210 kg of protein) were slight. It should be presumed that they were associated with adaptation to new enviromental conditions. Out of 1760 quarters evaluated, 96% showed no morbid symptomps. Acute and very acute inflammatory condition occured in 2,5% cases. Calvings were normal and the weight of healthy calves averaged 37,9 kg.

CATTLE PRODUCTION   [C]

Poster C3.47

Effect of breeds of the world on breeds of cattle small in number
O. Saveli, U. Kaasiku. Institute of Animal Science, Estonian Agricultural University, 51014 Tartu, Kreutzwaldi 1, Estonia.

During the 19th-20th century the Estonian Red breed and the Estonian Black-and-White (since 1998 Holstein) breed of cattle were formed in Estonia. In the last decades both Estonian cattle breeds have strongly been influenced by European and American breeds. The present investigation includes 219 bulls used for breeding of the Estonian Red breed, and 278 bulls used for breeding of the Estonian Holstein. Relative breeding value of each bull was calculated according to Animal Model. The bulls were grouped by blood content and the data were computed applying LSQ method. The Estonian Red breed was improved by 100% Ayrshire (BV=114; n=4), Swedish Red-and-White (SRB:BV=112; n=3), Brown Swiss (BV=109; n=19) and Red Holstein (RH:BV=108; n=26). Due to high genetic heterogeneity the gene pool of the Estonian Red breed needs stabilization. During 20 years the Estonian Holstein breed has been affected by the bulls of 100% Holstein blood from the USA and Canada, and by the bulls possessing less than 100% Holstein blood from Europe. The breeding value of the bulls with more than 75% Holstein blood (n=164) surpassed statistically significantly that of other bulls in milk yield (BV=316kg), milk fat production (BV=11.6 kg) and milk protein production (BV=8.4kg). The breeding value of bulls was decreased by the heterogenity of their genotype.

Poster C3.48

Analysis of testing series in the Retinto Beef Cattle Improvement Plan
F. Álvarez[1], F. Delgado[2], M. Valera[2]*, A. Molina[2] and A. Rodero[2]. [1]Asociación Nacional de Criadores de Ganado Vacuno Selecto de Raza Retinta. Madrid (Spain). [2]Departamento de Genética. Facultad Veterinaria. Universidad Córdoba. Avda. Medina Azahara, s/n 14005-Córdoba (Spain)

The testing center play an important role in the individual performance evaluation as a phase of the plan of selection for beef cattle. Even tough its advantages upon farm evaluation, it presents some inconveniences. In this work it has been analyzed the 28 testing series that have been performed in the autochthonous beef cattle Retinta Breed Improvement Plan. The results of the performance evaluation of 793 young bulls have been analyzed. The variables analyzed have been live weight at 12 months, average daily gain during the testing period, the morphological punctuation of the animal, scrotum circumference, and the feed conversion rate. The data were analyzed by BLUP Animal model that included as fixed factors the year (20) and the season (winter, summer and autumn) in which the series were celebrated, and the farm where the animals proceed from (97), and as random factors, the animal effect.
The connection between the series was possible thanks to the existence of 223 bull sires of 525 animals tested in different series.

CATTLE PRODUCTION   [C]   Poster C3.49

Evaluation of type classification in Limousin breed
A. Kovács[1], J. Tözsér[1], S. Balika[2], S. Bedö[1],
[1] Institute of Animal Husbandry Gödöllö University of Agricultural Sciences, H-2103 Gödöllö, Hungary. [2] Association of Limousine Breeders, H-1051 Budapest, Hungary

Type classification was introduced in 1986 in Hungary, which is based on 4 principal quality groups (utility score, score for length, score for width, score for muscularity) and included 22 type traits. Breeding Association of Limousin Breeders takes into consideration in their qualitication system has been worked out for type classification of cows. The investigations were carried out by using of 207 cows from two Limousin seedstock herds. In statistical procedure we have used a correlation analysis and a multiple regression analysis (backward stepvise). Correlation coefficients ($r > 0.65$) - have been accounted among single phenotypic scores show that lenght and width scores have been connected with muscularity. We have determined connections among traits inside the utility score of traits and other linear type trais which had a tendency in case of accounted correlation coefficients from height at withers to skeleton that it would be looser and looser. The estimated linear type traits characterized the final summit of phenotypic score by a multiple correlation coefficient ($R = 0.999$) and by a little evaluation deviation (0.369). The results for multiple regression analysis indicated that it is possible to reduce the number of recently estimated 22 type traits.

Poster C3.50

Association of growth hormone, κ-casein and β lactoglobulin gene polymorphism with meat production traits in Friesian bulls.
*Oprządek J *., Dymnicki E., Zwierzchowski L., Łukaszewicz M. Institute of Genetics and Animal Breeding, Polish Academy of Sciences, Jastrzębiec, 05-551 Mroków, Poland*

Genotypes of growth hormone (GH), κ-casein (CASK) and β lactoglobulin (BLG) of 138 Friesian bulls were analysed with PCR-RFLP technique. Following genotypes were identified (numbers of animals in parentheses): GH-LL (56), LV (50), VV (32), CASK - AA (84), AB (46), BB (8); BLG-AA (26), AB (52), BB (60). Animals were slaughtered and dissected at the age of 15 months; following carcass traits were measured: lean weight in valuable cuts, weight of lean, fat and bone in half-carcass. Interactions between GH, CASK and BLG genotypes and carcass traits were calculated. It was shown that GH genotypes significantly influenced lean weight in half-carcass ($P<0.01$). Animals with LL, LV and VV genotypes had lean weight in half-carcass 72,9, 72,2 and 69,7 respectively. The difference was approximately 3 kg in favour of LL and LV genotypes. No simple association of BLG and CASK genotype with meat production traits were found, however interactions between genotypes occurred. Combinations of GH x BLG genotypes influence carcass fat weight while the CASK genotype showed additive effect in combination with GH genotypes on lean weight in valuable cuts and in half-carcass. Animals with LL (GH) x BB (BLG) genotype had less fat then those with LL x AB or LV x BB genotypes. No such differences were found for other GH genotypes. CASK genotype showed additional effects on lean traits in combination with different GH genotypes, animals with LL x BB genotype having approx. 5 kg more lean in valuable cuts ($P<0,05$) then those with VV x BB genotype (50.3 kg *vs* 45. 6 kg).

CATTLE PRODUCTION   [C] Poster C3.51

Characteristics of browth and reproductive ability in Simmentalized and Hereford cattle of different ecogenesis.
G. Nezavitin, A.A. Permyakov, N.B. Zakharov. Siberian Livestock Research Institute of Novosibirsk Agrarian University, 160 Dobrolubov Str., 630039 Novosibirsk, Russia.

To improve aborigenal Simmentalized and Hereford cattle the genetic potential of their German and Canadian relatives is widely used. However, combining ability of the breeds under Russia conditions has not been sufficiently studied.
The efficiency to cross Siberian Simmentals with German Spots, Canadian Herefords and Simmentals has been studied.
When delivered Simmental cattle of different ecogenesis exceeded Canadian and aboriginal Hereford X Simmental crosses in live weight by 2.6-4.1 kg. The live weight of 1.5 year male and female animals reaches 496.9-500.8 kg and 488.3-494.5 kg, respectively.
Insemination index constituted 1.3 in Canadian Simmentals that is 0.3-0.5 more than that in Hereford X Simmental and German Spot first-cal-cows.
The results of the researches testify to the possibility to use Canadian and German animals along with aboriginal Simmental and Hereford ones.

Poster C3.52

The future model of high quality beef production in herds of milk cows on the way of emploing commercial croosing and transfering embryos of beef breeds.
H. Chmielnik, A. Sawa, University of Technology and Agriculture, Mazowiecka 28, 185-084 Bydgoszcz, Poland

On the basis of the carried out investigations, having the aim to obtain in milk herds from cows at the same time two calves: a crossbreed (commercial crossing) and a purebred beef calve (from a transferred embryo) an introductory monograph was elaborated for the future model of high quality beef production. It concerns partially cows of inferior genetical parameters, which may be embraced by the above program. The results concerning the fertility and fecunality indicators were presented at the 48th Annual Meeting of the EAAP in Vien 1997 which pointed out the different usefulness of meat breedfor this kind of progress. In this part of report fattening results and meatness of young cattle obtained from the commercial crossing and from the transferred embryos was presented. The investigation enfolded the growth and development period from birth to18 month of age and concerned the crossbreed yearling bulls and pure bred meat race. Meifers were served in other investigations. Toking into consideration together all the parameters having respect to fecundity and suitable for fattening and slaughter value the Limusin race obtained the best results, both in the care of using for commercial crossing and as for transfering embryos to Black and White (BW) breed cows.

CATTLE PRODUCTION   [C]   Poster C3.53

Carcass variability in seven Spanish beef breeds
J. Piedrafita[1]*, R. Quintanilla[1], C. Sañudo[2], J.L. Olleta[2], M.M. Campo[2], B. Panea[2], M.A. Oliver[3], X. Serra[3], M.D. Garcia-Cachan[4], R. Cruz-Sagredo[4], K. Osoro[5], M.C. Olivan[5], M. Espejo[6], M. Izquierdo[6]. [1]UAB, 08193 Bellaterra, Spain, [2]UNIZAR, 50013 Zaragoza, Spain, [3]IRTA-CTC, 17121 Monells, Girona, Spain, [4]ETC, 3770 Guijuelo, Salamanca, Spain, [5]CIATA, 33300 Villaviciosa, Asturias, Spain, [6]SIAEX, 06080 Badajoz, Spain.

A trial was designed to assess the genetic aptitudes to produce quality beef meat by several Spanish breeds in their specific production system. The breeds involved were Asturiana de los Valles, Asturiana de la Montaña, Aviléña-Negra Iberica, Bruna dels Pirineus, Morucha, Pirenaica and Retinta. During two consecutive years a total of 70 animals of each breed were studied. Fattening started at 5 to 7 months of age, calves being fed ad libitum concentrate meal. Slaughter weight was dependent upon the degree of maturation of the breed and local market preferences, ranging from 450 to 550 kg. Growth was monitored as well as carcass characteristics. After 24 hour of controlled chilling, standard carcass measurements were recorded, including carcass length, chest internal width, hindlimb length and diameter, area of the Longissimus muscle, kidney knob and channel fat, fatness score, conformation score and degree of ossification. The estimation of the tissue and regional composition was made from 6th rib. Considerable variability was found both between and within breed-production systems which suggests a great potential of the Spanish breeds to meet different market demands.

Poster C3.54

Relationships between carcass characteristics and ultrasonography measurements in Belgian Blue double muscle females : preliminary results
M. Evrard, J.L. Hornick*, A. Bielen, V. de Behr, L. Istasse. Nutrition B43, Faculty of Veterinary Medicine, University of Liège, 4000 Liège, Belgium.

Ultrasonic measures of fat thickness and of Longissimus muscle surface and thickness were taken at the beginning and at the end of a 3 mo fattening period on 21 culled females from the Belgian Blue double muscle type breed. After slaughter, killing-out proportion was recorded and carcass composition was estimated by the dissection of the 7th and 8th ribs, in order to separate lean meat, fat and connective tissues, and bones. Meat pH was also measured. Significant correlations ($P<0.01$) were found between surface of Longissimus dorsi and some carcass characteristics (carcass weight: $r = 0.61$; muscle weight in ribs: $r = 0.69$; weight of Longissimus dorsi muscle: $r = 0.71$). A negative correlation was also found between surface of Longissimus dorsi muscle and meat temperature 4h post-mortem. The higher fat thickness before slaughter corresponded to the higher increase in fat thickness during the experiment ($r = 0.41$). Bone proportion was negatively correlated with the muscle thickness ($r = -0.58$). This parameter showed also a significant relationships with carcass weight and weight of Longissimus dorsi ($r = 0.60$ and 0.57 respectively, $P<0.01$). A lower but significant correlation was found between killing-out percentage and the ratio of the length of the transverse apophysis to muscle thickness. Such results show the interest of ultrasonic measurements for carcass quality determination in the Belgian Blue breed and the needs for further researchs.

CATTLE PRODUCTION [C]

Poster C3.55

Effect of finishing system on fattty acid composition and tocopherol content of Simmental steers

F.J. Schwarz*[1], M. Timm[3], C. Augustini[2], K. Voigt[1], M. Kirchgeßner[1] and H. Steinhart[3]. [1]Lehrstuhl für Tierernährung, Technische Universität München, 85350 Freising, [2]Institut für Fleischerzeugung, Bundesanstalt für Fleischforschung, 95326 Kulmbach, [3]Institut für Lebensmittelchemie, Universität Hamburg, 20146 Hamburg.

Simmental steers (n=27) were either kept indoors and fed maize silage until a final live weight of 580 kg was reached (M), or grazed from 400 kg live weight for 180 d (P), or grazed for 120 d and were finished for a further 90 days on maize silage (MP). After slaughter the fatty acid patterns (total fat, phospholipids) and the tocopherol contents of the *musculus longissimus dorsi* (mld), the liver, the kidney fat and serum were measured. The i.m. fat content of the mld on treatment M was significantly higher that that on treatments P or MP. The proportion of the poly-unsaturated fatty acis (PUFA) of the total fat of the mld of animals on treatments P or MP was more than twice that for treatment M with inverse proportions of mono-unsaturated fatty acids (MUFA). The proportion of saturated fatty acids (SFA) did not differ between treatments. A tendency of a similar pattern (P and MP vs. M) was also found in the fatty acid composition of the total fat (liver and kidney fat) and in the phospholipids (mld, liver and serum). The α-tocopherol contents were significantly greater for treatment P than for treatments MP and M.

Poster C3.56

Effect of castration on the fatty acid profile and the incidence of yellow fat in bovine adipose tissue

P.-A. Dufey, Swiss Federal Station of Animal Production, 1725 Posieux, Switzerland

A fattening trial with 78 bulls and steers (Simmental crossbred) was carried out to evaluate the effect of castration on color measurements and fatty acid profile of the backfat. All animals were fed a maize (1/2)- grass silage (1/2) ration ad libitum. At the age of 14 months corresponding to a final live weight of 500 kg, the animals were slaughtered and the carotenoid content of the plasma and adipose tissue was analyzed. Only carotene are determinant in the bovine adipose tissue, since other carotenoid such as xanthophylls (lutein and zeaxanthin) could not be detected. Color measurements of the backfat were assessed by mean of a Minolta CR-300 and taxed also by an expert. The backfat of the steers and bulls was judged yellowish to yellow in 67 and 7 % of the cases, respectively. Concomitant, the chroma values ($P<.001$) and carotene levels of the backfat ($P<.001$) and plasma ($P<.005$) were higher for the steers. The b* values (yellow) and carotene levels were significantly correlated (r=0.38). The level of 16:1, 18:1, 20:1 was higher and of 18:0, 18:2, 18:3 lower in the backfat lipids of the steers compared to the bulls, which implies the significant effect of castration on the composition of the adipose tissue lipids.

CATTLE PRODUCTION [C]   Poster C3.57

Relationship between SEUROP conformation and fat grade and composition of carcasses of Belgian Blue slaughter bulls
G. Van de Voorde, S. De Smet*, M. Seynaeve and D. Demeyer. Department of Animal Production, University of Gent, Proefhoevestraat 10, 9090 Melle, Belgium

Classification of beef carcasses according to the SEUROP scheme for conformation and fat grade is obligatory in EU approved abattoirs. For grading systems to be a good tool for value-based marketing, they must relate to the carcass value. Carcass value is primarily determined by the relative content of meat, fat and bone, and by the conformation or shape of the joints. In the present study, the relationship between the SEUROP conformation and fat grades and the carcass composition of slaughter bulls of the Belgian Blue breed was examined.
A total of 786 carcasses were graded following the visual SEUROP scheme for conformation and fat grade by one expert classifier. Carcasses were graded in the S, E, U or R class, and in the fat 1, 2 or 3 class. The carcass content of lean meat, fat and bone was determined by dissection of a 1-rib sample and applying appropriate regression equations. Carcass dressing yield and lean meat content were clearly related to the SEUROP conformation grade, irrespective of the fat grade. Conformation grade relates thus not only to the shape of the joints, but also affects carcass lean content.
Bimodality of the data reflects the double-muscling phenotype.

Poster C3.58

Relationship between SEUROP-grading, genotype and meat quality in Belgian Blue slaughter bulls
S. De Smet*, E. Claeys, D. van den Brink, G. Van de Voorde and D. Demeyer. Department of Animal Production, University of Gent, Proefhoevestraat 10, 9090 Melle, Belgium

Little information is available on the relationship between SEUROP carcass grading data and meat quality traits in beef cattle. Whereas the SEUROP grading scheme relates to the conformation and the composition of carcasses, it is not clear to what extent meat quality traits vary across the SEUROP grade scheme.
In the present study, a total of 35 bulls of the Belgian Blue (BB) breed, both of the meat and dual-purpose type, was used. Carcasses were graded according to the SEUROP scheme, but only S, E, U and R carcasses were available. The genotype for double-muscling was also determined. Samples of five different roast muscles were taken at 1 day pm. Colour co-ordinates were measured. After 11 days of ageing, tenderness was determined by measuring shear force on raw and cooked meat samples. As expected, the genotype for double-muscling strongly affected SEUROP grading results. Meat of higher graded carcasses was paler. Shear force values of raw meat were lower in higher graded carcasses, reflecting the effect of the double-muscling genotype on meat collagen content. However, shear force values of cooked meat did not vary much across the SEUROP grading scheme, which does not correspond to earlier findings.

CATTLE PRODUCTION [C]

Poster C3.59

Slaughtering value and quality of meat in Polish BW cattle with regard to macro and microelement content
A. Litwińczuk[1], Z. Litwińczuk[2], J. Bartowska[1], M. Kędzierska[1]. [1] Subdepartment of Animal Materials Estimation and Utilization, Lublin Agricultural University, ul. Akademicka 13, 20-950 Lublin, Poland, [2] Department of Cattle Breeding,.Lublin Agricultural University, ul. Akademicka 13, 20-950 Lublin, Poland

Research involved 79 cattle from mass breeding including 22 cows, 33 heifers and 24 young bulls. The average body mass before slaughter amounted to 546.6 kg for cows, 506.2 kg for young bulls and 425.6 kg for heifers. The evaluated animals reached a relatively high hot carcass value: 49.1% for cows, 54.1% for heifers and 56.3% bulls. Meat content in carcasses fluctuated from 65.11% in cows, through 66.9% in heifers and 70.5% in bulls. Carcasses of evaluated young bulls also contained the heaviest sirloin. Meat of young bulls contained the least amount of fat: 1.05% in round of beef and 1.90% in sirloin and the greatest amount was found in heifers' meat - 1.72 % in round of beef and 2.48% in sirloin.
The results of macro (Na, Mg, Ca) and microelement (Cu, Mn, Fe, Zn) content analysis as well as Cd indicate their similar content in round of beef and sirloin. A significant difference in content of the evaluated element was noted only in the case of potassium and lead.

Poster C3.60

Histological structure of skin tissue of crossbreeds from Black-and-White cows x Italian beef bulls crossed with Aberdeen Angus
P. Zapletal, J. Szarek, A. Węglarz, Department of Cattle Breeding. Agricultural University of Cracow, al.Mickiewicza 24/28, 30-059 Kraków, Poland

Studies were conducted on the effects of crossing crossbreeds from Black-and-White (B) cows x Italian beef bulls (Piemontese - P, Marchigiana - M and Chianina - Ch) with Aberdeen Angus (AA) and also B and Simmental (S) cows with AA, on histological structure and substructure of skin tissue. The thickest grain sublayer (part of thermostatic layer) and the reticular layer were characteristic of bulls' skins in BxAA group (600μ, 6483μ). The angle of collagen fibres splicing formation was larger in skins of bulls BxAA (43^0) compared to those of S (30^0) and of BxS (33^0). A high proportion of the thermostatic layer in the leathers of S bulls (25%) and in BxS (21.5%) resulted in a considerable decrease in their tensile and tear strength parameters, which was verified by the results obtained. Preparations photographed under a scanning microscope revealed in the central part of the reticular layer a very well formed splicing of the collagen fibres and of a shorter diameter in skins of crossbreeds BxAA and (BxP)xAA compared to skins of S and BxS. Those findings were confirmed by the stained histological preparations. In skin tissue of the bulls BxAA and (BxP)xAA the least spacing between the bunches of collagen fibres and their least thickness were observed. A high proportion of collagen in skins of S bulls and BxS is indicative of its cross-linked structure.

CATTLE PRODUCTION [C]

Poster C3.61

The influence of interbreed crossing on the quality of hides in cattle husbandry
N.B. Zakharov. Siberian Livestock Research Institute of Novosibirsk Agrarian University, 160 Dobrolubov Str.,630039 Novosibirsk, Russia.

Variants of the crossing were studied: aboriginal Black-and-White and Simmental cows with foreign Rain (Italy), Limuzin (France) and Hereford bulls derived from English and Siberian breeding. The researches were carried out from 1972 to 1988.
6, 12 and 18 month Limuzin X Black-and-White animals exceeded same aged Black-and-White ones and Kain X Black-and-white crosses in the hide and skin thickness by 8.1 and 11.4% ($P>0.95$ and $P>0.999$), respectively. The weight of fresh hides was restricted to 32.1-41.3 kg in pure and cross breeds.Leather obtained from the hides of 1.5 year Simmental, Black-and-White, Hereford bulls and their crosses have high market technological characteristics and due to chemical composition, tensile strength, modulus of elasticity and other physical mechanical properties the leather fits manufacturing both tough and soft one.The highest leather output was obtained from Simmental, Hereford and Hereford X Simmental animals, the difference to their benefit in contrast with Black-and-White analogies constituting 6.4-10.2% ($P>0.95$).
It has been established that under west Siberian conditions the interbreed crossing of Black-and-White and Simmental with Limuzin and Hereford animals permits to increase the quality of cattlehides.

Poster C3.62

Évaluation des intrevalles velage-velage des vaches Limousines dans une élevage hongroise
M. Horvai Szabó[1], S. Balika[2], L. Gulyás[3], A. Kovács[1], J. Tözsér[1], [1]Institut de Zootechnie, Université des Sciences Agricoles, H-2103 Gödöllö, Hongrie [2]Association des Eleveurs hongrois de la race Limousine, H-1051 Budapest, Hongrie [3]Institut de Zootechnie, Université des Sciences Agricoles, Mosonmagyaróvár, Hongrie

Les performances individuelles de 126 vaches ont été enregistrées. Elles ont porté sur l'intervalle entre deux vélage (IVV), sur l' age de premier velage (APV), sur le poids au sevrage corrigé a 205 jours (POC), et sur les résultats du jugement de la conformation (notes sur 22 criteres et note synthése, NSYNT). Les valeurs moyennes de l'IVV, l'APV, le POC et le NSYNT ont été respectivement: de 428±51,46 jours, 34,1±5,31 mois, 225±17,19 kg et 55,0±7,69 notes. Dans l'analyse factorielle, l'ensemble des informations a pu se ramener aux 5 axes suivants: I longeur-largeur (valeur propre, VP: 9,701, contribution a la variation totale: CVT: 37,3%); II musculature (VP: 4,174, CVT: 16,1%); III grosseur des canons (VP: 2,296, CVT: 8,8%); IV age de premier velage (VP: 1,299, CVT: 5,0%) et V intervalle velage-velage (VP: 1,127, CVT: 4,3%). Les résultats ont confirmé que, l' IVV est totalement indépendent des caractéristiques envisagés (APV, POC, note de la conformation). Concernant des performances de 3 familles de demi-soeurs paternelle nous avons démontré les différences significatives ($p<0,05$). Les résultats de cette étude ont confirmé, que les éleveurs doivent continuer a une directe sélection pour l' IVV.

CATTLE PRODUCTION   [C] Poster C3.63

Belgium Blue cattle growth and reproduction as influenced by management
O. Gérard*[1], M. Evrard[1], V. de Behr[1], I. Dufrasne[2], L. Istasse[1], [1]Nutrition, [2]Expérimental Station, Faculty of Veterinary Medicine, University of Liège, 4000 Liège, Belgium.

The aim of this study was to evaluate the effect of age, sex and trimestre of birth to predict body weight of Belgian Blue cattle and to study the influence of growth rate on the reproduction performance of the breeding stock. Live weight was recorded four times a year on the growing stock. A total of 7006 live weight records was available on 1455 young cattle (from 1 to 80 weeks). To perform statistical analysis, data were separated according to variance analysis using trimestre of birth and sex as fixed factors. Live weights at first fecondation were calculated for fast (0.25 heaviest) and low (0.25 lightest) growing animals. The regression models showed that the average daily gain (ADG) was significantly different between males and females over the two periods of the growth curve (intersection point at 31-45 weeks). The ADG ranged between 0.80-0.85 kg/d and 0.91-1.24 kg/d for the males and 0.76-0.80 kg/d and 0.48-0.56 for the females. The determination coefficient (R^2) ranged from 0.76-0.88. In terms of ADG, the best trimestre of heifers birth were the third or the fourth while the second was less interesting with a difference of 60 kg at 80 weeks. For the low growth group, the ADG was 0.48 kg/d, the live weight at first fecondation was 362 kg (sd 51) and the heifers were 22.4 months old (sd = 3.4) vs 0.77 kg/d, 480 kg (sd = 42) and 18.2 months (sd = 3.1) for the fast growing animals.

Poster C3.64

Utility crossing using Belgian White Blue and Charolais semen in Hungarian Grey herds
K. Bölcskey,*[1] I. Bárány,[1] I. Bodó,[2] S. Bozó,[1] I. Györkös,[1] Ms. A. Lugasi,[3] J. Sárdi.[1] [1]Research Institute for Animal Breeding and Nutrition, H-2053. Herceghalom, HU; [2]University of Veterinary Sciences, Budapest, HU; [3]National Insitute of Food Hygiéne and Nutrition, HU.

Hungarian Grey (HG) cows were inseminated with the semen of three Belgian White Blue (BWB) and three Charolais (Ch) bulls. HG bulls were serving in the same herd. seven BWB x HG, 10 Ch x HG and 9 HG bull calves were taken into the same tied fattening system at 288.0, 265.6 and 251.9 days of age and 255.1, 242.9 i.e. 187.3 kg weight. After 344 days fattening average live weight corrected to the same age of the BWB x HG group was 633.7 kg, by 98 kg (18.3%, $P<1\%$) higher, than that of the HG, and by 14.3 kg (2.3%, $P>5\%$) higher, than that of the Ch x HG group. Average corrected weight of the Ch x HG group was higher by 83.5 kg (15.6%, $P<1\%$) than that of the HG group. Slaughter value of the groups was 61.6, 60.3, and 55.7%, respectively, the differences are all significant. Proportion of abdominal fat was less by 23% ($P<.1\%$) in BWB x HG as compared to HG and by 14.3% ($P<5\%$) as compared to the Ch x HG group. Boning data show 84.1 kg (46.4%, $P<.1\%$) more pure meat for the BWB x HG as compared to HG and 35.8 kg (15.6%, $P<1\%$) more as compared to the Ch x HG group.proportion of intermuscular fat was by about % less in the BWB x HG ($P<.1\%$) as compared to both the control and ChxHG groups. The proportion of bone was also the lowest, 16.1% in the BWB x HG group, 16.8% in the Ch x HG, and 17.8% in the HG groups. This way HG cows may produce offspring of excellent slaughter quality.

CATTLE PRODUCTION [C]

Poster C3.65

Commercial crossing with Belgian White Blue semen in Hungarian Simmental herds
K. Bölcskey,* I. Bárány, S. Bozó, I. Györkös, J. Sárdi. Research Institute for Animal Breeding and Nutrition, H-2053. Herceghalom, Hungary;

Beef-type Hungarian Simmental cows were inseminated with the semen of three Belgian White Blue (BWB) and three Hungarian Simmental (HS) bulls. Nine BWB x HS (F_1) and 10 HS (control) bull calves were put into fattening after weaning at the age of 232.7 days (272.6 kg), i.e. 218.4 days (246.8 kg) in free keeping under the same feeding conditions. After 265 days of fattening live weight of BWB x HS bulls was 587.2 kg, by 39 kg (7.1%, non-sign.) higher than that of the HS group. Average live weight production/life day was 1171, i.e. 1137 g/day for the whole time of fattening. Between days 238 and 265 of fattening the nearly similar gains between weightings became reduced, the F_1 group fall back to 975, the HS to 537 g/day, the difference is significant (P<5%). The results refer to identical fattening characteristics, but to a possibility for fattening to a higher terminal weight of F_1 animals. Seven - seven randomly chosen individuals were slaughtered at the age of 552, and 546 days, and a live of weight 643, and 644 kg, respectively. Average weight of cut halves was 394.6 kg for F_1, and 370.9 kg for HS (the difference is 6.4%, P<5%), dressing percentage was 61.4, i.e. 57.7% (P<0.1%). Bony meat production/day of life was high in both of the groups 714, i.e. 681 g/day, but the difference is non-significant. Differences of weight gain data are not significant, however slaughter quality was highly improved as the effect of BWB crossing.

Poster C3.66

Comparison of three multiphasic growth models for bulls of Czech Pied Cattle
H. Nešetřilová[1], J. Pulkrábek*[2]. [1]Department of Statistics, Czech University of Agriculture, Prague, Kamýcká 129, 165 21 Praha 6 - Suchdol, Czech Republic, [2] Department of Genetics and Animal Breeding, Přátelství 815, 104 01 Praha 10, Czech Republic.

It was found that the real growth data of breeding bulls of Czech Pied Cattle correspond with a multiphasic growth model better than a classical model with one inflexion point. Thus, an attempt was made to construct a multiphasic growth curve as a sum of two logistic curves. Such a multiphasic growth model is a non-linear regression model with six parameters which have to be estimated. Due to the statistical complexity of the task also two simpler models were considered (with four and five parameters) and the correspondence of the model to real data and the degree of nonlinearity of the model (which determines its mathematical properties) were studied. It was found that the most general model with six parameters is significantly better than the simpler models although the degree of its nonlinearity is higher than desirable.

CATTLE PRODUCTION   [C]

Poster C3.67

Double muscle cattle: Piemontese (PD) x Friesian (FR) and Blanc-Blue-Belge (BBB) x Friesian (FR) young bull crosses in Italy
M. Iacurto[1], C.Lazzaroni[2], S. Failla[1], S. Gigli*[1]. [1]*Animal Production Research Institute, Via Salaria 31 - 00016 Monterotondo (RM) - Italy.* [2]*Animal Production Department Via Leonardo da Vinci 44 - 10095 Grugliasco (TO) - Italy*

In the last years the use of BBB and PD sires as bulls on dairy cows for the crossbreed production increased in Italy because of the larger request of high quality meat by the consumer. We compare some trials carried out on young bull crossbreeds produced and reared in our Research Institutes for the characteristics at slaughter and at dissection. The final weight and the average daily gain were very similar: 511.1 and 519.3 kg, 1.110 and 1.119 kg/d respectively for PDxFR and BBBxFR; while the conversion index was better for PDxFR (5.76 vs. 6.35 MeatFU/kg). The BBBxFR showed higher net dressing percentage (68.58 vs. 66.16%). The carcass evaluation was similar with difference of 1/3 of class only ; the PDxFR had lower scores (4- vs. 4 and 2- vs. 2). The side weight was similar, 162 kg for PDxFR and 168 kg for BBBxFR, while the side composition was better in PDxFR with more meat (73.3 vs. 69.2 %) and less fat (8.5 vs. 10.3%). Two genotype examined are very similar for different weights and for carcass evaluation but PDxFR produces little more meat (+2%).

Poster C3.68

Meat productivity in cattle breeds of different ccogenesis
A.A. Permyakov, N.B. Zakharov, A.G. Nezavitin. Siberian Livestock Research. Institute of Novosibirsk Agrarian University, 160 Dobrolubov Str. , 630039 Novosibirsk, Russia.

Meat productivity was studied in Siberian principal cattle breeds during 1995-1997. Detailed observations were made in 4 groups of to 6-mounth bulls. The first group was built up by Hereford and Simmental crosses, the second one by Black-and-White bulls, the third by Simmentals, the fourth by Herefords. The animals were fed and managed in the same way. The bulls were managed loose indoors in winter and loose on pasture in summer. The diets contained 2093.2 forage units and 213.1 kg of digestible protein. The liveweight of Hereford and Simmental young animals was from 465 to 437 kg, that was 8.5% and 4.3% higher than that in Black-and-White animals of the same age. The carcass weight in Simmentals and Herefords was 237.3 and 223.3 kg, respectively. Hereford bulls and Hereford X Simmental crosses exceeded Simmentals and Black-and-White cattle in carcass, slaughter weight and meat output by 4.6-7.0%, 6.3-11.4% and 15.1-19.1% respectively. Herefords and the crosses had a much better ratio of muscular, fat and bony tissues and greater output of valuable meat putts. The meat and bone ratio in Herefords was 4.1% less than that in Black-and-White. The Hereford beef was evaluated by independent inspectors and proved more delicious and softer that than of Simmentals and Black-and-White.

CATTLE PRODUCTION   [C]   Poster C3.69

Influence of calving season and age of dams on rearing results of beef calves
Z. Litwińczuk, W. Zalewski, P. Jankowski, P. Stanek, B. Kuryło, Department of Cattle Breeding, Lublin Agricultural University, ul. Akademicka 13, 20-950 Lublin, Poland

Observations included 191 calving cases of beef cows including 88 heifers, 57 cows after second calving and 56 after third and subsequent. Two calving seasons were identified: winter lasting till the end of March and springs beginning on 1 April. After about 200 days of rearing with dams, the increases of calves from heifers were significantly lower (876 g) compared with calves reared at older cows (949g heifers and 1039g bulls). Calves from winter calving (before 1 April) came to pastures (with dams) with their body weight of 100 kg, reaching in the case of bulls daily increases lower by nearly 100 g. Significantly higher increases of calves from winter calving allowed them to achieve a relatively high mass at separating from dams (after the end of the pasture season), which amounted to 234 kg for heifers and 263 kg for young bulls.

Poster C3.70

The distribution of lactate dehydrogenase isocomponents in blood serum of bulls fattening with antidotes
J. Buleca[1], J. Szarek[2], D. Magic[1], V. Pavlík[3], Z. Vilinská[3], J. Mattová[1], [1]University of Veterinary Medicine, Komenského 73, 041 81 Košice, Slovak Republic, [2]University of Agriculture in Krakow, Poland, [3]Research Institute of Animal Production, Krmanova 1, Košice, Slovak Republic

The bulls fattening in the emission.-polluted environment were investigated. Antidotes eliminated the negative influence of magnesium in the tested groups: rape oil, rape cake. During the experiments the enzymological analysis was done especially of isoenzyme spectrum of LD. The results document the variability of values from the reference scale in individual phase of experiment. Biometric information confirms variability of serum components mainly in the comparation between initial and final phase of experiments.

CATTLE PRODUCTION   [C]   Poster C3.71

Effect of slaughtering age on carcass traits of "Tipo cebon" animals with the Rubia Gallega breed
L. Monserrat*[1], A, Varela[1], J.A. Carballo[1], L. Sánchez[2]. [1]C.I.A.M. Aptdo. 10. 15080. A Coruña (Spain), [2]Facultad Veterinaria. Aptdo. 280. 27002. Lugo (Spain)

The effects of slaughtering age on carcass quality were studied with the aim of produce "Tipo cebon" (bulls or steers of Rubia Gallega breed slaughtered between 18 and 36 mo.).12 bulls and 12 steers grazed in a rotational system and were finished indoors with maize silage *ad lib* and concentrate (4 kg/a./d.) for 3 months before slaughtering at 18 or 24 months. The data were analysed with ANOVA or χ^2 test. At the same slaughter age, bulls had heavier carcass than steers (376.5 kg. vs 316.8 kg and 477.0 kg. vs 382.7 kg. at 18 and 24 mo.) and higher dressing percentage (58.03 % vs 54.77% and 59.34% vs 56.05% at 18 and 24 mo.). Percentage of carcasses in "E" and "U" classes was grater for the animals slaughtered at 24 months, 20 vs 14% and 67 vs 29% for bulls and steers. Steer carcasses had better distribution of fore and hindquarters and worse distribution of high and low-priced cuts than bulls.The bulls had higher meat percentage(77.8 vs 73.9% and 79.5 vs 74.9% at 18 and 24 mo) and lower bone(16.8 vs 18.5 and 16.0 vs 18.3 at 18 and 24 mo.) and fat percentage (4.8 vs 7.5% and 4.5 vs 6.8% at 18 and 24 mo.) than steer carcasses.

Poster C3.72

The use of video image analysis in grading of beef quality.
T.Sakowski *1, M.Słowiński 2, J.Cytowski 3. 1 Institute of Genetics and Animal Breeding Jastrzębiec, 05-551 Mroków, Poland, 2 Warsaw Agricultural University, Poland, 3 University of Warszawa, Poland.

The aim of the investigations was to prove the ability of digitised image processing for estimation of beef quality. The information about colour, marbling, surface of MLD cross-section, which are highly correlated with other characteristics of slaughter productivity, could be a part of a fast objective method for grading beef carcasses and lean quality which could be next quickly selected for cooking or processing.
One hundred thirty one Black-and-White bulls were slaughtered and dissected at a mean body weight of about 450 kg. One hundred thirty one digitised pictures of the MLD cross-section were taken simultaneously in condition of homogenous lighting. The meat probes were physically and chemically analysed. Statistically significant correlations have been stated between colour components R,G,B (r = 0.56-0.70).
The results of this study showed, that the video image analysis could be useful as a grading method for beef processing. The following features: protein content (r=-.24), pH45 min.(r=-.26), pH24 hour (r=-.31), pH48 hour (r=-.37), water holding capacity (r=-.49), cooking loss (r=.42), density (r=.33), taste (r=-.29), and consistence (r=-.37) were highly correlated with R (one of the colour components) and confirmed its usefulness.

CATTLE PRODUCTION   [C]   Paper C4.1

Some thoughts about the development of dual purpose breeds-history, presence and future
G. Averdunk, Bayerische Landesanstalt für Tierzucht, D-85580 Poing-Grub, Germany.

The history of dual purpose breeds in Europe goes back to local breeds which were kept for triple purposes: milk, beef and draughting. Modern testing and breeding technics were introduced in most populations at an early stage in the last decades including AI, bull performance testing, embryo transfer and open MOET-schemes. In connection with a high organisational depth selection progress for milk constituents (fat and protein yield) and beef traits (net gain) in the past was moderate (~1 percent). Past breeding goals had an nearly equal weighting of milk and beef traits, which in reality meant higher weight on milk production through higher recording intensity. This led to changes in body proportions: taller animals with less pronounced muscling. Some populations also introduced genes from dairy populations, like Red Holsteins and Ayrshires. Future development depends mainly upon the development of price relations between milk and beef, leading to a general lower level of income. This requires a reduction of costs and increases the importance of functional traits, like fertility, calving ease, milkability, udder health and sound feet and legs, resulting in a long herd-life. Selection programs for the different breeds are adopted, resulting in lesser weight for milk and beef traits. There is a need for diversity in traits at high production levels (Simmental, brown and red breeds), if semen sexing enables systematic crossbreeding for milk and beef in commercial herds.

Paper C4.2

Total merit indices in dual purpose cattle
J. Sölkner[1], *J. Miesenberger*[1], *R. Baumung*[1], *C. Fürst*[2] *and A. Willam*[1]. [1]*Department of Livestock Science, University of Agricultural Sciences Vienna, Gregor-Mendel-Strasse 33, A-1180 Vienna, Austria,* [2]*Federation of Austrian Cattle Breeders, Universumstrasse 33/8, A-1200 Vienna, Austria.*

The economic efficiency of dual purpose cattle is influenced by a large number of traits which may be classified in groups of dairy, beef and functional traits. Only recently improvements in data collection systems and data analysis have led to a situation where estimated breeding values are available for bulls and cows on many of the traits of economic importance. The combination of these estimated breeding values in a total merit index, as long practised in some Scandinavian countries, is currently being implemented in a number of Central European states. Economic values for populations of dual purpose cattle in Austria and Switzerland derived from a deterministic herd model are presented. Traits in the dairy group are fat and protein yield; beef traits are daily gain, dressing percentage and cacass conformation; functional traits are longevity, persistency, fertility, calving ease, stillbirth and mastitis resistance. A rough average over populations of the relative economic importance of dairy vs. Beef vs. functional traits is 40:15:45 (traits are weighted with their genetic standard deviations, differences in expression of traits are taken into account). Due to the covariance structure of the traits most of the gain is expected for fat and protein yields (moderate heritabilities and high positive correlation of the two traits). The indices presented are compared with total merit indices used in other European countries.

CATTLE PRODUCTION   [C]

Paper C4.3

Les races à deux fins en France : situation actuelle, principaux systèmes d'exploitation, spécificités de leurs programmes de sélection
F. Arnaud[1], N. Bloc*[2], J. Pavie[3], J.L Reuillon[4]. [1234]Institut de l'Elevage, France, [1] Valparc, Valentin Est, 25048 Besançon Cedex ; [2]5 rue Hermann Frenkel, 69364 LYON CEDEX 07 ; [3]Chambre d'Agriculture de Normandie, 6 rue des Roquemonts, 14053 CAEN Cedex 4 ; [4] 12 avenue Marx Dormoy BP 455, 63012 Clermont-Ferrand Cedex 1.

Les quotas laitiers imposent de rechercher davantage une maîtrise économique globale du troupeau que des performances maximales par vache. En France, dans les régions herbagères et particulièrement dans celles à vocation fromagère, trois races « à deux fins » sont ainsi particulièrement valorisées : la Montbéliarde, la Normande, et la Simmental Française. Leurs objectifs de sélection ne sont pas exclusivement laitiers, le maintien d'un revenu viande reste non négligeable, et l'approvisionnement de leurs schémas de sélection privilégie les ressources nationales. De telles caractéristiques communes de sélection n'empêchent pas une large diversité de systèmes d'exploitation, allant du plus laitier au plus mixte suivant l'importance du quota et de la surface fourragère disponible par hectare. La part du co-produit viande dans le revenu est très variable. En outre, on peut également relever l'intérêt des races mixtes sur les plans :
- fromager : elles présentent un taux de matière utile élevé, particulièrement le taux protéique.
- des démarches de qualité : elles offrent une typicité et une identification régionale (AOC).
- de l'occupation du territoire : elles participent à un bon aménagement du territoire, spécialement dans les zones herbagères, fragiles ou difficiles.
- de la diversification : elles peuvent apporter un revenu complémentaire grâce à la large gamme des produits viande possibles (veaux, génisses, boeufs, jeunes bovins).

Paper C4.4

Use of Multiple-trait Across Country Evaluation (MACE) procedures to estimate genetic correlations in Austrian dual purpose Simmental breed
T. Druet*[1], J. Sölkner[2] and N. Gengler[1]. [1]Unité de Zootechnie, Faculté Universitaire des Sciences Agronomiques, B-5030 Gembloux, Belgium. [2]Department of Livestock Science, University of Agriculture, A-1180 Vienna, Austria.

In order to improve selection for dual purpose breeds, genetic correlations were estimated between production traits (dairy and beef traits) and 7 functional traits: survival, persistency, somatic cell count, fertility (maternal and paternal) and calving ease (maternal and direct). A multiple-trait across country evaluation procedure involving these traits was adapted to get a multiple-trait a posteriori evaluation procedure which could provide these genetic correlations. A set of bulls was selected from the Austrian Simmental population based on mean original reliabilities of at least 0.50 for yield traits or of at least 0.20 for other traits. Only breeding values above these limits were retained. The breeding values were deregressed assuming that they were obtained by single trait sire-maternal grand sire models. An Expectation-Maximization REML procedure based on the multiple-trait across country evaluation equations was used to compute genetic correlations among all these traits. Interactions between yield traits and the others were rather small except for maternal fertility (-0.20). The strongest relations concerning beef traits were their correlations with direct component of calving ease and persistency (-0.16 and -0.19 with daily gain). The most correlated functional traits were survival with persistency (0.23) and survival with maternal fertility (0.34).

CATTLE PRODUCTION [C] Paper C4.5

Ecological Total Merit Index for an Austrian dual purpose cattle breed
R.Baumung* and J.Sölkner; Department of Livestock Sciences, University of Agricultural Sciences Vienna, Gregor-Mendelstraße 33, A-1180 Vienna, Austria

Organic farming plays an important role in Austrian agriculture. Therefore it seems to be worthwhile to make suggestions about a more ecological breeding goal for cattle breeds. Because of the fact that Simmental is the most frequent dual purpose cattle breed in Austria, this breed is chosen to show the impact of selection under an ´ecological´ total merit index. With simple calculations the effect on selection response in milk production, beef production and functional traits under selection with the current economic total merit index and more ecolocigal index variants, e.g. with higher economic values for fitness and functional traits, is shown. As a basis for future decisions the efficiency of more or less ecological indices is compared. The results indicate that an increase of the current economic values of functional traits of about 50 percent does not present a great risk, expected selection responses for milk production traits are still high.

Paper C4.6

Future role of Simmental cattle in Austria: views of breeders and decision-makers
E. Gierzinger*, A. Willam, J. Sölkner, C. Egger-Danner. Department of Livestock Sciences, University of Agricultural Sciences Vienna, Gregor-Mendel-Strasse 33, A-1180 Vienna, Austria.

Counting about 220,000 herdbook cows, the Simmental breed is the by far dominant cattle breed in Austria. 13 regional breeding organisations serving 17,000 breeders are united in the parent organisation AGÖF. Bred as dual purpose cattle, the genetic gain in milk traits has been unsatisfactory in recent years. To be competitive in the future, changes in selection schemes and breeding programmes are planned.

Within the framework of a project for optimisation of the breeding programme, the general manager of AGÖF, the executive managers of the regional breeding organisations and official farmers' representatives were interviewed. The detailed interviews included questions about the organisational strucure of the breeding organisations, breeding goals, breeding programmes, marketing and finances. There were also questions about visions for the Austrian Simmental breed and expectations of future developments at the level of the breeding organisation and on-farm. Questionarios were sent out to all 17,000 breeders. Answers indicate a general agreement on centralisation of breeding decisions. The breeding goal of a dual purpose breed is not questioned and the total merit index is accepted as the main instrument to achieve this goal. Although the structure of farming is expected to change away from the current small units it is not anticipated that Simmental will lose its dominating role among Austrian cattle breeds.

CATTLE PRODUCTION [C] Paper C4.7

Relationships of polledness to production traits in the German Fleckvieh population
A. Lamminger[*1], O. Distl[1], H. Hamann[1], G. Röhrmoser[2], E. Rosenberger[2] and H. Kräußlich[3].
[1]Department of Animal Breeding and Genetics, School of Veterinary Medicine Hannover, Germany, [2]Bavarian Institute of Animal Breeding, Grub/Poing, Germany, [3]Gauting, Germany.*

The German Fleckvieh Cattle is a dual purpose breed showing the polled mutation (*P*) at a very low frequency. Due to a changing attitude towards animal welfare the interest in breeds which are genetically polled has increased. The objective of this work is to compare polled and horned animals with respect to traits regarded in the breeding programme like milk, fat and protein yield, fertility, calving performance, fattening and carcass traits. Pedigree and progeny information of the animals was used to differentiate further between the different possible genotypes causing polledness (*PP*, *Pp*, *pp*). The information available for milk production traits is based on 478 daughters of 23 polled or scurred sires and their contemporary herdmates from the same or random chosen herds. The model considered the effects of different herd groups, the sire genotype with respect to polledness and the additive-genetic value of the animal. Growth traits are tested in 90 sons of 10 polled or scurred sires. The results have shown that *pp*-sires exhibit higher breeding values for milk, fat and protein yield in the different lactation periods (milk: +450-600kg, fat: +20-25kg, protein: +16-22kg) compared to homozygous polled sires. Positive deviations of the *pp*-sires were also found in growth rate while the carcass traits were not affected by the genotypes for polledness.

Poster C4.8

Discussing lines of Simental bulls imported to Poland
J. Trautman, J. Gnyp, P. Stanek, J. Tarkowski. Department of Cattle Breeding. Agricultural University, ul. Akademicka 13, 20-950 Lublin, Poland

The paper attempted to compare the lines of Simental bulls brought to Poland from Switzerland, Germany and Austria. Female offspring of all 3 groups had similar efficiency which ranged from 4000-5000 kg of milk, 3.9-4.1% of fat and 3.3-3.4% of protein (maximum 7298-8726 kg of milk). Four German lines were distingusihed (Haxl, Hugo, Redad and Streik) as well as 3 Austrian lines (Polde, Senat and Saltus) and 4 Swiss lines (Notar-Aelpler, Sämu-Nebelick, Zimbo-Condor and Francoas-Rossli). All the lines mentioned above made a considerable impact on local stock in the Polish region of Simental race breeding.

CATTLE PRODUCTION   [C]   Poster C4.9

Difference in Economy of Czech Spotted and Holstein Breeds
P Šafus*, J.Přibyl. Res. Inst. Anim. Prod. Uhříněves - Praha 10, P.O.Box 1, 104 01, Czech Republic

Comparison of economic differences of a closed herd with all categories of Czech Spotted (Simmental type) (C) and Holstein (H) cattle was performed from the data file covering 237 of C and 291 of H cows. Data for both breeds are from an identical conditions of the same farm in the years 1994 - 1998. Individual recording is for the milk production - number of lactations, milk yield, fat and protein content and payment for the milk; the meat production - number of bulls and the number of negatively selected animals from all categories with slaughter weight; the costs for the reproduction - number of inseminations and natality; and the costs of feeding. Whole period was divided into the shorter seasons with identical conditions. Differences between breeds for an income and costs were calculated within seasons. Whole difference between breeds was determined as the weighted average according the effective No. of observations. The higher price for milk has C, higher income for milk H. The higher income for beef has C. The profit was determined as difference between income from milk and meat production and costs for reproduction and feeding. The preliminary partial difference of profit (without fattening of bulls) was 637.70 CZK in advance of H (1.34 % of year income).

Poster C4.10

The effectivity of cows of dual purpose in comparison with dairy breed
V. Kudrna, J. Přibyl*, P. Lang. Res. Inst. Anim. Prod., Uhříněves - Praha 10, P.O. Box 1, 104 01 Czech Republic

Dairy cows of Czech Spotted (Simmental type) (C) are compared with Holstein (H) and Black Spotted breed of an former type (N) in an average level of production 6350 kg of milk. High pregnant heifers (48), calved in the same stable and in the same seasons, were involved in the study. Under conditions of an exact individual feeding with two feeding techniques (divided and total mixed rations) were practised. The composition of feed rations was the same for all breeds. An evaluation was carried out with linear model considering calving season, order of lactation, the kind of feeding technique, breed, individuality of the cow, and age at calving. Majority of all effects were not statistically significant. During firs three lactation (after correction on the other effects), was the average difference in feed dry matter intake of one lactation between H a C breeds approx. 10 %. The difference in FCM between H vs N is 4 % and H vs. C 13,6%. For milk protein production differences are 3,5 and 10,5 %. Differences among breeds decreased with order of lactation. There is not to expect high difference in economy of milk production among watched breeds under conditions of similar level of production.

CATTLE PRODUCTION   [C]

Poster C4.11

A crossbreeding system for cattle improvement in developing countries
A.K.F.H. Bhuiyan[*1], Gerhard Dietl [2] and Gunter Klautschek[2]. *Research Institute for the Biology of Farm Animals, D-18196 Dummerstorf, Germany[2], AvH Fellow, Dept. of Animal Breeding & Genetics, Bangladesh Agricultural University[1].*

Recurrent selection can be a useful tool for the development of new breeds that better fit specified production environment. With this view, genetic trends were estimated for various traits within different layers of a beef breeding population developed during 1975 to 1990 in the eastern part of Germany. Single trait animal model REML and BLUP methodologies were used respectively for variance component and breeding value estimation. Estimated breeding values were regressed on year of birth to obtain estimates of genetic trend for each trait. Estimated trends were: in beef cows, -0.38, 0.24, 0.50, 0.32, -0.35, 1.10 and 0.82 kg respectively for body weights at 6,12,15,18 months, first, second and third calving; in beef bulls, ranging from -0.003 to 0.12 cm for nine body measurement traits; 3.69 g and -3.00 NE (net. energy) respectively for test period average daily gain and energy efficiency; for body weight of beef bulls at the end of the test period 1.18, 0.59,1.52 and 0.96 kg respectively for overall, industrial farms, breeding farms and testing stations; for birth difficulty respectively -0.002 and -0.0002 in beef bulls and beef x dairy animals; 0.025 kg for birth weight in beef x dairy animals. Estimates were of 0.02 to 0.18, 0.07 to 0.33, 0.01 to 0.11 % of the population mean respectively for reproduction, production and body measurement traits. Nearly all estimated trends were significant, in expected direction, of indirect nature, based on taking only the additive genetic merit of animals into account and has indicated the effectiveness of the breeding system used and may be more suitable for cattle improvement in situations with low inputs e.g. developing countries.

Poster C4.12

Milk yield of crosses of Slovak Simmental and Holstein Red and White breeds in dependence on sire's breed
J. Huba, M. Oravcová, L. Hetényi[*], D. Peškovičová, P. Polák. Research Institute of Animal Production, 949 92 Nitra, Slovak Republic.

Milk yield of crosses of Slovak Simmental and Holstein Red and White breeds in dependence on sire's breed was studied. Daughters of purebred sires (mainly German Fleckvieh) produced in the same herds and in the same period as daughters of crossbred sires (Slovak Simmental with Holstein Red and White < 50%). The 1st - 3rd lactations were analysed separately using GLM method.
Statistically significant ($P < 0.01$) influence of herd, year of calving, lenght of lactation, age at first calving and previous calving interval on milk, fat and protein production were found. Statistically significant influence of herd x sire breed interaction was found only in fat and protein percentage. Daughters of crossbred sires produced more milk per lactation than daughters of purebred Fleckvieh sires (11 - 90 kg according to parity). No statistical significance was observed. Daughters of crossbred sires achieved more favourable results in fat production (154.6 resp. 151.0 in the 1st lactation, 177.4 resp.170.9 in the 2nd lactation and 198.4 resp. 186.3 in the 3rd lactation) and fat content (4.24 resp. 4.14 in the 1st lactation, 4.30 resp. 4.23 in the 2nd lactation, 4.32 resp. 4.12 in the 3rd lactation). These differences were statistically significant.

CATTLE PRODUCTION   [C]

Poster C4.13

In vivo carcass value evaluation of Simmental bulls by ultrasonographic method
P. Polák[1], K. Słoniewski[2], T Sakowski[2], L. Hetényi[1]*, J. Huba[1], D. Peškovičová[1], E.N. Blanco Roa[1]. [1]Research Institute of Animal Production, Hlohovská 2, Nitra 949 92, Slovakia. [2]Institute of Genetics and Animal Breeding Jastrzębiec, 05-551 Mrokow, Poland.

An ultrasonographic method was used for slaughter value analysis. The data from 50 dual purpose Simmental bulls at the age of 450 days were used in the analysis. The thickness of muscle including fat and skin were taken on five positions (os scapula, 7th and 13th thoracis vertebrae, 5th lumbalis vertebra, os ischia). The correlations among traits mentioned above and slaughter traits (hot carcass weight, cold carcass weight, weight of meat in carcass and weight of meat in valuable cuts) were studied. Weight before slaughter was highly correlated with slaughter traits (r=0.78 - 0.94), while their correlations with sonographic measurements were low (r=0.30 - 0.55). Sonographic measurements were used to construct linear regression models for slaughter traits using stepwise procedure. The coefficients of determination were high (R^2=0.60 - 0.89) in case of using weight before slaughter as independent variable together with sonographic measurements. The models did not fit very well (R^2=0.13 - 0.43) when weight before slaughter was excluded.

Poster C4.14

Frequency of umbilical hernias in German Fleckvieh and their effect on meat performance
R. Herrmann[1], J. Utz[2], E. Rosenberger[2], R. Wanke[3], K. Doll[4] and O. Distl[1]
[1]Institute of Animal Breeding and Genetics, School of Veterinary Medicine Hannover, [2]BLT-Grub, 85586 Poing, [3]Institute of Animal Pathology, University of Munich, [4]Clinics for Ruminants, University of Giessen, Germany.

The data of 53,105 German Fleckvieh calves were collected on 77 livestock auctions to investigate the frequency of umbilical hernia. The incidence of umbilical hernia was in male calves 2.2% and in female calves 1.5%. In twins the incidence significantly increased to 3.2% (males) and 3.0% (females), resp. Among sires significant differences were evident and there could be not indentified any sire without affected progeny. A monogenic autosomal recessive inheritance seems not very probable from these figures. In addition to this field study 62 males calves affected by umbilical hernia were tested on station according to the protocol of the progeny test together with a control group of 82 unaffected male half sibs/Animals showing in the beginning a hernial opening of 1 to 2 fingers thick developed a complete occlusion in most cases (%), whereas larger hernial openings closed only in 2 of 6 calves. No differences could be observed among these two groups for mortality, .. and carcass traits. The pathological examination could not detect significant deviations in tissue structure among affected and healthy control calves.

CATTLE PRODUCTION   [C]   Poster C4.15

Inheritance and Introgression of Polledness in the German Fleckvieh Population
H. Hamann*, A. Lamminger, O. Distl, Department of Animal Breeding and Genetics, School of Veterinary Medicine, Hannover, Germany

At the experimental farm in Schwaiganger, Bavaria, a herd of genetical polled Fleckvieh bulls and cows was built up during the last 25 years. Since then data from about 1500 animals are recorded. Among these animals phenotypes observed were horned, polled and scurred. Using the data collected in Schwaiganger it was possible to examine different models of inheritance (Brem *et al.*, 1982; Long and Gregory, 1978). These models assume epistatic interactions between two independently segregating loci and both consider different phenotypic expressions of the polled and the scurs status between the sexes. Cases not to be explained by the model of Brem *et al.* (1982) were dams with scurs which have polled sons. The model of Long and Gregory (1978) does not explain the case of polled sires with scurred daughters and other horned offspring. A model was developed which also takes the origin of the different alleles (paternal or maternal) into account and overcomes the inconsistencies encountered by the other two models. Based on this model different strategies of spreading polledness among the Fleckvieh population were tested with respect to breeding goals. It is supposed that polled animals are inferior in milk yield and other traits compared to horned animals. One of the strategies is driven by the primary interest of getting homozygous polled animals while in another one heterozygous animals were also selected in order to keep the high level in the production traits. The efficiencies were compared by simulation studies.

Poster C4.16

Comparaison of Holstein and Meuse Rhine Ysel lactation curves with three mathematical models
A. Félix*, J. Detilleux and P. Leroy. Department of Genetics, Faculty of Veterinary Medicine, University of Liège, Belgium.

Three mathematical models were used to compare lactation curves in Holstein (HF) and Meuse Rhine Ysel (MRY) cows. Data consisted in monthly test-day milk yields from 1136 first and 1443 later lactations on 1165 MRY cows in 494 herds and from 2676 first and 3865 later lactations on 2736 HF cows in 912 herds. The models included a gamma function, an inverse quadratic polynomial function, and a regression model of yields on day in lactation (linear and quadratic) and on log of 305 divided by day in lactation (linear and quadratic). For each lactation and each breed separately, coefficient of determination, squared deviations of predicted versus actual weights, and correlation between predicted and actual weights (Table 1) were used to compare the 3 models.

Table 1. Correlation between predicted and actual weights per breed, parities and model.

Model	HF 1st parity	HF later parities	MRY 1st parity	MRY later parities
Gamma	0.50	0.69	0.57	0.73
Inverse	0.45	0.57	0.51	0.58
Regression	0.47	0.68[a]	0.56[a]	0.74[a]

[a]several regression coefficients of the model were not significantly different from zero.
Regression coefficients were different between HF and MRY cows and between parities. Thus, new models should be applied to describe lactation curves in HF and MRY cows.

CATTLE PRODUCTION   [C]   Paper C5.1

Herd health schemes - problems and potential
T.R. Wassell*[1], B.R. Wassell[2], R.J. Esslemont[3], [1]*Estates Office, Scottish Agricultural College, Auchincruive, Ayr, KA6 5HW.* [2]*DAISY-National Milk Records, Chippenham, Wiltshire.* [3]*Dept of Agriculture, Earley Gate University of Reading, Reading, RG6 2AT.*

This paper discusses the development of the Herd Health Scheme concept as used by many veterinarians and dairy farmers in the UK. Herd Health Schemes have had a slow uptake by both the dairy farmer and the veterinary surgeon. A few attempts have been made over the last twenty years to increase veterinary and farmer awareness of this approach to preventative medicine, but unfortunately these have had little long-term impact. Most of the academic research into the use of herd health schemes has concentrated on the development of computerised recording schemes rather than the logistics of setting up and operating such schemes.

In its current form, the HHS consists of three elements: data handling, processing and utilisation. At present however, the HHS still focuses on a fairly narrow range of criteria linked to veterinary intervention. An alternative approach to Herd Health takes the concept one step further. The Integrated Herd Health Scheme encompasses the traditional aspects (physical records and the veterinarians' input) with managerial, feeding, quality assurance and statutory records in a 'one-stop' recording environment. This allows experts from these areas, through sharing this common pool of information, to give a co-ordinated response to the farmer.

Paper C5.2

Quality monitoring of milk recording
P.J.B. Galesloot*[1], G. de Jong[1] and G. van de Boer[2]. [1]*Research and Development department and* [2]*Quality department, NRS, P.O. Box 454, 6800 AL Arnhem, The Netherlands.*

Milk recording data are the basis for dairy herd management and dairy cattle improvement. Therefore the data collected has to be reliable. Currently the NRS has a system of repeated tests to guarantee the quality of test day data. Per year about 1.7% of the farms are visited unexpectedly at the day after testing, in order to execute a control test. This is a labour intensive system. To save costs and to be able to check more test day data the possibilities for a monitoring system are investigated. The research aimed to develop parameters to detect tests which could be incorrect. Test day data and data of milk deliveries to milk factories in the period '97-'98 were analysed. A suitable parameter to detect incorrect tests is the difference between milk production/cow/min by day and milk production/cow/min by night. If daily yields measured during test day concern more than 24 hours, milk production/cow/min by day is higher than milk production/cow/min by night. The relationship between this parameter and the difference between the recorded milk, fat and protein yield and the delivered yield to the factory was estimated.

CATTLE PRODUCTION   [C]   Paper C5.3

A dairy farm budget program as a tool for planning and evaluating
F. Mandersloot[*,1], A.T.J. van Scheppingen[2] and J.M.A. Nijssen[2]. [1]Research Station for Pig Husbandry, P.O. Box 83, 5240 AB Rosmalen, The Netherlands. [2]Research Station for Cattle, Sheep and Horse Husbandry, P.O. Box 2176, 8203 AD Lelystad, The Netherlands.

Before setting up or changing a dairy farm operation, it is higly preferred to explore the consequences of the lay-out and changes planned. To calculate technical and economic consequences a dairy farm budget program has been developed. This program is used for strategic and tactic planning purposes in studies as well as on commercial farms.

The set up of three research dairy farms has been planned using the farm budget program. One of these farms (De Marke) emphasises the integration of nutrient flows. This farm is in operation for seven years now. Results show that losses of nutrients are reduced to the planned levels. However, costs are considerably. Two other farms (the Low-cost and the High-tech farm at the Waiboerhoeve) aim at a reduction of the cost price of milk. The Low-cost farm is using simple management tools on herd level, where the High-tech farm is highly automated on individual cow level. First results indicate that a reduction in cost price by either low investments or producing at a larger scale may be achieved within the next years.

Based on the farm budget program also tools at operational level become available. A tool to support the grazing management is being developed.

Paper C5.4

Modelling lactation curves of Holstein Friesian cows on large scale farms in Malawi
M.G. Chagunda[*,1,2], Z. Guo[1], E. Bruns[1] and C. Wollny[2]. [1]Institute of Animal Breeding and Genetics, Georg-August-University of Göttingen, Germany, [2]Department of Animal Science, Bunda College of Agriculture, University of Malawi, Malawi.

Six alternative mathematical functions describing the lactation curve were examined for Holstein Friesian cows performing on large scale farms in Malawi. The functions which could be categorised as three-parameter and five-parameter functions were: Wood's function, Wilmink II function, Mixed-Log-Model I, Ali and Schaeffer model, Modified Wilmink II function and Mixed polynomial function. Although a little more demanding, five parameter functions had a better fit than three parameter ones ($p<0.001$) across parities and farms. Both season and year of calving had a significant influence ($p<0.001$) on the function of best fit. Milk yield predictions done by applying such functions need to take into consideration the importance of environmental differences prevailing in different situations.

CATTLE PRODUCTION   [C]   Paper C5.5

Field performance recording in smallholder dairy production systems in India
F. Schneider[1], *M.R. Goe*[2], *C.T. Chacko*[3], *M. Wieser*[4], *H. Mulder*[5]. [1]*Swiss College of Agriculture, CH-3052 Zollikofen, Switzerland,* [2]*Jonas Furrerstrasse 45, CH-8400 Winterthur, Switzerland,* [3]*Kerala Livestock Development Board, Trivandrum - 695004, Kerala, India,* [4]*RuralConsult, CH-2525 Le Landeron, Switzerland,* [5]*Intercooperation, P.O. Box 6724, CH-3001 Bern, Switzerland.*

An estimated 70 million smallholders in India own livestock - 60 million of them own cattle, buffalo or both. Field performance recording (FPR) is an effective management tool for monitoring livestock performance (milk, growth, reproduction, etc.) at the smallholder level. Since 1977 FPR has been a component of Swiss support to the livestock sector in India as part of sire evaluation and breed improvement programmes through milk recording and field progeny testing schemes for crossbred cattle, buffalo and goats. FPR is carried out by government and semi-government institutions, dairy co-operatives and nongovernmental organisations in selected areas of Andhra Pradesh, Gujarat, Kerala and Rajasthan States. FPR can help farmers increase productivity and genetic merit of their animals, manage available resources more efficiently and, on a wider scale, assist governments in developing sustainable livestock strategies. This paper reviews approaches used for designing and implementing FPR activities in India. Strengths and weaknesses are highlighted and important lessons learned regarding prerequisite institutional, technical and human resources for sustaining FPR as an integral part of an overall strategy for livestock development are discussed.

Paper C5.6

Etude du stress thermique dans les conditions tunisiennes à travers le calcul de l'index température-humidité
R. Bouraoui[1], *A Majdoub*[2] *et M Djemali*[2]. [1]*ESA Mateur 7030 Mateur.* [2]*INAT 43 Avenue Charles Nicole 1082, Cité Mahrajène Tunis.*

L'étude de l'effet de la température sur la production laitière bovine est toujours liée à l'humidité sous forme d'index THI (température- humidité index). Les données bibliographiques montrent qu'à partir de la valeur 72 de cet index l'effet du stress thermique se fait sentir sur l'ingestion volontaire, la production laitière, la reproduction aussi bien que sur la santé de l'animal. C'est dans ce cadre que cette étude a comme objectifs : la détermination des THI pour toutes les régions de la Tunisie, l'établissement de cartes de THI par saison, le classement des zones sur la base des valeurs THI en précisant la relation production laitière - THI et la détermination de la place de la Tunisie dans le cadre mondial sur la base de THI.
Pour ceci, des données mensuelles de température et d'humidité de 20 stations climatiques durant 10 ans ont été analysées. Il se dégage de ce travail que : Le stress thermique se manifeste en été début automne (Valeurs THI : 72 - 78), au fur et à mesure qu'on s'éloigne du littoral le stress thermique est plus marqué, la courbe d'évolution des valeurs THI en Tunisie se rapproche de celle de l'Egypte en Afrique et de l'Arizona en Amérique du Nord. Cette courbe montre que la Tunisie a un climat méditerranéen à été chaud et sec avec des effets négatifs sur la production laitière.

CATTLE PRODUCTION [C]

Paper C5.7

INFOLEITE - Educational Programs, Research and Sustainability within a Dairy Production System
M. M. Guilhermino. Departamento de Zootecnia, Universidade Federal do Ceará, P.O. Box 12.168, Fortaleza, Ceará, Brasil.

INFOLEITE is a multidisciplinary project aiming at: Education - Developing intellectually demanding and practically oriented educational programs for dairy producers objecting at economically feasible and sustainable production systems. Research - Setting up a reliable and dynamic data set to assist research in all areas of milk production to help students in their postgraduate courses at the University of Ceará. Production - Implementing new technologies, management techniques and dairy farm planning according to the farmers progressiveness and economical situation. Sustainability - Recommending actions that allow farmers to produce milk in a sustainable way so that future generations can use the same systems as does the current generation. Social Agricultural - Awareness about personal and domestic aspects of hygiene, implementation of horticulture, fruit crop and medicine plants program. Localisation - On the opposite to globalisation, the knowledge and the local culture is valued to contribute in the management decisions and to allow the knowledge to pass through all the generations. Nowadays, ten dairy farms participate in the project and their herds are being monitored in terms of health, fertility, production and costs. Management recommendations are made by a team of specialists of the University and the execution of the recommendations are made by the technicians employed by the municipality of Quixeramobim in Sertão Central of the state of Ceará in the Northeast of Brazil.

Poster C5.7

Breeding efficiency in evaluation of reproductive performance of dairy cows
E. Szücs*[1], K. Bódis[1], Gy. Látits[1], J. Tözsér[1], I. Györkös[2], and A. Gáspárdy[3]. [1] University of Agricultural Science, Páter K. u. 1, Gödöllö, [2]Research Institute for Animal Breeding and Nutrition, Gesztenyés u. 1, Herceghalom, [3]University of Veterinary Science, István u. 2, Budapest, Hungary

Accuracy of BE_{Tomar} was tested in a simulation study where age at first calving (AFC) and calving interval (CI) were varied. Estimated R^2=0.99 and SE= 0.81-0.40 for 1-5 parities reveal accuracy of BE_{Tomar} depending on AFC and sum of CI. Applicability of BE_{Tomar} has been tested on data set of 28631 lactations from purbred Holstein (HF) and Hungarian Red Spotted (HRS) x HF crossbred cows with 96.8, 93.7, 87.5, 75.0 and 50.0 % HF genes. Overall mean and SD for days in milk, milk yield, butterfat as well as protein yield and percentage, conception rate, calving interval and BE_{Tomar} were 320±47, 6460±1775kg, 234.9±62.2kg, 3.67±0.47%, 216.1±66.5kg, 3.31±0.37, 1.63±0.97, 385±47, and 91.7±7.31%, respectively. Even though statistical differences among means in BE_{Tomar} for genotypes, parities (1-5), years (6), months of calving (12), age of dam (2-14 years) and parity x month of calving interaction were established, order of them seemed to be low. Means for 1-5 parities were 91.8, 92.9, 94.6, 94.7 and 96.4%, respectively, a phenomenon which might be due to quadratic regression of actual AC on BE_{Tomar}. No effect for month of calving on BE_{Tomar} was shown. Distribution of BE_{Tomar} changed among conception rate classes depending on parity.

CATTLE PRODUCTION [C] Poster C5.8

Application of software programme „Herd hierachy" (for Windows) for beef cattle herds kept on pasture
K. Chudoba, A. Dobicki*
Agricultural University of Wrocław, Kożuchowska 7, 51-631 Wrocław, Poland

Programme „Herd hierarchy" (for Windows) (Chudoba, 1996) conform to ISO 9001 standards as well as to Polish standards PN-ISO 9000-3 (1994). Programme analyzes grouping of animals in the space (objects: bull, cow, cow + calf, cow + 2 calves, heifer, bullock) according to two parameters: (1) sum of rank score from empirical observations, (2) variability in rank position. It is possible to incorporate additional data as: site, time of observation of event, other circumstances occuring during observations. Input of data: (1) basic documentation and graphic data (ex. exterior traits, colour pattern as pictograms) (2) empirical observations - added to file in succession. Printout of herd hierarchy is based on dendrite (1- or 2-sided) and presents distribution of subgroups of the herd. Results of pilot trials of using „Herd hierarchy" in beef suckler herd of mixed genotypes (Charolaise, Red-and-White) - 82 cows grazing mountainous pasture - were encouraging. Distinct printouts of clear cut subgroups: leader, dominant, submissive, follower and marginal animal groups were obtained.

Poster C5.9

Dairy marketing trends in the suburban perimeter of Kenitra city, in the heart of the irrigated zone of Northern Morocco
M.T. Sraïri*. Department of Animal Productions, Institut Agronomique et Vétérinaire Hassan II, P.O. Box 6202, 10101 Rabat, Morocco.

There has been a dramatic decrease in raw milk quantities annually collected by dairy factories in the suburban perimeter of Kenitra city (from more than 28.4 million kg in 1992 to less than 10.8 million kg in 96). Many reasons can explain this trend. Climatic variations with massive drought in 95 and flood in 96 can be mentioned. However, these causes are not the only responsible for this decrease. Several surveys at milk collecting centres have revealed that these structures have been suffering from major activity crisis : many dairy producers refuse to deliver to them their milk anymore, because they consider that the price they get from the factories is too low (2.9 Dirhams/kg $\approx$ 0.3 US \$/kg), as they prefer to sell it at a much more attractive price (4 Dirhams/kg $\approx$ 0.4 US \$/kg), directly to consumers in Kenitra. The number of dairymen who deliver their milk in eight centres situated in a 30 km distance of Kenitra, has fallen from 540 to less than 140 in an interval of 5 years (92/97). Thus, the informal circuit of milk marketing is getting very active, with a intense peddlers organisation. However, quality control is reduced. This is why perspectives for the development of milk production remain unclear, and this rises numerous questions for the dairy sector durability in suburban perimeters in Morocco.
Key-words : Milk collecting centre, Milk factory, Peddler, Morocco, Suburban perimeter.

CATTLE PRODUCTION   [C]   Poster C5.10

The role of the very low somatic cell count on clinical mastitis: a review
*W. Suriyasathaporn[*1] and Y.H. Schukken[2]. [1]Division of Farm Animal Health, Utrecht University, Yalelaan7, 3584CL Utrecht, the Netherlands, [2]Department of Population Medicine and Diagnostic Sciences, Cornell University, Ithaca, NY, 14853 USA.*

Somatic cell count (SCC) is a number of cells, mainly leukocytes, presenting in milk. SCC is used to define the subclinical mastitis, the bulk milk SCC is an indicator for the success of the mastitis control program. However, the role of SCC on clinical mastitis (CM) is not clearly understood. Therefore, this study was conducted to review the role of SCC on CM. Mastitis is caused by the intramammary infection (IMI). In response to IMI, both bacteria and local leukocytes release chemoattractants causing the influx of neutrophils into the infected udder. Quicker and more neutrophils accumulated in udder results in less severity of mastitis. Therefore, very low SCC (VLSCC) indicating the fewer number of udder leukocytes could related to the lower capacity of the udder defence. In supported by experimental-mastitis studies, VLSCC cows showed higher severity of mastitis after experimental infection with the certain amount of bactaria than cows with higher SCC. However, results in epidemiological studies still showed that VLSCC was related to the lowest incidence of CM. Since the CM incidence is based on pathogens, udder defence, and management, the lowest CM incidence in VLSCC cows may be caused by the fewer time of IMI rather than the impairment of udder defence mechanisms.

Poster C5.11

The effects of various milk feeding periods on growth performances of Black Pied calves
*Ö. Sekerden[*1], M. Sahin[2]. [1]Mustafa Kemal University, Faculty of Agriculture, Department of Animal Science, 31034, Hatay, Turkey, [2]Hatay State Farm, Hatay, Turkey.*

This research was conducted in order to determine the most economic weaning age and amount with milk in Black Pied calves reared in Hatay State Farm. The material of the study was formed by various data belong to male and female calves of the cows calved first and second times. The calves born were classified into 3 groups in birth order as follows; The calves fed with 243 kg milk in 2 months: 1st, the calves fed with 343 kg milk in 2.5 months: 2nd, The calves fed with 378 kg milk in 3 months: 3rd group. Live weight, height at withers, body length, chest depth, chest girth, chest width, shin girth were determined at birth, in 15, 30, 45, 60, 75, 90, 105, 120, 135, 150, 165 and 180 days in trial animals. In 1-5 age groups, the effect of birth season, 6-13 age groups the effects of birth season and the group of feeding with milk on each investigated characteristic were determined using Least Square Analysis Method. Standardization were applied where is necessary. The average values were calculated belong to each characteristic in every ages. Following results were concluded; 1) Milk feeding groups of male calves are different from each other statistically from the point of view of chest depth and shin girth at the 6 months age. 2) In male calves feeding with milk for 8, 10, 12 weeks do not effect the live weight and various body measurements reached at the 6 months age significant statistically.

SHEEP AND GOAT PRODUCTION   [S]

Paper S1.1

Mountain sheep farming in the British isles
A. Waterhouse, SAC, Hill and Mountain Research Centre, Food Systems Division, Auchincruive, Ayr KA6 5HW, SCOTLAND

Hill and mountain sheep systems are the dominant land use in much of the north and western British Isles on land with difficult topography and poor soils. The relatively low altitude of the mountains (< 1,000 metres above sea level) and the temperate climate has enabled the creation of annual grazing systems by free-ranging sheep. These systems make heavy use of poor quality semi-natural grazing, typically heather moorland and acid grassland mozaics. Limited supplementation is provided. Environmental and nutritional conditions are harsh and output is typically well below the biological capacity of the hardy sheep breeds used. Hill and mountain grazing, which may be common land or used by a single farm, is usually associated with land in the lower valleys close to the farmhouse and buildings. This land is managed more intensively, divided into fields and often used strategically for lambing, mating and for better nutrition of weaned lambs and ewes with twin lambs. Silage or hay may be produced in these areas. Whilst virtually all hill and mountain farms have a sheep enterprise, only a proportion have a beef cattle enterprise. Managed game shooting often co-exists with sheep farming. Land is of high landscape interest with considerable areas designated for nature conservation and tourism plays a crucial role in the local economic viability. Sheep farming has a direct impact upon the landscape and nature conservation, but arguably also forms a key role itself as part of the cultural landscape and heritage.

Paper S1.2

Sheep and goat production in Norwegian wet mountain areas
T. Ådnøy*, G. Steinheim, and L.O. Eik. Department of animal science, Agricultural University of Norway, Box 5025, N 1432 Ås, Norway.

Of mainland Norway (324000km^2, 59-71°N), 3% is cultivated, and 22% is forests. Over half is mountain range lands - a cheap summer feed resource, giving 45% of total intake in sheep production. Quality is often good. One million sheep are intensively fed indoor 5-8 winter months, in expensive housing. High lamb production per ewe is sought. Fertility is 2,0 lambs born to sheep two years and older. Ewes often lamb at one year of age. Lambing is at end of indoor feeding - slaughtering before. Out of season lambing is not usual. Lowland pasture spring and autumn often a limiting factor on sheep farms. Income is from meat and wool - no milk.
Breeding goals include lamb growth, two lambs per mature ewe, good mothering ability, and non-declining wool quality and amount.
Free range grazing is in conflict with mounting numbers of protected carnivores: wolverines, lynx, eagles, wolfs, and bears give losses.
Sheep units are often part-time. Whereas sheep farming is not, dairy goat farming is labor intensive. Some 56000 goats give 23 tons of milk per year. A quota system limits realized production. Alternatives to traditional goat whey cheese are sought and have been developed. Meat production in goats is not much exploited.
Use of grazing animals to control shrubs and trees is considered.

SHEEP AND GOAT PRODUCTION   [S]   Paper S1.3

Sheep and goat production in wet mountain areas of Switzerland
R. Lüchinger Wüest[1], *M. Schneeberger*[1], *A. Zaugg*[2]. [1]*Swiss Sheep breeders association*[1], *P.O. Box, 3360 Herzogenbuchsee,* [2]*Swiss goat breeders association*[2], *P.O. Box, 3360 Herzogenbuchsee*

38 % of the total surface of Switzerland (41'285 km^2) is agricultural area, 14 % is used for mountain grazing. Sheep production is economically important especially in the climatic wet, pure grassland areas (1000-3000 mm/yr rainfall) of the mountains (Alps and Jura). The climatic conditions are not suited for keeping of fine-wool-breeds (Merino). The main-product of Swiss sheep production is lamb-meat, creating 98 % of the income. The main lambing season is spring. After grazing on valley pasture, about 50 % of Swiss sheep stay, for about 90 days, on alpine pastures situated up to more than 2000 m above sea level. The lambs reach the weight for slaughter of 40 kg with an average daily gain of 150 to 200 g after the alpine pasture in mid-September. Goats, too, are mainly kept on pure grassland farms in the wet and rainy areas. The most important goat breed produce 750 kg of milk with 2.7 % protein and 3.2 % fat in 260 days of lactation. Goats are kept on mixed farms, mostly beside of cattle, and represent an interesting niche.

The Sheep as well as the goat flock increased during the last 30 years. Today, the sheep flock consists of 440'000 heads, the goat flock of 55'000 heads. The sheep are kept in small flocks with 20 animals, the goats in flocks with 5 animals on average. The increasing importance of sheep and goat production in Switzerland is due to structural changes in agriculture, with increasing numbers of farms being run by part time farmers, and surplus in cattle production (milk and meat).

Paper S1.4

Sheep production systems under the hard conditions in alpine regions in Austria
F. Ringdorfer. Departement of Small Ruminants, Federal Research Institute of Agriculture in Alpine Regions, BAL Gumpenstein, A-8952 Irdning, Austria

In austria, sheep production is a small part of total animal production. About 20.000 farmers keep 380.000 sheep. That means the flocks are very small, in the average about 18 sheep The main objective is production of lambs for slaughter. Liveweight at slaughter is 15-25 kg (milk lamb) or 35-45 kg. Fattening heavy lambs should be finnished with an age of 4 to 5 months.

57 % of the agricultural area are in mountain regions. More than 60 % of the sheep were kept in this area. This regions are characterized in more than 1000 mm precipitation, cold temperature in winter and steep grassland. About 180 days of the year the sheep and all other animals are kept in stable. The main fodder is hay and grass silage in winter and pasture in summer. From may, june to september sheep are on pasture in the mountains. In this time no or less lambs are born, because in the alps the rearing rate is very low.

The best breed for this aerea is the austrian mountain sheep. This breed is very fertile (240%), have no specially breeding season, get lambs twice a year and can climb very good in the mountains. Meat performance is not so good, so the mountain sheep is used in crossing with meat breeds like suffolk, black head or texel.

Problems in lamb meat production are the small flocks (different quality) and high costs of production (high price of meat). An other problem are the cheap imported lambs.

SHEEP AND GOAT PRODUCTION [S] Paper S1.5

Attempts of describing the influence of some climatic conditions on the effects of sheep breeding production in mountainous regions of Poland
J. Ciuruś[1] and R. Niżnikowski[2], [1]Mountain Sheep Farming Research Station in Bielanka, 34-723 Sieniawa, Poland, [2]Sheep and Goats Breeding Department, Warsaw Agricultural University, Przejazd Street 4, 05-840 Brwinów, Poland.*

Sheep breeding in mountains remained only in some parts of Carpathians. Sheep stay about 5 months on pasturage i.e. from May till September in the open air under the supervision of shepherds. Polish mountain sheep come from a large race group called "cakiel", it is good adapted to severe conditions of environment in mountains. The pasturelands, on which sheep is recently pastured, are placed at an altitude between 500 and 1500 m. The number of sheep being recently pastured in mountains is estimated at about 70.000.

The annual rainfall at an altitude of 1000 m fluctuates between 1200 and 1400 m with the temperature of 4,5 °C, whereas in the period of pasturage it is 600 m with the temperature of 11,6 °C (results collected during past 20 years). Experiments proved that in the period of heavy rainfalls lactation of sheep kept 24 hours on the pastureland fell down to 13-40%. Sheep kept nights in a building were characterized by 8,4-13,8% higher daily lactation and 8,3-11,3% complete production per season, in comparison to sheep kept the whole night in the open air. What was found out was the regrowth and the rendement evaluation of wool. However the influence on weight and wool production was not stated.

Paper S1.6

Sheep raising in wet mountain areas of Czech Republic : a three-level system
Mátlová, V. Research Institute of Animal Production, 104 00 Praha 10 - Uhříněves, Czech Republic

Different systems of sheep raising in mountain grasslands on three extensity levels are examined : extreme E1, (0 pasture improvement, 0 shelter, 0 hay/concentrate), medium E2 (overnight housing, hay supplementation) and mild E3 (improving lamb pastures, overnight housing, hay/concentrates lamb supplementation). In E1 system native Šumavka (S) and Merinolandschaf (ML) breed are used to produce maternal population for E2 and E3 system, which is based on cross-breeding with meat rams for intensive lamb production. For E2 and E3 system East Friesian (VF) breed is examined too. Out of grazing season ewes are moved into somewhat kindly range, lambing both indoor and outdoor. Fertility rate in ML, S, and VF reaches 119-135, 120-122 and 172-189% respectively. Growth rate of ML 210-day ewe lambs ranges 114 -178 g.day^{-1} according to their birth date (January - June), and that of 100-day meat crossbred lambs 210 -228 g.day^{-1}. In S ewe lambs, S x Texel and S x Oxford Down crossbred meat lambs growth rate ranges 185, 221 and 209 g/day. Based on economic indices, required gain structure (lamb production/ landscape servicing/ state subsidies) in the three systems is estimated.

SHEEP AND GOAT PRODUCTION   [S]   Paper S1.7

Alpine lambs out of optimal reproduction and grazing techniques
K. Emler*, C. Marguerat, H. Leuenberger, N. Künzi. Institute of Animal Science, Swiss Federal Institute of Technology, CH-8092 Zurich

The objective of this project is to improve competitiveness of sheep production by optimal combination of reproduction and fattening traits with sustainable usage of alpine grassland area. Important criteria were year-around lambing (with exception of the alpine grazing period) and number of litters per ewe within 24 months. A preliminary analysis included 829 litters with 1284 lambs, originating from 2 breeding herds comprising 100 ewes each and consisting of Brown Alpine Sheep (BSB) as well as Charollais x Brown Alpine Sheep (CHAxSBS) and White Alpine Sheep x Brown Alpine Sheep (WASxSBS), respectively. Lambings took place all over the year, with increased frequency in late winter and fall. 60% of all litters sired by CHA rams were born in February and March, 65% of all litters sired by WAS rams were born from October to January, whereas litters sired by SBS rams were unequally distributed all over the year. Purebred SBS ewes had 2.8 litters within 24 months with 1.7 born-alive lambs per litter, with a rearing rate of 91% by 90 days. Crossbred ewes (CHAxSBS) had 2.5 litters with 1.7 lambs per litter, with a rearing rate of 93%. Crossbred lambs had significantly higher birth weight and daily gain compared to purebred SBS lambs. Fattening performance of backcrosses with Charollais was significantly worse compared to WAS and SBS. Lambs born in spring had significantly higher daily gain. Data collection will be completed by the end of 1999.

Poster S1.8

Use of N-alkane technique to determine herbage intake of sheep from natural mountain pasture in Sudetes.
P. Nowakowski*, K. Aniołowski, K. Wolski, A. Ćwikła
Agricultural University of Wrocław, 51-631 Wrocław, Kożuchowska 7, Poland

Natural mountain pasture in Sudetes at 470 - 500 m above sea level (average precipitation = 696 mm/year) consisted with 17 species of grasses (79,4 % of total DM in October, with dominance of *Festuca rubra* and *Agrostis vulgaris*) 6 papilionaceous (8,3 % of total DM in October) and 21 species of herbs. No lime nor fertilisers were applied. Herbage dry matter yield, crude protein, crude fibre and n- alkanes contents of herbage were estimated. There was in average 12 dt herbage DM/ha in October. Eleven 2-year old ewes of two genotypes: Wrzosówka and Merino x Romanov cross (body weight from 33,0 to 54,0 kg and average body condition score of 3,43 ±0,374) were dosed with 130 mg of C_{32} alkane for two weeks in October and individual collections of feaces were done in the second week to evaluate DM intake according to method described by Mayes et al. (1986). Average intake of DM when using C_{32}/C_{31} alkane pair for calculations was 60,6 ±7,20 g/$W^{0,75}$. The lowest calculated intake was 34,1 and highest 93,5 g/$W^{0,75}$/day due to variations between animals and days. These data are in good agreement with values quoted by ARC (1980).

SHEEP AND GOAT PRODUCTION   [S]   Poster S1.9

Small ruminants grazing as a management tool for controling bush invasion of natural mountain pastures
A. Ćwikła, J. Gawęcki, W. Łuczak, P. Nowakowski*
Agricultural University of Wrocław, Kożuchowska 7, 51-631 Wrocław, Poland

Not fully utilised natural mountain pastures in Sudetes at 470 - 500 m above sea level (average precipitation = 696 mm/year) are invaded nowadays with *Alnus incana, Crataegus oxyacantha, Fraxinus excelsior, Prunus avium and Prunus spinosa*. Three species of small ruminants: goats, sheep and fallow deer were kept all year out on wild pasture covered with up to 50 % by bushy growth. Nongrazed - control plots were kept intact. The seasonal pattern of browsing on shoots and bark was observed. During the vegetation season all animals browsed shoots of *Crataegus* and *Rosa* limiting yearly growth of bush crowns by up to 60 %. From late autumn to early spring when use of sward was limited (due to snow cover) and despite continous access to meadow hay animals were barking all species except *Prunus avium*. Main trunks and branches in animals mouth reach were barked. Up to 90 % of *Fraxinus*, 60 % of *Prunus spinosa*, 40 % of *Crataegus* and 30 % of *Rosa* were barked. Samples of shoots and bark of invading pasture species were collected during the peak season of browsing for chemical analysis.

Paper S2.1

The effect of Nigella Sativa supplementation on sheep milk composition and cheese manufacture
S.T.Abd El-Razek[1]* and M.A.A.EL-Barody[2]. [1]*Dairy Science Department, Fac. of Agric., Minia Univ.-Minia Egypt.* [2]*Animal production department,Fac. of Agric., Minia, Univ.-Minia-Egypt*

Fourteen pregnant Ossimi ewes, averaged 48 kg of body weight were used in this study. They were divided into two equal groups. The first group was fed concentrate mixture (0.5 kg/head/day) and supplemented with Nigella sativa by 10 % and the second group (control,fed concentrate mixture 0.5 kg/head/day) only. The feeding period starts on the last month of pregnancy period and continues for two months post parturition. Milk yield was recorded once weekly over 28 weeks period, starting from the third day post parturition. Milk samples from does were analyzed for fat, protein, total solids and ash percentages. Two batches of domiati cheese were made from milk of sheep of the two groups. Cheese samples were scored for flavor, body and texture and appearance. Nigella sativa had improved the persistency for producing milk at the 27 th week by 28.56 % vs.24.23 % of g1 and g2, respectively. Nigella sativa additive had no effect on milk composition (fat, protein, total solid, lactose and ash). Nigella sativa enchanced appearance score. The cheese made from milk of (g1) had higher scored 15 vs. 13 of (g2). Nigella sativa had improved the flavour, body and texture scores. In conclusion, herbs additive (Nigella sativa) has a beneficial effect in improving milk production and cheese manufacture.

SHEEP AND GOAT PRODUCTION   [S]   Poster S2.2

Comparison of matings using Moroccan Timahdit and D'man purebreds, first and terminal crosses. 1. Ewe productivity, lambs survival and growth performances
M. El Fadili* [1,2], C. Michaux[2], J. Detilleux[2], P.L. Leroy[2]. [1] Département de Zootechnie, INRA, B-415, Rabat, Morocco, [2] Département de Génétique, ULg, 4000, Liège-Belgium.

Timahdit (T), D'man (D) and 3 meat breeds (M) were involved to study (1) ewe performances: litter size at lambing (LSL) and at weaning (LSW), litter weight weaned per ewe exposed (LWW), and fleece weight (FW), (2) total lambs survival to weaning (TS) and (3) growth performances: birth weight (BW), weaning weight at 90 d (WW), average daily gains from 10 to 30 (ADG1) and 30 to 90 (ADG2). Matings realized were purebred D, and T, DxT, MxT and MxDT crossbred. Records on 1,187 ewes, 1,586 born lambs and 1,346 weaned lambs were analyzed with linear fixed models including non-genetic and mating effects. Least-squares means for matings are:

Matings	LSL	TS	BW	ADG1	ADG2	WW	FW	LSW	LWW
	(Lamb)	(%)	(kg)	(g)	(g)	(kg)	(kg)	(lamb)	(kg)
D	2.27[a]	68[a]	2.65[a]	146[a]	162[a]	17.26[a]	1.07[a]	1.45[a]	24.59[a]
T	1.17[b]	87[b]	2.95[b]	153[a]	171[a]	17.35[a]	2.06[b]	1.01[b]	20.13[b]
D x T	1.11[b]	85[b]	3.07[c]	153[a]	184[a]	18.56[b]	2.06[b]	0.98[b]	18.82[b]
M x T	1.14[b]	81[b]	3.48[d]	193[b]	210[b]	21.63[c]	2.09[b]	0.94[b]	22.62[a]
M x (DT)	1.74[c]	85[b]	3.27[e]	187[b]	201[c]	20.76[d]	1.60[c]	1.40[a]	28.75[c]

Results from table show that DT offers a potential for increasing productivity, and that M(DT) should be taken into account to improve sheep production in Morocco.

Poster S2.3

Carcass, meat and fat quality of Suffolk and Charmoise sired crossbred lambs compared with pure-breds of three common types of German breeds
M.R.L. Scheeder*[1], C. Sürie[2] and H.-J. Langholz[2]. [1] Institute of Animal Science, ETH Centre / LFW, CH-8092 Zurich; [2] FOSVWE, University of Göttingen, D-49364 Vechta

On four farms differing in stocking rate, use of N-fertiliser and soil conditions a crossbreeding experiment was carried out during two years in northern Lower Saxony. Texel (TX), Whiteheaded (WH) and Blackheaded Mutton (BH) ewes were all kept on the farm practising intensive pasture management. The other three farms performed an extensive grazing system only keeping ewes of one of these breeds. The ewes were mated either to a ram of the same breed or to Suffolk (SF) or Charmoise (CH) rams producing SFxTX, SFxWH and CHxBH crossbred lambs. Carcass characteristics of 370 lambs and meat and fat quality traits of a sample of 185 lambs were measured. CHxBH showed a more favourable lean to bone ratio over BH, but higher amounts of fat in carcass, hind leg and muscle. Flavour intensity was also higher but panellists judged the overall acceptability slightly lower. SFxTX also showed a higher fat deposition compared with TX. TX showed highest proportion of lean, highest max shear force and lowest intramuscular fat, but having high sensory acceptability. SFxWH grew slightly faster than WH but hardly showed any difference in carcass or meat and fat quality traits. Subcutaneous and intermuscular fat of the more extensively reared, older lambs showed higher melting and crystallisation temperatures. Thus enabeling slaughter at an earlier age yielding carcasses of desirable fat quality would be an advantage of crossbred lambs under extensified conditions.

SHEEP AND GOAT PRODUCTION [S]   Poster S2.4

Mapping quantitative trait loci causing the muscular hypertrophy of Belgian Texel sheep.
F. Marcq*[1], J_M. Elsen[2], J. Bouix[2], F. Eychenne[2], M. Georges[1] and P.L. Leroy[1]. [1]Department of Genetics, Faculty of Veterinary Medicine, University of Liège, Sart-Tilman B43, 4000 Liège, Belgium ; [2]INRA, Station d'Amélioration Génétique des Animaux, BP 27, 31326 Castanet Cedex, France.

Texel sheep, particularly the Belgian strain, are characterized by a generalized muscular hypertrophy sometimes referred to as « double-muscling ».
We initiated a linkage analysis in a (Texel X Romanov) F2 pedigree, including 5 families and 49 F2 individuals. Preliminary results obtained with 3 microsatellite markers flanking the *myostatin* gene showed lodscores > 3 for 2 of the 39 continuous traits included in the analysis, providing strong evidence that a QTL with considerable effect on muscular development maps to proximal 2q in sheep.
Since this study was performed, new F2 lambs (n=62) were born. Using the whole pedigree material, we will : 1. confirm the mapping of the QTL on proximal 2q ; 2. initiate a total genome scan using a subset of 200 polymorphic microsatellite markers.
QTL mapping is performed using a software package developed in our laboratoty and based on a maximum likelihood multilocus algorithm. Results of this study will be presented.

Poster S2.5

Repeatability of the breeding value of growth traits in the Merino sheep.
A.C. Sierra[1], J.V. Delgado*[1], A. Rodero[1], A. Molina[1], F. Barajas[2] y C. Barba[1]. [1]Departamento de Genética. Universidad de Córdoba. Av. Medina Azahara, 9. 14005 Córdoba. Spain. [2]ANCGSM. 06300 Zafra. Badajoz. Spain

We are developing a breeding plan in the Spanish Merino Sheep based on the use of a testing station for the genetic correction of the herds integrated in the improvement plan. We have employed an animal model with maternal effect genetic in the evaluation of sires, dams and lamb. 11.932 animals have been enclosed in the evaluation, 142 of those were sires, 4.361 dams and the remand lambs.
The weight at four different ages, and the daily gain in three periods were the traits studied here.
We have obtained clearly different levels of repeatability between the additive maternal and direct effects such lower in the first. Also some characters such us the birth weight have shown a low correlation between the number of controlled descendants and the level of repeatability, it mean that these characters are bad defined from the sampling points of view.
The rank of repeatability calculated in these character oscillated between 0,20 to 0,90 with 50 lams in the progeny for direct additive value, and 0.20 to 0,40 with the same progeny for additive maternal value.
The conclusion is that these methods are efficient in system were the artificial insemination is not possible, but it must be improved to obtain better results in the evaluation of additive maternal effects.

SHEEP AND GOAT PRODUCTION   [S]   Poster S2.6

Sire breed effects on glucose tolerance and plasma metabolite levels in Merino cross lambs
A.R. Bray*, D. O'Connell and S.R. Young. AgResearch, P.O. Box 60, Lincoln 8152, New Zealand.

Lambs from Merino dams and six sire breeds which were reared in a common flock underwent a glucose tolerance test 12 weeks before slaughter at nine months of age. There were 16-40 lambs per sire breed from 6-9 sires of each of Merino, Texel, Oxford Down, Border Leicester, Poll Dorset and Suffolk breeds. Glucose, β-hydroxybutyrate, urea and creatinine were measured in plasma samples collected 0, 20, 40, 60 and 90 minutes after intravenous administration of a dose of glucose (0.3 g/kg liveweight). Weight of carcass and measures of carcass fatness (GR) and muscularity (eye muscle A and B, and leg F, T and G) were recorded at slaughter.

At the mean carcass weight of 18.2 kg, Merino sired lambs had highest glucose and β-hydroxybutyrate, and lowest urea levels in plasma. These features were associated with lowest AxB and GR measurements and highest F, T and G measurements. Border Leicester sired lambs had the lowest glucose half-life and high F, T and G measurements. Poll Dorset and Texel sired lambs had high urea and creatinine levels and high GR measurements. The difference between breeds in glucose half-life was highly significant, but breed means bore no obvious relationship to GR nor muscularity measurements.

Poster S2.7

Réponse à la sélection d'un élevage ovin à viande dans les conditions du semi-aride
S. Bedhiaf *1, A. Ben Gara 2, M. Ben Hamouda 2 et M. Djemali 3. 1 Institut National de la Recherche Agronomique de Tunisie, Laboratoire ROC, 2049 Ariana, Tunisie.
2 ESA, Mateur 7040, Tunisie,..3 INA Tunisie, Mahrajène 1002.

Les objectifs de cette étude est d'exploiter la mine de données, relative aux contrôles de croissance des ovins à viande, que possède l'INRA.Tunisie depuis les années soixante, et ce afin, de fournir un critère optimal pour le choix des reproducteurs.L'étude a porté sur un total de 10409 agneaux de race Barbarine, qui sont nés entre 1968 et 1997.

L'évaluation génétique des troupeaux étudiés est faite par l'application de la théorie générale du BLUP au cas du modèle animal multicaractère.

Les résultats de ce modèle individuel montrent qu'il existe une différence de 5,5 Kg à 3 mois d'âge, entre les agneaux nés simples et ceux nés doubles, en faveur des premiers; Quant, aux agnelles, la supériorité des simples est de l'ordre de 5 Kg.

Les agneaux issus d'agnelages tardifs (décembre) gagnent 1 Kg de poids de plus à 3 mois d'âge, par rapport à ceux nés en octobre.

Les agneaux issus de mères âgées de 5 ans gagnent presque 2 Kg de poids de plus à 90 jours d'âge, par rapport à ceux issus de brebis primipares.

L'évolution du niveau génétique des animaux est presque stable dans le temps, avec une légère supériorité des index femelles par rapport aux index mâles.

SHEEP AND GOAT PRODUCTION   [S]   Poster S2.8

Genetic evaluation of sheep to improve carcass quality in Ireland
O.J. Murphy*, E. Wall, E.J. Crosby, D.L. Kelleher, V. Olori. Department of Animal Science and Production, University College Dublin, Belfield, Dublin 4, Ireland.

Four studies were carried out to improve the system of genetically evaluating sheep in Ireland with the long-term aim of improving carcass quality. The studies were: (i) study the effect of age at measurement on the error in predicting liveweight, muscle depth and fat depth in pedigree Charollais and Texel lambs; (ii) study of the balance-connectedness between the sires and flocks; (iii) estimation of genetic parameters (heritabilities, genetic and phenotypic correlations) of the performance traits; and (iv) recommended genetic evaluation procedure. This study found that an improved method of adjusting an animals liveweight to a fixed age in a situation where the lambs are recorded only once is to use that animals birth weight to calculate a linear growth rate correction factor. Poor genetic connections between flocks affect across-flock evaluations by reducing the precision of the comparison of animals. These genetic connections need to be created so that the comparison of animals is not adversely affected by the need to estimate flock effects. Heritability estimates, for the Charollais and Texel breeds respectively, were 0.24(±0.03) and 0.19(±0.03) for liveweight, 0.21(±0.02) and 0.11±(0.02) for ultrasonic muscle depth, 0.17(±0.03) and 0.08(±0.02) for ultrasonic fat depth. Findings of the first three studies were used to develop a new sheep genetic index using multi-trait Individual Animal Model BLUP.

Poster S2.9

Using of French Alpine and Nubian breeds to improving performance of Egyptian local Baladi
M.T. Sallam*; K.M. Marzouk; A.A. El.Hakeem and M.M. Abd Alla, Dept. of Anim. Prod., Faculty of Agric., Minia University, Minia, Egypt.

This study aimed at improving the reproductive and growth performance of local Baladi goats (B) by crossing them with French Alpine (FA) and Nubian (N)breeds. Differences in fertility, litter size at birth and at weaning of B and FA purebreds, F1 (crosses 1/2FA 1/2B and 1/2N 1/2B) goats were studied under Middle Egypt condition. Growth performance of pure B, FA and crossbred kids (1/2FA 1/2B, 3/4FA 1/4B, 1/2N 1/2B & 3/4N 1/4B), were also studied. The study was carried during the period from 1994 till 1997, at the Farm of Animal Production Department, Faculty of Agriculture, Minia University, Minia, Egypt. Genotype of does significantly (p<0.05) affected fertility and litter size at birth and at weaning. The F1 crossbred (1/2FA 1/2B) does were more fertile (0.89) followed by B (0.87) and FA does (0.71), while the F1(1/2N 1/2B) gave the lowest value (0.68). On the other hand, the F1 (1/2FA 1/2B) cross performed generally better than local B in both fertility and litter size at weaning. Also, genotype of kids born had a significant effects (p<0.05 or p<0.01) on body weights at birth, weaning and six months of age, as well as, the pre- and post weaning daily gain and survival rates at weaning. Year and season of kidding significantly (p<0.05 or p<0.01) affected all studied traits except survival at weaning.

SHEEP AND GOAT PRODUCTION   [S]   Poster S2.10

Phenotypic and genotypic variability of slaughter indexes in interbreeding
S.Sh.Mirzabekov, M.A.Yermekov, Kazakh State Agrarian University, 480013 Alma-Ata, Prospect Abaya 28, Kazahstan, CIS

Kazakh mutton crossbred-wool sheep are of importance in deserts and semideserts of Southern Kazakhstan. The reproductive crossings in the animals pre-selected for adaptivity to all year round pasture climatic conditions resulted in Chuisk intrabreed type of KMCW sheep with valuable biological characters preserved. The KMCW ewes' pasture plant consumption and rest periods are 4,3 - 26.9% higher and 17 - 78% lower than those of the crosses, respectively. 1.5 year lambs and ewe lambs reach 94.1 and 97.4% of the adults'indexes for the withers hight, respectively. The research in exterior showed that the young sheep differ in body type: the ewe lambs of the "best" year have shorter extremeties, wide and deep breast, longer body and developed mutton shapes. In the "best"and "worst"years h^2 of the living weight was 0.256 (P<0.001) and 0.076(P.0.05), respectively.

Phenotypic and genotypic variability of slaughter indexes in interbreeding testified to significant genotypic differences of KMCW tupping rams and was studied in the offsprings of Southern Kazakh Merino ewes and KMCW rams different in wool fineness. When slaughtered at 4.5 months, the KMCW offsprings and SKM control tupping rams were determined to exceed the pure breed lambs of the same age in body and slaughter weight by 0.7 kg but lagged 1.4 kg behind them for pre-slaughter weight.

Poster S2.11

Genotypic differences of southern Kazakh merinoes of Merkensky intrabreed type
S.Sh.Mirzabekov, Kazakh State Agrarian University, 480091 Alma-Ata, Shevchenko St. 80-22, Kazakhstan, CIS

To improve the market looks and technological characteristics of wool four 1/4 blood tupping rams from "Merkensky"state breeding farm were used in the introductory crossing of adult SKM with Australian merinoes that had been obtained from 23 and 18 month first class ewes, sheep herds 1 and 2, respectively. The control consisted of Australian sires of "strong"type preevaluated at 4-4.5 months in 1998 and SKM acknowledged as the best improver. The offspring of the 1/4 blood rams truly lagged bihind both pure and half blood ewe lambs averaged for the living weight. Heritability of the wool clippings from the ewe lamb offspring of the 1/4 SKMxAM tupping rams constituted h^2=0.4, F=20.0, h^2=0.34, F=16.5. The offspring was found to have the possibility to inherit high output (%) and washed wool clippings (kg) from the crossed MI 1/4 SKMxAM rams. The 1/4 blood offspring of "M"SBF exceeded the standard requirements of group "B"productivity mutton-wool sheep to elite animals in the washed wool clippings by 5.0- 34.0% for herd 1and 2, respectively, whereas SKM had the priority (8.0%) but only for herd 1.Regarding the economically valuable characters and their summary index of the offsping class composition there were revealed the improvers of selected characters.

SHEEP AND GOAT PRODUCTION   [S]   Poster S2.12

Role of small holder in improvement of management of small ruminant in Egypt
A.M.S. Al khbeer[1] *and K.M. Marzouk*[2]. [1]*Dept. of Home Economics Sci.,Faculty of Specifical Education,* [2]*Dept. of Anim. Prod., Faculty of Agric., Minia Univ.Egypt.*

Data were collected from 82 flocks owned by small holder located in Minia Governorate (247 km south Cairo). This study amid to evaluate the knowledge level of sheep and goats management technological package.The results revealed that sheep and goats represented 95% and 5% from small ruminant population, respectively. The sheep breeds raised in this area mainly were Ossimi and crossbreds with other native breeds, sheep are yarded,sometimes even housed, at night and shepherded by day onto stubble, road and canal verges. The prevalent production system is characterized with free matings (all-year round), insufficient disease control and low external input.Feeding, from December to May, depends mainly on Egyptian clover (Berseem) while over the rest at the year animals scavenge crop residues and stubble beside household. Family size is the main social factors of positive effect on flock size, number of adults, in the family working in agriculture and farmers land in use had no significant effects on flock size. Owing to the small holder has not a recording system, breeding work is absence. More attention must be done for recording, veterinarian, nutrition and technology transfer to improve the productivity of sheep flocks owned by small holder.

Poster S2.13

Improvement of sheep and goat raising - development project for the Republic of Mali, West Africa
Fantová, M.[1] *and Mátlová,V**[2]. [1] *Institute of Tropical and Subtropical Agriculture, Czech University of Agriculture, Prague, Czech Republic,* [2] *Research Institute of Animal Production, Prague 10 - Uhříněves, Czech Republic*

In addition to veterinary and nutrition tasks, introduction of advanced methods in reproduction (targeted mating, recording, selection) is the necessary condition for sheep/goat raising improvement in the agropastoral joint grazing system. Non-surgical castration of excess males is unknown and due to ritual grounds not accepted in every situation. Growth intensity and carcass yield in castrated (rubber rings) and intact Toronke rams and Sahelian bucks was observed, to encourage farmers to apply this technique. During 240 days of trial both bucks (B) and rams (R) castrated in various age (60-240 days) grew better (B: 46-35 g.day^{-1}, R : 56-36 g.day^{-1} according to the age) than the intact ones (B: 37-33 g.day^{-1}, R: 43-27 g.day^{-1}). Carcass yield of castrates was higher (B: 45,1% at 17 kg liveweight, R: 42,4% at 23 kg) than that of intacts (B: 42,9% at 15 kg, R: 41,3% at 23 kg). Only slight differences in bone : meat ratio in both rump and shoulder section between castrates and intacts were found (B rump: 0.40 and 0,42; B shoulder 0,53 and 0,52; R rump: 0,38 and 0,38; R shoulder 0,48 and 0,47).

SHEEP AND GOAT PRODUCTION   [S]   Poster S2.14

A new lamb feeding method
I. Filya[*1] *and A. Karabulut*[1]. [1]*Uludag University, Faculty of Agriculture, Animal Science Department, 16059, Bursa-Turkey.*

There are various lamb production and growth methods in Mediterranean countries. However, intensive feeding level is low in this methods.

One of the best methods to improve intensity is to use only concentrate feeds as adlibitum. A "creep" area is provided to four weeks old lambs which they can pass but their mothers can not. The lambs completely weaned at six weeks age. The concentrate diet for the lambs, which are fed adlibitum, should be based on whole cereals. To ensure optimum feed conversion and to prevent digestive problems, the cereals must not be processed in any way. Any cereal can be used, but best results are obtained with high energy cereals such as barley, wheat, sorghum and maize. The oil seed meals are used as a protein source in the ration. If automatic feeders are used they prevents the lambs from contaminating the feed with their feet, which results in low feed intake and the spreading of coccidiosis and other infectious diseases. In addition to these, required labour is decreased.

We have obtained positive results from a number of studies by using this method in Turkey. According to the traditional growth methods, lamb fattening performance and carcass quality are improved significantly.

Poster S2.15

Observations on the appearance of mastitis in milked Merino ewes and in their crossbreds with prolific breeds
T. Pakulski, B. Borys and M. Osikowski. National Institute of Animal Production, Experimental Station Kołuda Wielka, 88-160 Janikowo, Poland.

The examination of the appearance of clinical and subclinical states of mastitis and of the contents of somatic cells and bacteria in the milk was carried out for 4 years in milked Polish Merino ewes [Mp], their F_1 crossbreds with prolific breeds: the Booroola Merino [BM], the Finnsheep [FM] and the Romanov sheep [RM], as well as on the prolific Merinofinn line [Mf-40], 530 heads in all. The ewes were milked from the 56th day [weaning] till the 6th month of lactation. The appearance of subclinical states of mastitis was investigated periodically by means of three indirect methods: the CMT test, the test paper and the electric conduction measurement. The somatic cell content was marked on the Fossomatic meter, while the bacteria content was measured by means of inoculation. An influence of the genotype, the year and the stage of lactation on the appearance of clinical and subclinical states of mastitis and the contents of somatic cells and bacteria in the milk was confirmed.

SHEEP AND GOAT PRODUCTION [S]   Poster S2.16

Milk fat composition in goats representing different genetic variants of α S1 casein
B. Reklewska[1], Z. Ryniewicz[2], M. Góralczyk[1], A. Karaszewska[1] and K. Zdziarski[1]
[1] Warsaw Agricultural University, 05-840 Brwinów, Przejazd 4, [2] Institute of Genetics and Animal Breeding, Polish Academy of Sciences, Jastrzębiec, 05-551 Mroków

The aim of the study was to compare milk fat composition in goats representing strong (S) and weak (W) genetic variants of α S1 casein. Milk protein genotypes were determined by PAGE at the beginning of lactation and two groups of 20 goats identified as S or W variants were selected from a herd. Milk samples were collected between 2nd and 7th month of lactation. Milk fat was fractionated by TLC and then fractions and fatty acid (FA) composition were quantified spectrometrically or with GC. Milk of goats representing S variants of α S1 casein contained significantly (P<0.01) more of protein and fat than W phenotypes. Besides, due to significantly lower FFA concentration, the milk fat of the S α S1 casein variants was less susceptible to lipolysis compared with the W phenotypes. The effect of genotype was highly significant (P<0.01) for saturated FAs only in case of capric acid $C_{10:0}$. In MUFA strong variants of α S1 casein contained significantly more (P<0.01) palmito-oleic acid $C_{16:1}$ than W variants. The most important differences occurred among PUFA. The S variants of α S1 casein were associated with a significantly higher (P<0.01) content of arachidonic acid, but considerably lower (P<0.001) content of $C_{22:5}$; $C_{22:6}$ and conjugated linoleic acid (CLA) than W variants.

Poster S2.17

The growth and reproductive characteristics and milk yield of Karakaş sheep in rural farm conditions
Ö.Gökdal*, M.Bingöl, A.Çivi, Y.Aşkın, F.Cengiz. Yüzüncü Yyl Üniversitesi, Ziraat Fakültesi, Zootekni Bölümü,Van, Turkey.

This research has been carried out to determine the growth and reproductive characteristics and milk yield of Karaka_ sheep, known as a variety of Akkaraman, in rural farm conditions. In general, information related to the performance of native sheep breeds of Turkey under rural farm conditions is insufficient.
The means of live weights of lambs at birth, 90, 134 (at weaning) and 180 days of age were 3.69, 18.93, 27.19 and 27.33 kg, respectively. The average daily gains from birth to weaning and from birth to 180th days were 0.170 and 0.128, respectively. Infertility, abortion, parturition and twinning rates were found to be 5.26 %, 5.26 %, 89.47 % and 8.82 %, respectively. Fecundity and litter size were found to be 0.97 and 1.08, respectively. The survival rates of lambs until 7th and 134th (until weaning) days were 94.60 % and 87.84 %, respectively. The means of lactation length and lactation milk yield for ewes were 197.8 days and 54.75 l, respectively.

SHEEP AND GOAT PRODUCTION   [S]   Poster S2.18

Milk protein polymorphism in Portuguese sheep breeds: αS1-casein and β-lactoglobulin
A. M. Ramos[*,1], P. Russo-Almeida[1], A. Martins[1], F. Simões[2], J. Matos[2], A. Clemente[2], T. Rangel-Figueiredo[1] 1- Dep. Zootecnia, UTAD, 5001 VILA REAL Codex, Portugal 2- INETI/IBQTA/DB/BQII, Est. Paço do Lumiar, 1699 LISBOA Codex, Portugal

Because of the influence of some variants on milk production traits and cheesemaking properties, milk protein polymorphism is considered as a potential tool for selection of dairy sheep. For this purpose, the knowledge of the allelic frequencies for each breed and the effects of the variants, on both the dairy traits and milk processing properties, is essential. For αS1-casein 5 allelic variants are determined (A, B, C, D, and E) while for β-lactoglobulin 3 variants are referred (A, B, and C). Studies carried out so far have shown contradictory results. While for some variants a marked effect on milk traits was observed, for others the results indicate that there are differences between breeds.
DNA samples were amplified by PCR and digested with the restriction enzymes *Mbo*II and *Mae*III (αS1-casein) and *Rsa*I (β-lactoglobulin), allowing the identification of the variants. Genetic variation was found and the frequencies for each breed were determined.

Poster S2.19

The influence of different lymphocytes subpopulations in milk on the health state of udder in sheep
W.P. Świderek[1], A. Winnicka[2], Wł. Kluciński[2], K.M. Charon[*,1], [1] Department of Genetics and Animal Breeding, Warsaw Agricultural University, Przejazd 4, 05-840 Brwinów, Poland, [2]Faculty of Veterinary Medicine, Warsaw Agricultural University, Grochowska 272, 03-849 Warsaw, Poland.

The investigations were conducted in the Warsaw Agricultural University Experimental Farm on Polish Lowland Sheep of Żelazna variety. Health condition of the mammary gland of 40 ewes was examined in the first month of lactation on the basis of somatic cell count (SCC) in milk samples taken from each half of udder separately.
The subpopulations of lymphocytes, isolated from milk, were analysed using the flow cytometry method with the help of FACStrak (Becton Dickinson) apparatus. The specific monoclonal antibodies were used, anti: CD2, CD4, CD8, CD19, WC1-N2. Final analysis was performed by the use of SimulSet programme.
The increased frequency of CD2 lymphocytes and decreased frequency of other subpopulations of lymphocytes in milk of mastitic ewes comparing to the healthy ones was observed.

SHEEP AND GOAT PRODUCTION   [S]   Poster S2.20

The milk production of Finnish landrace sheep and their crosses with meat-wool breeds
M.R.Nassiry[1], V.P.Shekalova[2], E.A. Karasov[1], [1]Department of Sheep breeding, Timiryazev Agriculture Academy, Timiryazevskaya st. 49, 127550, Moscow, Russia, [2].All-Union Research Institute of Animal Husbandry RF-142012 P.O.Dubrovitsy, Podolsk District, Russia

The experiment was carried out on 76 lambs born from 42 Finnish landrace and their crossed ewes. The objective was to evaluate milk and daily milk production of ewes during their first (20)days of lactation by indirect method. The animals descended from the following genetic groups :1) Finnish Landrace (purebred), 2)Crosses of Romney Marsh and Finnsheep (1/2 Rom x1/2 Finn) ,3)Three-breed crosses (2/8Rom x 1/8Lincoln x 3/8Finn) and also 4) four-breed (2/8Rom x 2/8Lin x 1/8Texel x 5/8Finn). Significant differences were in BW and DG between genotypes. There were non significant differences in MP and DM between genotypes ,but the highest value of MP(32.74 kg) was observed in four-breed crosses and the lowest in Finnish Landrace (27.64 kg). In the other hand the MP of four-breed crosses increased by 15.58 % and 11.24 % more than that of purebred and three breed crosses, respectively. Significant difference was found between the different olds of ewes for MP and DM traits. The highest values of MP and DM traits were observed in ewes four and six years old. Ewes which suckled triple produced higher amounts of MP and DM as ewes which nursed singles and twins, ewes which suckled twins produced higher as ewes which nursed singles.
Results of this experiment show that individual differences in MP and DM Finnsheep and their crosses of ewes testifies to opportunities of selection.

Poster S2.21

The effect of ration's maize gluten meal -in substitution of soy bean meal- on yield and composition of ewes' milk in early lactation
D. Liamadis*, and Ch. Milis. Department of Animal Nutrition. Aristotle University of Thessaloniki 540 06 Thessaloniki. Greece.

A total of 30 ewes of «Thessaloniki» breed in early lactation (1st - 4th and 5th - 8th week of lactation) were used in a 2 X 2 factorial design (two periods X two rations). They were allocated into two groups of 15 animals each, equivalent in terms of mean live weight and milk yield. Animals were fed in groups, strictly to their needs, with alfalfa hay, maize silage and one of two isocaloric-isonitrogenous rations having a different main protein source. The 1st ration contained Soy Bean Meal (group SBM) and the 2nd Maize Gluten Meal 60% (group MGM). The analysis of variance indicated that: 1) the stage of lactation (period) significantly affected milk yield, Fat Corrected Milk (FCM) yield and milk fat content (2.26 and 1.97 kg of milk, 2.52 and 2.11 kg of FCM, 6.89 and 6.59%, for the 1st and 2nd period respectively), 2) the nature of the main source of protein (SBM or MGM) had also significant effect on milk yield, FCM yield as well as on the milk protein content (2.25 and 1.98 kg of milk, 2.47 and 2.17 kg of FCM, 5.45 and 5.36% for the SBM and MGM groups respectively), 3) daily produced fat, protein, lactose and solids-non-fat of the milk were also significantly affected by both period and source of protein, while the milk lactose and solids-non-fat content remained unaffected. In conclusion, the maize gluten meal cannot totally replace the soy bean one in ewes diet in early lactation, since it mainly decreases the milk yield as well as the milk protein content, due to low lysine content. However, the addition of lysine to increase the feeding value of gluten and consequently establish this substitution feasible, is subject to research.

SHEEP AND GOAT PRODUCTION   [S]   Poster S2.22

Correlation between the udder health state, its dimensions and milk productivity in the milking hybrids ewes F_1 East Friesian x Polish Merino.
S. Mroczkowski[1]*, B. Borys[2], D. Piwczyński[1]. [1]Academy of Technology and Agriculture in Bydgoszcz, Faculty of Animal Science, ul. Mazowiecka 28, 85-084 Bydgoszcz, Poland, [2]National Research Institute of Animal Production in Cracow, Experimental Station Ko3uda Wielka, 88-160 Janikowo, Poland

The research was carried out on 95 ewes, crossbreeds F_1 East Friesian and Polish Merino in first to fourth lactation, in Experimental Station at Kołuda Wielka situated in the lowlands of central Poland. The ewes were milked mechanically twice a day during the period of four months [May-September in 1997 and 1998] after the weaning of 8-week-old lambs. The wholesomeness of the udder was estimated on the basis of the results of clinical examination of sheep as well as the somatic cell counts (SCC). The estimates of SCC, udder dimensions, yield and chemical composition of milk were taken in the milking period, in the 10th, 16th and 22nd week of lactation. No clinical mastitis was recorded in any ewe. The SCC were quite variable, ranged from 12×10^3 to 3229×10^3, and the highest at the beginning of the milking period. The phenotypic correlation between SCC and the udder dimensions, milk yield as well as main milk components were general low and insignificant. In the studies were also analysed wholesome state of udder according to the ewe's age and phase of lactation

---

Poster S2.23

Competitiveness of lamb meat quality in Hungary
G. Molnár. Department of Animal Breeding and Nutrition, Debrecen University of Agricultural Sciences, Debrecen Böszörményi u.138. H-4032, Hungary

The author investigated 153 lambs of nine genotypes originating from breeding flock, and 50 lambs originating from production flock. The investigations were performed between 1995 and 1998. The author discussed the evaluation of comformation and fat cover according to EUROP qualification. The author also investigated the proportion ratio of valuable meat in case of the different genotypes, and - out of the internal value indicators - the dry-matter content, the protein content, the fat content, the connective-tissue content, and the hermin content, and compared the flavour, the aroma, the tenderness, and the oven loss of the different genotypes.
The following should be mentioned among the achievements and findings:
- The Hungarian Combed Merino breed should be improved, as - according to EUROP qualification, more than 70 % of them were rated as quality "R".
- The Hungarian fattening technology has to be preserved, as the lambs reach the desired slaughter weight within a short period of time, and without over-fattening.
- Readiness for slaughter, typical of each genotype, has to be defined, and slaughter in proper weight has to be achieved.
- It has to be re-evaluated, whether the Hungarian Combed Merino is the only breed which can be used in Hungary, as none of the investigations really proved the special characteristics and significance of this breed.

SHEEP AND GOAT PRODUCTION   [S]   Poster S2.24

A study of performance traits of Charolaise sheep imported from France to Poland
S. Czarniawska-Zając, W. Szczepański, M. Oczapowski *University of Agriculture and Technology, 10-719 Olsztyn, Poland*

In recent years in Poland the interest of breeders in meat breeeds of sheep has been increased. The evaluated Charolaise sheep were imported to the region of Warmia and Mazury in 1997 due to their splendid meat characters, early maturing and high prolificity. Another reason for importing these sheep was their ease to acclimatize to new climatic conditions.

The study included a Charolaise pedigree flock kept in the region in 1997 and 1998. The first stage of the study was concerned with the evaluation of the animals. The flock imported from France numbered 50 ewes and 3 rams. There was no relationship between them and they were large. The mean body weight of the ewes at the age of eight months was 64.34 kg. In October 1997 the ewes were mated and after lambing, the lambs were weighed every 21 days till 70 days of age. At this age, the rams weighed 25,10 kg and ewes 23,8 kg, on an average. Body weight increases of rams and ewes were higher between 30-70 days than between 10-30 days of age. Reproduction index for the ewes of this breed in the first year was high and amounted to 155%. The mean annual gain of grease wool was 2,36 kg (in ewes); staple length was 6,41 cm and thickness was 23,37 μm.

Poster S2.25

The influence of weaning and movement to a slaughter-house on the level of cortisol, glucose and haematocrit in blood of lambs
J. Sowińska[1], H. Brzostowski[2]. Z. Tański[2]. *[1]Department of Animal Higiene, [2]Department of Sheep, University of Agriculture and Technology, 10-718 Olsztyn, ul. Oczapowskiego 5/108, Poland.*

Levels of cortisol, glucose and haematocrit in blood of 50-day-old rams of different genotypes: Pomorska (P), Ile de France (IF) and Pomorska x Ile de France (P x IF) were determined.

Blood samples was collected from the jugular vein: before waning of lambs, after 15-hours weaning and after movement to a commercial slaughter-house located 250 m away.

After weaning and movement to a slaughterhouse a significant ($P \leq 0.01$) increase in the hormone leves from 21.85 to 44.06 and 42.09 nmol/l were found.

No significant relationship between the glucose (3.12 , 3.09 and 3.46 mmol/l, respectively) and haematocrit levels (0.29; 0.02 and 0.30, respectively), and time of blood sampling were found.

SHEEP AND GOAT PRODUCTION [S]

Poster S2.26

The influence of age at castration on carcass quality in sheep
R. Süß*, U.E. Mahrous and E. von Borell. Institute of Animal Breeding and Husbandry with Veterinary Clinic, Martin-Luther-University Halle-Wittenberg, A.-Kuckhoff Str. 35, D-06108 Halle/Saale, Germany

Castration of male lambs is worldwide practised for different reasons. A behavioural and physiological study concerning animal welfare aspects of castration was completed by investigating the effect of method and age of castration on carcass quality. 72 Fox lambs were randomly allocated to 8 groups when 10 days old. Beside two control (C) groups, each 27 lambs were castrated by different methods at an age of 21 (E) and 56 days (L) respectively. When about 240 days old, 36 lambs were selected for determination carcass value including meat and fat quality traits. The mean final weights of the three groups doesn't differ ($39,8\pm1,1$; $40,1\pm1,5$ and $39,6\pm0,9$ kg for C, E and L resp). Within the statistical model, the effect of birth type, castration method and age, day of slaughter and final weight as covariate were taken into consideration. In comparison to the entire ram lambs castrated ones showed a higher fat-meat ratio of the loin, fat thickness, kidney fat, total non carcass fat and an increased percentage of high valueable cuts. Fatty acid number of subcutaneous as well as kidney fat was reduced. Early castration resulted in a significant higher percentage of non carcass fat and breast and reduced percentage of high valueable cuts. L- and a-value of the meat colour were also reduced. No relevant differences were found in any of the involved traits related to the method of castration.

Poster S2.27

Effect of castration method and age on behaviour and cortisol in sheep
E. von Borell*, U.E. Mahrous and R. Süß. Institute of Animal Breeding and Husbandry with Veterinary Clinic, Martin-Luther-University Halle-Wittenberg, Adam-Kuckhoff-Str. 35, 06108 Halle, Germany.

Nine Fox lambs (from a total of 72) in each of 4 groups were either subjected to castration with a Burdizzo emasculator (B), Burdizzo emasculator combined with a local anaesthesia (BA), surgical castration (S) at 2 ages (d 21 and 56) or handled as controls (C). Blood samples for cortisol were collected before and every 30 min for 3 hrs after castration. Behaviour was recorded continuously by a video system for one day before and after each treatment day. Data were analysed by two way analysis of variance using SAS. All methods of castration resulted in a significant elevation of cortisol with a higher response ($P<0.05$) in younger lambs. (S) lambs showed the highest, (BA) lambs the lowest response to castration. 21 d old (S) treated lambs suckled less frequent and in shorter bouts ($P<0.05$) compared to all other treatments, while 56 old (S) treated lambs showed a reduced feeding time [$P<0.01$ vs. (C) and $P<0.05$ vs. (B)]. Most of the other behaviours observed (resting, locomotion, comfort and exploration) were affected by castration, however, (S) treated lambs deviated most from (C). In general, no clear cut conclusions can be drawn from age x treatment effects on behaviour. It can be concluded, however, that (BA) treated lambs were behaviourally and physiologically least affected, presumably by minimising pain associated with castration.

SHEEP AND GOAT PRODUCTION [S]

Poster S2.28

Relationship between SEUROP grading, composition and value of lamb carcasses
G. Van de Voorde, S. De Smet* and J. Depuydt. Department of Animal Production, University of Gent, Proefhoevestraat 10, 9090 Melle, Belgium

Grading systems for carcasses are a tool for value-based marketing. Carcass value is mainly determined by the relative composition of meat, fat and bone, and by the conformation or shape of the joints. Grading according to the SEUROP scheme for conformation and fat grade is voluntary for lamb carcasses, whereas it is obligatory for beef carcasses in EU approved abattoirs. Few studies have been done in lamb carcasses relating SEUROP grading results to the carcass composition and the relative amount of high-value cuts.

During the last years, a total of 95 lamb carcasses have been dissected at our department. The carcass content of lean meat, fat and bone was determined by dissection of the whole carcass or by dissection of a 3-rib sample and applying appropriate regression equations. The relative amount of high-value cuts was also determined. In comparison with beef carcasses, relatively small differences in dressing yield, in carcass composition and in the relative amount of high-value cuts of lamb carcasses were found between the SEUROP conformation classes. It appears that the higher value of carcasses of better conformation is mainly related to the better shape of the cuts and much less to the carcass composition.

Poster S2.29

Ovine omental and subcutaneous adipocytes display depot and breed differences in culture and in vivo
B. Soret, A. Arana*, P. Eguinoa, J.A. Mendizábal, A. Purroy. ETSIA. Universidad Pública Navarra. Campus Arrosadía. 31006 Pamplona. Spain.

Lipogenic enzyme activities *in vivo* were studied in suckling lambs (12 kg live weight, LW) of the Spanish breeds Lacha (L) and Raza Navarra (RN). It was found a higher synthesis of triglycerides, estimated by glycerol 3-phosphate dehydrogenase (G 3-PDH) activity, in RN lambs than in L lambs in both omental (OM) and subcutaneous (SC) adipose tissues. OM adipocytes showed higher G 3-PDH activity than SC adipocytes ($p<0.05$).

In order to elucidate breed and depot specific differences an *in vitro* system was used in which OM and SC preadipocytes from lambs (12 kg LW) of L and RN breeds were allowed to proliferate and to differentiate in culture. Differentiation was achieved under the effect of either a lipid mixture (1%) or dexamethasone (10nM) or a combination of both and it was monitored by the activities of the enzymes G 3-PDH, fatty acid syntetase (FAS), glucose 6-phosphate dehydrogenase (G 6-PDH), malic enzyme (ME) and isocytrate dehydrogenase (ICDH). Breed and depot specific differences obtained in vitro displayed a similar pattern to those found in vivo: RN preadipocytes showed higher differentiation (G 3-PDH, FAS, G 6-PDH, EM and ICDH activities) than L preadipocytes ($p<0.001$) and G 6-PDH in OM preadipocytes was higher than in SC ones ($p<0.05$).

SHEEP AND GOAT PRODUCTION   [S]

Poster S2.30

An investigation of fatty acid composition and meat quality in lambs from different breed and production system backgrounds
E. Kurt*[1], J.D. Wood[1], M. Enser[1], G.R. Nute[1], L.A. Sinclair[2] and R.G. Wilkinson[2]
[1]Division of Food Animal Science, University of Bristol, Langford, Bristol BS40 5DU UK, [2]Harper Adams University College, Edgmond, Newport, Shropshire TF10 8NB UK

There is interest in whether certain sheep breeds produced in different ways have characteristic flavours in the meat due to genetic or nutritional effects on muscle fatty acid composition. In this study, 4 breed x feeding system groups of lambs, representing extremes in the UK sheep industry were compared: purebred Welsh Mountain produced on upland grass; purebred Soays produced on low land grass; and Suffolk crosses from the same flock produced on either grass or concentrates.

The three groups fed grass all had high concentrations of n-3 PUFA in muscle in contrast to the Suffolks fed concentrates which were high in n-6 PUFA. The Soays, which were extremely lean, had high concentrations of n-3 and n-6 PUFAs. The two high n-3 grass fed groups (Welsh Mountain and Suffolk) had similar flavour profiles as identified by the taste panel and the highest overall liking scores. Soays and the concentrate fed Suffolks had a similar flavour profile and low 'overall liking scores'.

The results show that there are important differences between sheep types in the content of PUFA and associated flavour characteristics. Fatty acid composition also affected shelf-life as determined by lipid oxidative stability and colour.

Poster S2.31

Quality of lamb's meat and fat as effected by crossing of Polish Merino with prolific and meat breeds
B. Janicki[1], B. Borys[2], M. Osikowski[2], E. Siminska[1]. [1]Academy of Technology and Agriculture - Bydgoszcz, 85-084 Bydgoszcz, Mazowiecka 28, [2]National Institute of Animal Production, Experimental Station Ko3uda Wielka, 88-160 Janikowo, Poland

The aim of the investigation is evaluation of the effect of lamb genotype and sex on select features of muscular tissue, and distribution and quality of their fat tissue. The experiment is performed in two repetitions, together on 64 lambs of Polish Merino breed, and second degree hybrids from crossing of Suffolk rams with F_1 ewes Finn x Merino, Romanov x Merino and Booroola x Merino. The lambs are fattened intensively with all-mash feed from weaning at the age of 56 days to the obtained live weight: ram-lamb 35-40 kg and the ewe-lamb 30-35 kg. Analysed is carcass conformation, fatness, and tissue composition, distribution of fat in carcass and select traits of muscular and fatty tissue quality, including composition of fatty acids and cholesterol content in external, intermuscular and intramuscular fat. The results of the first repetition of investigation show that Merino lambs, in respect of the analysed parameters, were generally a little better than triple crosses. The ewes-lambs in comparison with ram-lambs were characterized by higher content of valuable cuts, however at greater carcass fatness, with worse muscle-fat ratio, and with smaller content of polyunsaturated fatty acid in fat.

SHEEP AND GOAT PRODUCTION [S] Poster S2.32

Growth, carcass and fleece quality traits in Swedish sheep breeding
A. Näsholm*. Department of animal breeding and genetics, Swedish University of Agricultural Sciences, P. O. Box 7023, S-750 07 Uppsala, Sweden.

Genetic parameters were estimated and genetic trends were studied by using information from the progeny testing of rams of different Swedish breeds. Almost 70,000 lambs were included and traits studied were 110-day live weight and fleece quality. The animals were from flocks that used AI or were participating in the progeny testing of rams. The lambs were born during the period 1991-1998. The analyses were done with use of an animal model. An analysis of variance on carcass data from the slaughterhouses was also performed. Around 20,000 lambs were included. Heritabilities for fleece quality traits varied between 0.22 and 0.33. Heritability of 110-day weight was for the Swedish peltsheep high with an estimate of 0.52 and for the other breeds this value was 0.33. An annual genetic progress among lambs of the rams included in the progeny testing scheme of about 1.4 % in live weight was noticed and for fleece quality the improvement was around 0.8 %. For all lambs in the study these figures were lower with values og 0.6 - 1.1 % for live weight and 0.3 % for fleece quality. Considerable variation both between and within breeds for net weight gain and carcass grades was shown. For improvement of the genetic evaluation scheme it was concluded that carcass traits should be included.

Poster S2.33

Field testing of progeny of breeding rams on growth and slaughter value
M. Momani Shaker*, I. Šada, F. Vohradský and Ghassan Agil. Czech University of Agriculture, Prague, Institute of Tropical and Subtropical Agriculture, Suchdol, Kamýcká 129, 165 21, Prague 6, Czech Republic.

The aim of this study was the observation of growth of lambs of both sexes, fattening ability and slaughter value of ram lambs of five breeding rams included into pen-fold grazing system. Besides the effect of father, the effect of sex of lambs and litter frequency was observed. 119 lambs were included into the observation and their birth weight and the weight at the age of 130 days were taken. At the end of the experiment five ram lambs of each breeding rams were selected for control slaughter. Mean live weight of lambs at birth was 3.74±0.05 kg and at the age of 130 days 38.38±0.50kg. Live weight of lambs at birth was influenced by the father's genotype ($P \leq 0.001$), however, significant differences between the breeding rams on the growth were not found. Differences in live weight and weight gains between sexes at the age of 130 days were significant ($P \leq 0.05$). Litter frequency influences weights gains and live weight at birth and at the age of 130 days very significantly ($P \leq 0.05$). The highest dressing percentage in warm stage (53.8±1.17 %) had the ram lambs of the ram Chinin, the lowest value (47.9±0.72) of the ram Chevalier and the differences between the breeding rams in dressing percentage were significant ($P \leq 0.05$). Significant differences in the share of leg musculature between the rams were found on the level ($P \leq 0.05$). The effect of father was revealed in the weight of kidney fat on the level ($P \leq 0.05$).

SHEEP AND GOAT PRODUCTION   [S]   Poster S2.34

Conformation data of different Tsigai types in Hungary
A. Gáspárdy*[1], F. Eszes[1], L. Jávorka[1] and T. Keszthelyi[2], [1]Dept. of Anim. Husb. of the Univ. of Vet. Sci., H-1400 Budapest, István 2., [2]Dept. of Mechanics of the Gödöllö Univ. of Agr. Sci., H-2103 Gödöllö, Páter 1., Hungary.

The traditionally multipurpose Tsigai breed is registered within the sheep group of mountain origin. In the course of its spreading in the Carpathian-basin the lowland-type has been developed. In this investigation the differences in body measurements and indices between the 4-5 years old ewes of lowland-type (n=38) and mountain-type (n=46) are revealed.
Animals of lowland type comparing to the other were significantly ($p<0.001$) larger in each measurements (heart girth: 92.4/88.5, canon girth: 9.2/8.1; withers height: 72.8/65.7, body length: 81.7/71.8, body depth: 38.1/33.5, body width: 25.8/23.6 and rump with at the trochanter: 26.5/22.8 cm) excepting the length of head and the body weight.
The body indices of narrowness, deepness and length were higher by 3-4% in the lowland-type than in the mountain-type where the indices of stubbiness, fullness, depth and the compactness have been found higher by 6-9%.
Comparing the present data to the earlier ones by 60-70 years it can be concluded that the body sizes have slowly increased in both Tsigai-types, while the body proportions in each type have still remained.

Poster S2.35

The carcass traits of Finnish Landrace sheep and their crosses with Meat - wool breeds
M.R.Nassiry[1], V.P.Shekalova[2], [1]Department of Sheep Breeding ,Timiryazev Agriculture Academy, Timiryazevskaya st. 49 127550, Moscow, Russia, [2]All-Union Research Institute of Animal Husbandry RF-142012 P.O.Dubrovitsy, Podolsk District, Russia

The experiment was carried out on lambs (N=24) of 5-6 month of age during 60 days (fatting period) .The animals descended from the following genetic groups :1) Finnsheep, 2)Crosses of Romney Marsh and Finnsheep (1/2 Rom x 1/2 Finn) ,3) 3-breed crosses (2/8Rom x 1/8Lincoln x 3/8Finn) and also 4) 4-breed (2/8Rom x 2/8Lin x 1/8Texel x 5/8Finn). Difference in bone in the carcass between the different genotypes was significant ($p<0.05$) , since bone content was heavier in purebred than in 3-breed and 4-breed crosses by 14.22 % and 10.15 % respectively , whereas meat content was heavier in 4-breed crosses than in purebred by 6.64 % . Significant differences ($p<0.05$) were found between genotypes for soft drops weight (heart, liver, and spleen) and LIL (Large intestine long) but significant differences ($p<0.05$) weren't found between genotypes for kidneys, lungs with trachea, stomachs and intestines plus and without digest , hard drop[head, Skin , and feet] , Kidney fat , pericardial fat and omental and mesenteric fat weight . Purebred averaged 33.2 % and 6.8 % higher BFTHO (thickness of back-fat on the "eye" muscle)than did 4-breed and 3-breed crosses respectively. The high values for meat in the carcass and LIL and low values for bone in the carcass and BFTHO for cross breeds showed the effect of crossbreeding on carcass quality and internal fat.

SHEEP AND GOAT PRODUCTION [S]

Poster S2.36

The weaning weight and measurement traits of young Finnish Landrace and their crosses with Meat-wool breeds

M.R.Nassiry[1], V.P.Shekalova[2], E.A. Karasov[1], [1]*Department of Sheep Breading,Timiryazev Agriculture Academy, Timiryazevskaya st. 49, 127550, Moscow, Russia,* [2]*All-Union Research Institute of Animal Husbandry RF-142012 P.O.Dubrovitsy, Podolsk District, Russia*

Weaning weight (WW) and ten body measurement traits were recorded on 86 sire and female lambs of 5-6 month of age, born in 1997 in the experimental basis of VIGH institute. The animals descended from the following genetic groups: 1) Finnsheep, 2) Crosses of Romney Marsh and Finnsheep (1/2Rom x 1/2Finn), 3) 3-breed crosses (2/8Rom x 1/8Lincoln x 3/8Finn) and 4) 4-breed (2/8Rom x 2/8Lin x 1/8Texel x 5/8Finn). WW did not differ among genotypes and between single and twins of birth type, but showed a highly significant among sex ($p<0.01$). Analysis of variance showed that, genotype effect on measurement traits was not significant but dam age, sex and birth type effects on several traits were significant. Regression estimation of WW and measurement traits on suckling days were high significant ($p<0.01$). There were significant differences in the several measurement traits between genotypes. The correlation between WW and the linear dimensions and among the dimensions themselves, show that there is a marked degree of independence in the size of the separate parts. The coefficients ranged from 0.11 to 0.80. In the first and fourth indexes (meatiness indexes) the highest value was observed in 3 - breed crosses (155.77% and 67.02% res.) and the highest value of WW indicated in pure and 3-breed animals (26.37 and 25.80 res.). It is evident that 3-breed cross and pure breed are important in meat production than the other genotypes.

Poster S2.37

The growth and the slaughter value of Pomeranian ewes and their crossbreds with meat rams

Z. Tanski[1], H. Brzostowski[1], J. Sowinska[2]. [1] *Departament of Sheep-farming, Hunting and Goat Breeding,,* [2]*Department of Animal Hygiene, University of Agriculture and Technology, 10-718 Olsztyn, ul. Oczapowskiego 5, Poland.*

The investigations were carried out on 50 days old rams of Pomeranian race (P) and crossbreeds of Pomeranian ewes with Blackhead (PxCz) and Teksel (PxT) rams.
Cross-breeding of Pomeranian ewes with Blackhead rams influenced profitably on the body weight of the newborn and 50 days old lambs as well as on the growth rate during the period 21-50 days. However crossbreeds of Pomeranian ewes with Teksel rams did not give expected effects in day increase of weight and growth rate during the period till 50 days.
Cross-breeding of ewes with Blackhead and Teksel rams increases slaughter value and percentage participation of meat in the gammon saving simultaneously its high quality.

SHEEP AND GOAT PRODUCTION [S]

Poster S2.38

Bestimmung der körperzusammensetzung bei lebenden Merinoschafen mit hilfe des computertomographen
G. Pászthy[1], A. Lengyel[2]. Bábolna AG, H-2943 Bábolna, Mészáros u. 1; Pannon Agrarwiss. Universität, Fakultät Tierproduktion, H-7401 Kaposvár, POB.16

200 Ungarische Merinolämmer mit einem Lebendgewicht zwischen 20-45 kg wurden mit CT hinsichtlich ihrer Körperzusammensetzung untersucht. Die CT-Aufnahmen erfolgten an den Rückenwirbeln IX, XI und XIII, den Lendenwirbeln II, IV und VI sowie am Femurkopf. Von den einzelnen Aufnahmen wurden die Größe der Muskel- und Fettfläche sowie deren Breite und Tiefe dokumentiert. Nach der CT-Untersuchung wurden die Lämmer auf dem Schlachthof geschlachtet und lt. Vorschriften der Fleischindustrie zerlegt. Für die einzelnen Teilstücke wurden Fleisch-, Knochen- und Fettmenge festgestellt. Mit 5 mathematischen Modellen wurde versucht, die Muskel- und Fettmenge für die einzelnen Körperteile zu bestimmen. Die r^2-Werte für den Fleischgehalt des Schlachtkörpers in den 5 Modellrechnungen waren: 0.79, 0.86, 0.80, 0.84, 0.84. Die r^2-Werte für den Fettgehalt in der gleichen Reihenfolge waren: 0.59, 0.62, 0.53, 0.49, 0.57. Mit Probeschlachtungen wurde die praktische Anwendung der Schätzungsgleichung bewertet. Für die praktische Zuchtarbeit werden 2 Schätzungs-gleichungen empfohlen. Die Schätzung des Gewichtes der einzelnen Körperteile auf Grund der CT-Aufnahmen kann bei den ungarischen Merinobeständen mit Berücksichtigung der Selektionsparameter verwendet werden.

Poster S2.39

The meat quality of Pomeranian and Ile de France rams and their crossbreds.
H. Brzostowski[1], Z. Tański[1], J. Sowińska[2]. [1] Departament of Sheep-farming, Hunting and Goat Breeding, [2] Departament of Animal Hygiene, University of Agriculture and Technology 10-718 Olsztyn, Oczapowski str. 5, Poland.

The investigations were carried out on the quadriceps muscle of the thigh (*m. quadriceps femoris*) and the longest back muscle (*m. longissimus dorsi*) in 50 days old lambs of Pomeranian (P), Ile de france (IF) rams and their crossbreeds F_1 (PIF). Samples of the muscles were studied in respect of chemical constitution, fatty acid composition of the intramuscular fat, energy value as well as physical properties of the meat. There were no radical differences in the meat quality with respect to the indices studied. The genotype of the animals was differentiated only by pH of the meat 48h after slaughter, $C_{16:0Izo}$ acid content and the thickness of muscle fibres. The thinnest muscle fibres were observed in IF rams (17,24μm) while the thickest ones occurred in P rams (19,31μm). The quadriceps muscle of the thigh had the higher dry matter, crude protein and fat content, more irregular post-mortem glycolysis (18h after slaughter), higher percentage participation of $C_{16:0Izo}$, and lower $C_{14:0}$, better crispness, juiciness, taste and aroma than the longest back muscle.

SHEEP AND GOAT PRODUCTION   [S]

Poster S2.40

Effect of muscle location and sex on the evaluation of meat quality in sheep
Katy Heylen*, R. Suess, R., G. v. Lengerken. Institute of Animal Breeding and Husbandry with Veterinary Clinic, Martin-Luther-University Halle-Wittenberg, Adam-Kuckhoff-Str. 35, D-06108 Halle, Germany, e-mail: heylen@landw.uni-halle.de

Muscle and location within muscle has been found to influence the measured values for pH, electrical conductivity, intramuscular fat content (IMF) and sensory assessment in pigs and cattle. Considering the difference in the physiological maturity between sexes at a same age and/or weight it seems therefore important to study this problem also in lambs. The investigation took place in 47 male and 45 female lambs of the German Longwool Merino breed at 3 locations in musculus longissimus dorsi and in musculus semimembranosus. According to German marketing requirements final weights ranged between 35 and 45 kg. PH-value, electrical conductivity, meat colour and IMF content were determined.

The investigation indicates a significant effect of measuring point on the obtained values for most all meat quality traits. This differentiation occurs in the same pattern in lambs of both sexes. From a methodological point of view therefore sex effects can be neglected when evaluating meat quality in lambs. Beside IMF content there are only small differences in meat quality traits between male and female lambs at a comparable range of weight.

Poster S2.41

Meat production characteristics of three goat genotypes kept under intensive (stable) and extensive (biotope) feeding conditions
Pera Haumann[1], H. Snell*[2] and E.S. Tawfik[1]. [1]Department of International Animal Husbandry, University of Kassel, Steinstr. 19, 37213 Witzenhausen, Germany, [2]Institute of Agricultural Engineering, University of Göttingen, Gutenbergstr. 33, 37075 Göttingen, Germany.

Twenty twin pairs of kids allotted to three genotypes - Boer x Alpine (F1), F1 x Cashmere (F2), Cashmere - were weaned from their mother at the age of 95.4 ± 5.3 d. Whereas one of each twin was fed with concentrate ad lib. in the stable, the other was kept outside on infertile grassland for the purpose of biotope conservation. After the grazing period (77 d) they were finished with concentrate ad lib. in the stable. All of the kids were slaughtered at the age of 261.4 ± 5.3 d.

The daily weight gain (LSM) of the three genotypes compared was 160^a, 154^a and 96^b until weaning, 104^a, 98^a and 56^b during the next period and 120^a, 103^{ab} and 71^b g/d with all kids kept in the stable. The performance was strongly influenced by the feeding conditions, daily weight gain in the intensive and extensive group conducted 134^a and 38^b g/d during the grazing period. No compensatory growth could be observed, body mass development in the mentioned groups amounted to 103^a and 93^a g/d in the last period. The percentage of valuable meat cuts, espec. of the gigot, was lower in the intensive group (42.9^b vs. 44.3^a, 26.7^b vs. 28.5^a %). However meat quality was satisfying with both feeding systems.

SHEEP AND GOAT PRODUCTION [S] Poster S2.42

Contemporary orientation of utility type of breeding of sheep in Czech Republic
L. Štolc, L. Nohejlová, V. Dřevo, V.Nová. Department of Cattle Breeding and Dairying, Czech University of Agriculture Prague, 165 21 Prague 6, Czech Republic

Conditions of market economy considerably influenced the system of utility orientation of sheep breeding in CR. Formerly prefered the wool performance was in the long-term improvement programme in all breeds raised replaced by fertility, milking capacity and mutton production.

In the year 1999 number of sheep kept on the area of Czech Republic is 93 000 head, including 72.3 % of sheep with combined wool-mutton breeds (Merinolandshaft, Mutton Merino, Kent, Tsigai, Improved Walachian, Šumavka and Bergshaft), 26.9 % of mutton breeds (Charollais, Texel, Suffolk, Oxford Down) and the rest of population (0.8 %) is formed by the milk breeds and prolific breeds (East Friesien, Romanov, Finish sheep).

Since the year 1992 the prolificacy per one ewe lambed increased from 115 % to 151 %. Rams and gimmers from plentifuls litters were included into breeding. In the entire population of sheep bred an intensive up-grading on mutton utility takes place. The pen-fold grazing system is continuously extended with spring lambing.

Poster S2.43

Hay and silage in the feeding of pregnant and lactating ewes
R. Sormunen-Cristian. Animal Production Research, Agricultural Research Centre of Finland, 31600 Jokioinen, Finland.

A comparison was made between hay and silage made from timothy/fescue sward and fed to pregnant and lactating ewes. Hay was harvested at later stage of growth than silage. Intake of hay (H), silage (S) and hay/silage (HS), and performance of 30 Finnish Landrace ewes carrying twins, triplets and quadruplets and suckling either twins or triplets was measured during the last 8 weeks of pregnancy and 6-week lactation. There was no difference in dry matter intakes between ewes during pregnancy nor neither between H and S ewes during lactation. The H, S and HS ewes consumed metabolizable energy on average 12.8, 14.3 and 13.4 MJ per day (H vs. S, $P< 0.05$) in pregnancy, respectively, and 17.5, 20.5 and 18.0 MJ per day (H vs. S, $P< 0.001$) (S vs. HS, $P< 0.001$) in lactation, respectively. The total daily protein intakes of H, S, and HS, calculated in terms of amino acids absorbed in the small intestine (AAT) were 92, 110 and 102 g (H vs. S, $P< 0.01$) in pregnancy and 156, 187 and 161 g (H vs. S, $P< 0.001$) (H vs. HS, $P< 0.001$) in lactation, respectively. The ewes consistently performed equally when fed hay or silage. Ewes given silage alone, however, produced most offspring. Lamb mortality and the number of artificially reared lambs were highest when ewes were fed both hay and silage ad libitum. Lamb growth on S and HS diets was nearly equal, and clearly higher than on H diet. Concerning winter feeding of ewes, grass silage compared favourably with ordinary hay and therefore hay could be fully replaced by grass silage during late pregnancy and early lactation.

SHEEP AND GOAT PRODUCTION   [S]

Poster S2.44

Effect of selenium and zinc supplement on fattened lambs
J. Kuchtík, G. Chládek, V. Koutník and M. Hošek. Departement of Animal Breeding, Mendel University of Agriculture and Forestry, Zemedelska 1, Brno 613 00, Czech republic

The study was carried out on 24 lambs, F_1- crosses of German Longwool sheep and Oxford Down sheep. The lambs were distributed into three groups of 8. Groups 1 and 2 were fattened in semi-intensive systems with their feed ration based on protein concentrate supplemented with vitamins and minerals, on crushed grain, hydro-thermally treated toasted soya beans, lucerne hay and a supplementary pasture. In addition, the ration fed to the group 1 contained the optimal amount of selenium, provided in organic forme. The ration fed to the group 2 contained optimal amounts of selenium and zinc, all selenium and at least two thirds of the zinc provided in organic form. The group 3 was fattened on the pasture only. The highest daily weight gain in the period from birth until weaning (200 g) was found in group 2 which was given selenium and zinc. This group also showed the highest daily weight gain during the period from weaning until slaughter (235 g) and during the overall period from birth until slaughter (213 g). Dressing percentage varied from 39.69 - 42.43 %, however no significant differences were found between groups. The proportion of a leg weight of the total carcass weight after slaughter was highest in group 2 (33.68 %) and lowest in group 1 (33.30). The highest proportion of muscle in the leg was found in the group 2 (on average 1.75 kg, which is 68.18 %). The proportion of fat in the leg ranged from 8.70 to 11.47 % between groups.

Poster S2.45

Nutrient consumption and body mass development of suckling goat kids of the production genotypes of milk, meat and fibre
H. Snell[1] *and E.S. Tawfik*[2]. [1]*Institute of Agricultural Engineering, University of Göttingen, Gutenbergstr. 33, 37075 Göttingen, Germany,* [2]*Department of International Animal Husbandry, University of Kassel, Steinstr. 19, 37213 Witzenhausen, Germany.*

In the goat production kids are frequently reared together with their dams for a period of several month. In this case the growth performance is essentially determined by the nutrient supply with the goat's milk. In the context of this investigation relevant breed differences should be studied.
The milk consumption, as determined with the kid suckling method over 63 days, of kids of the genotypes Alpine (BDE), Boer, Cashmere, BDE x Boer, Boer x BDE and R_1 amounted to 1134[b], 789[a], 815[a], 1001[ab], 1239[b] and 814[a] g/d. The milk composition was analysed from milk samples collected at four lactation stages (4th, 24th, 45th and 62nd d). With the aid of this data an ingestion of 94[ab], 77[a], 78[a], 98[ab], 104[b] and 77[a] g/d solids-not-fat could be calculated. The corresponding values for milk fat conducted: 18[a], 17[a], 21[ab], 27[b], 22[ab] and 19[ab] g/d. Due to the insufficient udder evacuation, gaining the milk samples, the milk fat data represent the lower bound of actual conditions. The correlation between the consumption of solids-not-fat and the daily weight gain amounted to 0.85. The investigations indicate, that goat breeders have to strive for a sufficient milk yield even with non-dairy goats. Additionally a valuable concentrate should be supplemented under all feeding conditions.

SHEEP AND GOAT PRODUCTION [S]

Poster S2.46

Effect of weaning age and ration energy level on performance, carcass quality and growth rate of quadrolocular stomach of fattening lambs

D. Liamadis*, S. Milioudis, A. Hatzikas and Ch. Milis. *Department of Animal Nutrition, Aristotle University of Thessaloniki, 540 06 Thessaloniki, Greece.*

One 2 X 3 factorial experiment (two weaning ages X 3 rations of different energy level) was carried out in order to examine the effect of the weaning age (30 or 45 days) and the dietary energy level (12.0, 10.8 and 9.9. MJ/kg DM) on Daily Live Weight Gain (DLWG), feed efficiency ratio, carcass quality and growth rate of the total volume and weight of quadrolocular stomach. 24 male fattening lambs were used for this purpose, which were slaughtered at their final body weight of 32 kg. The obtained data showed that: 1) DLWG was not significantly affected by the weaning age, but the high energy ration obtained better results ($P<0.05$), 2) feed efficiency and lean content were increased by increasing energy ration, 3) warm and cold carcass yield were affected only by the weaning age, 4) kidney and channel fat were not influenced by the weaning age; however the lambs with the high energy ration had more kidney and channel fat than the other ones, 5) warm carcass without internal fat (%) and lean content of cold carcass without internal fat (%) were not significantly affected by the 3 rations, but they were influenced by the weaning age, 6) the total volume of quadrolocular stomach was higher in 30-day weaning than the other ones, yet it was not affected by the 3 rations, 7) the total weight of quadrolocular stomach was, also, higher in the 30-day weaning lambs fed with the low energy ration.

Poster S2.47

Effects of dietary by-pass glucose on plasma metabolite concentrations in lambs

S. Abbas*[1], T. Matsui[2] and H. Yano[2]. [1]*Department of Animal Production, Faculty of Agriculture, Assiut University, Assiut, Egypt.* [2]*Division of Applied Agriculture, Kyoto University, Kyoto 606, Japan.*

Eight wether lambs (30 kg) were randomly divided into two groups after 6 days of preliminary period. In the supplementation period, each group was additionally given 2 kinds of by-pass glucose at 200 g/head/day for 9 days. One by-pass glucose consisted of 40% glucose, 30% calcium salt of fatty acids, 20% beef tallow and 10% CaCO3. The other by-pass glucose consisted of 50% glucose, 20% calcium salt of fatty acids and 30% beef tallow. Blood was collected just before morning feeding 0 and 2, 4 and 8 h after morning feeding on the last day of the preliminary period and the third, sixth and ninth day of the supplementation period. The supplemented groups of by-pass glucose significantly increased plasma glucose and triglyceride concentrations on day 3. However, plasma glucose and triglyceride became less than the levels of preliminary period on day 6. Each supplement decreased plasma urea nitrogen on the day 3 of treatment but it was recovered on day 9. Both the by-pass glucose supplements affected plasma metabolites concentrations in lambs but these effects were temporal. The findings suggest that the influx of a large amount of glucose changed glucose and lipid metabolism. These changes may be beneficial for productivity of ruminants.

SHEEP AND GOAT PRODUCTION   [S]

Poster S2.48

Conception rate and embryonic mortality in Booroola * German Mutton Merinos depending on the time of mating in relation to ovulation

K.-H. Kaulfuß*[1], S. Moritz[2], Institute of Animal Breeding and Husbandry with Veterinary Clinic, Martin-Luther-University Halle-Wittenberg [1], Ambulatory and Obstetrical Veterinary Clinic, University of Leipzig[2]; Germany.

The aim of this study was to determine the influence of a mating before, during and after the ultrasonographic determined ovulation on the conception rate or the level of embryonic mortality (e.m.) in sheep. The examination included 38 ewes of Boorool * Mutton Merino crosses. One time a day with a teaser ram the natural oestrus was determined. Immediately after it an ultrasonographic ovary diagnosis took place. Between day 7 and 12 after mating the ovulation rate (o.r.) and between the 29th and 31st day after mating the number of living embryos were diagnosed also by transrectal scanning. By a mean o.r. of three 16 ewes finished the ovulation short before mating (g1) and 12 ewes showed preovulatoric and ovulated follicles together (g2). Only 10 ewes were in a preovulatory situation and the ultrasonographic clear nonechogenic follicles had an diameter from 4.2 ± 0.8 mm (g3). There was no influence of the group on the level of the total e.m. ($\approx 30\%$). The number of ewes with a total loss of embryos or which not conceive was nearly the same (g1: 18.8%; g2: 16.7%; g3: 20.0%) but there was a clear trend (non significant) for an increasing number of ewes with a partial loss of embryos when mating during or after ovulation took place (g1: 43.8%; g2: 50.0%; g3: 20.0). Summarizing we couldn't certify that a fertil mating is unpossible after ovulation.

Poster S2.49

Synchronization of oestrus in goats: Dose effect of progestagen

J.P.C. Greyling* and M. van der Nest, University of the Free State, P.O. Box 339, Bloemfontein, 9300, Republic of South Africa

The effect of dose progestagen on the efficiency of synchronization in the breeding season was evaluated in Boer and Indigenous goats. Group 1 served as control, while Group 2 and 3 were treated with 60 mg MAP or halved MAP intravaginal sponges plus 300 IU PMSG. AI was performed at fixed times with fresh diluted semen. Time to oestrus was not significantly different between breeds or treatments. Oestrous response varied between 65 and 85% in Boer and 70 and 75% in Indigenous goats. The duration of the induced oestrous period was significantly ($P<0.05$) longer in both treatment groups. Duration of oestrus could not be related to conception rate. There was no significant difference in the mean serum progesterone levels for Boer and Indigenous does following sponge withdrawal (1.21 ± 0.44 ng/ml and 1.41 ± 0.45 ng/ml for Boer and Indigenous does respectively). The mean serum LH level was significantly ($P<0.05$) lower in the 60 mg MAP group following sponge withdrawal (0.59 ± 0.22 ng/ml vs 0.84 ± 0.66 ng/ml vs 1.9 ± 0.49 ng/ml for the 60 mg, control and halved MAP groups respectively). No detrimental effect on reproductive performance was observed with higher doses. Halving dose of progestagen holds no real advantage. Intravaginal progestagen gave acceptable synchrony and conception rates (mean 75%) in Indigenous and Boer goat does, irrespective of the dose.

SHEEP AND GOAT PRODUCTION   [S]   Poster S2.50

An attempt to find an optimum time for mating in Anglo-Nubian goats in and outside breeding season based on luteinizing hormone (LH) concetration

J. Udała[1], B. Błaszczyk[1], M. Baran[1], K. Romanowicz-Barcikowska[2], B. Barcikowski[2], [1]Department of Animal Reproduction, Agricultural University, 6 Judyma St., 71-460 Szczecin, [2]Institute of Animal Physiology and Nutrition, Polish Academy of Sciences, 3 Instytucka St., 05-110 Jabłonna n. Warsaw, Poland.

Keeping in mind a dominant role of LH in the course of ovalation, it was decided to investigate its secretion during synchronised oestrus in and outside sexual activity season. The experiment was carried out in autumn (October) and spring (April) on 12 goats that were treated with vaginal sponges (Chrono-Gest, 40mg) inserted them for 12 days. After sponge removal, animals were given a 500 i.u. PMSG. Every two hours the level of LH in blood serum with RIA method was determined. As the result of hormonal stimulation applied, the LH peak concentration was noted that occurred after approximately 39 hours after sponge removal in the breeding season, while being approximately 43 hours outside it. The highes concentration of that hormone in autumn was 19,8ng/ml on the average, whereas in spring 13,5ng/ml. Taking into account that ovulation in goats takes place about 20 hours after the LH release, an approximate time of ovulation occurred on 59 hour after sponge removal in the breeding season, and on 63 hour outside it. The most convenient timing for mating goats is about 10 hours prior to expected ovulation. In this connection, the time of mating in goats falls between 47-49 hour after sponge removal and PMSG administration in the breeding season, while between 51-53 hour outside it.

Poster S2.51

Der Einfluß von Selenzusatz in monatlichen Abständen auf die Spermaqualität von Schafböcken im Verlauf des Jahres

B. Seremak, J. Udala, B. Lasota. Landwirtschaftliche Universität Szczecin, Wissenschaftsbereich Fortpflanzung der Tiere, ul. Judyma 6, 71-466 Szczecin, Polen.

Die Untersuchungen wurden an 12 Jungböcken der Rasse Polnisches Langwolliges Schaf, die in eine Versuchs (VG)- und eine Kontrollgruppe (KG) unterteilt und unter gleichen Umweltbedingungen gehalten wurden, durchgeführt. Die Versuchstiere erhielten intramuskulär jeden Monat 2 ml 1%-er Natriumselenitlösung. Dreimal im Monat wurde von den Böcken Sperma entnommen und folgende Parameter ermittelt: Ejakulatsvolumen, Spermienkonzentration, Massenbewegung, Dichte und Anteil an vorwärtsbeweglichen und morphologisch veränderten Spermien. In der Versuchsgruppe wurden bei allen Spermaqualitätsparametern bessere Ergebnisse festgestellt, besonders außerhalb der Reproduktionssaison, in der die Spermaqualität schlechter ist. Signifikante Unterschiede traten bei der Spermienkonzentration, beim Anteil an vorwärtsbeweglichen und morphologisch veränderten Spermien auf. Die Spermienkonzentration im Ejakulat betrug im Jahresdurchschnitt 3,2 Mio. je 1 mm^3 in der Versuchsgruppe und 2,95 Mio. je 1 mm^3 in der Kontrollgruppe. Beim Anteil an vorwärtsbeweglichen Spermien betrugen die Werte in VG und KG entsprechend 75,5% und 69,2%. Demzufolge ist eine zusätzliche Versorgung von Schafböcken mit Selen, insbesondere auf den Selenmangelgebieten des Pommerns, zweckmäßig.

SHEEP AND GOAT PRODUCTION [S]

Poster S2.52

Dynamics of estradiol and progesterone blood concentration changes in Anglo-Nubian goats with stimulated ovulation in and outside breeding season
J. Udała, B. Błaszczyk, Department of Animal Reproduction, Agricultural University, 6 Judyma St., 71-460 Szczecin, Poland.

The experiment was carrired out in atumn (October) and spring (April) on 12 Anglo-Nubian goats 1,5-2 years old. Oesrtus synchronisation and ovulation stimulation was made using vaginal sponges (Chrono-Gest, 40mg) inserted them for 12 days. Immediately after sponge removal, goats were given a 500i.u. PMSG (Serogonadotrophin) intramuscular injection. Blood samples were drawn every two hours from 24 to 68 hour after sponge removal. The level of hormones was determined with RIA method. It was showed that the average concentration of progesterone in autumn was 0,46ng/ml (0,37-0,60ng/ml) being significantly higher than in spring - 0,30ng/ml (0,22-0,37ng/ml). Most likely, these differences resulted from luteal hypofunction due to the lack of functioning corpora lutea on the ovary outside breeding season. The secretion rhytm of estradiol (E_2) in both experimental periods was similar. Differences referred to average initial and maximum concentrations. The highest value of E_2 in atumn was 31,3ng/ml on the average, whereas in spring - 20,85ng/ml. These divergences, despite of similar rythm of E_2 secretion, point to a different sensitivity of the system controlling the course of reproduction processes in goats in and outside the breeding season. It might be a reason of the weakened growth of ovarian follicles and reduced secretion of esrtradiol.

Poster S2.53

The prolactin blood level in Anglo-Nubian goats with synchronised oesrtrus in and outside breeding season
B. Błaszczyk, J. Udała, Department of Animal Reproduction, Agricultural University, 6 Judyma St., 71-460 Szczecin, Poland.

One of hypotheses says that prolactin inhibits or stimulates ovary functions in response to changes in melatonin secretion, and thus playing a significant role in controlling the seasonality of reproduction. Therefore, in the study were determined trends in the level prolactin in periovulatory period in and outside the breeding season in goats with synchronised oestrus. Our experiment included 12 goats aged 1,5-2 years that were treated with vaginal sponges (Chrono-Gest, 40mg) in autumn (October) and spring (April) inserted them for 12 days. After their removal, animals were given a 500 i.u. PMSG (Serogonadotrophin). Blood samples were drawn every two hours up to 68 hour after sponge removal. The prolactin concentration was determined with RIA method. In the experiment carried out well-marked differences were noted in the seasonal liberation of prolactin. The average concentration of prolactin in the breeding season was 21,5ng/ml in the periovulatory period analysed, whereas it was almost four times higher outside the breeding season and amounted to 81,7ng/ml. The highest level of prolactin in spring was 164ng/ml, while its peak liberation in autumn did not exceed 40ng/ml. Differences observed in dynamics of changes in the prolactin concentration between autumn and spring point to a contribution of this hormone in controlling seasonal changes of sexual activity in goats.

SHEEP AND GOAT PRODUCTION   [S]   Poster S2.54

Performances Zootechniques á la Jeunesse Ovin, (l' anné deuxieme), en Fonction de Fertilisation
I. Scurtu, Inst. des Reche. et de Prod. sur la Culture Prairies, 5, rue Cucului, cod 2200 Brasov, Roumanie

En vue d'etudier le potentiel de valorisation des prairies temporaires, on a utilisé les resultats optenus pendant plusieus années (1996-1998). La prairie est formée de un melange simple dans quatre systems de fertilisation caracterises par l'apport de la fumure minérale, fumure minerale + fumier (mixte), fumier et parcage. On a forme quatre les lots des aminaux homogene du point de vue du poids corporel. Observations effectuées: matiére seche (MS), proteine brute (PB), cellulose brute (Cel. B), analyses botaniques, problemes de pollution, le gain poids vif (g/j), fibre de laine, etc.

L'interpretation statisque a été l'analyse de la variance, corelations et regretions. Ces premiers resultats optenus menttent en evidence la qualité du fourage administre aux animaux. Le gain de poids vif moyen par jour pour la fertilisation organique (fumier et parcage) a été de 112 g et pour la fertilisation minerale a été 101 g. Pourcentage de legumineuses a été plus élevée, sur les parcelles fertilisées organiquement par rapport a celles fertilisées chimiquement. Le contenu en PB a été plus élevé pour la fertilisation organique, par rapport á celle chimique (20.8% par rapport á 18.1%). La quantité d'herbe données aux animaux durant toute la periode a été de 7.5 t MS/ha pour la fertilisation organique et de 8.8 t MS/ha pour la fertilisation chimique.

Poster S2.55

Nutritional evaluation of complex phosphate (Ca, Mg and Na) in growing lambs
F. Meschy. Laboratoire de Nutrition et d'Alimentation INRA INA-PG, 16, rue Claude Bernard 75231 Paris cedex 05, France.

Almost all ruminant basal diets are P deficient and must be supplemented. This supplementation is more efficient and less pollutant when high quality feed phosphates are used. The aim of this trial was to compare new processed complex phosphate (*Pcom*) to monodicalcium phosphate (*Mdp*) for phosphorus supply; and to magnesium phosphate (*Pmag*) and two magnesium oxides (*Ox1*, *Ox2*) for magnesium supply. The experiment was conducted with 4 groups of 3 growing lambs (Lacaune breed, 30 kg LW and 250g ADG) fed by semi-synthetic diets. Two 5 days-balance trials were performed.

The apparent absorption and retention of P as well as its true absorption (calculated by the AFRC equation for endogenous losses) were significantly higher for magnesium phosphates compared to *Mdp*, but no difference between *Pmag* and *Pcom* was observed.

Apparent absorption of Mg was significantly higher for phosphates compared to oxides but no difference was observed neither between phosphate nor between oxides. Nevertheless, global retention of Mg was only higher for *Pmag*, the *Pcom* group showed a high and unexplained Mg urinary excretion. In our experimental conditions, results showed that *Pcom* could be considered as a high quality feed phosphate allowing an efficient supplementation of animal diets. It is more difficult to conlude in terms of magnesium.

SHEEP AND GOAT PRODUCTION   [S]   Poster S2.56

Mean values of milk yield and its components in Awassi x Barki crossbred ewes
A.A. El Shahat[1] and Y.S. Ghsnem[2], [1]Anim. Prod. Dept. NRC, Dokki, 12622, Cairo, Egypt, [2]Anim. Prod. Dept. Desert Res. Inst., Mataria, Cairo, Egypt.*

The main objective of the present study was to determine milk production and milk constituents in crossbred Awassi x Barki ewes. For estimating milk production of the experintal animals, hand milking method was applied at weekly intervals for period of 18 weeks. The ewes were hand milked twice, at 7.00 a.m. and 4.00 p.m.

The results revealed that the average milk production was 39,7 kgs. The mean percentages of total solids, fat, solids non-fat, protein, lactose and ash were 16.8, 4.6, 6.1, 9.7, 4.0 and 0.875, respectively. The results will be discussed on the basis of animal nutrition and sheep production sciences.

Paper SN3.1

The energy and protein requirements of lactating goats - the AFRC report
J.D.Sutton, Department of Agriculture, University of Reading, Reading RG6 6AT, UK.

In 1998 an AFRC Working Party reported the results of a review of The Nutrition of Goats (CABI, Wallingford, UK). One important aim was to determine whether the metabolisable energy (ME) and metabolisable protein (MP) requirements established earlier for lactating cattle (AFRC 1993) could be applied to goats. For milk production, goats were divided into Saanen/Toggenburg (S/T) and Anglo-Nubian (A-N). Estimates of body weight changes relied heavily on INRA (1988). For energy it was concluded that: fasting metabolism is very similar to cattle at 315 kJ/kg$^{0.75}$ so maintenance ME requirement is 441 kJ/kg$^{0.75}$ at ME/GE (q) = 0.6; efficiencies (k) are the same as for cattle; at milk yields of 6 kg/day, ME requirements at q=0.6 are 4.8 and 5.7 MJ ME/kg milk for S/T and A-N goats respectively; the energy value of live-weight (LW) change is 23.9 MJ/kg; LW loss is 1.0 kg/week for the first 4 weeks, yielding 4.6 MJ ME/day, then a gain of 1.2 kg/month from the fourth month, requiring 1.5 MJ ME/day. For protein, it was concluded that maintenance requirements and efficiencies are the same as for cattle. For milk, requirements are 38.4 and 47.7 g MP/kg for S/T and A-N goats respectively. LW loss in the first month yields 30 g MP/day at 143 g net protein/kg LW while LW gain later in lactation requires 4 g MP/day. Requirements for pregnancy with twins increase from 1.35 to 6.15 MJ ME and from 7.2 to 32.2 g MP daily over the last 3 months. Areas of uncertainty are activity costs and the extent and composition of LW losses and gains.

SHEEP AND GOAT PRODUCTION [S]   Paper SN3.2

Recent progress in the assessment of mineral requirements of goats
F. Meschy. *Laboratoire de Nutrition et d'Alimentation INRA INA-PG, 16, rue Claude Bernard 75231 Paris cedex 05, France.*

For many years, mineral requirements of goats have been considered as half way between cows and ewes ones. During last years, advances in the particular goat nutritional research allow more specific mineral recommendations. Values of the parameters used in the factorial approach seem to be slightly different for goats. Endogenous losses of Ca and P might be more related to the level of DMI than to the Live-weight (LW) of the animals. Ca and P deposit in conception products are higher because of the frequency of twin (or more) bearing goats while the net requirement for growth is lower at least than the calf one (5-6 g of P and 9-10 g of Ca / kg gain). True absorption coefficient of P is probably higher for goats (70-75 %) than for the other ruminant species. Nevertheless, Ca and P content of milk are very closed to cow ones (1.3 g/L and 0.9 g/L respectively). Goat requirements for other macro-elements (Mg, Na and K) do not significantly differ from those for the other ruminants. A special attention must be given to S according to specific fibre production needing. Several results indicate that goats are less sensitive to copper toxicity than sheep and can tolerate higher levels of Cu in their diets. Moreover, goats can tolerate Mo levels 10 times higher than sheep. On the other hand, the iodine requirement, or sensitivity to deficiency seems to be higher in goats. Particular attention must be given to Se and Cu status of pregnant goats and/or new-born kids according to metabolic disorders as white muscle disease or swayback.

Paper SN3.3

Concentrate supplies of dairy goats on pasture
V. Fedele[1] and Y. Le Frileux[2] [1] *Istituto Sperimentale per la Zootecnia, Viale Basento 106, 85100 Potenza, Italy.* [2] *Lycée Agricole olivier de Serres, Station Experimentale Caprine, Domaine du Pradel, 07170 Mirabel, France*

Seasonal variation of herbage availability and quality not always meets goat requirements. In situation where concentrate supplies appear justifiable, their nutritive composition could be complementary at best herbage quality.
In Mediterranean conditions, where pasture is very rich in legumes, the concentrate supplies should be low in protein and high in energy. Protein supplements reduce herbage intake and increase milk production less than energy supplements. The increase of one percentage point in crude protein content reduced herbage intake by 30 g, while 0.1 MFU by 60 g. From late autumn to winter, when fermentable sugars in herbage are very high, goats prefered high NDF concentrate. In opposite, when sugars are low, they prefered rapidly degraded starch and medium protein concentrate.
In lactating goats, the substitution rate herbage/concentrate decreased as concentrate supplements increased. With 150, 500 and 900 g of concentrate, substitution rate decreased from 1.4 to 1.1, and to 0.7 respectively.

SHEEP AND GOAT PRODUCTION [S] Paper SN3.4

Influence of the intensity of forage production on nitrogen nutrition of dairy goats
R. Daccord*, Federal Research Station for Animal Production, CH-1725 Posieux, Switzerland

4 different hays were produced in a long range trial studying the interactions of 3 N fertilisation levels (0, 150, 300 kg N/ha/year) and 3 cutting intensities (2, 3 and 5 cuts/year). These hays were completed with concentrates so that the rations were iso-nitrogenous and almost iso-energetic. Each ration was distributed ad libitum to 4 dairy goats. The nutrient balances were determined.

The goats consumed greater quantities of hay from treatments N300/C5 and N150/C5 than from treatments N150/C3 and N0/C2. With the C5 rations, 65 % of the ingested crude protein came from hay, whereas this proportion was only 30 % with the C3 and C2 rations. The mean milk production was 5.75 kg/day with the C5 rations, i. e. 21 % higher than with the C3 and C2 rations. The highest production of proteins, fat and lactose in the milk was obtained with the hay N150/C5.

The N losses in the feces were greater with the C5 rations. In contrary, the N losses in the urine were greater with the C3 and C2 rations. The global N utilization was best with the hay N150/C5. The cutting intensity of the meadow had a greater influence on the hay quality than the N fertilisation level. The low hay quality could not be compensated by the concentrates.

Paper SN3.5

Relationship between feeding and goat cheese quality: present days problems
P. Morand-Fehr*[1], R. Rubino[2] and Y. Le Frileux[3] [1]Laboratoire de Nutrition et Alimentation, 16 rue Claude Bernard, 75231 Paris cedex 05, France, [2]Istituto Sperimentale Zootechnica,Viale Basento 106, 85100 Potenza, Italy, [3]Lycée Agricole Olivier de Serres, Station Expérimentale Caprine, Domaine du Pradel ,07170 Mirabel, France

Presently, it appears two main trends for goat cheeses market: keeping them in a position of festive products with high goats characteristics and vulgarizing them as expendable products in extended markets. Since some years, modern and intensive management and particularly feeding methods could reduce the quality of goat cheese for consumers (lacks of unctuousness and goaty taste). These defects are mainly due to low fat percentages or reversing of fat and protein contents in goat milks. We can partly correct these defects by adding an lipid source of high quality and improving the level of fibrosity in the high energy diets. On the other hand in various production basins, goat milk producers want to keep productions of typical local cheeses in protected marked (labels…). For that, they are endeavouring to organize committees to support their cheeses of high quality or associations as « Fromaggi sotto il cielo » in Italy to support the cheeses made with milks of goats reared on pasture. In these cases they ask to find typicity parameters to discriminate their products. Presently feeding experiments were carried out in Italy and France to compare cheeses made with goats fed in goat-houses or on pasture. The present results are not yet clear : it seems difficult to discriminate goat cheeses only by feeding management because other factors can influence the typicity of goat cheeses.

SHEEP AND GOAT PRODUCTION [S]   Paper SN3.6

Effects of recombinant bovine somatotropin administration to lactating goats
K. Kyriakoy, S. E. Chadio, G. Zervas, C. Goulas, J. Menegatos, Agricultural University of Athens, Department of Animal Science, 75, Iera Odos, 11855 Athens, Greece.

Eight multuparus lactating goats, matched for age and time of parturition were used in order to evaluate the effects of recombinant bovine somatotropin (rbST) administration on milk yield and composition, as well as on certain metabolic parameters. The experiment was a switch-back design, with three periods lasting 28 days each. Animals were divided in two groups (n=4) and treatments consisted of subcutaneous injection of 160 mg rbST at 14-d interval. Controls remained uninjected. Supplementation with rbST increased significantly over the entire experimental period (1506±512 vs 1337±518 ml/d, $P<0,05$). Milk fat percentage was increased during the treatment with rbST, (3,26±0,67 vs 2,9±0,55, $P<0,05$), while protein percentage did not differ significantly between treated and control animals. Lactose content (%) also increased after rbST administration (3,78±0,19 vs 3,65±0,24, $P<0,01$). Plasma NEFA as well as b-hydroxybutyrate levels were not affected by treatment. Yields of short-chain fatty acids were higher for goats in the rbST group, indicating probably an enhanced de novo synthesis in these animals. In conclusion, rbST administration increased milk yield and the percentage of milk fat and lactose, while had no effect on milk protein content.

Poster SN3.7

The influence of stimulating level of feeding Carpathians goats on their milk production
V. Tafta and C. Neacsu, University of Agriculture and Veterinary Sciences, Bd. Marasti 59, Bucharest, Romania

In order to study at extent the stimulating feeding of Carpathians goat (that produces an average of 200...220 litres of milk during the lactation period) influences the total milk amount, 35 goats have been fed from the second day after birthing until the end of nursing the kids (60 days) with a fodder composed of: 1 kg lucerne, 0.3 maize, o.5 kg barley, 0.35 kg combined nutrition, 0.0012 kg salt and 0.0010 kg concentrated fodder that represents 2.11 kg dried substance, 1.97 UN and 248 PBD.
After a bimonthly control one recorded a production of 387.23 kg milk, over 270 days of lactation, with 13.92% dried substance, 3.75% proteins and 4.17% grease, which means the double amount compared to the normal situation when the same goats are fed on low yield pastures under extensive maintaining conditions. The peak of lactation curve, with 80.10 kg milk, is placed within the second month, and during the first three months 54.76% of the total production is obtained. After this, the monthly production decrease continuously up to 17.20 kg in September.The recorded prolificity was of 130% with the kid weight at birth of 3.68 kg for males and 3.16 kg for the females, and the weight at weaning of 15.58 kg and 13.35 kg respectively.
The daily weight gain was of 172.42 g with a maximum of 230 g, for males and 147.68 g with a maximum of 175 g for females, respectively. The maternal milk demand for a kilo of weight gain was of 8.6 kg.

SHEEP AND GOAT PRODUCTION [S] Poster SN3.8

Effect of rumen-protected methionine and lysine in diets with different protein content on N balance of lactating goats
Rapetti*, L. Bava, A. Sandrucci, A. Tamburini, G. Galassi and G.M. Crovetto.
Istituto di Zootecnia Generale, via Celoria 2, 20133 Milano, Italy.

Six primiparous Saanen goats (50 kg LW) were paired and fed *ad libitum*, in a 3x3 Latin Square design (one week adaptation, one week collection), three isocaloric diets with different CP content (% DM) added or not with rumen-protected methionine and lysine: 15.5% CP (diet C); 15.5% CP + AA (diet 15+AA); 13.0% CP + AA (diet 13+AA).

Diet		C	15+AA	13+AA
DM intake	g/d	2258[a]	2231[ab]	2146[b]
Milk yield	g/d	3754[A]	3697[A]	3423[B]
Milk protein	%	2.83[ab]	2.85[a]	2.75[b]
Faecal N	% ing. N	31.7[B]	29.8[B]	36.0[A]
Urinary N	% ing. N	17.1[A]	19.5[A]	13.2[B]
Milk N	% ing. N	29.6[B]	29.8[B]	33.1[A]
Retained N	% ing. N	21.5[a]	21.0[a]	17.7[b]

A, B: $P \leq 0.01$; a, b: $P \leq 0.05$

The data obtained suggest that the reduction of the protein content and the concomitant addition of the two amino acids is not convenient in terms of milk production, and it is questionable as a key to reduce N excretion effectively.

Poster SN3.9

Milk performance and milk composition of goats under conditions of restriction as well as realimentation of energy and nutrient supply
V. Manzke, H. Münchow and Susam Dündar. Animal Nutrition Section, Institute of Animal Sciences, Faculty of Agriculture and Horticulture, Humboldt University Berlin, Philippstr. 13; 10115 Berlin, Germany

Multiparous goats of two breeds [Bunte Deutsche Edelziege (BDE); Kashmir (KMZ)] were fed with pelleted feed and hay semi ad libitum considering the milk yield (period I: 30 days). That was followed by periods of energy restriction (II: 24 days about 65 % of I) as well as of realimentation (III: 30 days ad libitum). At these nutritional levels the pellet : hay-ratios were almost constant (75 : 25).
Under the experimental conditions of ad libitum (I, III) and restrictive (II) feeding there were no significant breed-specific differences in milk yield. Regarding the FCM and fat output BDE goats reacted more moderate to energy and nutrient restriction than the KMZ goats. Concerning the milk fat and protein concentration the breeds responded in divergent manner to the reduction of energy supply. With BDE the fat content increased noticeably, whereas the protein concentration decreased. With KMZ these parameters were only slightly changed.
Within each breed the variation of the energy and nutrient ingestion affected the relations of the fatty acids of milk fat. At each nutritional level there were also breed-specific differences in the fatty acid composition of milk fat which reflected the measured pattern of free fatty acids of blood serum.

SHEEP AND GOAT PRODUCTION   [S]   Poster SN3.10

Investigations of rumen and intermediary metabolism of goats under conditions of restriction as well as realimentation of energy and nutrient supply
H. Münchow, V. Manzke and Susam Dündar. *Animal Nutrition Section, Institute of Animal Sciences, Faculty of Agriculture and Horticulture, Humboldt University Berlin, Philippstr. 13; 10115 Berlin, Germany*

In investigations of compensatory abilities of lactating goats to marginal nutritional situations reactions of rumen and intermediary metabolism to a changing energy and nutrient supply of these animals have been examined. In three successive periods of 30 (I and III) and 24 days (II) of duration all goats of two breeds got a ration equal in quality (75 % concentrate, 25 % hay) in varying amounts (I: ad libitum; II: restrictiv, app. 65 % of I; III: ad libitum). Under these experimental conditions of ad libitum (I, III) and restrictive (II) feeding there were no significant breed-specific differences concerning parameters of rumen and intermediary metabolism. With both breeds in the restriction period (II) versus ad libitum levels (I, III) a change of molare proportions of volatile fatty acids could be measured (more intensive break down of fibre fraction). Regarding parameters of intermediary metabolism due to the malnutrition situations no excess of metabolism was to register. The differences between the ad libitum feeding level of period I versus the marginal conditions of period II reflecting the catabolic situation (temporary decrease of triacylglycerol concentrations as well as activities of selected enzymes; increase of keton bodies in blood and milk). These changes could be compensated with both breeds in the following realimentation period (III).

Poster SN3.11

Effect of physiological stage on the feeding behaviour of dairy goats fed at the trough
J.A. Abijaoudé, P. Morand-Fehr [*], J.Tessier and Ph. Schmidely. *Laboratoire de Nutrition et Alimentation, 16 rue Claude Bernard, 75231 Paris cedex 05, France*

Two experiments were carried out to analyze the effect of different physiological stages of Alpine and Saanen goats on the main parameters of feeding pattern. The experiments I and II were conducted during 13 weeks (from late pregnancy to early lactation) and 12 weeks in mid lactation respectively. Five physiological stages were considered according to the variations of behavioural parameters : 1 : end of gestation, 2 : 2-4 weeks of lactation WL, 3 : 5-7 WL, 4 : 8-10 WL, 5 :20-32 WL.
Jaw movements and the weights of feed containers were recorded over 48 hours. All goats were fed mixed diets composed of lucerne hay, dehydrated lucerne, beet pulps, barley, soybean meal, minerals and vitamins.
DM intake increased from late gestation to 10 weeks of lactation. DM intake in main meals increased beginning from 5 weeks after parturition. The number of secondary meals taking place between feed distributions tended to decrease during lactation. Daily eating and ruminating durations seemed to stabilize during late gestation and lactation but chewing time per kg of feed decreased clearly in early lactation with a tendency to increase from early to mid lactation. During lactation, the higher the level of intake, the lower the eating and ruminating times per unit of feed.

SHEEP AND GOAT PRODUCTION   [S]   Poster SN3.12

Use of Decoquinate to improve the growth of female goats and milk production of one year old goats
P. Morand-Fehr*[1], A. Richard[2], J.Tessier[1] and J. Hervieu[3], [1]Laboratoire de Nutrition et Alimentation, 16 rue Claude Bernard, 75231 Paris cedex 05, France, [2]Alpharma, Silic 411, 3 Impasse de la Noisette, 91374 Verrières le Buisson, France, [3]INRA, Domaine de Brouessy, 5 rue Paul et Jeanne Weiss, 78114 Magny les Hameaux, France

Coccidiosis is one of the main causis of growth slow down in young female goats. On two groups of young female goats of our experimental flock during five years we compare the group S received 3 times an sulfadimérazine treatment for three days, one before and two after weaning and the group D received 1 mg Decoquinate per kg LW for 30, 75 or 120 days after weaning. At 7 months, the live weight of group D was always higher than group S from 1.4 to 4.8 kg. The 75 and 120 days treatment appeared to be more efficient than 30 days treatment. In all the cases the feed efficiency was improved. The 100 or 200 days milk production of the first lactation was improved with 75 or 120 days treatments but not 30 days treatment. The effect on milk production would be due to heavier live weight at mating and parturition of group D goats. Consequently Decoquinate treatments around 75 days would be efficient to improve growth and milk production of young female goats.

Paper S4.1

Genetic resistance in sheep to scrapie, Salmonella and gastrointestinal nematodes
J.M. Elsen*[1], F. Lantier[2], J Bouix[1], J Vu Tien[1],C. Moreno[1], L. Gruner[3]. [1]SAGA, INRA, 31326 Auzeville, France, [2]PII, INRA, 37380 Nouzilly, [3]PAP, INRA, 37380 Nouzilly.

This paper is based on exemples coming from experiments in sheep by pathologists and geneticists of INRA about salmonellosis, scrapie and gastro-intestinal strongylosis.
Evaluation of resistance may be based on either direct or indirect criteria. Direct criteria are recorded either under field conditions or during a challenge with pathogen. They are related to number of pathogens in target organs, clinical signs, pathogenetic effects, host immune response.... Indirect criteria are potential predictors independant of exposure to the pathogen (humoral response to a specific or non-specific challenge, genetic markers...). Some example of genetic determinism are given : polygenic inheritance (gastro-intestinal parasitism), major locus (scrapie) or oligogenic heredity (Salmonellosis). Identification of QTL (Quantitative Trait Locus) is now possible with the help of genetic markers
Breeding strategies have to be set up according to the characteristics of each situation : production losses associated with the disease under consideration, existence of carriers of pathogen without clinical signs, risk for human health, efficiency and drawbacks of alternative approaches...
Inclusion of an additional selection criterion concerning resistance in a preexisting breeding program require to define accurately the objective, to know the genetic variability of the resistance trait and its genetic relationships with production traits and to optimize the selection scheme organisation.

SHEEP AND GOAT PRODUCTION [S]   Paper S4.2

Selecting sheep for resistance to gastrointestinal nematode parasites
S.C. Bishop*[1] and M.J. Stear[2]. [1]Roslin Institute (Edinburgh), Roslin, Midlothian EH25 9PS, UK. [2]Department of Veterinary Clinical Studies, Glasgow University Veterinary School, Bearsden Road, Glasgow G61 1QH, UK.

This paper considers genetic selection as an alternative to chemotherapy for controlling nematode parasite infections in sheep. Faecal egg count (FEC), an indicator on the degree of infection, has been shown in many studies to be moderately heritable and extremely variable between animals, thus amenable to selection. In 6-month old Scottish Blackface lambs the heritability of FEC is 0.33, with the component traits of worm fecundity and worm number having h^2s of 0.55 and 0.14. FEC shows strong associations with IgA response and specific alleles at the MHC DRB1 locus. Modelling studies demonstrate that selection for decreased FEC results in selection responses up to twice that predicted by quantitative genetic theory, due to the epidemiological benefits of sheep with enhanced resistance excreting fewer eggs - thus decreasing the pasture larval contamination. These studies also demonstrate that this reduced parasite prevalence can lead to large correlated responses in growth rate, and they show how the genetic relationships between resistance and productivity depend upon the level of parasite challenge. The feasibility of selection is now being investigated on commercial farms using FEC, IgA and MHC measurements. The next challenge is to design selection indices incorporating resistance to nematode parasites that account for the severity of the parasite challenge faced by the flock.

Paper S4.3

Effect of expected degree of parasitism on the economic value of genetic host resistance to internal parasites
P.R. Amer*[1], S. J. Eady[2] and J.C. McEwan[1]. [1]AgResearch Invermay, Private Bag 50034, Mosgiel, New Zealand. [2]CSIRO Div. of Animal Production, Chiswick, Armidale, NSW 2350, Australia.

Incorporation of selection for host resistance to internal parasitism into commercial sheep breeding programmes in New Zealand and Australia is expected to result in a reduced dependence of sheep farming on chemotherapy. However, selection for genetic resistance to disease necessitates a trade-off in the amount of progress that can be achieved by selecting for improvements in typical production traits such as growth rate, wool production and prolificacy. A method of calculating an economic value to facilitate optimal weighting of selection for disease resistance in a breeding programme will be described. It is shown that the economic value is highly dependent on the expected degree of parasitism in the commercial flock in which genetic improvement is expressed. Growing incidences of parasite resistance to currently available chemotherapy's and increased consumer pressure to reduce chemical use in sheep farming are expected to increase the degree of parasitism on commercial sheep farms, and therefore also increase the economic value of disease resistant sheep.

SHEEP AND GOAT PRODUCTION [S] — Paper S4.4

Evidence for breed differences in resistance to nematode parasitism
J.P. Hanrahan* and B.A. Crowley, Teagasc, Research Centre, Athenry, Co. Galway, Ireland.

Gastrointestinal parasitism is a major concern in the management of sheep in lowland production systems. It is known that *Oestertagia* are the predominant nematode species in the summer/autumn period. A study of breed differences in resistance to nematode infection was undertaken with particular reference to the Suffolk and Texel breeds. Faecal samples were collected, beginning in September, from 7-month old ewe lambs which had been managed together from birth and had not received anthelmintic treatment for at least 2 months. Faecal egg counts (FEC) were significantly (P<0.01) higher in Suffolks. The animals received a standard anthelmintic treatment (oxfendazole) and were re-sampled 8, 21 and 35 days later. On all these sampling dates Suffolk lambs had significantly (P<0.01) greater FEC values than Texels. The study was repeated and gave the same results. A follow-up study of young ewes of these breeds also yielded significantly lower FEC values for Texels. FEC values for a sample of lambs representing Suffolk, Texel and Belclare breeds confirmed the difference between Suffolk and Texel and showed no difference between Texel and Belclare. The repeatability of log (FEC + 1) values, for samples obtained at least 21 days after anthelmintic treatment, were 0.5 to 0.8 (P<0.01). These results provide strong evidence that the Texel breed is more resistant to nematode infection than the Suffolk.

Paper S4.5

Genetic parameters for Eimeria resistance followed natural infections in Merinoland lambs
K.J. Reeg[1], M. Gauly[1]*, C. Bauer[2], R. Beuing[1], M. Kraus[1] and G. Erhardt[1]. [1]Department of Animal Breeding and Genetics, [2]Institute of Parasitology, Justus-Liebig-University, Ludwigstr. 21B, D-35390 Giessen, Germany

The aim of this study was to estimate genetic parameters of resistance to *Eimeria* infec-tions in German Merinoland lambs, to calculate interactions with other traits of economic value and to describe the first incidence and prevalence of different *Eimeria* species. The experimental flock were the offspring of 10 Merinoland rams (n = 235). Faeces (n = 5581) and blood was sampled regularly from all lambs. Individual oocysts counts were carried out using the McMaster method. The *Eimeria* antibody titer were monitored using an ELISA-test. The *Eimeria* species were differentiated. *Eimeria* oocysts have been in 75.5 % of all samples. In 86.5 % of the positive samples between 2 and 5 different *Eimeria* species were recorded. The species were: *E. faurei, E. ovinoi-dalis, E. weybridgensis/crandallis, E. bakuensis, E. pallida, E. parva, E. ahsata, E. intricata, E. granulosa. E. ovinoidalis* was dominant in most of the samples. It showed the highest prevalence on day 41. The ram effect on the number of oocysts was not significant. The phenotypic correlation were between number of oocysts and weight at 40 days - 0.33 (p < 0.01). The estimated heritability for total number of oocysts and different species was between 0.03 and 0.06.
Acknowledgement: This study was supported by the German Research Foundation (SFB 299).

SHEEP AND GOAT PRODUCTION   [S]   Paper S4.6

Genetic parameters of gastrointestinal nematodes resistance in German Rhön sheep
M. Gauly*, H. Mathiak, K .Hoffmann and G. Erhardt. Department of Animal Breeding and Genetics, Justus-Liebig-University, Ludwigstr. 21B, D-35390 Giessen, Germany

The aim of this study was to estimate genetic parameters of resistance to gastrointestinal nematodes in German Rhön sheep, and to calculate interactions with other traits of economic value. The experimental flock were five groups of 20 ewes mated with 5 rams. Ewes lambed in January and February 1998 (n = 131). They were kept on pastures from April to November. Lambs were weaned at an age of 4 to 6 months. Faeces and blood was sampled regularly from all lambs. Individual FECs were carried out using the McMaster method. The gastrointestinal nematode infection was predominantly *Cooperia*, *Strongyloides* and *H. contortus* according to the infective larvae collected from faecal cultures. There were small numbers of *Trichostrongylus spp.*, *Chabertia* and *Nematodirus* larvae. Faecal egg counts were lower in male than in female lambs. Neither the type of birth nor the age of the ewe had an effect on FECs. The ram effect was highly significant on FECs (range 57 - 408). Genetic correlations were - 0.60 between FEC and daily gain and - 0.33 between FEC and hematocrit. The phenotypic correlation between daily gain and hematocrit was 0.33.
Acknowledgement: This study was supported by the German Research Foundation (SFB 299).

Paper S4.7

The relation between nematode parasite infection and frequency of lymphocytes subpopulations in blood of the Heath Sheep lambs
K.M. Charon*[1], R.Rutkowski[1], B. Moskwa[2], A.Winnicka[3], W.P.Siderek[1], [1] Department of Genetics and Animal Breeding, Warsaw Agricultural University, Przejazd 4,05-840 Brwinów, [2] Institute of Parasitology, Twarda 51/55,00-818 Warsaw, [3]Faculty of Veterina-ry Medicine, Warsaw Agricultural University, Grochowska 272 03-849 Warsaw,Poland.

Polish Heath Sheep lambs, 3-7 months old, were examined during period of half a year (September 1998 to March 1999). lambs from the Warsaw Agricultural University Experimental Farm were tested. Faecal samples were taken every two month and the number of gastrointestinal nematode eggs per gram of faeces (EPG) by a modified McMaster method was estimated. The analysis of frequency of leukocytes populations and lymphocytes subpopulations using the flow cytometry method was performed. The following monoclonal antibodies were used: anti- CD2, CD4, CD8, CD19, WC1-N2, MHC I and MHC II. The individual differences, from 0 to 1250, in nematode parasite egg output were observed. The highest number of nematode eggs in faeces of lambs was noticed in September (after grazing season) then the decrease of EPG was observed. The changes in the frequency of analysed lymphocytes subpopulations depending on the EPG were noticed. The CD2 and CD8 lymphocytes were the most frequent in blood samples tested in September, while MHC II in blood taken from older lambs.

SHEEP AND GOAT PRODUCTION   [S]   Paper S5.1

The genetic improvement of Angora goats in France
D. Allain*[1] and J.M. Roguet[2]. [1]Institut National de la Recherche Agronomique, Station d'Amélioration Génétique des Animaux, BP27, 31326 Castanet Tolosan, France, [2] Caprigène France section Angora, Agropole, Route de Chauvigny, 86550 Mignaloux Beauvoir, France.

The angora breeding of goat began in France about 1980. Today, about 7500 animals from 180 different farms, produce annually 30t of mohair. The entire French production is collected, graded and processed under the control of farmer's cooperatives, then marketed directly to consumers by farmers in the form of finished products, under a common label "Le Mohair des Fermes de France".

A national selection scheme was developped with breeders in order to improve both quality and quantity of mohair produced by Angora goats. This selection scheme is based on a performance recording system, a national genetic database and a central testing station for bucks raised on a common standardised environment.

The mean production data and the effects of some non genetic factors (sex and age of animals, physiological staus of females: nursing or not, year and breeding) on the production of mohair are described. Heritability of the main fleece traits average to strong and lies between 0,12 and 0,56. The weight and the homogeneity of the fleece are in unfavourable genetic correlation with mean fibre diameter (0,37 and - 0,53 respectively), but favorable with the kemp score (0,61 and - 0,51 respectively). These results are taken into account to consider the breeding value of animals.

Paper S5.2

Genetic improvement of cashmere goats
S.C. Bishop*[1] and A.J.F. Russel[2]. [1]Roslin Institute (Edinburgh), Roslin, Midlothian EH25 9PS, UK. [2]formerly: Macaulay Land Use Research Institute, Craigiebuckler, Aberdeen AB15 8QH, UK.

This paper summarises research in Scotland into the genetic control of fibre production in cashmere goats and considers breeding options designed to match the cashmere produced to the market specifications. Research has concentrated on improving the fibre characteristics of kids, taking measurements on patch samples midway through their first fibre growing season. The population was a composite of feral goats and strains imported from Siberia, New Zealand, Tasmania and Iceland. Measurements included weight of cashmere in the sample, fibre diameter and variability, fibre length and live weight. From these measurements annual cashmere production was estimated. All fibre traits were strongly inherited, with heritabilities greater than 0.5, and extremely variable. Fibre weight and diameter were strongly antagonistically correlated. Selection designed to increase cashmere weight yet hold diameter constant was successful, and responses in cashmere weight over 6 years (compared to an unselected control) were greater than 7%/year. Conversely, selection to reduce fibre diameter had a lower rate of response (1.25%/year) and resulted in a marked decline in cashmere weight. Research is now focussing on fibre shedding patterns, which affect harvesting and goat husbandry requirements, and traits influencing cashmere processing, e.g. crimp frequency, staple structure and fibre colour.

SHEEP AND GOAT PRODUCTION [S] Paper S5.3

Cashmere production on Norwegian goats
L.O. Eik*, T. Ådnøy and N. Standal, Department of Animal Science, Agricultural University of Norway, P.O. Box 5025, N-1432 Ås, Norway.

The potential for production of down fibre (cashmere) on dairy goats, cashmere goats, and crosses in Norway is investigated.

In the current production system, milk is the major product of the goat industry and goat kids not needed for replacement are usually disposed of right after birth. Kidding normally takes place in February. The dairies receive peak deliveries of milk from April to September and only small quantities during mid winter. An even distribution of high quality milk is essential for the production of fresh goat cheeses. The study shows that changing time of kidding to spring combined with suckling of goats during months of surplus would improve annual distribution and quality of goats milk and increase farm incomes through production of kid meat and cashmere.

Norwegian dairy goats produce moderate quantities of down. Cashmere might be included the national breeding program to give added income and to keep goats warm in cold and wet weather. Does not needed for breeding purposes may be crossed with cashmere bucks and pure cashmere goats could be used for fibre and meat production. Encroachment of former grazing land is a problem on the Norwegian countryside. Mixed grazing with sheep and goats could become a cost efficient management method for such areas.

Paper S5.4

Genetic parameters for wool traits in Finnsheep
M-L.Puntila*[1], A.Nylander[1] and E-M .Nuutila[2], [1]Agricultural Research Centre of Finland, Animal Production Research, FIN-31600 Jokioinen, [2] Department of Animal Science, P.O.Box 28, FIN-00014 University of Helsinki.

Entire data was obtained from the flocks of the Fine Finnwool-project (EU 5B program) for two years (1997-98) and from the Finnsheep nucleus flock (1986-98) and contained the following traits; fineness grade (FG), crimp frequency/3cm (CF), staple length (SL), density (DS), staple formation (SF), lustre (LT), fleece uniformity (FU) and fleece weight (FW). For 180 wool samples fiber diameter (FD) was measured by the Optical Fibre Dia- meter Analyser (OFDA), which was introduced in 1998. Variance components were esti- mated by several multitrait analysis (AI-REML) using different subsets of the data, depending on the involved traits.The fitted animal model contained birth/rearing type, age of dam, sex and flock by year as fixed effects, age/weight as covariates and the random
animal effect. Heritability estimates for wool quality traits ranged from 0.16 to 0.50. For FW estimated heritability was 0.32. Genetic correlation between FG and CF was highly positive, whereas both traits' genetic correlation with SL were highly negative. Genetic correlations for FW with SL and FG were moderately high. Subjectively scored traits as DS, FU, SF and LT showed lower association with related wool traits. The average fibre diameter was 25.6 µm (CV 8.2%), 26.2 µ(CV 7.6%) for white and coloured lambs, respectively. Phenotypic correlation between FD and CF was - 0.43. Next step will be the estimation of breeding values for the most important traits.

SHEEP AND GOAT PRODUCTION   [S]   Paper S5.5

Main aspects of fine wool production in Hungary
S. Kukovics[1] and A. Jávor[2]. [1]Department of Sheep and Goat Breeding, Research Institute for Animal Breeding and Nutrition, Gesztenyés u. 1. Herceghalom, 2053 Hungary, [2]Department of Animal Science, Debrecen University of Agriculture, Böszörményi út 138. 4032 Debrecen, Hungary

Considering the Hungarian sheep breed structure Hungary might be in a very exceptional position in fine wool production. More than 90% of the total sheep population are belong to Merinos, and the dairy, meat as well as ancient breeds represent only a couple of %, respectively. At the beginning of the 1970's the wool gave more than 70% of the total income in sheep industry. By the second half of the 80's the relative value of the wool decreased to 25%, and nowadays the wool could give only 5-8% of the gross income in sheep sector.

As one of the consequences of the political changes in the last decade the wool processing industry almost disappeared from Hungary and the fine Merino wool was exported mainly as raw grease wool.

The grease wool prices (85-95 cent per kg in 1998) were covering the shearing cost and gave limited income but from management point of view the wool became as a necessary wrong one in sheep industry. However, the wool quality was still the basic condition in determining estimated breeding value with Merinos, not so many interest were paid to fibre diameter during the 90's.

The changes of wool traits and prices as well as the role of wool products on the Hungarian market are evaluated in the paper.

Paper S5.6

The production, wool trade and wool quality in Poland
A. Radzik-Rant, D. Sztych, R. Niżnikowski*. Department of Sheep and Goats Breeding, Warsaw Agricultural University, st. Przejazd 4, 05-840 Brwinów, Poland.

The changes in the sheep population and the wool production since the years before Second World War to the present period have been analysed. This analysis contains also the wool production per sheep and changes of the breed structure adjusted to desire purpose.

The level and principles of the wool sale and haw, the domestic clip meet requirements of the industry have been presented. The particular attention has been paid on the considerable achievements in the clip preparation and in the improvement of objective measurements of wool quality. On example of the auction system sale of wool during 1990-1994 years the practical application these achievements was showed.

The wool quality of the Polish sheep breeds and sheep breeds imported to Poland was recorded on the base of the exact studies conducted in our unit. The worsening of wool quality, caused by the change dual meat-wool purpose to meat purpose was indicated, too.

The main way of utilisation of domestic wool before the change of the Polish economy has been reported. The perspectives and other possibilities of utilisation of domestic wool considering the present situation were also defined.

SHEEP AND GOAT PRODUCTION [S]   Paper S5.7

Speciality wool production in Europe
M. Merchant*[1] and A.J.F. Russel[2]. [1]*Macaulay Land Use Research Institute, Craigiebuckler, Aberdeen, AB15 8QH, UK,* [2]*Newton Bank, Frankscroft, Peebles, EH45 9DX, UK.*

The social and environmental impact of keeping approximately 120 million sheep in Europe is enormous. These sheep are bred primarily for the production of meat and milk but new initiatives are underway to increase the value of the wool to levels where it makes a significant contribution to the economic viability in a range of production systems. Systems specialising in wool production are potentially suited to extensive systems of management in some of the less favoured areas of Europe where nature conservation objectives require to be taken into account in the development of multi-objective systems. Success will depend on appropriate management protocols, the production of high quality wool from superior fine and speciality wool genotypes, the establishment of grading and marketing structures and the production of a diverse range of high quality products. To achieve this, research, organisation and commitment will be required. Wool quality is the most important criterion. This involves not only relating the characteristics of the wool produced to the end product, but also taking into account the efficiency of processing. In general diameter is the main determinant of quality and the value of the fibre increases exponentially below 21mm. Other quality traits which need to be considered include colour, medullation, fibre length and tensile strength, yield and degree of contamination. Improvements in wool quality can be achieved through specifically designed cross-breeding and selection programmes such as those in progress with the Bowmont sheep in the UK, the White Merino in Portugal and the Finnsheep of Finland, Sweden and Denmark.

Poster S5.8

Economic analysis of hair goat breeding farms in Antalya province in Mediterranean region in Turkey
Ilkay Dellal*[1], Ahmet Erkus[1]. [1]*Department of Agricultural Economics, Faculty of Agriculture, University of Ankara 06110 Ankara, Turkey.*

In this research, hair goat breeding farms, their assets and annually farm results were examined in Antalya province, in Mediterranean region in Turkey. The data were collected from selected 89 sample farms by random sampling methods in November and December 1996, and these selected farms divided into 4 size groups according to goat flock size as 1-10 Large Animal Unit (LAU), 11-25 LAU, 26-50 LAU and 51 and more LAU.

According to research results, hair goat breeders and their family live in scrub, brush, mountain and forest area, 64,04% of them was semi-nomad. The number of hair goats for per farm was 221,63 head, lactation period was 95,18 days, economical life was 6,92 year, pasture days were 12 month. 20,22% of total farms was landless farms, the area for per farm was 39,47 decare. In the small farm size, fixed assets had more portion of total assets, in the big farm size operating assets had more portion. And in all groups net worth had the more portion into liliabilities and net worth. Into total gross production value, animal production and hair goat production values had the biggest portion in all groups.

SHEEP AND GOAT PRODUCTION   [S]   Poster S5.9

Histomorphological characteristics of the skin structure of grey blueKarakul lambs from Talassky sheep stud
G.S.Dunayeva, Kazakh State Agrarian University, 480072 Alma-Ata, Seyfullin St. 534 - 75, Kazakhstan, CIS

Karakul quality as well as the quality and quantity of wool products are first and foremost defined by the inheritance characteristics of skin histostructure, fixed skin samples have been taken for biopsy from the sacrum of the first class grey blue lambs of jacket type, middle curl (3 ewe lamb groups from Ilyich breeding farm; 3 ewe lamb groups and 3 lamb groups from "Talassky" sheep stud, Kazakhstan) obtained from different variants of selections for Karakul sheep colour.

Talassky SS grey blue Karakul lambs of jacket type, are distinguished by the following characteristics: high absolute indexes of piliar layer (Xd = 140.1 - 403.8 mkm; η^2_x = 0.3 - 0.7) with their absence throughout the general thickness of skin, depth of hair follicles (Xd = 233.8 mkm - 242.7 mkm, η^2_x = 0.6), absolute (mkm) and relative (%) margins (amplitude) of sweat and adipose glands location; 3-folded indexes of the diameter of the secretory region of sweat glands.

PIG PRODUCTION   [P]

Paper P2.1

The nutritional significance of "dietary fibre" analysis
K. E. Bach Knudsen[*1]. [1]*Department of Animal Nutrition and Physiology, Danish Institute of Agricultural Sciences, Research Centre Foulum, P.O. Box 50, DK-8830 Tjele, Denmark.*

The term "dietary fibre" (DF) is in most recent animal literature used for cell wall or storage non-starch polysaccharides (NSP) and lignin. DF can be measured as soluble, insoluble and total DF by enzymatic-gravimetrically methods or as soluble, insoluble and total NSP by enzymatic-chemical methods and lignin by gravimetry. NSP comprise 70-90% of the plant cell wall, while the remaining parts are lignin, protein, fatty acids, waxes etc. Plant cell wall NSP is a diverse group of molecules with varying degrees of water solubility, size and structure, which may influence the digestion and absorption process to a variable degree. The action of NSP in the stomach and small intestine is essentially a physical one in which the plant cell either acts as barrier to the release of nutrients or increases the viscosity of the liquid phase, and slow and restrict their absorption. Studies with pigs show that 80-90% of fed NSP is recovered in ileal effluents; the remaining being lost due to microbial degradation in stomach and small intestine. The NSP will be degraded to a variable degree in the large intestine by anaerobic fermentation. The degradation of NSP in the large intestine depends of the degree of lignification, solubility and structure of the polysaccharide. The degree of lignification and water solubility provides important information's about the degradability of NSP in the large intestine, while the effect of NSP on the digestion and absorption processes in the small intestine is more difficult to predict from any of the chemical parameters measured.

Paper P2.2

Role of dietary fibre in the digestive physiology of the pig
Caspar Wenk, Animal Sciences, Nutrition Biology, ETH Zürich, Switzerland

Dietary fibres are usually defined as the sum of plant polysaccharides and lignin that are not hydrolysed by the enzymes of the mammalian digestive system (non starch polysaccharides and lignin). The amount and composition of the dietary fibres varies over a wide range between but also within feedstuffs and a sharp distinction between dietary fibres and starch is far from easy . Therefore the analysis as well as the physiological function of that fraction in the digestive tract of the pig can considerably vary. Furthermore the age of the pig interacts with the digestive processes and there is a adaptation over time of the animal to fibrous diets.
Dietary fibres are generally considered as a fraction with a low energy content. This diluting effect of the diet is used to increase the feed intake in periods with low performance of the animals.
Dietary fibres reduce in the upper and increase in the lower digestive tract the transit time and therefore decrease the digestibility of almost all nutrients and energy. They increase on the other hand microbial growth in the gastrointestinal tract. This can lead to an increased excretion of nutrients in the faeces.
Finally dietary fibres beneficially influence the pigs wellbeing and health. Fibrous feedstuffs give the pigs the opportunity to chew the feed over a longer time. More short chain fatty acids are produced and undesired micro organisms excluded. A regular peristaltic avoids the possibility of constipation.

PIG PRODUCTION [P] Paper P2.3

Effect of dietary fibre on the energy value of feeds for pigs
J. Noblet* and G. Le Goff. INRA, Station de Recherches Porcines, 35590 St-Gilles, France

Dietary fibre (DF) is an inevitable component of organic matter in pig feeds since it is present in most ingredients and to a high extent in by-products (wheat bran, for instance) or forages which are commonly used. In growing pigs, digestibility coefficients of DF average 40 to 50% but they range from about 0% in high lignin and water-insoluble DF sources (wheat straw, for instance) to 80-90% in fibre sources with high pectin or water-soluble DF levels (sugar beet pulp or soybean hulls, for instance). But, even it is partly digested, DF provides negligible amounts of digestible or metabolisable energy to the growing pig in connection with increased endogenous losses and interactions between DF and other dietary components. Digestive utilisation of DF improves with body weight of the pig, the improvement being dependent on the DF botanical origin. Consequently, DF has a positive contribution to energy supply in adult sows. Digestion of DF is also associated with energy losses as methane. The efficiency of utilisation of metabolisable energy for net energy is poorer when it originates from DF (50 to 60%; 80% for starch). Finally, the actual contribution of dietary fibre to energy balance of the pig can be affected by climatic conditions (heat increment of DF is used for thermoregulation) or changes in behaviour of pigs (lowered physical activity with higher DF supplies).

Paper P2.4

Effect of dietary fibre on ileal digestibiliy and endogenous nitrogen losses
W.B. Souffrant. Department of Nutritional Physiology "Oskar Kellner", Research Institute for the Biology of Farm Animals, Justus-von-Liebig-Weg 2, D-18059 Rostock, Germany

Dietary fibre, by definition, is a heterogeneous mixture of structural and non-structural polysaccharides and Lignin. A large number of investigations were carried out to study the effect of dietary fibre on digestibility and endogenous losses in pigs. Most of the authors reported that fibre content of the diet can impair apparent ileal digestibility of dietary nutrients, endogenous nitrogen and amino acid secretion and losses. Particularly the investigations within the last years have shown that the effect of dietary fibres differ with the source and nature of the fibres and relates to their chemical composition as well as to their physico-chemical properties. The effect of pure cellulose is rather low in contrast to other types of fibre (i.e. hulls, bran, inner fibres or pectin). In our experiments the endogenous nitrogen losses were almost twice higher after feeding barley inner fibre instead of barley hulls (331 vs 180 mg N per 100 g DMI). It is difficult to ascertain which physical or chemical properties are responsible for the observed effects. The data presented in this review illustrate that dietary fibre solubility, viscosity, water-holding capacity and chemical-binding of the final feed seems to be the most important factors influencing nutrients ileal digestibility and endogenous losses in pigs. The latter is also affected by protein sources.

PIG PRODUCTION   [P]

Paper P2.5

Effect of dietary fibre on the behaviour and health of the restricted fed sow

M.C. Meunier-Salaün*[1], S.A. Edwards[2] and S. Robert[3]. [1] INRA Station Recherches Porcines 35380 Saint-Gilles, France, [2] University of Aberdeen, 581 King Street, Aberdeen AB24 5UA, UK, [3] Research and Development Centre, 2000 Road 108 East PO Box 90 Lennoxville J1M 1Z3, Canada.

The typical restricted level of gestation feeding, whilst adequate to maximise health and economic performance, might not fulfil behavioural needs of the sow. Hunger and frustration of feeding motivation have been linked to the occurrence of stereotypic activity, and accentuate aggression and feeding competition in group-housing systems. Incorporation of fibre in diets to increase bulk, without changing the daily dietary energy content, has been shown to result in at least a doubling of eating time, a 20% reduction in feeding rate, a 30% reduction in operant response in feed motivation tests, a reduction of 7-30% in stereotypic behaviour, and a decrease in general restlessness and aggression. Results suggest a reduced feeding motivation, which is only effective if nutrient requirements are met. Investigations of circulating glucose, insulin and volatile fatty acid levels in sows fed fibrous diet indicate a more constant nutrient absorption and greater microbial fermentation in the gut, which should increase satiety. There is inadequate information on effects of dietary fibre on physiological stress and health indicators. In addition to welfare considerations, a range of technical requirements and economic factors must be considered when making decisions on the use of fibre during gestation.

Paper P2.6

Dietary fibre for pregnant sows; - effect on performance and behaviour

Viggo Danielsen*[1] and Ellen-Margrethe Vestergaard[2]. Danish Institute of Agricultural Sciences, [1]Department of Animal Nutrition and Physiology and [2]Department of Animal Health and Welfare, Research Centre Foulum, P.O. Box 50, Dk-8830 Tjele, Denmark.

The effect of high fibre levels in diets for pregnant sows was studied in a long-term experiment. Three diets were compared: A control diet (C) based on mainly barley and soy bean meal contained 17.6 per cent dietary fibre, a second diet with inclusion of 50 per cent sugar beet pulp (SBP) contained 44.6 percent dietary fibre and a third diet in which a 50 per cent inclusion of mixed fibre sources (MFS) as dried grass meal, wheat bran and oat hulls resulted in a content of 34.4 per cent dietary fibre. The SBP-diet and the MFS-diet had relatively high contents of soluble dietary fibre (SDF) and insoluble dietary fibre (IDF), respectively. The diets were used in a production trial combined with a behaviour study on sows. The first comprised 120 sows allocated to the three different pregnancy diets for three consecutive reproduction cycles, and the latter included 54 sows representing one gestation period each. Irrespective of diet, sows were fed similar levels of estimated daily net energy. Weight gain of sows during pregnancy was significantly higher for SBP and MFS than for C. Litter size at birth and at weaning was not affected by diets, but weight of piglets at birth was negatively influenced by SBP. Weaning weight of piglets was not significantly different. Eating time for sows was increased and the time spent on rooting was reduced for sows fed SBP and MFS. Aggression in sows was reduced by the SBP-diet.

PIG PRODUCTION   [P]

Paper P2.7

Sugar beet pulp silage as dietary fermentable carbohydrate source for group-housed sows: effects on physical activity and energy metabolism
M.M.J.A. Rijnen*[1], J.W. Schrama[1], M.J.W. Heetkamp[1], M.W.A. Verstegen[1] and J. Haaksma[2], [1]Wageningen Institute of Animal Science, Wageningen Agricultural University, P.O. Box 338, 6700 AH Wageningen, The Netherlands. [2]Institute for Sugar Beet Research, Bergen op Zoom, The Netherlands.

In this study the dose response effect of fermentable NSP on physical activity in relation to metabolic rate in sows was examined. Twelve groups of each six dry sows were fed one of four experimental diets with 0, 100, 200 or 300 g/kg DM sugar beet pulp silage (SBPS). Diets were identical except for starch and NSP content. The sows were group-housed in climatically controlled respiration chambers. The study consisted of a 5-week adaptation and a 1-week experimental period. Heat production, nitrogen and energy balances and faecal digestibilities were measured. Intake of digestible starch and NSP decreased and increased, respectively, with increasing dietary SBPS content ($P < .001$). Heat production and energy retention were unaffected. The exchange of NSP for starch however lowered energy expenditure on physical activity. Sows were more quiet when dietary NSP content increased. Based on heat production data and apparent digestibility of crude protein, crude fat, and NSP, the estimated net energy value of fermented NSP was 15.2 kJ/g. This relative high energy value of fermented NSP was only partly related to the lowered energy expenditure for physical activity, 1.7 kJ/g of fermented NSP. In this study sows were capable of using energy from fermented NSP as efficiently as energy from digested starch.

Paper P2.8

Ileal amino acid digestibility of high fibrous oil cake meals in growing pig
C. Février*[1], Y. Lechevestrier[2] and Y. Jaguelin-Peyraud[1]. [1]Station de Recherches Porcines, I.N.R.A., 35590 Saint-Gilles, France, [2]Central Soya France, BP 108, 78191 Trappes Cedex, France.

The quality of cotton and palm kernel meals, widely used of In African countries, presents very large differences according to the origin, preparation and preservation. Their main characteristic is a high fibre content, also varying in a large range. Five cotton meals (CA to CE) and four palm kernel meals (PA to PD) were tested in several ileal digestibility trials. From CA to CE, the NDF content ranged from 27.3 to 47.5 % /DM and the crude protein content from 38.4 to 50.6 but not correlated with NDF. However, true standardised ileal digestibility (tsID) of lysine varied from 69.7 to 44.7 % and for methionine from 78.7 to 33.3 %, from CA to CE, Digestibility of lysine is highly correlated (.95) with NDF content, more than with ADF content or KOH solubility. From PA to PD the NDF content ranged from 59.2 to 70.7 %/DM and the crude protein content from 15.1 to 17.7 % independently from NDF. The tsID of lysine varied from 86.4 to 18.1 % and for methionine from 86.5 to 49.1 % from PA to PD, variyng also with the association or not with sodium-caseinate, in both circumstances lysine digestibility is also highly correlated (.97) with NDF content. Residual fat content also improves the tsID for palm kernel meals.

PIG PRODUCTION [P]

Paper P2.9

The use of sugar beet-pulp in the diet of heavy pigs
R. Scipioni. *Department of Morfofisiologia veterinaria e Produzioni animali. University of Bologna. Via Tolara di Sopra, 50 - 40064 Ozzano Emilia (Bologna), Italy.*

Data deriving from several studies carried out, since 1988, at our facilities on pressed beet pulp silage (PBPS) in the diet of heavy pigs (160 kg l.w.) are reviewed and discussed. The results can be summarised as follows. Large amounts of PBPS (50% on dry matter basis) in the diet are tolerated by pigs and determine a reduction of gastric lesions. PBPS can be used in practical heavy pig diets at the inclusion rate of 15-20% (on dry matter basis) without any adverse effect both on growing parameters and slaughtering performances. The use of PBPS does not modify the weight losses of raw hams (*Parma ham*) during their seasoning process (lasting for this typical product 12 months); similarly the acidic composition of cured ham fat is not influenced by the dietary inclusion of beet pulps. From an "environmental" point of view, increasing levels (up to 24% on dry matter basis) of PBPS can reduce nitrogen excretion with urine (N-balance trial in metabolic cages). The proved trophical effect on large intestine mucosa and the high quality of PBPS fibrous fractions may explain the positive role of this by-product in pigs feeding and justify the possibility of partially replacing cereals (barley in our case) with PBPS in feed formulations.

Paper P2.10

The effects of sugar beet pulp silage added with vinasse in heavy pigs feeding
G. Martelli*, L. Sardi, P. Parisini, A. Mordenti, R. Scipioni. *Department of Morfofisiologia veterinaria e Produzioni animali. University of Bologna. Via Tolara di Sopra, 50 - 40064 Ozzano Emilia (Bologna), Italy.*

Sixty Landrace x Large White pigs of the initial live weight of about 55 kg were divided in three groups fed as follows: T1 control group in which pigs were fed a "traditional" diet based on cereals and soybean meal; T2 in which pressed beet pulp silage (PBPS) added with 5% vinasse partially replaced barley for 10% on d.m. basis and T3 in which PBPS added with 5% vinasse partially replaced barley for 20% on d.m. basis. Pigs were fed at the rate of 9% of their metabolic l.w. up to a maximum of 3.2 kg per pig per day. Animals were regularly weighed and feed intake recorded to calculate average daily gain and feed conversion rate. Pigs were slaughtered at around 160 kg l.w. and the main qualitative parameters of carcasses (dressing out percentage, muscle percentage, lean and fatty cuts yield) and meat (pH and colour) were collected. Dietary treatment did not significantly influence either growing performances or post-mortem parameters although pigs on T3 diet (20% PBPS) ate less food and showed some tendencial modifications such as a worsening of daily weight gain (548 g/d vs. 563 g/d), an improvement of muscle percentage (50.6% vs. 49.4%) and of lean to fatty cuts ratio (2.06 vs. 1.95).

PIG PRODUCTION   [P]

Paper P2.11

The performance response of growing and finishing pigs fed differing proportions of oatfeed as a dietary fibre source

S.A. Chadd*[1] and D.J.A. Cole[2]. [1]Royal Agricultural College, Cirencester, Gloucestershire GL7 6JS, ,England, [2]University of Nottingham, School of Agriculture, Sutton Bonington, Loughborough, Leicestershire LE12 5RD.

The fibrous by-product oatfeed, was incorporated into eight pig diets at 7.2 to 36.1% inclusion rates. Experimental consideration was given to the interaction of such a 'bulky' material with appetite fulfilment, the regulation of food intake (VFI), together with growth and carcass characteristics. Sixty-four (Large White x Landrace) x Large White progeny were fed *ad libitum* from 25 to 120kg live weight. Diets contained f9.50 to 14.75 MJ DE/kg fresh weight in determined crude fibre (CF) levels of 118.0 to 34.6 g/kg. VFI of all pigs decreased as dietary energy density was increased from 10.25 MJ DE/kg and CF level was reduced from 107.0 g/kg in both growing (25 to 60) and finishing phases (60 to 120kg), (p<\<>0.001). VFI of boars resulted in similar daily intakes of DE across dietary treatments with gilts less consistent. Pigs fed the highest level of oatfeed (360.8g/kg) were not able to compensate for nutrient intake and the diet appeared to represent a physical constraint on feed intake capacity. Overall, results demonstrated the capability of the modern hybrid pig to produce satisfactory daily live weight gains and carcass leanness when offered diets containing a considerable range of fibre levels and taken to heavy slaughter weights.

Paper P2.12

Influence of energy supply on growth characteristics in pigs and consequences for growth modelling

N. Quiniou*[1], J. Noblet[2], J.-Y. Dourmad[2] and J. van Milgen[2]. [1]Institut Technique du Porc, BP 3, 35650 Le Rheu,France, [2]Station de Recherches Porcines, INRA, 35590 Saint-Gilles,France

The aim of the present paper is to investigate the effect of energy supplies on growth characteristics in pigs when protein supplies are non-limiting. An experimental program was carried out at INRA between 1992 and 1994 and involved 60 pigs from three types fed at five energy levels. The effect of energy intake on protein deposition (PD) and lean gain on one hand and on lipid deposition (LD) and fat gain on the other hand were found to be very similar which is logical with the fact that more than 55% of PD and 80% of LD are deposited in lean and fat tissues, respectively. Our results indicate that PD increases with energy supplies according to a linear-plateau relationship, which slope (ßp) and maximum value are influenced by genotype, sex, and stage of growth. From our results and those of literature, ßp is reported to vary between 2.9 and 6.1 g per MJ digestible energy (DE) intake over the 45-100 kg body weight (BW) range. Simultaneously, LD increases linearly with energy level but neither the genotype nor the sex influence significantly the slope of the relationship (14.5 g/MJ DE between 45-100 kg BW). Differences in growth response to energy intake between types of pigs result in an important variation in BW gain and its composition associated to energy intake. Therefore, characterisation of relationships between growth components and energy intake specific to each type of pigs is necessary for modelling response to energy intake.

PIG PRODUCTION [P]

Poster P2.13

Ad libitum feeding of loose housed pregnant sows with fibre rich diets
Gunner Sørensen* and Brian N. Fisker. The National Committee for Pig Breeding, Health and Production, Axeltorv 3, 1609 Copenhagen V, Denmark

Ad libitum feeding with a fibre rich diet compared to restrictive feeding with an ordinary grain based diet for pregnant sows has been tested in Denmark. The ad libitum diet contained 70 per cent Sugar Beet Pulp per kg DM and 3 MJ NE per kg feed. The sows were housed in stable groups (12-20 sows per pen) after service. Both diets contained the same amount of nutrients per energy unit.

The test involved 1900 litters and showed that the number of liveborn pigs per litter was reduced by 0.3 pigs, the pregnancy period was one day shorter and the weight at birth was reduced by 100 gram per pig, when the sows were fed ad libitum. The reason for these statistical significant results is probably that the feed intake during the first 10 days and the last 2-3 weeks of the pregnancy period was insufficient according to the sows energy need.

The low feed intake is presumed to be due to an adaptation of the sows gastro/intestinal system to a larger feed intake and because the intestinal volume in this period was reduced due to the size of the horn of uterus.

Ad libitum feeding in the pregnancy period resulted in an increased feed intake and a lower weight loss in the lactation period - both parameters are statistical significant.

Poster P2.14

Effect of fibre rich diets for loose housed pregnant sows
Brian N. Fisker* and Gunner Sørensen. The National Committee for Pig Breeding, Health and Production, Axeltorv 3, 1609 Copenhagen V, Denmark

A Danish test with a total of 2443 litters showed that litter size can be increased up to 0.7 total born piglets (live + dead born) if sows housed in stable groups are mixed 4 weeks after service. The effect of not mixing the sows in the implantation period may be due to a lower level of stress caused by establishing rang order and/or because the sows were restrictively fed.

A test with approx. 2400 litters was carried out in order to see if it was possible to reduce the aggression between the sows and thereby decrease the effect of mixing by feeding the sows semi ad libitum with a fibre rich diet. In group 1 and 2 sows were restrictively fed with an ordinary diet for pregnant sows. Sows in group 3 were fed semi ad libitum with a fibre rich diet containing 60 per cent dried sugar beet pulp. The sows in group 2 and 3 were mixed just after service, sows in group 1 were mixed 4 weeks after service. All sows were housed in stable groups with 14-16 sows per pen. Reproduction results showed that feeding a fibre rich diet semi ad libitum did not reduce the mentioned effect of mixing the sows in the implantation period. The amount of total born piglets in group 2 and 3 was reduced by 0.3 and 0.5, respectively, compared to group 1.

The same effect of mixing the sows in the implantation period has been seen when the sows where fed ad libitum with a diet containing wet sugar beet pulp but has not been found with sows housed in large dynamic groups fed restrictively.

PIG PRODUCTION   [P]   Poster P2.15

Non-starch polysaccharides from soybean hulls and beet pulp in combination with Ca-sulphate affect digestion and utilisation of nutrients and manure quality in pigs
Z. Mroz, A.W. Jongbloed, K. Vreman, J.Th.M. van Diepen, and Y. van der Honing*. Institute for Animal Science and Health, P.O. Box 65, 8200 AB Lelystad, The Netherlands

Synergistic/antagonistic effects of non-starch polysaccharides (NSP) in combination with different acidifying salts as interactive factors altering dietary buffering capacity (BC), ileal/faecal digestibility (ID/FD) and retention of nutrients and indoor NH_3 volatilisation have been poorly evaluated. Therefore, 12 cannulated and 12 intact pigs of 60 kg initial BW were used to measure the effect of BC (high vs low, as obtained by replacing Ca-bicarbonate for Ca-sulphate) and NSP source (tapioca, soybean hulls and sugar beet pulp) in a 2 x 3 factorial arrangement. The ID and FD of N and other proximate nutrients was reduced ($P<\!\!<\!\!0.05$) by adding 25% of soybean hulls (SH) and beet pulp (BP), but not by adding 0.54% of Ca-sulphate. However, the amount of urinary N in pigs fed SH and BP was approximately 8% N lower than in those fed tapioca, and therefore, N retention remained similar. This 'shift' of N from urine into faeces in combination with a lowered faecal pH at presence of SH and BP (due to a higher VFA content) resulted in slowing-down the indoor NH_3 volatilisation by 26 and 13%, respectively. Ca-sulphate lowered urinary pH by up to 0.6 units. In consequence, in pigs fed tapioca-based diets, manure pH and, thereby, NH_3 emission were reduced by 0.7 mmol/g N ($P<\!\!<\!\!0.05$). However, despite acidifying urine with Ca-sulphate in pigs fed SH and BP, NH_3 emission from manure was not affected due to a sufficient buffering ('masking') potential of faeces.

Poster P2.16

Ammonia emmision from finishing pigs fed 15 percent pelleted sugar beet pulp
N.M. Sloth*[1] and H.B. Rom[2]. [1] The National Committee for Pig Breeding, Health and Production, Udkjærsvej 15, Skejby, DK-8200 Aarhus N, [2] Research Centre Bygholm, P.O. Box 536, DK-8700 Horsens

160 pigs (53 to 98 kg BW) were allocated to either a diet with 15% pelleted sugar beet pulp or a pure cereal-soyabeanmeal-based diet (control) with similar content of nutrients per energy unit. The experiment consisted of two replicates of airquality and 8 replicates of productionresults. Both diets contained 19% crude protein and 15% digestible crude protein. The feed was given as dry pellets ad libitum.
The results showed a pH-decrease in slurry from 6.9 to 6.6 and a 12.4% reduction in ammonia emmision (1.9 grams/pig/day), but a 9% decrease in daily gain (92 grams) and a 9.5% increase in feed conversion ratio (0.26 Feed Units equal to 0.24 kg feed per kg gain) for the pigs fed sugar beet pulp. As a result of the poorer feed conversion ratio, the loss of nitrogen to faeces increased, which neutralized a small decrease in ammonia emmision in percent of N in faeces. Thus, calculated per kg produced pork, pelleted sugar beet pulp did not reduce ammonia emmision in this experiment.
It is stressed, that this experiment couldn't meassure the impact on production results precisely as the two diet of obvious reasons couldn't be fed in the same house. The impact on production results will be tested in a bigger scale later.

PIG PRODUCTION [P]

Poster P2.17

Impact of de-inking sludge used for bedding on aluminium, copper and HAP concentrations in blood, and in liver, fat, meat and urine of pigs.
C. J. Beauchamp[1], R. Boulanger[1] and G. St-Laurent[2]. *DÈpartements de [1]Phytologie et des [2]Sciences animales, FSAA, UniversitÈ Laval, Sainte-Foy, QuÈbec, Canada, G1K 7P4.*

In the non-cereal producing areas, there is not enought straw produced to fill the amount required for animal bedding. However, pulp and paper mills produce organic residues. Primary sludges consist mainly of cellulose, hemicellulose and lignin fibers. The de-inking sludge is produced by de-inking mills and differs from primary sludge by the presence of paper coating, fiber clay, ink and various chemicals added to dissociate these materials from the pulp fibers. The majority of potentially toxic inorganic and organic compounds are at their detection limit in de-inking paper sludge. Among detected compounds, some mineral elements and organic molecules are detected such as, aluminium and copper, and the polycyclic aromatic hydrocarbons (PAHs). Twelve piglets were grown on de-inking sludge and wood shavings bedding for about 120 days. Half of them were analysed. The presence of aluminum and copper were evaluated in the blood, whereas the presence of PAHs were evaluated in the liver, fat, meat and urine of pigs. The concentration of aluminum in the blood has not been detected at a threshold of 40 µg/l. For copper, the average value in the blood of pigs on a litter of wood shavings was 1,19 mg/l (± 0,08) and 1,16 mg/l (± 0,11) for pigs on litter of de-inking sludge. The presence of PAHs in the liver, fat, meat and urine of pigs grown on the de-inking sludge or wood shavings litter have not been detected.

Poster P2.18

Leistungsstabilisierung beim absetzferkel durch einsatz von nahrungsfasern
H. Münchow, V. Manzke und L. Hasselmann . Fachgebiet Tierernährung, Institut für Nutztierwissenschaften, Landwirtschaftlich-Gärtnerische Fakultät, Humboldt-Universität zu Berlin, Philippstr. 13, 10115 Berlin, Germany

Vorrangig bei frühem Absetztermin stellen Leistungsdepressionen der Ferkel, die mit Störungen im Gastrointestinaltrakt verbunden sind, ein Problem dar. Obwohl bisher durch fütterungsprophylaktische Maßnahmen über den Einsatz von Futterzusatzstoffen (Anti-, Chemo-, Probiotika) eine Verbesserung dieser Situation erreicht werden konnte, ergeben sich aus dem neuerlichen Einsatzverbot einiger dieser Verbindungen Fragen nach deren nutritiver Ersatzmöglichkeit. In Untersuchungen an Absetzferkeln wurde daher die diätetische Wirkung eines fermentierbaren Faserträgers (HCl-behandeltes Strohmehl, 5 - 10 % der Tr.-Subst.) ohne ergotropen Futterzusatz bzw. kombiniert mit einem Chinoxalin-dioxid-Derivat (COD, Dosierung: 100 mg/kg Ferkelfutter) auf einige Leistungskriterien sowie auf mikrobielle Umsetzungsprozesse im Darmtrakt geprüft. Unter zusatzstofffreier Ernährung führte der Faserträger gegenüber der Kontrolle (ohne Faser) zu merklicher Leistungssteigerung und Reduzierung des Durchfallgeschehens. Ebenso stellte sich ein additiver Effekt der Nahrungsfaser unter den Fütterungsbedingungen des COD-Einsatzes heraus. Die erhobenen gastrointestinalen Kenndaten spiegelten positive Wirkungsmechanismen des Faserträgers deutlich wider (pH-Wert-Senkung im Magen, FFS-Konzentrationsanstieg, Verringerung des NH_3-Gehaltes sowie pathogener Keimgruppen im Darm).

PIG PRODUCTION   [P]

Poster P2.19

The use of expeller copra meal in grower and finisher pig diets
M. P. McKeon, M.G. Dore*, A.B.G. Leek and J.V. O'Doherty, Lyons Research Farm UCD, Ireland.

Two experiments investigated the nutritive value of copra meal (copra) for grower and finisher pigs (40-95kgs). Experiment 1 determined the nutrient digestibility of copra, when included at 200 and 400 g/kg in the diet. In experiment 2, productive performance was determined in group fed pigs (n=360) offered diets containing a control diet (0 copra) (T1), 100g/kg copra (T2) and 200g/kg copra (T3) formulated as a direct replacement for barley and 100 g/kg copra (T4) and 200 g/kg copra meal (T5) formulated on a least cost basis. The control and least cost diets were formulated to have similar DE and ideal protein. Nutrient digestibilities were higher for the control diet than the 400 g/kg copra diets ($P<0.05$). The nutrient digestibility of the 400 g/kg copra meal diet was higher in the finisher ($P<0.05$) than in grower period. Gains (kg/day) of 0.886, 0.884, 0.843, 0.862 and 0.897 (s.e.m. 0.020: $P<0.05$), intakes (kg/day) of 2.43, 2.35, 2.16, 2.33, 2.29 (s.e.m. 0.045: $P<0.05$) and FCR of 2.64, 2.57, 2.46, 2.59, 2.47 (s.e.m. 0.071: $P<0.05$) were recorded for T1 to T5 respectively. The inclusion of copra in the diet decreased killout proportion as did the method of formulation. In conclusion, 200 g/kg copra meal can be used in the diet of grower-finisher pigs but its performance will depend on the method of formulation used.

Poster P2.20

Lipoic acid and vitamin C used in rations of pregnant sows
N.A.Nosenko* Siberian Research and Technological Institute of Animal Husbandry, Krasnoobsk, 633128, Novosibirsk region, Russia.

Grain ration in sows of Large White breed was balanced in accordance with the detailed norms of 1985 by adding of protein vitamin-and-mineral additives (PVMA). The difference between PVMA was the following: in the 2-nd group the lipoic acid and vitamin C in ratio 1:19 were added to PVMA two weeks before coupling, then within the first 30 days and the last 14 days of the **pregnancy** period. Within the other days PVMA similar to those used in the control group No.1 were fed to sows. In each group 8 sows were coupled. To the 100-th day of **pregnancy** period the live body weight of sows in the second group was 20,1 kg more in comparison with live body weight of sows in the 1-st group because sows in the second group were more calm and at the similar feed consumption the assimilation of feed was much better. The average period of farrowing was 114,25 days (from 110 up to 116). **At farrow** from every sow in the 2-nd group we had in average 12,62 live piglets, whose safety to the 21-st day of their life constituted 87,1% whereas the similar indices were 10,62 and 85,9 in the 1-st group, correspondingly. To the 21-st day of lactation the milk capability of sows in the 2-nd group was 50, 40 kg and the loss of live body weight constituted 21, 18 kg or 8%; whereas in the 1-st group the same indices were equal, correspondingly, to 42,82; 16,95 kg or 7%.

PIG PRODUCTION   [P]   Poster P2.21

The efficacy of feeding fattening pigs with meal prepared from discarded peanuts
J. Koczanowski, J.B. Pyś, W. Migdał, C. Klocek, A. Gardzińska and A. Siuta
- Department of Pig Production Agricultural University of Kraków, al. Mickiewicza 24/28, 30-059 Kraków, Poland

Growning gilts (weighing from ca. 50 to 96 kg BW) were divided into two groups: control group in which gilts were fed on the balanced concentrate mixture containing 10% of repeseed meal and the treatment group in which gilts were fed on the balanced concentrate mixture containing 10% of meal from discarded peanuts.

This tested by-product was prepared from peanuts which were discarded during the production process (e.g. sorting, scarching or packing) or/and were rejected because of inappropriate colour or they fell down on the production floor.

The addition of discarded peanuts increased significantly the average of the daily gain of growing pigs (803 g vs. 736 g). Moreover, the feeding efficiency was lower in the treatement group (3.2 kg/kg BW) comparing to the control group (3.6 kg/kg BW) but the backfat thickness increased (ca.3 mm) and the muscling was 3% lower in pigs which were fed on the peanuts supplement.

Paper P3.1

Fattening and meat qualities of different purebred and crossbred pigs raised in Lithuania
J. Remeikiene. Department of Animal Breeding and Genetics, Lithuanian Veterinary Academy, Tilzes 18, Kaunas 3022, Lithuania.

The data of 6769 fattening pigs (5117 purebred and 1679 crossbred) have been analyzed. Purebred pigs of 2 local and 5 imported breeds, crossbred pigs of 7 different crosses of Lithuanian White with imported breeds were included. They were fattened and slaughtered in 5 Control Fattening Stations during period of 1993 - 1998. The data included indexes of fattening and meat qualities.

The average means and standard deviations of most important indexes of fattening (feed consumption, daily gain, growing rate) and meat qualities (carcass length, carcass percentage, area of longissimus dorsi, ham weight, average back-fat thickness) have been computed.

The best fattening qualities were established for purebred German Landraces and crossbred Lithuanian White x Duroc pigs. The best meat qualities were established for purebred Norwegian Landraces and German Landraces, for crossbred Lithuanian White x German Large White pigs. Local Lithuanian White pigs had average fattening and meat qualities, just back-fat thickness was quite big (30.74 mm on average).

PIG PRODUCTION [P]

Paper P3.2

Comparison of different pig combinations by using data from Piglog 105
A. Tänavots, T. Kaart. *Department of Animal Breeding, Institute of Animal Science, Estonian Agricultural University, Kreutzwaldi 1, Tartu 51014, Estonia.*

Aim of research was to investigate effect of foreign breeds on carcass quality of local breeds, compare meat traits between different breed combinations. 6538 pigs were tested with ultrasonic equipment Piglog 105 in 1996-1997. Seven groups of breed combinations were used in research: purebred - Estonian Large White (ELW), Estonian Landrace (EL) and Hampshire (H); crossbred - H♂ x ELW♀, ELW♂ x EL♀, EL♂ x ELW♀, ELW♂ x (EL/ELW)♀. The traits taken under observation were testing weight, backfat thickness at last (X1) and 10th (X3) rib, diameter of loin eye (X2) and lean meat percentage (Y). Least-square means were calculated for all breed combinations, years and seasons. Highest backfat thickness and lowest lean meat percentage were calculated for EL (X1=18.72mm, X3=18.09mm, Y=55.70%) and ELW♂ x EL♀ (X1=18.72mm, X3=18.09mm, Y=56.00%). Largest influence on backfat thickness was observed in EL sows. However diameter of loin eye was the same in both breeds. Low backfat thickness, large diameter of loin eye and high lean meat percentage showed H breed (X1=10.75mm, X3=10.77mm, X2=49.56mm, Y=62.42%) and its combination H x ELW (X1=12.98mm, X3=13.89mm, X2=47.76mm, Y=59.81%). Year had significant influence on all traits. Significantly lower backfat thickness, larger diameter of loin eye and lean meat percentage were in autumn. To improve local pig's carcass quality, coloured breeds must use.

Paper P3.3

Heterosis effect demonstrated as an increase of the testes size and an improvement of the semen traits of cross-breed boars between Duroc and Pietrain breeds
R. Czarnecki[1], M. Różycki[2], J. Udała[1], M. Kawęcka[1], M. Kamyczek[2], A. Pietruszka[1], B. Delikator[1], *Department of Pig Breeding, University of Agriculture, Dr. Judyma 10 Street, 71-460 Szczecin, [2]Institute of Animal Science, Kraków, Poland*

This study was based on a total of 262 young boars (77 of D breed; 50 of P breed; 78 ♀D x ♂P and 57 ♀P x ♂D), born and brought up in a single experimental farm. At the age of 180 days they were weighed, the size of their testes was determined as well as the libido parameters, and their semen was evaluated (based on three consecutive ejaculations). The crossbreed boars of both combinations in most cases exceeded the pure-breed boars of both breeds. They were better in the following features: the body weight on day 180 (by 6.28 kg), daily increment up to 180th day (by 45.94 g), surface area of testes (by 8.53 cm^2), testes volume (by 56.72 cm^3), age of the first sampling of semen (by 13.15 days), time to first mounting (by 15.31 s), ejaculation time (by 23.81 s), ejaculate volume (by 20.59 cm^3), percentage of progressive spermatozoa (by 1.31%), percentage of spermatozoa with major defects (by - 4.13%), percentage of spermatozoa with minor defects (by - 3.38%), percentage of spermatozoa with normal acrosome (by 7.56%), activity of AspAt enzyme in the sperm fluid (by - 6.88) and ORT test of the semen (by 3.44). The boars of both crossbreed combinations had similar content of meat in their bodies (58.27% and 58.61%).The present study fully confirmed the better growth rate and better breeding performance of the crossbreed boars, compared to the pure-breed ones.

PIG PRODUCTION   [P]   Paper P3.4

Monitoring of hybrid pig performance in commercial herd thorough the field testation.
M. Šprysl, R. Stupka, J. Èitek, M. Pour; *Department of Pig and Poultry Science, Czech of Agriculture University, Prague, 165 21 Prague - Suchdol, Czech Republic*

Field tests were undertaken in large-scale swine production herd. Their objective was to determine performance level and profitability of 4 genotypes which are being used by those farm on the base of using profitability criteria and determine the importance of the effects affecting individual performance traits. The reproductive performance was determined on the base of 546 litters of sows BuxL mated purebred and crossbred boars. The fattening capacity and carcass value of 4353 offsprings was determined as well. Based on the results found, the following conclusions can be drawn.
- Thanks to asserting of genetic factors on main reproductive traits there is possibility to use a genetic progress gained in the sphere of breeding to the production herd. With regard to reproductive sow performance expressed by number of reared piglets per litter and sow there is necessity to rise it.
- The genotypes slighty affected the level of carcass value which suggested low level of slaughter value traits in mothers and sires positions. The effect of sex is significantly reflected in sales of farm.
- Thanks to higher uniformity of fattened pigs the three-bred combination appears more suitable for the establishment concerned.

The field tests shows whether there are reserves in the large-scale production herds for improvement of environmental conditions and whether a genetic potential of the animals is manifested.

Paper P3.5

Risk factors and genetic variance components of pre-weaning mortality in piglets
R. Röhe* and E. Kalm. *Institute of Animal Breeding and Husbandry, Christian-Albrechts-University of Kiel, D-24098 Kiel, Germany*

Risk factors and variance components of pre-weaning mortality were estimated using generalized linear mixed models. Data were from 12727 piglets born alive of 1338 litters recorded at the swine breeding farm of the University of Kiel from 1989 to 1994. Deviance reductions due to risk factors and their odds ratios were estimated using generalized linear models with binomial errors and a logistic link. Variance components of sire, dam and litter effects were estimated using a logit or probit link function as well as a linear model for which estimates were transformed to the underlying continuous scale.

Highest reduction in deviance was obtained after exclusion of individual birth weight (1206) from the model, followed by year-season (217), parity-farrowing age or interval (58), genotype of piglets (56), sex (39), total number of born piglets (18) and gestation length (16). Odds of pre-weaning mortality was 2.0 times higher in piglets from German Landrace dams than in piglets from Large White dams. Piglets with individual birth weight of less than 1.7 showed a rapid increase in odds ratios of pre-weaning mortality with 2.5, 6.6 and 15.1 for 1.5, 1.2 and 1.0 kg, respectively. Estimates of heritability for pre-weaning mortality on linear observed, transformed underlying, logit and probit scale were low with 0.02, 0.06, 0.07 and 0.07, respectively. Therefore, selection for individual birth weight, the main risk factor for pre-weaning mortality, is expected to improve survival of piglets during the nursing period more efficiently than direct selection for the binary trait itself.

PIG PRODUCTION   [P]

Paper P3.6

Productive characterization of varieties of the Iberian pig branch
C., Barba[1], J.V. Delgado[*1] *and E. Dieguez[2]. [1] Departamento de Genética. Universidad de Córdoba. Av. Medina Azahara, 9. 14005 Córdoba. Spain. [2] AECERIBER. Av. Antonio Chacon, 7. 06300 Zafra. Badajoz. Spain.*

We have developed the characterisation of the diverse varieties forming the Iberian Pig from three point of view : morphological, genetics and productive. In this paper we are presenting the results obtained from the statistical analysis of 18 productive variables grouped in three categories : antemorten preweaning (3 variables), antemorten postweaning (5 variables) and postmortem (10 variables), ones the 7 varieties of the Iberian Pig.
Firstly we have applied an analysis of the descriptive statistist with the purpose of the description of the productive characterises of each variety. Secondly, we have developed an ANOVA using the general lineal to detect homogeneities and differences among varieties. Finally we have developed a multivariate analysis using discriminate and factorial techniques (discriminant analysis, Mahalanovis distance and principal components analysis) with a view to define the relationship among varieties.
Our results have shown a clear specialisation of several varieties on different productive aspect, some have growing specialisation in early periods while other have shown better qualification for the production of the expensive pieces of the carcass such us arms, ham and Lomb.

Poster P3.7

The evaluation of the level of utility traits and breeding steps in pigs by means of correlation analysis
J. Fiedler, L. Houška, J. Pavlík, J. Pulkrábek *
Research Institute of Animal Production, CZ 104 01 Praha-Uhříněves, Czech Republic.

Utility traits and breeding steps were checked in the population of Large White breed. In every of 37 breeding herds the level of reproduction, feeding capacity and carcass value were observed together with an average selection intensity used in the rearing of young breeding animals. The relations between individual utility traits and between utility traits and breeding steps in the herds were evaluated by means of correlation analysis. Closed relations between the litter weight and average daily gain in the gilts ($r=0.42$) and in young boars ($r=0.49$) in the period of rearing were observed. The herds with high level of average daily gain in gilts in the period of rearing showed high level of carcass traits, too ($r=0.32$ to 0.49) unlike in young boars. Selection intensity however is highly influenced by the level of traits from station test especially by carcass traits and in boars by traits of feeding capacity, too. Selection intensity in young boars is influenced as by carcass value traits from the field test (especially backfat thickness, $r=0.52$ and lean percentage, $r= -0.44$) as by traits from station test especially consumption of metabolised energy ($r= -0.36$) and average backfat thickness ($r=0.40$). It is evident that much higher emphasize in the current way of selection is put on the traits of feeding capacity and carcass value from the station test than from field test.

PIG PRODUCTION [P]

Poster P3.8

Relationship between the breeding value of boars and fertility of their sisters.
R. Czarnecki[1], J. Owsianny[1], M. Różycki[2], M. Kawęcka[1], B. Delikator[1], K. Dziadek[2], [1]Department of Pig Breeding, University of Agriculture, Dr. Judyma 10 Street, 71-460 Szczecin, [2]Institute of Animal Science, Kraków, Poland

The study was based on a total of 1000 of young boars representing three genetic groups and 195 their full sisters. The sisters gave birth to three consecutive litters (I to III). This offspring were arranged in a total of 747 pairs (brother-sister). Between day 70 and 180 of their life the boars and sows were reared in the same standard conditions, fed on individual basis, and evaluated in the same experimental farm. At the age of 180 days the young boars had their testes measured and their libido and semen evaluated: volume of testes (555,19 cm^3), time to effective mounting (199,66 s), number of leaps to effective mounting (1,21), time of ejaculation (253,33 s), ejaculate volume after filtration (126,79 cm^3), concentration of spermatozoa in cm^{3}x10^6 (187,05), total number of spermatozoa x 10^9 (22,28), spermatozoa with major defects (13,63%), spermatozoa with minor defects (13,49%) and with defects of acrosome (7,26%), average number of piglets born with third first litters (9,83). Very low values of the correlation coefficients were obtained between the size of testes, parameters of libido and traits of the semen of the boars and fertility of their sisters (r_p = from 0.01 to 0.085). The only higher, positive correlation values were observed between the overall semen volume (and the volume after filtration) of the boars-and the number of piglets delivered in individual litters by their sisters (r = 0.143** and 0.140**).

Poster P3.9

Perinatal mortality in the pig in relation to genetic merit for piglet vitality
J.I. Leenhouwers*[1], E.F. Knol[2] and T. van der Lende[1]. [1]Animal Breeding and Genetics Group, Wageningen Institute of Animal Sciences, Wageningen Agricultural University, P.O. Box 338, 6700 AH Wageningen, The Netherlands, [2]Institute for Pig Genetics, P.O. Box 43, 6640 AA Beuningen, The Netherlands.

Data were collected on a nucleus pig farm located in Brouennes, France. Records of 123 purebred litters, produced by 113 sows (parity 1-8) from two lines were used. All these litters had known breeding values for piglet vitality. Perinatal mortality was divided in stillbirth and mortality of live-borns until first check up after parturition (£ 12 h after birth). Stillbirth was defined as piglets which showed no signs of decay and were found dead lying behind the sow at first check up after parturition. Autopsy was performed on stillborn piglets to distinguish piglets which died before, during or immediately after parturition.
The statistical model consisted of fixed effects line, parity and breeding value for piglet vitality and covariables average birth weight of the litter and variation in birth weight within the litter. Significance was tested by logistic regression, using stepwise elimination of non-significant (p>0.05) effects. Litters with a high breeding value for piglet vitality had significantly lower mortality immediately after parturition and significantly lower mortality of live-borns. These results show that mortality occurring in the first hours after parturition might have a genetic basis.

PIG PRODUCTION   [P]

Poster P3.10

Reproductive characteristics of different types of Large White boars bred in Siberia
P.G. Kharchenko* Department of Pig Breeding, Siberian Research and Technological Institute of Animal Husbandry, Krasnoobsk, 633128, Novosibirsk region, Russia.

Reproductive characteristics of different types of Large White boars belonging to different zonal types (European, Siberian) bred at the breeding farm "Kudryashovskoe" (Novosibirsk region) were studied. Fertilization of sows of both types took place at the same period. By the sperm of 29 boars of Siberian type, 106 sows of the same zonal type were fertilized. 26 boars of European zonal type were crossed to 87 sows of the same type. The degree of fertilization by the boars of Siberian type was 80,19%, litter size 11,70; still birth 1,10; litter birth weight 14,7 kg. In European zonal type, degree of fertilization was 78,16%, litter size 11,8; still birth 1,23; litter birth weight 14,5 kg. Under similar exploitation of boars of different types, the fertilization of sows of Siberian zonal type significantly (P(0,05)) exceeded that in European zonal type. It may be explained by better adaptability the animals of Siberian type to Siberian environment. The other parameters were shown to be similar in Siberian and European zonal types.

Poster P3.11

Field testation - a source of enhancing of hybrid pig performance in commercial herds
R. Stupka, M. Šprysl, J. Èítek, M. Pour. ; Department of Pig and Poultry Science, Czech of Agriculture University, Prague, 165 21 Prague - Suchdol, Czech Republic

The objective of this work was to assess performance of (PCxL)xSL and (PCxLW)xSL crossbred combinations in a large-scale operation using a field test. The reproduction performance was assessed in 35 resp. 30 litters. During period of growing and finishing 213 animals of both combinations has been monitored and carcass indicators were recorded. In order to examine more precisely the phenotype values of performance of genotypes used, and the effects affecting this one, linear models with fixed and random effects were used as well as profitably formula for determine the profitability of used genotypes.
Based on the results found, the following conclusions can be drawn.
– Reproduction performance of the both combinations was nearly the same. This, except the length of farrowing interval, is not affected by genetic effects. This situation indicate the mistakes in the husbandry practice in swine category inside the operation.
– As regards growth, fattening and carcass indicators in relation to genotype, there is no significant differencies in pigs performance except the weight achieved, which is obviously caused by low number of animals in the trial.
– (PC x BU) x SL combination had better results in carcass value assessment, especially as regards lean meat share.
From the economic point of view, the genotype of (PC x BU) x SL pigs appears more suitable for the establishment concerned.

PIG PRODUCTION   [P] Poster P3.12

Genetic correlations between weight of tissues from different carcass cuts of pigs
Barbara Orzechowska, Marian Różycki
National Research Institute of Animal Production, 32-083 Balice, Poland

Genetic correlations were found between the weight of meat and all carcass cut except ribs. No statistically significant correlations were found between the weight of meat of ham and belly, and loin and belly. The other coefficients of genetic correlations were high and ranged from $r_G=0.407$ to $r_G=0.739$

Loin fat is highly correlated genetically with fat of neck ($r_G=0.463$), shoulder ($r_G=0.555$) and ham ($r_G=0.335$). The weight of fat of neck is correlated with the weight of fat of ham muscle ($r_G=0.682$), leg ($r_G=0.564$), belly ($r_G=0.650$) and ribs ($r_G=0.452$). Significant correlations were also found between fat content of ham and shank ($r_G=0.504$) and between ham and belly ($r_G=0.538$).

The weight of fat of belly was significantly correlated with the weight of fat of shoulder ($r_G=0.458$), shank ($r_G=0.565$) and ribs ($r_G=0.509$).

Genetic correlations for the weight of bones were observed between four carcass cuts, i.e ribs, shoulder, neck and ham. High genetic correlations were found from all combinations of these traits, ranging from $r_G=0.412$ to $r_G=0.627$. Correlations between the content of bones were also found for shank and shoulder ($r_G=0.462$) and shank and ham ($r_G=0.646$).

Poster P3.13

The age dynamics on selected hematological and immunological parameters of different breed types of pigs
J. Buleca,[1] J. Szarek,[2] I. Mikula[1], L'.Tkáčiková[1] and J. Mattová[1]. [1]University of Veterinary Medcine, Komenského 73, Košice, Slovak Republic, [2]University of Agriculture in Krakow, Poland

The selected blood parameters of the hematological and immunological profile of various breed and their cross-breeds were investigated: Large White (LW), Vietnam (V) and Large White x Landrass (LWxL). The statistically significant differences were found ($P < 0.05$, $P < 0.01$) in the parameters of the hematological and immunological profile in favour of the LW breed and cross-breeds LWxL in comparison with Vietnam breed. The values of the metabolic activity of the INT test are statistically significant ($P < 0.01$) on day 42 and 56 in favour of Vietnam pigs compared to the LW breeds and their cross-breed.

PIG PRODUCTION   [P]

Poster P3.14

The profitable Iberian pigs crossed effect at reproductive and maternal ability traits
J. Benito *, C. Vázquez-Cisneros, J.L. Ferrera, C. Menaya y J.M. García-Casco
Sección de Porcino, Servicio de Investigación y Desarrollo Tecnológico
Junta de Extremadura. Apdo 22. Badajoz. Spain

The reproductive parameters and maternal ability was evaluated in three Iberian Pig genotypes: *Valdesequera* and its 50% crossed with the *Torbiscal* and *Guadyerbas* strains. 55 sows per group along two consecutive farrows were involved in that trail. About reproductive characters, the following data were analysed: Fertility, Prolificity, Total born alive and neonatal mortality. Considering maternal ability characters, piglets were weighted from birth up to 21 days of live. Although first farrow did not show any statistical differences about reproductive characters, these are sensible in the second one. Higher total born and total born alive numbers up to 1.4 and 0.8 piglets respectively were obtained in *Valdesequera* x *Torbiscal* sows respect to pure strain. However, the stillbirth decreased 0.5 piglets in shorter litters as it was established in the pure strain. There are not statistical significance in any farrow about the Neonatal mortality character. The *Valdesequera* x *Torbiscal* genotype show higher growing pig index, around 0.100 and 0.500 Kg/piglet at birth and 21 days respectively. These preliminary results show the profitable heterosis effect for the Iberian Pig breed, mainly in prolificacy and maternal ability characters.

Poster P3.15

The genetic differences in the pigs reaction to ultrasound treatment
T.N. Nicolaeva*, V.L. Petukhov, O.S. Korotkevich Institute of Veterinary Genetics of Novosibirsk Agrarian University, 160 Dobrolubov Str.,630039 Novosibirsk, Russia

The ultrasound influence (0.1-0.2 W/cm^2 intensity for 1 minute, on the keel bone area) on the organism of 1.5 month piglets was investigated. The aim of the study was to assess the genetic determination of the pig reaction to ultrasound. Four family groups of littermates were formed. Every group was divided in two subgroups (the experimental and the control). Among the measurements characterising the reaction of pigs to ultrasound influence, the lysozyme level in blood serum was analyzed. The significant difference in the average lysozyme level between the experimental and control littermates of one family group was found (P<0.01). The lysozyme level in the experimental and control subgroup piglets was 3.0 ± 0.099 % and 16.4 ± 1.85 % relatively. In other family groups the similar exceeding of the lysozyme level was not found. Thus, the genetic determination of the pigs reaction to the ultrasound treatment was established.

PIG PRODUCTION   [P]

Paper P3.16

Feeding strategies for meeting lysine requirements of the grower-finisher pig
A.B.G. Leek* and J. V. O'Doherty. Lyons Research Farm, UCD, Ireland

Thirty six males and 36 females, were used to examine the merit of phase feeding compared with a constant lysine:energy ratio from 38kg to slaughter. Treatments were (**SR**) single ration (0.86g lysine/MJDE from 38kg to slaughter), (**HS**) high split (0.86g lysine/MJDE for 35 days (d) and 0.74g lysine/MJDE to slaughter) (**MS**) medium split (0.8g lysine/MJDE for 35d and 0.68g lysine/MJDE to slaughter), (**LS**) low split (0.74g lysine/MJDE for 35d and 0.62g lysine/MJDE to slaughter), (**HP**) high phase (0.86g lysine/MJDE for the 17d, 0.8g lysine /MJDE for 18d, 0.74g lysine/MJDE for 14d and 0.68g lysine/MJDE to slaughter) and (**LP**) low phase (0.8g lysine/MJDE for the 17d, 0.74g lysine/MJDE for 18d, 0.68g lysine/MJDE for 14d and 0.6g lysine/MJDE to slaughter). Growth of 898, 796, 790, 743, 814 and 780 g/d (sem = 19.5 g/d, $P<0.001$) and FCR of 2.54, 2.62, 2.70, 2.78, 2.58 and 2.77 (sem = 0.050, $P<0.01$) were recorded for treatments SR, HS, MS, LS, HP and LP resp. Pigs fed the SRC diet had a higher protein deposition rate ($P<0.05$). There was an interaction between treatments and sex ($P<0.05$). Boars fed SRC and HS diets had higher growth rate than gilts ($P<0.01$). There was no difference in growth rate between sexes fed the remaining diets.

Paper P3.17

Relationships between *in vivo* conformation measurements and the composition of growing and finishing pigs
D.J.Abrutat*[1,2], C.P.Schofield[2], J.D.Wood[1], A.R.Frost[2] and R.P.White[2]. [1]Division of Food Animal Science, Department of Clinical Veterinary Science, University of Bristol, Langford, Bristol BS18 7DY, UK. [2] Bio-Engineering Division, Silsoe Research Institute, Wrest Park, Bedford, MK45 4HS, UK.

The conformation or physical shape of a pig can differ markedly both between and within breeds. Pigs with good conformation have well rounded hams and thick loins. Previous research links conformation to compositional differences in the levels of muscle, fat, bone and other effects such as nutrition and environmental factors. Trials in which *in vivo* conformation measurements were taken using Video Image Analysis with a feeder mounted camera system have shown distinct contrasts between the development of specific body areas of 36 gilts of two extreme conformation genotypes. Pureline Large White and Large White/Pietrain cross gilts were grown from weaning to 100kg slaughter weights on an *ad libitum* diet. Preliminary results indicate that levels of subcutaneous fat in the hams of the Large White/Pietrains were actually significantly higher, 14.6%, than that found in the pureline Large Whites, 11.8% ($P<0.05$). Lean percentage in the ham was greater in the pureline Large Whites, 65.6%, than the Large White/Pietrains, 62.4% ($P<0.05$). Correlating dissection data with *in vivo* conformation measurements could make it possible to grade carcass quality and predict lean and fat content before slaughter.

PIG PRODUCTION   [P]   Poster P3.18

Low phytic acid barley for growing pigs: pig performance and bone strength.
T. L. Veum*[1], D. R. Ledoux[1], and V. Raboy[2]. [1]Department of Animal Sciences, University of Missouri, Columbia, MO, USA; [2]USDA-ARS, P. O. Box 307, Aberdeen, ID, USA.

Crossbred barrows (n=35) averaging 13.5 kg were used to evaluate a low phytic acid mutant barley (MB) containing the lpa1-1 allele compared to a near-isogenic normal hybrid barley (NB) in a 35 day experiment. The MB and NB, respectively, contained 0.35 and 0.35% total phosphorus (tP) and 0.21 and 0.11% phytate free (estimated available) P (aP). The five treatments (T) were: (1) a NB diet containing 0.14% aP and 0.32% tP, (2) a MB diet containing 0.22% aP and 0.32% tP, (3) diet 1 with monosodium phosphate added to increase aP to 0.22% to equal the aP in diet 2, (4) a NB diet containing 0.57% tP, 0.30 aP and 0.65% Ca, and (5) a MB diet containing 0.50% tP, 0.30 aP and 0.65% Ca. Diets 1 to 3 were supplemented with whey protein concentrate, blood cells and ground limestone to bring Ca to 0.50%. Diets 4 and 5 were supplemented with soybean meal, ground limestone and dicalcium phosphate. Pigs were housed in individual pens and fed to appetite. Pig performance, and metacarpal and radius bone breaking strength were similar ($P>0.04$) for T2 and T3, and for T4 and T5. In conclusion, pig performance and bone strength confirmed our estimate of 0.21% aP in the low phytic acid barley.

Poster P3.19

Low Phytic Acid Barley for Growing Pigs: Calcium and Phosphorus Balance.
T. L. Veum*[1], D. R. Ledoux[1], and V. Raboy[2]. [1]Department of Animal Sciences, University of Missouri, Columbia, MO, USA; [2]USDA-ARS, P. O. Box 307, Aberdeen, ID, USA.

Thirty five crossbred barrows averaging 13.5 kg were used in a 35-day experiment to evaluate a low phytic acid mutant barley (MB) containing the lpa1-1 allele compared to a near isogenic normal hybrid barley (NB). The barleys and the five dietary treatments are described in our companion abstract that summarizes the pig performance and bone strength results. The MB reduced ($P<0.03$) fecal P excretion 55% compared to NB (1.8 vs 4.0 g/d) when barley was the only source of phytate in the diet. With soybean meal as the protein source, fecal P excretion was reduced ($P<0.14$) 15% by MB compared to NB (3.7 vs 4.4 g/d). The absorption and retention of P (g/day) were higher ($P<0.05$) for the diets containing MB compared to NB. Calcium absorption was also increased ($P<0.05$) with MB compared to NB in the diets where barley was the only source of phytate. In conclusion, MB significantly reduced fecal P excretion and increased P absorption and retention compared to NB. The absorption of Ca was also increased by MB compared to NB.

PIG PRODUCTION   [P]

Poster P3.20

The performance of large white and local Malawian pigs fed rations based on cowpeas (vigna unguiculata), soyabeans (glycinemax), or pigeon peas (cajanus cajan)
V. Simoongwe[2], J.P. Mtimuni[1], R.K.D. Phoya[1] and C.B.A. Wollny[1]*
[1]University of Malawi, Bunda College of Agriculture, Animal Science Department, Box 219, Lilongwe, Malawi[3]; [2]PDTI, Box 50199, Lusaka, Zambia

Thirty Large White (LW) and 24 local Malawian (LOM) entire weaned male and female pigs of 8 weeks of age were compared based on average daily gain (ADG) when consuming rations similar in nutrient composition and containing roasted cow-peas (ROCO), roasted soybeans (ROSO) or pigeon peas (ROPI) as protein sources. Ten LW and 8 LOM pigs were randomly allocated to ration, replicating twice for each breed. Weights were recorded every week for 15 weeks when LW pigs weighed 60 kg. Average daily gain (ADG) was 181.2 ± 18.7, 406.9 ± 17.5, and 64.1 ± 9.6 for LW, and 180.6 ± 17.8, 312.8 ± 18.7 and 46.3 ± 10.7 g/day for LOM pigs fed ROCO, ROSO and ROPI respectively. Treatment differences within breed were highly significant ($p < 0.001$). Weight gain varied inversely with trypsin inhibition levels. LW pigs fed ROSO grew faster ($p < 0.05$) than LOM, whereas breed performance on ROCO and ROPI rations was similar. The experiment showed that under sub-optimal feeding conditions of locally available cowpeas the local pig performs equally well as the exotic pig. However roasting techniques were ineffective to reduce trypsin content to an acceptable level. The use of pigeon peas in pig rations cannot be recommended using conventional roasting.

Poster P3.21

Performance of local pigs under village conditions in the rural areas of Malawi
C.G. Mulume, C.B.A. Wollny*, J.W. Banda and R.K. Phoya. *Animal Science Department, Bunda College of Agriculture, P.O. Box 219, Lilongwe, Malawi, Africa*

Productive performance of local pigs kept under traditional village conditions under harsh environmental conditions was monitored in the Shire Valley in Malawi, Southern Africa. Fifty-seven farmers were interviewed and reproductive and productive performance of randomly chosen 32 sows and 212 piglets was monitored. All animals were described as indigenous pigs to Malawi. No interventions or extension services were provided. Pigs were kept in small herds of two to three breeding sows with modest or no housing provided in a mixed farming system by smallholder farmers. No interventions were offered during the course of the study. The major constraints according to farmers were feed problems (lack or unavailability of feed - 77% of responses), lack of any extension services (77%) and high incidence of diseases. Season of birth and parity of sow affected ($P < 0.05$) body-weight of piglets at first weighing (1-14 days of age) averaging 1.42 kg (SD 0.52kg). At the age of 50 to 64 days average live-weight was 7.26 kg (SD 2.38kg). 20 weeks after birth the mean live-weight was 19.04 kg (SD 5.65kg) yielding an average daily gain of 140g. Possible interventions may focus on basic changes of management practices resulting in a more efficient and sustainable utilisation of the adapted local pig population in an adverse production environment.

PIG PRODUCTION [P]

Poster P3.22

Suitability of crambe cake and crambe meal in pig feeding
D. Kampf, H. Böhme and R. Daenicke*, Federal Agricultural Research Centre (FAL), Institute of Animal Nutrition, Bundesallee 50, 38116 Braunschweig, Germany

Due to the high erucic acid content *crambe abyssinica* is grown as a renewable resource and there is a potential for feeding the by-products after pressing (cake) or extraction (meal). In the balance experiments and the feeding trial involving 100 pigs in the live weight range 25 - 120 kg, a crambe cake and a meal were used having a residual fat content of 14.4 % or 2.6 % in DM. The crambe oil was found to have an erucic acid concentration of 56 % related to total fatty acids. Glucosinolate contents were found to be 55 or 77 µmol/g DM in the cake or in the meal respectively.

Due to the digestibility of crude nutrients the energy concentration of crambe cake and meal were estimated to be 10.6 or 9.3 MJ ME/kg DM. Inclusion of 10 % crambe cake or 10 % meal in a soybean/cereals based diet resulted in significantly decreased live weight gains 742 g/d or 752 g/d versus 782 g/d in controls. Feed conversion ratio increased correspondingly. Influences on carcass quality as affected by feeding crambe cake or meal were not detected, but a tendency towards reduced carcass dressing percentage was observed, due to the enlarged organs. The increased intake of glucosinolates associated with feeding crambe by-products resulted in enlarged thyroid glands, weighing 27.1 g (10 % cake) or 24.9 g (10 % meal) as compared to 8.8 g (controls). Feeding 10 % cake during the whole growing-finishing period the erucic acid content in backfat was found to be 1.5 % of total fatty acids.

Poster P3.23

Choice and phase feeding methods of growing pigs: use of diets containing yellow lupin (*Lupinus luteus L.*)
J.Falkowski*, W.Kozera and D.Bugnacka, Faculty of Animal Bioengineering, Olsztyn University of Agriculture and Technology, 10-718 Olsztyn-Kortowo, Poland

28-day experiment using 16 crossbred (Polish Landrace x Polish Large White) pigs (23,5 kg average initial live weight) was conducted to evaluate the effectiveness of two methods of feeding. Eight pens of two pigs (barrow and gilt) were assigned to:
(I) conventional phase feeding - C (17,7% crude protein diet) and (II) choice feeding with the use of high - protein diet (21,0% CP) and low-protein diet (14,5% CP). The experimental diets contained 25,0; 44,0 and 8,0% of ground sweet yellow lupin of Polish var. Juno, respectively.

Specification		Phase feeding	Choice feeding
Weight gain	(g d^{-1})	798	864
Feed intake	(kg d^{-1})	1,99	2,09
Feed/gain	(kg kg^{-1})	2,51	2,42
Protein intake	(g d^{-1})	343	360
Protein/gain	(g kg^{-1})	435	419

There was no significant effect of feeding method on pig performance. Choice feeding method was not an effective one under conditions of the described experiment.

PIG PRODUCTION [P]

Poster P3.24

The effect of rapeseed oil and tallow supplement in a diet on the reproductive performance in sows
J.M. Paschma*. Departament of Technology and Livestock Ecology, National Research Institute of Animal Production, 32-083 Balice, Poland.

The purpose of this study was to evaluate the effect of dietary rapeseed oil and tallow supplement on reproductive performance in sows. Three experiments were carried out on 136 Polish Large White multiparous sows whose diets were fat supplemented during late gestation (91 to 110 d.) and lactation. In the lst experiment 4% rapeseed oil, in the 2nd - 4 and 8% rapeseed oil, and in the 3rd experiment - 4% rapeseed oil and tallow were added to the diets of sows in experimental groups. A total of 348 litters were evaluated.

It was found that supplementing the diets with rapeseed oil or tallow during late pregnancy and lactation had no clear beneficial effect on the number of piglets born alive, their growth during lactation or other reproductive characteristics of sows. Fat supplements increased fat content of colostrum and in most of the trials also milk fat content. Tallow was slightly better than rapeseed oil as fat supplements to the sows diets.

Poster P3.25

The influence of feeding methods on the results of fattening in crossbred gilts and castrates
J.Koczanowski, W.Migdał, C.Klocek. Department of Pig Production, Agricultural University of Cracow, Al. Mickiewicza 24/28, 30-059 Cracow, Poland.

The experiments were conducted on 30 crossbred gilts and 30 crossbred castrates a(Polish Landrace x Polish LargeWhite) x `(Duroc x Hampshire). Animals were fat-tened from 30 to 100kg body weight using feed which contained 12,5MJ ME and 158g CP. Number of heads have been divided into two grups. The pigs of the first grup were given limited feeding(2,4kg/day). The pigs of the second group were given at libitum.

Characteristics	Gilts		Castrates	
	2,4kg	at libitum	2,4kg	at libitum
Daily geins(g)	7,14±77,8	740±85,4^x	744±80,1	801±88,4^{xx}
Fat thickness(cm)	1,58±0,22	1,82±0,37^x	1,92±0,30	2,34±0,34^{xx}
Eye muscle area(cm^2)	50,93±5,17	52,80±4,98	48,60±5,25	49,27±5,03
Meat content in carcass cut(kg)	21,60±1,98	21,75±2,01	20,64±1,93	20,92±2,17
Carcass leannes(%)	57,5±5,1	56,8±5,2	59,6±5,26	50,0±5,17
Quantity of carcass in E class(%)	71,4	72,1	28,0	17,7

The obtained results show that during fatten of gilts may be use full feeding whereas castrates must be fatten by using limited feeding.

PIG PRODUCTION   [P]

Poster P3.26

Efficiency of various balanced protein-vitamin supplements to pig ration
V.A. Bekenev* and V.G. Pilnikov. *Department of Pig Breeding, Siberian Research and Technological Institute of Animal Husbandry, Krasnoobsk, 633128, Novosibirsk region, Russia.*

At the breeding farm CAC "Kudryashevskoe" of the Novosibirsk region, a set of experiments was performed in order to study the pig productivity of Large White breed of Siberian selection. The comparison was made between the productivity of pregnant, lactating sows and the weaning piglets fed with rations balanced with protein concentrates produced by Dutch firm "Provimi" and with other balanced supplements according to the standards adopted in Russia. The productivity of sows (N=180) fed during the periods of pregnancy and lactation by the supplements SP and SL was compared to that of the control group (N=195). Litter size equaled to 11,1 piglets (9,9 - in control); the average piglet weight at the age of 21 days and 2 months- 1,42 and 17,8 kg, respectively (in control, 1,31 and 16,0 kg, respectively); the average litter weight at 21 days - 57,1 kg (52,6 kg in control), the viability of piglets - 86,2% (78,3% in control). The cost of the weaning piglet was 17% less in the experimental group than in control. Feeding of the weaning piglets (N=930) for 75 days with the ration enriched by the "starter" and "grower" supplements produced by "Provimi" has enabled to obtain the average daily gain 524 g (380 g in control, N=850). The cost of 1 kg average daily gain was 17% less in the experimental group than in control.

Poster P3.27

Evaluation of the fat score by NIR-spectroscopy
D. Schwörer*, D. Lorenz und A. Rebsamen. *Swiss Pig Performance Testing Station, 6204 Sempach, Switzerland*

The fat score (modified iodine value) as a quality parameter relates to the firmness and the oxidative stability of the fatty tissue. In most slaughterhouses of Switzerland the fat score is introduced as a quality securing instrument and is integrated in most payment systems for pork carcasses. The fat score is usually measured by titration or can be derived from the gaschromatographic evaluation of the fatty acids. As the titration method is time consuming and in the slaughterhouses only batches of pigs can be analysed, the Swiss Pig Performance Testing Station searched for a quick method in regard to selection purposes at the Testing Station.
A total of 1457 fatty tissue samples (backfat, leaf, belly) were used for the calibration with an InfraAlyzer 450 (Bran+Luebbe). The fat score was evaluated by gaschromatography. The calibration range was 42 to 66 fat scores. From 19 filters 4 were used: 1680 nm, 1734 nm, 1778 nm, 1818 nm. Calibration results: $r = 0.93$, SEE = 1.75. The validation of the calibration was done with 505 independant test samples: $r = 0.93$, SEP = 1.76. Results are showing that NIR can be a valuable, supplementary technique in evaluating fat scores for the mentioned purposes.

PIG PRODUCTION   [P] Poster P3.28

A comparison between different methods to measure lean meat content in carcasses of pigs
K. Eilart, A. Põldvere. *Institute of Animal Science, Estonian Agricultural University, Kreutzwaldi 1, Tartu 51014, Estonia.*

In 1998 experiment was carried out to compare basic meat and fattening performance traits of pigs of Estonian Landrace and Large White breed. Goal of this study were to compare meat characteristers of carcasses of the pigs (Ultra-FOM 100 and ZP-method). The data of 451 pigs of the Estonian Landrace breed and of 121 pigs of the Large White breed were used. At slaughter live weight of investigated pigs was from 95 to 105 kg
1. The lean meat content measured by lean meter FOM was on the average 2,3% higher than that obtained by ZP method (P<0.001). Therefore the ZP-method should be recommended to small meat industries.
2. Great variability of meat characteristics creates the necessity for applying lean meter and ZP-method in pig breeding.
3. Pigs shall be selected on a basis of genetical breeding value of the directly determined traits. The correlation analysis showed that besides the lean percentage of a carcass more attention must be paid to measurements of backfat thickness. In case of both methods the lean percentage depends most of all on backfat, less on lean meat measurements.
4. 1 mm increase in backfat thickness reduces the share of lean meat in a carcass by 0.2-0.8% whereas 1 cm^2 increase in loin eye area increases its share by 0.2 %.

Paper P3.29

Piglet mortality in a research herd - causes and effect of parity of the sow
S. Stern* and I. Wigren. *Department of Animal Breeding and Genetics, Swedish University of Agricultural Sciences, Funbo- Lövsta, S-755 97 Uppsala, Sweden*

Piglet mortality was studied with data from a research herd. A total of 18,753 piglets (mainly pure bred Large White) born in 1,732 litter during 1983 -1996 were included in the study. Piglets were individually weighed at birth. For the 3,662 pigs that died within 9 weeks of birth, age, weight and cause of death were recorded by the technical staff. Parity of sow and litter size (total born) were also recorded.
Total piglet mortality, including stillbirths, was 19.5% between birth and 9 weeks. Stillborn was the main cause of death (34.4%), followed by crushed (25.2%), runts (6.6%), alimentary tracts diorders (5.8%), injury (4.8%), weakness (4.3%), infections (2.9%), savaged (2.1%) and others (13.9%).
The mean litter size was 10.8. Litter mortality was 2.1 piglets per litter. However, a large variation in mortality was found between litters, and in 23% of the litters no pigs died. Piglets destined to die before 9 w. were lighter at birth than surviving piglets (1.13 vs 1.44 kg, p<0.001) Piglet birth weight was 1.42 in first parity sows vs 1.47 kg in later parity sows (p<0.001). Crushing and runt-associated were more common causes of death in later-parity litters than in gilt litters, whereas the reverse was true for savaging-related deaths. Overall mortality was slightly higher in later parities but more piglets survived since the litters were larger.

PIG PRODUCTION   [P]   Paper P3.30

Influence of extended photoperiod during lactation on behaviour and performance in domestic sows and piglets
S.G. Deligeorgis[*1] *and P.N. Nikokyris*[2]. [1]*Department of animal breeding and husbandry,* [2] *Department of animal nutrition, Agricultural University of Athens, Iera Odos 75, Athens 118 55, Greece*

The influence of photoperiod stimuli, considered biologically relevant for lactating sows, on nursing and suckling behaviour and piglet physiology and growth was investigated. The two light regimes utilized were 10h light (L) : 14h dark (D) and 20h L : 4h D. A total of 12 sows were used. The behaviour of sows and piglets were video recorded for 24 h on days 0, 1, 5, 14 and 19 post partum. The weaning weight of piglets kept on 20h L was higher than those kept on 10h L. On day 0 the number of milk ejections tented to be higher for sows on 20h L than those on 10h L regime; however, milk ejections per h were more (1.39±0.03 v. 1.28±0.03) during the experimental period and duration of milk ejections was longer (62.1±0.4 v. 60.0±0.4) on 10h L than 20h L regime. The active piglet massaging of the sow's udder was higher before and after (61.1±0.7 v. 56.8±0.7 sec) milk ejection for the 10h L than for 20h L group. Blood profile of piglets was different between the two treatments. Piglets on 20h L regime had more platelets, lower mean hemoglobin concentration and higher numbers of white blood cells (14.6±1.1 v. 10.1±1.2 x10^3). The results of the present experiment indicate that extended photoperiod affect the nursing and suckling behaviour of sows and piglets as well as the growth rate of piglets, which is related to significant alternations of blood profile and their immunological status.

                                                         Poster P3.31

Influence of the number of piglets born on the composition of sow's colostrum milked immediately after parturition
J. Csapó, Zs. Csapó-Kiss, Z. Házas, P. Horn & T. Németh[1]. *PANNON Agricultural University Faculty of Animal Science, H-7401 Kaposvár. P.O.Box 16. Hungary.* [1]*Kajtorvölgye Agricultural Cooperative, Aba, Hungary*

On determination of the content of the first colostrum of 25 Hungarian Large White x Dutch Landrace F_1 sows with respect to dry matter, total protein, whey protein, casein, non-protein nitrogen, ash and macro- and microelements (potassium, sodium, calcium, phosphorus, magnesium, zinc, iron, copper and manganese) it was established that most of the components examined were present in the highest concentrations where litter size was around 10 to 12, and that a decrease or an increase in litter size is accompanied by a decrease in the concentrations of the components studied. The changes, following a maximum curve dependent on litter size, observed in the concentration of these components are attributed partly to the effect of placental lactogen (in the section of the curve indicating increase), and partly to physiological overburden on the mother (in that indicating decrease).

PIG PRODUCTION   [P]

Poster P3.32

Behaviour and performance of sows and piglets in crates and a Thorstensson system
R. H. Bradshaw and D. M. Broom*. Department of Clinical Veterinary Medicine, Madingley Road, Cambridge CB3 OES, United Kingdom

Ten sows were analysed from a group-house deep-straw Thorstensson system (mean parity: 1.90 ± 0.40) and compared with eight sows from crates (mean parity 1.62 ± 0.26). 24h videos allowed a farrowing day to be established for each sow. During the 24h following this day (12.00h to 20.00h) the number of each lying event was noted. Also the number of piglets within 0.3 m of the sow was noted, expressed as a proportion of litter size and two indices were calculated based on 10-minute scans and all lying events. Production parameters were collected throughout. Mean total lying events were greater in the Thorstensson system (crate: 6.50 ± 1.26; Thorstensson: 12.80 ± 1.38; $p < 0.01$) due to increased sternal to lateral movements (crate: 3.62 ± 0.73; Thorstensson: 6.10 ± 0.78; $p < 0.05$) and rollover events (crate: 0.37 ± 0.18; Thorstensson: 2.20 ± 0.55; $p < 0.05$). In the Thorstensson system piglets aggregated closer to the sow at lying (Crate: 0.30 ± 0.07; Thorstensson: 0.54 ± 0.05; $p < 0.05$) and piglet mortality was higher due to crushing (crate: 0.25 ± 0.16; Thorstensson: 1.90 ± 0.48; $p < 0.01$). The Thorstensson system allowed sows freedom of movement but this resulted in increased piglet mortality due to crushing.

Poster P3.33

Sows behaviour after regrouping
K.Zhuchaev*, V.Koshel. Department of Pig Breeding, Novosibirsk State Agrarian University, 630039, Russia

Observations were realized on sows landrace - 1 month post-service, 5 animals in a group, two repetitions - on young sows (Y) and adult (A) sows. The period of recording comprised about 7 hours every time.
The most share of this period sows rested (34-54% at the day of regrouping and 66-70% 4-6 days after it, respectively in Y and A-groups). Stress-effect of regrouping was revealed in increasing of the frequency of feeding and water consumption. Amount of aggressive interactions increased 10-15 times. The general time of feeding decreased to 2-3%. A-sows had higher number of exploratory reactions than Y-sows, that might be result of the stronger stress of young animals.

PIG PRODUCTION   [P]

Poster P3.34

Inducing of exploratory activity in sows after regrouping
V.Koshel, K.Zhuchaev*. Department of Pig Breeding, Novosibirsk State Agrarian University, 630039, Russia

The behaviour of sows (landrace, 1 month post-service) in small groups after regrouping was investigated (5 animals in a group, two repetitions: on young and adult animals). Control groups were housed in stall without bedding. Experimental groups have got about 30dm^3 sawdust in the center of the stall.
The sawdust induced the growth of exploratory activity at the day of regrouping (5,5 times in young sows, 2,5 times in adult sows). Animals in experimental groups spent resting more time (50-57%) in comparison to control groups (34-54%). The number of aggressive interactions decreased among experimental animals to 30-50%. The time of stability fixing in groups was similar: 4 days for young sows and 6 days for adult sows.To that time the frequency of exploratory actions increased in control groups to the level of experimental groups (8 times in young and 1,8 times in adult sows). The gain of the such activity in stall with sawdust made up 24 and 7% respectively.
Inducing of exploratory activity allowed to decrease the agonistic behaviour of sows in considerable degree.

Poster P3.35

The fattening pigs welfare in various housing system
J. Walczak*. Departament of Technology and Livestock Ecology, National Research Institute of Animal Production, 32-083 Balice, Poland.

A group of 240 fattening pigs of synthetic line 990 was examined. Pigs were kept for 90 days in litter (A), non-litter (B), deep-litter (C) and self-cleaning (D) systems. The behaviour, fitness, basal level of ACTH, cortisol, T4, adrenaline, noradrenaline were compared. The twenty-four-hour recordings of ECG, pulse and temperature were taken by means of telemetric method. The lowest basal cortisol concentration was recorded in pigs penned in system D (37.5 nmol/l) ($P \leq 0.05$). There were no significant differences among the adrenaline and noradrenaline concentration. The lowest level of aggression was observed in D pens (6.2 x/day), and the highest - in C ones (13.7 x/day). The highest number of stereotypic behaviour was recorded in B pens (10.2 x/day). Systems D and C proved to be the cleanest, and with the best functional zoning. The highest average pulse level was measured among the B penned pigs (112.6 x/min.), while the lowest in A penned ones (81.8 x/min.), ($P \leq 0.05$). The lowest average skin temperatures were recorded in the B system (34.8^{0}C), the highest in the D system (37.5^{0}C), ($P \leq 0.05$). Some minor irregularities of ECG were noticed among animals in the B system. Summarizing, the best and the poorest relifare the D and B penned pigs had, respectively. The A system was characterized by the standard welfare. The C system did not provide clear enough evidence, for it seems to be suitable for more numerous groups on technical terms.

PIG PRODUCTION   [P]

Poster P3.36

Influence of tropical climate and season on energy balance and chemical body composition in young growing pigs
D. Rinaldo[1], G. Saminadin[1], G. Gravillon[1] and J. Le Dividich[2], INRA, Guadeloupe[1], Saint Gilles[2], France.

Although in the tropics, low performance of growing pigs is said to be related to their difficulty to dissipate increased heat production in a warm environment, their energy balance is unknown. A trial involving 24 Large White pigs fed *ad libitum* from 15.0 kg to 35.2 kg body weight (BW), was conducted in Guadeloupe (F.W.I.) to investigate the influence of climate and season on energy balance. Individually penned animals were subjected to one of the 3 following treatments : tropical climate during the warm season (WTC, average temperature= 27.1 °C) or the cool season (CTC, average temperature= 24.6 °C) in a semi-open room, or constant ambient temperature of 20°C and relative humidity of 75% in a climatic room (CE). Energy balance and body chemical composition, measured by the comparative slaughter method, were not significantly different in CTC and CE pigs. During the warm season, a 6 % reduction in metabolizable energy intake ($P < 0.05$) and a 9 % decrease in heat production ($P < 0.05$) were observed in WTC animals, as compared to CE pigs. Fat deposition was decreased by 12% ($P < 0.10$) in WTC animals relatively to CE pigs, with no significant variation in total energy retained. The net energy content of feed was assessed at 9.84 MJ.kg^{-1} in WTC and 9.30 MJ.kg^{-1} in CE ($P < 0.05$). Present data suggested that tropical climate may not be on the whole a warm environment, with minimal heat production and improved efficiency of dietary energy in WTC.

Poster P3.37

The change of piglets'biochemical indices under laser influence
O.S.Korotkevich*. *Institute of Veterinary Genetics and Selection of Novosibirsk Agrarian University, 160 Dobrolubov Str.,630039 Novosibirsk, Russia.*

One of the most ecologically pure methods to influence animal organisms,that can be applied in veterinary practice, is laser reflexotherapy.The investigation of blood serum was done in 40 bronchopneumonia piglets at the age of 21 days in Kemerovo region. The animals were divided into two groups (experimental and control). Biologically active points (BAPs) of piglets' lung meridian were exposed to laser radiation produced by "Mustang" apparatus. The power of the laser radiation was from 2 to 5 W, with the frequency - from 80 to 600 Hz. 5 procedures were performed. The level of creatinine in blood serum of the experimental animals decreased by 15.5%($P<0.01$) as compared to the control ones that demonstates the activity of metabolism and elimination of toxic substances in piglet organisms. While the content of alkaline phosphatase increased by 30%, the level of acid phosphatase decreased by 13% in the experimental group.It may result from hyperfunction of osteoblasts. Thus, the laser influence on lung meridian BAPs activizes compensatory mechanisms and help restore homeostasis.

PIG PRODUCTION   [P]   Poster P3.38

Effect of ultrasound influence on pigs' biologically active points (BAPs)
O.I.Sebeghko*, O.S.Korotkevich, D.A.Odnoshevsky, G.N.Korotkova. *Institute of Veterinary Genetics and Selection of Novosibirsk Agrarian University, 160 Dobrolubov Str.,630039 Novosibirsk, Russia.*

Ultrasound is a catalyst of physico-chemical and biophysical processes. The experiment was carried out in 20 bronchopneumonia piglets at the age of 1.5 month that were divided into two groups: experimental and control. Before and after treatment the total blood protein and its fractions were stidied. High frequency ultrasound 0.88MHz with intensity 0.2-0.4 W/cm^2 in 2ms impulse-wave was used for 1-2 minutes to affect piglets' lung meridian BAPs. After the treatment in the experimental group the level of total blood protein, α- and γ-globulins increased 2.1 times (from 52.6 to 109 g/l),1.2 and 1.5 times, respectively. The level of albumins and β-globulins decreased 1.7 and 1.2 times, respectively. Thus, the influence of ultrasound on BAPs leads to biosynthesis of the total blood protein and its fractions, ultrasound was used as a regulater of metabolism in piglets' organisms and the increase in γ-globulins promoted the activity of immune system.

Poster P3.39

Stress resistance of Siberian North pigs
V.G.Kuznetsov*, O.S. Korotcevich, V.L. Petukhov. *Institute of Veterinary Genetics of Novosibirsk Agrarian University, 160 Dobrolubov Str.,630039 Novosibirsk, Russia*

Only 3 pig breeds: Siberian North, Kemerovskaya and New Meat were bred for the whole history of animals husbundry in Siberia. It necessary to pay more attention to stress resistance under Siberian pig selection for productivity. With the help of halothane test the Siberian North piglets were investigated at the age 40-60 days in winter and summer. The apparatus "Narkon II" and mask made by Hart V.V. were used in the experiment. The piglets were anesthetized by the mixture of ftorothane and oxygen for 2-3 minutes. 147 piglets were investigated, where 82±3,2% of the animals were stress resistant (HAL -) and 18±3,2% of them were susceptible (HAL +). There were tremor and convulsions that evidenced exstrapyramidal hyperkinesia. If in winter these signs were observed in the all investigated piglets, while in summer only in 9.6% animals. The results of the investigation showed, that the periods of exposure and sleep indices of were higher in stress resistance piglets, than in HAL+ ones. Resulted from the lysosomic-catione test the indices of non-specific resistance were 1.13±0.056 (HAL -) and 1.09±0.073 (HAL +).

PIG PRODUCTION [P]

Poster P3.40

The relationships between foot measurements and body weight in piglets
K.N. Kotomin*, V.L. Petukhov, T.N. Nicolaeva, V.G. Kuznetsov Institute of Veterinary Genetics of Novosibirsk Agrarian University, 160 Dobrolubov Str., 630039 Novosibirsk, Russia

The investigation was carried out on the farm of West Siberia. The data of the 150 early-ripe meet strain piglets at the 3 month of age were used. The claw height, claw length, diagonal, length of the dorsal border, claw width and cannon circumference were measured.
The relationship between the foot measurements and the body weight was positive and high. The phenotypic correlations lay in the range between 0.60 and 0.94 for these parameters. The influence of the sex on the foot dimensions in animals has been found. The problem of the some measurements application for the genetic improvement of the foot soundness is being discussed.

Poster P3.41

The investigation into relationship between cholesterol, sex steroids concentrations and fertility in gilts
C. Klocek*, J. Koczanowski, W. Migdal, Department of Pig Production, Agricultural University of Cracow, Al. Mickiewicza 24/28, 30-059 Kraków, Poland.

Cholesterol as a well-known precusor of many steroid hormones (e.g. progesterone, oestradiol, androgens) could influence on the reproductive performance in gilts.
Blood samples were collected from 20 Polish Landrace x Large White gilts on Days 17 and 30 after mating. Gilts were divided into two experimental groups after estimation of cholesterol concentrations. Cholesterol concentration in blood samples was ≤ 100 mg/dL in Group 1 (8 gilts) and > 100 mg/dL in Group 2 (12 gilts). All gilts were slaughtered on Day 30 of pregnancy and the number of corpora lutea and living embryos was determinated. The number of living embryos in gilts group 1. was 10.44 and 9.17 in the second group.
The corelation cosficients between cholesterol concentration and the living embryo number was -0.81; between oestradiol level and the living embryos number was -0.59 and between progesterone level and the living embryo number was 0.69. On the other hand it was 0.71 between progestrone and oestradiol level and 0.56 between progesterone and cholesterol concentration.

PIG PRODUCTION [P]

Poster P3.42

Sexual behavior of Polish Landrace sows and their actual fertility
A. Stasiak, A. Walkiewicz, P. Kamyk. *Department of Breeding and Production Technology of Swine, Agricultural University, Akademicka 13, 20-950 Lublin, Poland.*

Sows of PL breed (202 animals) at the age of 150 days were under studies. Their sexual activity in the third estrus - in presence of a boar - was estimated. They were classified into two groups: 1) very clearly showing the estrus and fully tolerating the boar; 2) weakly showing the estrus and difficult to be mated. After farrowing, number of alive and dead-born piglets, number of piglets reared up to the 21st day, litter and piglet weight at the 21st day were estimated. Among 202 sows introduced into the basic herd as replacement ones, 74.2% showed the lordosis reaction very clearly and well tolerated the boar; 25.8% were classified as difficult for mating. Sows well manifesting the lordosis reaction in the first litter, were characterized by higher reproduction performance factors as compared with weakly reacting to the boar's presence and difficult for mating. Significant differences as regarding to born piglets number (1.3 piglets - 12.6%) and reared ones up to the 21st day (1.2 piglets - 12.2%) were found between those groups.

Poster P3.43

Reproduktionsleistungen von zuchtsauen unter praxisbedingungen bei zusätzlicher β-carotin-zufuhr
V. Manzke, H. Münchow und H. Fechner. *Fachgebiet Tierernährung, Institut für Nutztierwissenschaften, Landwirtschaftlich-Gärtnerische Fakultät, Humboldt-Universität zu Berlin, Philippstr. 13, 10115 Berlin, Germany.*

Die Reproduktionsleistung von Zuchtsauen bestimmt die Ökonomie des Produktionszweiges erheblich. Erhebungen zu nutritiven Einflüssen auf diesen Leistungskomplex gehen auch von einer eigenständigen, zum Teil widersprüchlichen Positivwirkung von β-Carotin aus. In vorliegender Untersuchung wurde der Einfluß einer β-Carotin-Supplementation (400 mg/Tier x d ab dem 85. Trächtigkeitstag über die folgenden 80 Tage) in einem Bestand mit 600 Sauen auf die Fruchtbarkeits- und Aufzuchtleistung der Tiere geprüft (lebend geborene und abgesetzte Ferkel, Gewichtsentwicklung und Verluste der Ferkel, Besamungsaufwand u.a.m.). Bei mit β-Carotin versorgten Jungsauen konnten gegenüber den unsupplementierten Probanden signifikant mehr lebend geborenen Ferkel sowie eine tendenziell höhere Wurfmasse zur Geburt registriert werden. In den entsprechenden Erhebungen an Altsauen, die im vorherigen Reproduktionszyklus schon einmal β-Carotin erhielten, ergaben sich dagegen keine Anhaltspunkte für eine positive Wirkung des Provitamins. Ebenso blieben die Trächtigkeitsrate nach Erstbesamung sowie der Gesundheitsstatus der Ferkel von der β-Carotinversorgung der Mütter unbeeinflußt, obwohl die Zulage den β-Carotin- und Vitamin A-Gehalt der Sauenmilch steigerte.

PIG PRODUCTION [P]

Poster P3.44

Comparison of the semen quality in pure-bred and hybrid boars based on ORT test
J.Udała[1], R.Czarnecki[2], M.Kawęcka[2], M.Różycki[3], M.Kamyczek[4], [1]*Department of Animal Reproduction,* [2]*Department of Pig Breeding, Agricultural University, 6 Judyma St., 71-460 Szczecin,* [3]*National Research Institute of Animal Production, Department of Pig Breeding, 32-083 Balice,* [4]*Zootechnical Experimental Station, 64-122 Pawłowice, Poland.*

In the study were determined trends in the value of spermatozoa osmotic resistance test, ORT, in which the percentage of spermatozoa with normal acrosome is evaluated during a 15 and 120 minutes incubation in a 300 and 150mosm/kg BTS diluent, in groups of pure-bred: L990 x L990 (L), Duroc x Duroc (D), Pietrain x Pietrain (P), and hybrid: LxD, DxL, LxP, PxL, DxP, PxD, boars. Starting from 70-180 day of life boars were maintained and fed individually. On the average, 195 ejaculations were evaluated in each group of 65 boars. In most purebred boars the value of ORT test was lower (63,6; 61,2; 61,1%) when compared with hybrid ones (61,1; 64,9; 62,0; 66,5; 62,2; 66,0%). It is observed that boars having a higher ORT test value distinguished themselves by a higher percentage of spermatozoa with intact acrosome in fresh semen (72,7-75,9%), greater volume of testes on 180 day (596,0-744,9cm^3), earlier age of obtaining the firs ejaculation (238,4-279,5 days), and a lower AspAT semen plasma activity (38,6-92,8μU/ml). Hybrid boars were also of larger body weight on 180 day (105,1-111,4kg) than pure-bred ones (104,2-10,1kg). The results obtained confrim in great measure usefulness of ORT test to evaluate the quality of semen, which is higher in hybrid boars than in pure-bred ones.

Poster P3.45

Relationship between the growth rate and the thickness of the back fat of young boars and the size of their testes as well as the parameters of their libido and sperm.
R. Czarnecki[1], M. Różycki[2], M. Kawęcka[1], B. Delikator[1], K. Dziadek[2], A. Pietruszka[1].
[1]*Department of Pig Breeding, University of Agriculture, Dr. Judyma 10 Street, 71-460 Szczecin,* [2]*Institute of Animal Science, Kraków, Poland.*

The present study was based on a total of 1000 young boars (792 represented line 990, 111-line 890, and 97-Duroc breed). After finished half year age evaluated their libido and semen traits that amounted: time to effective mounting (199,66 s), number of leaps to effective mounting (1,21), time of ejaculation (253,33 s), ejaculate volume after filtration (126,79 cm^3), concentration of spermatozoa in cm^{3}x10^6 (187,05), total number of spermatozoa x 10^9 (22,28), spermatozoa with major defects (13,63%), spermatozoa with minor defects (13,49%) and with defects of acrosome (7,26%). The acquired correlation values between the growth rate and the thickness of the back fat of the boars and their parameters of libido and sperm were low. Boars with thicker back fat had higher percentage of progressive spermatozoa (0.103**), higher spermatozoa concentration (0.121**) and slightly lower percentage of spermatozoa with major defects (- 0.082*), and with altered acrosome (- 0.096**). The growth rate was positively correlated with the size of testes (0.270**), while the thickness of back fat was not correlated with this trait of the boars (- 0.034).

PIG PRODUCTION   [P]   Poster P3.46

Relationship between the sexual activity of young boars and the traits of their semen.
R. Czarnecki[1], M. Kawęcka[1], M. Różycki[2], B. Delikator[1], K. Dziadek[2]. [1]Department of Pig Breeding, University of Agriculture, Dr. Judyma 10 Street, 71-460 Szczecin, [2]Institute of Animal Science, Kraków, Poland.

The present study was based on 1000 young boars older than half a year. Out of this number 792 represented line 990, 111-line 890, and 97-Duroc breed. They were reared in uniform, standard conditions (from day 70 to day 180 they were kept and feed on individual basis). Their average body weight on day 180 was 115.81 kg and their libido parameters were as follows: time to effective mounting 199,66 s; number of leaps to effective mounting 1,21; time of ejaculation 253,33 s; and traits of semen: ejaculate volume after filtration 126,79 cm^3; concentration of spermatozoa in cm^{3}x10^6=187,05; total number of spermatozoa x 10^9=22,28; spermatozoa with major defects 13,63(%); spermatozoa with minor defects 13,49(%) and with defects of acrosome 7,26(%). Obtained very low and nonsignificant relationship between mentioned on the beginning two traits of libido and all studied traits of the semen (r=from 0,002 to -0,090). Time of ejaculation, however, had statistically significant ($P \leq 0.01$): positive effect on ejaculate volume after filtration (0.573**), negative effect on the spermatozoa concentration (- 0.291**), and also negative effect on % of spermatozoa with minor defects (- 0.102**). It had no effect on the overall number of spermatozoa in the ejaculate, % of spermatozoa with major defects, and % of spermatozoa with altered acrosome.

Paper PGN4.1

What is pork quality?
Henrik J. Andersen, Department of Animal Product Quality, Research Centre Foulum, Danish Institute of Agricultural Sciences, P.O.Box 50, DK-8830 Tjele, Denmark

At present significant variations for most quality parameters occur between pork carcasses. Unfortunately, up to now only few in the pork industry have realised or cared about this problem. Only lean meat content has been in focus. However, dramatic changes in the international market place over the past ten years, caused by the changing lifestyles and requirements of consumers in Europe and around the world, have started to require high standards of quality assurance regarding diversity, quality and safety of products and the environmental and animal welfare aspects of their production. Consequently, the concept "pork quality" is developing and it includes beside composition and size, eating quality, nutritional quality, technological quality, health and hygienic quality and ethical quality. The present presentation will especially focus on technological and eating quality aspects expected to be of outmost importance for the customers (industry/consumers) of pork meat now and into the next millennium.

PIG PRODUCTION   [P]   Paper PGN4.2

Influence of genetics on pork quality
A.G. de Vries*[1], L. Faucitano[1], A. Sosnicki[3] and G.S. Plastow[2], [1]PIC Europe and [2]PIC Group, Fyfield Wick, Abingdon, OX13 5NA, UK; [3]PIC Americas, P.O. Box 348, Franklin, KY 42135-0348, USA.

The aim of this paper is to discuss the opportunities of gene technology in relation to the quality of pork. After dealing with breed effects and within-breed variation, an overview of major genes and DNA technology is given. It is demonstrated that some of the breed effects can be fully explained from the presence of a single gene with major effect. Within breeds, there is considerable genetic variation in relevant meat quality traits like waterholding capacity and intramuscular fat. Again, part of this variation is due to major genes. As a result, DNA marker technology can play an important role to improve meat quality. Selective breeding based on this technology will also increase the uniformity of the final product. Furthermore, the exploitation of major genes can be highly relevant for differentiation of breeding populations for specific markets.

Paper PGN4.3

Genetic parameters of meat quality traits in station tested pigs in France
T. Tribout*, J.P. Bidanel, Station de Génétique Quantitative et Appliquée, I.N.R.A., 78352 Jouy-en-Josas Cedex, France

Genetic parameters of traits recorded on slaughtered animals controlled in French central test stations were estimated for the Large White (LW) and French Landrace (FL) breeds using a restricted maximum likelihood procedure applied to a multiple trait animal model. The data consisted of, respectively, 13000 and 5600 records in LW and FL breeds, collected from 1988 to 1999. Four performance traits - average daily gain (ADG), feed conversion ratio (FCR), dressing percentage (DP), carcass lean content (CLC) - and 8 meat quality traits - ultimate pH of Adductor femoris (PHAD) and Semimembranosus (PHDM) muscles, two reflectance measurements (REFL and L*), water holding capacity (WHC), a visual note (NOTE), two meat quality indexes (MQI1, MQI2) - have been considered. In both breeds, heritabilities of performance traits were close to previous estimates. Heritabilities of meat quality traits were low to moderate. Large positive genetic correlations were obtained between the two pH and the two reflectance measurements. Reflectance, pH, NOTE and the two MQI had large favourable genetic correlations. Genetic correlations between WHC and the other meat quality traits were lower, but remained favourable. All meat quality traits, except WHC, had unfavourable genetic correlations with CLC and FCR. Conversely, meat quality traits were genetically almost independent from ADG.

PIG PRODUCTION   [P]   Paper PGN4.4

Dissection of genetic background underlying meat quality traits in swine
J. Szyda[*1], E. Grindflek[2] and S. Lien[2]. [1]Department of Animal Genetics, Agricultural University of Wrocław, Kożuchowska 7, 51-631 Wroc3aw, Poland, [2]Department of Animal Science, Agricultural University of Norway, 1432 Ås, Norway.*

The primary goal of the study is to localise the quantitative trait loci (QTL) responsible for meat quality in swine. The additional information on the family structure and on sex- and family specific differences in recombination rate can be incorporated into the statistical model in order to provide better insight into the genetic background underlying analysed traits. Available traits comprise taste panel data (16 traits), water-holding capacity, intramuscular fat, fatty acid content, protein in muscle, connective tissue protein and carcass characteristics, scored on 320 individuals from Duroc and Norwegian Landrace breeds. Genotype information is available for three generations: parental, F1 and F2. Statistical procedures applied for the analysis involve (i) the composite interval mapping (Zeng, 1994), and (ii) a modified regression approach originating from the method of Haley and Knott (1992).

Paper PGN4.5

Skeletal muscle fibres as factors for pork quality
A. Karlsson[* 1], R. E. Klont[2] and X. Fernandez[3]. [1]Danish Institute of Agriculture Sciences, DK-8830 Tjele, [2]ID-DLO, NL-8200 AB Lelystad, [3]INRA, Theix, FR-63122 Saint Genès Champanelle.*

Interactions between muscle fibres characteristics (fibre type composition, fibre area, metabolism, and glycogen and lipid contents), *post-* and *peri-mortem* energy metabolism and different environmental factors determine *post-mortem* transformation from muscle to meat. Muscle fibres are not static structures, but easily adapt to altered functional demands, hormonal signals, and changes in neural input. Their dynamic nature makes it difficult to categorise them into distinct units. Some properties may change without affecting others or without changing histochemical appearance of a given fibre, and within a muscle a fibre type may show a continuum of structural and functional properties which overlap with other fibre types. Therefore, a distinction between specific types must strictly refer to the method that has been used for the typing.
The literature indicate possibilities to include muscle fibre characteristics in breeding schemes for improved meat quality, while preserving optimal production traits. In order to use muscle fibre characteristics in a beneficial way for future breeding programmes, further investigations are needed to better understand the physiological mechanisms. Selection experiments based on biochemical and histochemical characteristics determined in biopsies may possibly provide better tools to study these relationships.

PIG PRODUCTION   [P]

Paper PGN4.6

Selection progress of intramuscular fat in Swiss pig production
D. Schwörer*[1], A. Hofer[2], D. Lorenz[1] und A. Rebsamen[1]. [1] *Swiss Pig Performance Testing Station, 6204 Sempach, Switzerland,* [2] *Institute of Animal Sciences, Swiss Federal Institute of Technology (ETH), CLU, 8092 Zürich, Switzerland*

The Swiss Pig Performance Testing Station started in the year 1980 with the evaluation of intramuscular fat (IMF: amount of free fat in the 10th rib section of the M. longissimus dorsi) by Near Infrared Reflection. Since the year 1985 all full sibs of the Testing Station were routinely analysed on IMF. Until the end of 1998 a total of 57975 animals were tested. Inclusion of IMF in the combined selection index began in the year 1989 (desired selection progress per generation: 0.03 % IMF). Since the year 1997 IMF is also considered in the BLUP evaluation system.
Selection for higher IMF has proved to be effective. In Swiss Large White (SLW) IMF increased phenotypically from 0.8 % in the year 1987 to 1.9 % in the year 1998, and in Swiss Landrace (SL) from 1.0 % to 1.6 %. The variance remained constant. The BLUP breeding values were used for the estimation of genetic trends. In the years 1991-1998 IMF increased in SLW genetically by 0.3 % and in SL by 0.2 %. In the same period a genetic increase in premium cuts of 1.3 % in SLW, of 1.7 % in SL and in daily gain of 18 g in SLW, of 25 g in SL were estimated. The genetic improvement in feed conversion was in both breeds 0.1 kg/kg.

Paper PGN4.7

Nutritional and genetic influences on meat and fat quality in pigs
K. Ender; K. Nürnberg; G. Kuhn; U. Küchenmeister
Division of Muscle Biology & Growth, Research Institute for the Biology of Farm Animals, D-18196 Dummerstorf

The results of three experiments show the influence of nutrition and genetics on meat and fat quality. N-3 fatty acid enriched diet to pigs did not affect the meat quality (Exp. 1), however this diet increased the percentage of n-3 fatty acids in backfat, in neutral and polar fractions of muscle and heart. A study with Saddle Back pigs (SB) and German Landrace pigs (GL) indicated differences in meat and fat quality (Exp. 2). The larger fat content of SB pigs was related to an increased intramuscular fat concentration of *longissimus* muscle. The saturated fatty acid concentration was higher in SB pigs and that of the polyunsaturated fatty acids significantly lower than in GL muscle. Heterocygotes (NP) of SB showed higher values for drip loss, and meat colour compared to homocygote (NN) SB. It can be concluded that a higher fat content is no guarantee for PSE free meat. A mutation of the calcium release channel causes an increased metabolism post mortem resulting in cell injury and inferior meat quality in Pietrain pigs (Exp. 3). Malignant hyperthermia susceptible pigs (MHS) indicated a higher concentration of n-3 fatty acids in muscle phospholipids, enhanced peroxidation of lipids and an insufficient $SR-Ca^{2+}$ pump in comparison to stress resistant pigs. The results lead to the recommendation to exclude MHS pigs from breeding programs.

PIG PRODUCTION   [P]

Paper PGN4.8

Food waste products in diets for growing-finishing pigs
N.P. Kjos* and M. Øverland. Department of Animal Science, Agricultural University of Norway, P.O. Box 5025, N-1432 Ås, Norway

Food waste products (FW) can be defined as the edible waste from food production, distribution and consumption, and may constitute as much as 20% of the total human food supply. Forty-eight growing-finishing pigs (28.3 kg average initial weight) were used to study the effect of FW in diets on growth performance, carcass merits and meat quality. Treatments consisted of six levels of FW (0, 200, 400, 600, 800, 1000 g kg^{-1} diet) in combination with a barley-soybean meal diet. The FW averaged 21.4% dry matter, 4.9% crude protein, 3.3% crude fat, 0.5% crude fiber, 11.5% NFE and 1.2% ash. Increasing dietary levels of FW reduced (linear, P < 0.001) average daily intake of feed dry matter (ADFI) and feed DM:gain ratio, but had no effect on weight gain. The ADFI was significantly lower for pigs fed the diets containing 600 g kg^{-1} FW or higher. Increasing dietary levels of FW gave a linear reduction (P < 0.01) of fat firmness, and lightness (L*) of backfat and loin meat. The proportion of saturated and polyunsaturated fatty acids in backfat decreased (linear, P < 0.001), and increased (linear, P < 0.001), respectively. Sensory quality of loin muscle, fresh or stored at - 20°C for six months, was not affected by the dietary treatment. To conclude, food waste products could be used in diets for growing-finishing pigs without adversely affecting growth performance, carcass merits and sensory quality when used in moderate levels.

Paper PGN4.9

Conservation and development of the bísaro pig. Characterization and zootechnical evaluation of the breed for production and genetic management
J. Santos e Silva*[1], J. Ventura[2], P. Albano[3] and J. S. Pires da Costa[4]. [1]Estação Experimental de Produção Animal, Direcção Regional de Agricultura Entre Douro e Minho, S. Torcato, 4800 Guimarães, Portugal, [2] Universidade de Trás-os-Montes e Alto Douro, Secção de Zootécnia, Apt. 202, 5000 Vila Real, Portugal, [3]Centro de Estudos de Ciência Animal da Universidade do Porto, Campus Agrário de Vairão, 4480 Vila do Conde, [4]Estação Zootécnica Nacional, Departamento de monogástricos, Fonte Boa, Vale de Santarém, 2000 Portugal.

This study is part of an extensive plan of conservation and recovery of an ancestral genetic resource, the bísaro pig, and has the objective of reversing the lack of variability that has been observed for the past 50 years, as well as establishing technical and economic references which are necessary for the development of high quality production involving this breed and its production systems. This study shows that the bísaro breed maintains the zootechnical characteristics of its ancestors: slow growth (550 g/day), low feed efficiency (3.77), poor conformation (NE), low prime cut yield, high bone yield, average quantity of external fat (20 mm), best post mortem pH values and water retention capacity than other breeds studied at the same time. The criteria for breed conservation and breeder selection are: molecular identification of the halotane gene, CRC locus, (revealing the secretion of allele n (Nn) which has negative effects on sensorial quality and meat technology), mean values of post mortem pH observed in the families, as well as A.D.G., FC and backfat.

PIG PRODUCTION   [P]

Paper PGN4.10

Meat consumption and health
M. Zimmermann. Laboratory for Human Nutrition, Food Science Institute, Swiss Federal institute of Technology, Zürich, Switzerland.

Public health concerns in the industrialized countries where obesity, coronary heart disease and cancer are common have led to dietary recommendations to the public to change their diet. These include a reduction in fat and cholesterol consumption, particularly saturated fat. Because consumption of meat in industrialized countries typically supplies about a quarter of the saturated fatty acids in the diet and approximately a third of the cholesterol, dietary guidelines often suggest a reduction in red meat consumption. In addition there are public concerns about residual pesticides, hormones and growth promoters in meat, and the transmission of human diseases from meat. However, meat is an important source of protein, the B-vitamins, and many minerals. In particular, meat is a rich source of highly bioavailable iron and zinc. Therefore, although meat is not an essential part of the diet and intake of fatty meats should be minimized, meat consumption supplements and complements a diet based on plant foods and often helps to ensure nutritional adequacy.

Paper PGN4.11

Who eats meat: factors effecting pork consumption in Europe and the United States
W. Jamison. Worcester Polytechnic Institute, 100 Institute Road, Worcester, MA 01609-2280, USA

The pork industry is typified by excellent parameters of production and quality. Pork is leaner and more nutritious than at any other time in the history of intensified pork production, and the industrial production of pork has been characterized by increasing levels of production, increasing levels of efficiency, and decreasing cost to consumers. With these advantages, the pork industry should expect to enjoy increasing market share and profitability. Nonetheless, that is not necessarily the case; consumers in Europe and the US are increasingly making purchasing decisions on factors other than the cost, nutritive value and availability of pork. Indeed, a significant percentage of consumers are making their purchasing decisions based upon parameters that may be unfamiliar to pork producers. Some consumers consider the ethical treatment of pigs, the impact of pork production on the environment, and other extraneous issues to be paramount in their purchasing decisions. Thus, animal scientists and pork industry officials who believe that by increasing efficiency and nutritive value, when combined with decreasing costs to consumers, they will be able to increase market share and enjoy perpetual profitability may be mistaken. Data indicate that some consumers in Europe and the US no longer view those industry standards as important to their choice of food. This paper examines the possible implications of shifting consumer attitudes upon modern pork production.

PIG PRODUCTION   [P]   Poster PGN4.12

Breed effect on meat quality of Belgian Landrace, Duroc and their reciprocal crossbred pigs
G. Michalska, J. Nowachowicz, B. Rak, W. Kapelanski*. University of Technology and Agriculture, Mazowiecka 28, 85-084 Bydgoszcz, Poland

The breed effects on meat quality in crossing are not ease to foresee and ought to be studied in the respective crossing wariants. In the study there were compared the meat quality traits of the Belgian Landrace (BL), Duroc (D), BLxD and DxBL crossbred pigs (sire breed being on the first position). Each group was consisted of 30 gilts slaughtered at 185 days of age. The determined meat traits were: pH_1, colour lightness and soluble meat protein content which is assumed to be close related to water holding capacity of meat. Mean pH_1 value for BL was significantly lower than for D, BLxD and DxBL pigs (5.92 vs 6.25, 6.15 and 6.13, resp.). Meat colour was the highest in BL pigs (28.6 vs 25.8, 23.6 and 24.8%, resp.) and soluble protein content was significantly lower for BL than for the Duroc and crossbred pigs (7.60 vs 8.03, 8.15 and 8.15%, resp.). The heterosis effect for pH_1 was minimal (1.15 for BLxD and 0.82% for DxBL) but for colour lightness was distinct (-13.11 for BLxD and -8.76% for DxBL) and was 4.35% for soluble protein content in meat of the two crossbred group of pigs.
It is worthy to note that the progeny of BL boars and Duroc sows was characterized by the best meat quality.

Poster PGN4.13

Genetic and energy effects on pig meat quality
Z. Gajic*[1] and V. Isakov[2], [1]University of Belgrade, Faculty of Agriculture 11081 Beograd, [2]A.D. "29. Novembar, Subotica, Yugoslavia

Direct and interaction effects of breed (Mangalitsa, Yugoslav Meat Breed and Belgian Landrace) and levels of diet energy (low, medium and high) on meat quality traits were examined in a 3 x 3 factorial experiment. Each of 9 subclasses consisted 20 pigs all taken off test at 100 kg for slaughter.
The three breeds differed regarding some of the meat quality parameters. The Belgian Landrace had significantly lower pH_1 in *M.semimembranosus* (6.12) and nearly the same for Mangalitsa and Yugoslav Meat Breed (6.43 and 6.46, respectively). A similar tendency was observed in *M. longissumus dorsi*. The meat from Belgian Landrace pigs had a lower water holding capacity, observed as higher wetness value in *M. longissimus dorsi* (4.14 cm^2) than Mangalitsa (3.75 cm^2) and higher Yugoslav Meat Breed (3.56 cm^2). Differences in dry matter and crude protein content between the breeds were significant. Percent of the intramuscular fat in the meat show that Mangalitsa ranked highest (2.93), followed by Yugoslav Meat Breed (1.93) and than Belgian Landrace (1.67). The differences are highly significant.
From these comparative investigations it may be concluded that the Mangalitsa, Yugoslav Meat Breed and Belgian Landrace greatly differ in terms of the meat quality traits studied. No trait was significantly affected by diet energy level. Breed x energy levels interactions were not significant for all meat quality traits except dry matter content.

PIG PRODUCTION   [P]

Poster PGN4.14

Genetic parameters for fattening traits in the Belgian Piétrain population
D. Geysen*[1], S. Janssens[1], W. Vandepitte[1]. [1]Centrum voor Huisdierengenetica en selectie, Department Animal Production, K.U.Leuven, Minderbroederstraat 8, 3000 Leuven, Belgium.

Genetic and phenotypic parameters for fattening traits were estimated for a centrally tested purebred stress susceptible Piétrain population. Data (from 1984 to 1998) consisted of individual records on growth (18,755) and percentage of lean meat (7,908). Growth was expressed as the number of days (DAYS) needed to grow from 25 kg to 100 kg liveweight. Lean tissue growth (LTG) was calculated as the product of average daily gain (ADG) and percentage of lean meat (LM%). REML variance components were estimated using an animal model that included fixed effects for sex (barrow or gilt) as well as the random direct genetic effect of the animal, the random permanent environmental effect of the litter (pigs are housed in pens of four littermates) and the random effect of herd-year-season. Data on DAYS were precorrected for initial and final weight. For the traits LM% and LTG initial and final weight were included in the model as covariates nested within sex. Total number of animals (including pedigree information) was 30,136. Resulting heritabilities were 0.403, 0.363 and 0.393 for DAYS, LM% and LTG respectively. Genetic correlation between DAYS and LM% is moderately unfavourable (0.329), between DAYS and LTG is highly negative (-0.949) and is not significantly different from zero between LTG and LM% (-0.063). These results indicate that selection for increased lean tissue growth will lead to faster growing Piétrains without loosing meatiness.

Poster PGN4.15

Genetic parameters for colour traits and pH and correlations to production traits
S. Andersen* and B. Pedersen. National Committee for Pig Breeding, Health and Production, Axeltorv 3, DK-1609 Copenhagen V, Denmark.

Meat colour in m.longissimus dorsi was measured as Minolta L, a and b values and subjectively scored on the Japanese scale (Jap). Ultimate pH was also measured in m.longissimus dorsi. In total 4,902 boars from three breeds, free from the Hal^n and the RN$^-$ alleles, were measured. Average daily gain (dg), lean meat percentage (%lm) and feed efficiency (FE, from individual feed intake recordings) were also measured. Heritabilities for Jap; L, a, b, pH were estimated to 0.30, 0.18, 0.52, 0.27, 0.17 and correlations to Jap were cor(Jap; L, a, b, pH) = (-0.68, 0.55, 0.03, 0.22)
Correlations to production traits were low, cor(Jap; dg, %lm, FE) = (0.06, 0.00, 0.01), cor(L; dg, %lm, fe) = (-0.01, -0.12, 0.09) cor(a; dg, %lm, FE) = (-0.11, 0.00, 0.10).
The possibility to construct a colour index for Jap from measurements of L, a and b are discussed. The expected correlated response in Jap from selection on an index of production traits was found to be moderate.

PIG PRODUCTION   [P]

Poster PGN4.16

Halothane gene effect on carcass and meat quality by use of Duroc x Pietrain boars
H. Busk[*1], A. Karlsson[1] and S.H. Hertel[2]. [1]Danish Institute of Agricultural Sciences, Dept. of Animal Product Quality, Research Centre Foulum, P.O.Box 50, DK-8830 Tjele, Denmark. [2]PIC Denmark A/S, P.O.Box 7021, DK-9200 Aalborg SV, Denmark.

The halothane gene in pigs has in the heterozygote state (HAL-Nn) a positive effect on production and carcass traits. Concerning meat quality there are still problems with the water-holding capacity of meat in some breeds and breed-crosses. Duroc pigs and crosses are known to have a good meat quality compared to other breeds. Therefore, the aim of this project is to combine production traits with improved meat and carcass quality using halothane heterozygote cross-breed boars of Pietrain and Duroc (PxD).
PxD boars (PIC 409) were mated to LxY sows. The offspring selected for the trial were equally distributed with Halothane negative homozygotes (HAL-NN) and heterozygotes (HAL-Nn). In total, 196 pigs were slaughtered at a high strees level. The results showed no differences in daily gain between the genotypes. Concerning carcass quality, the area of m. longissimus dorsi was 2.0 cm^2 less, the sidefat thickness 1.9 mm thicker and percentage of lean meat in the carcass 1.9 percent less for HAL-NN than for HAL-Nn. Concerning meat quality, the pH values at 0 min, 45 min and 24 hours were higher, and the drip loss 3.7 percent units lower for HAL-NN than for HAL-Nn, but there were no differences in color. In summary, the results showed no effect of the HAL- gene on daily gain, but a positive effect on carcass quality and a negative effect on the meat quality expressed as water-holding capacity.

Poster PGN4.17

In vivo and post mortem changes of muscle phosphorus metabolites in pigs of different malignant hyperthermia genotype
M. Henning[*1], U. Baulain[1], G. Kohn[1] and R. Lahucky[2]. [1]Institute of Animal Science and Animal Behaviour (FAL), Mariensee, 31535 Neustadt, Germany. [2]Research Institute of Animal Production, 94992 Nitra, Slovakia.

Stress susceptibility in swine causes a significant economic loss due to poor meat quality and preslaughter death. Different malignant hyperthermia genotypes - stress susceptible Piétrain (nn), heterozygous German Landrace (Nn) and homozygous non stress susceptible German Landrace (NN) were investigated by means of non invasive ^{31}P-NMR spectroscopy. Changes of inorganic phosphate (P_i), phosphocreatine (PCr) and ATP in M. biceps femoris were analysed during halothane exposure. Intracellular pH and magnesium (Mg^{2+}), which is required for many cellular functions, were estimated from the spectra. As expected, nn genotypes showed dramatic changes in their muscle metabolism after administration of halothane. A significant decrease of PCr and a corresponding increase of P_i were observed. Intracellular pH declined from 7.1 to 6.4 and Mg^{2+} increased by more than 100%. Nn and NN pigs did not show a reaction after administration of halothane. At rest intracellular Mg^{2+} was significantly higher in nn pigs as compared with NN and Nn. Mg^{2+}-concentration of heterozygotes was intermediate but closer to NN. Post mortem studies of muscle metabolism in NN and Nn genotypes showed significant differences in time course changes of PCr and P_i as well as in the rate of pH decline.

PIG PRODUCTION   [P]

Poster PGN4.18

Interactive effects of the *HAL* and *RN* major genes on carcass quality traits in pigs
P. Le Roy*[1], C. Moreno[1], J.M. Elsen[2], J.C. Caritez[3], Y. Billon[3], H. Lagant[1], A. Talmant[4], P. Vernin[4], Y. Amigues[5], P. Sellier[1], G. Monin[4]. INRA, [1]SGQA, 78352 Jouy en Josas cedex, [2]SAGA, BP27, 31326 Castanet Tolosan cedex, [3]Domaine du Magneraud, 17700 Surgères, [4]SRV, Theix, 63122 Saint Genès Champanelle, [5]LABOGENA, 78352 Jouy en Josas cedex, France.

An experiment was set up in order to estimate the interactive effects of the two major genes *HAL* (*N* and *n* alleles) and *RN* (*rn*[+] and *RN*[-] alleles) on carcass quality traits in pigs. Forty pigs from each of the nine combined *HAL-RN* genotypes were recorded for growth, body composition and meat quality traits. The results showed that major effects of *HAL* and *RN* are essentially additive for carcass composition traits but significant *HAL* by *RN* interaction effects occur for most of the meat quality traits. In the latter case, two situations were encountered, either a « snowball » interaction effect when the effect of one mutation, *n* or *RN*[-], is increased by the other one, or a « lessening » interaction effect when the effect of one mutation is decreased by the other one. For example, the adverse effect of the *n* allele on pH_1 (or drip loss) was greater in carriers than in non-carriers of *RN*[-], whereas the adverse effect of the *RN*[-] allele on Napole technological yield (or cooking loss) was smaller in *nn* than in *NN* pigs.

Poster PGN4.19

Meat quantity to meat quality relations when the RYR1 gene effect is eliminated
J. Kortz[1], W. Kapelanski[2]*, S. Grajewska[2], J. Kuryl[3], M. Bocian[2], A. Rybarczyk[1]. [1]Agricultural University, Dr Judyma 24, 71-460 Szczecin, [2]University of Technology and Agriculture, Mazowiecka 28, 85-084 Bydgoszcz, [3]Institute of Genetics and Animal Breeding, Polish Academy of Science, 05-551 Mrokow, Jastrzebiec, Poland

The comparison of the overall (r_o) and intragroup (r_i) correlations between carcass lean content (CLC) and meat quality traits was performed on three groups of pigs tested in respespect to RYR1 gene status (33 NN, 27 Nn and 30 nn). It was assumed that the intragroup correlation eliminates the genotype effect. The evaluated carcass traits were: backfat thickness (BF), loin eye area, (LEA) and CLC measured by UFOM. The correlated meat quality traits were: total quality score composed of eight meat characteristics (Q), pH_1 and meat colour lightness.
It is worthy to note that the overal correlations of Q with BF, LEA and CLC were high (r_o = 0.46**; -0.56**; -0.58** and -0.53** resp.) whereas the intragroup correlations were not significant (r_i = 0.14; -0.19; -0.17 and -0.14 resp.). However, the intragroup correlations of pH_1 or colour values with carcass traits were not diminished so distinctly: pH_1 vs BF r_o = 0.55** and r_i = 0.26**; pH_1 vs LEA r_o = -0.64** and r_i = -0.26**; pH_1 vs CLC r_o = -0.63** and r_i = -0.31**. Colour vs BF r_o = -0.43** and r_i = -0.17; colour vs LEA r_o = 0.50** and r_i = 0.21*; colour vs CLC r_o = 0.50** and r_i = 0.22*.
The results suggest that the significant negative relation between the qantity and quality of meat depends upon the linked effect of the nn genotype on meat deposition and on meat quality.

PIG PRODUCTION   [P]

Poster PGN4.20

Correlations between growth rate, slaughter yield and meat quality traits after the elimination of RYR1 gene effect
W. Kapelanski[1]*, J. Kortz[2], J. Kuryl[3], T. Karamucki[2], M. Bocian[1]. [1]University of Technology and Agriculture, Mazowiecka 28, 85-084 Bydgoszcz, [2]Agricultural University, Dr Judyma 24, 71-460 Szczecin, [3]Institute of Genetics and Animal Breeding, Polish Academy of Science, 05-551 Mrokow, Jastrzebiec, Poland

Several productive traits in pigs are under the genetic control and some of them are influenced by RYR1 gene status of pigs. That mainly concern to meat quality and in some extent to growth rate. The study was carried out on pigs of three RYR1 genotype: 33 NN, 27 Nn and 30 nn. The applied method and calculation of the overall (r_o) and intragroup (r_i) correlations between the compared traits allow to eliminate the group efffect, i.e. RYR1 genotype, on the assessed relationships. The av. daily gain (ADG) and fattening period (30 up to 103 kg l.w.) were not significantly differ between the NN, Nn and nn groups, whereas the slaughter yield (SY) differed at $P < 0.01$ (79.3 vs 80.3 vs 82.5%, resp.). The higher ADG was negative correlated with meat pigment content ($r_o = -0.51**$; $r_i = -0.51**$) and positive with greater free water content in meat ($r_o = 0.44**$; $r_i = 0.46**$). Similar relations refer to fattening period. The another relations however, were stated between r_o and r_i values for SY and meat traits such as colour lightness ($r_o = 0.37**$; $r_i = 0.10$), thermal drip ($r_o = 0.40**$; $r_i = 0.13$), free water ($r_o = 0.30**$; $r_i = 0.12$), pH_1 ($r_o = -0.49**$; $r_i = 0.08$), total quality score Q ($r_o = 0.48**$; $r_i = 0.18$). The results suggest that the significant negative relation between SY and meat traits exist only as an effect of RYR1 gene on the meat quality formation, whereas the relation between growth rate and meat quality exists per se as an unfavourable dependence.

Poster PGN4.21

Detection of the RYR1 locus variants and its implications in the Portuguese pig breeding programs
A.M. Ramos*[1], F. Simões[2], J. Matos[2], A. Clemente[2], T. Rangel-Figueiredo[1] 1- Dep. Zootecnia, UTAD, 5001 VILA REAL Codex, Portugal 2- INETI/IBQTA/DB/BQII, Est. Paço do Lumiar, 1699 LISBOA Codex, Portugal.

The RYR1 locus variants in the Portuguese swine breeds, Bísaro and Porco alentejano, were studied. Genotyping of the individuals was carried out using a method based on PCR. DNA samples from 70 Bísaro and 88 Porco Alentejano pigs, were amplified and digested with the restriction enzyme *Hha*I, allowing the determination of the variants.
Genotype and allelic frequencies for each breed were determined, as well as the genotype distribution among boars and sows. All the 3 possible genotypes were found and the recessive allele was found in both breeds and in both sexes.
For the purpose of providing the Portuguese traditional pork products industry with high quality meat, an efficient detection of the carriers of the recessive allele must be achieved, in order to eliminate these animals from the breeding schemes. This objective can be reached using the method described.

PIG PRODUCTION [P]

Poster PGN4.22

Effect of RYR 1 gene on meat quality in pigs of Large White, Landrace and Czech Meat Pig breeds

R. Bečková, P. David. Research Institute of Animal Production in Prague - Uhrineves Department of Pig Nutrition and Meat Quality in Kostelec n. O. Czech Republic

From the meat traits we found the values of pH_1, pH_{24}, in the LD muscle, the content of intramuscular fat (IF) and percentage of lean meat (by two point method) Genotypes of RYR 1 were determined by the method of the DNA test and their frequencies were as follows: Large white (LW) 26 N/N and 8 N/n, Landrace (L) 20 N/N and 8 N/n, Czech Meat Pig (CVM) 13 N/N, 21 N/n and 9 n/n. The differences in BU pigs between N/N a N/n genotypes were not significant. PSE meat has been found at N/n genotype only 5,88 %. The differences in L pigs between N/N and N/n genotypes were not significant again but more pronounced, except pH_1 ($P < 0,05$). The occurrence of PSE meat was much higher (17,8 %, N/N - 80 % and N/n 20 %). Significant differences between N/N, N/n and n/n genotypes were found in meat traits in CVM breed, where all three genotypes were found: pH_1 (P<0,05). The occurrence of PSE meat was much higher (17,8 %, N/N - 80 % and N/n 20 %). Significant differences between N/N, N/n and n/n genotypes were found in meat traits in CVM breed, where all three genotypes were found: pH_1 ($P < 0,05$), IF (N/N and n/n - $P< 0,05$), the percentage of lean meat (N/N and n/n - $P< 0,05$). The frequencies of genotypes correspond to the occurrence of PSE meat in this breed 39,53 % (N/N - 0 %, N/n - 58,8 % and n/n - 41,2 %). The results obtained proved that the pigs of N/N genotype showed a higher resistance to stress, performance of PSE meat, higher content of IF, and a lower percentage of lean meat. The heterozygous N/n genotype showed this meat traits to be medium.

Poster PGN4.23

Development of a highly accurate DNA-test for the *RN* gene in the pig

Chr. Looft[*,1], D. Milan[2], J.T. Jeon[3], S. Paul[1], C. Rogel-Gaillard[4], V. Rey[2], A. Tornsten[3], N. Reinsch[1], M. Yerle[2], V. Amarger[3], A. Robic[2], P. Le Roy[4], E. Kalm[1], P. Chardon[4] and L. Andersson[3].*
[1]*Institute of Animal Breeding and Husbandry, Christian-Albrechts-University, Kiel, Germany,* [2]*INRA, Toulouse, France,* [3]*Swedish University of Agricultural Sciences, Uppsala, Sweden,* [4]*INRA, Jouy en Josas, France*

The porcine *RN* gene, previously mapped to chromosome 15, has a large effect on meat quality. Four research groups have now joined their efforts in an attempt to identify the *RN* gene.
Reference families comprising altogether about 1000 backcross animals have been collected for precise linkage mapping of *RN*. The map shows that *RN* is located in a short interval between microsatellites Sw2053 and Sw936. FISH mapping of YACs containing Sw2053 or Sw936 allowed a regional assignment to chromosome 15q2.5. Pig markers have been developed for nine genes assigned to the corresponding region in humans and mapped using a pig radiation hybrid (RH) panel. A contig of more than 1.2 Mb containing the *RN* gene is constructed by screening YAC and BAC libraries. Additional markers are isolated by sequencing BAC ends. With these resources we developed a marker based test to determine the *RN*-status with a high accuracy. BAC clones will allow us to identify the *RN* gene by comparative sequencing and cDNA-selection in the near future.

PIG PRODUCTION   [P]

Poster PGN4.24

Performances of the Piétrain ReHal, the new stress negative Piétrain line.
P.L. Leroy[1], V. Verleyen [1] Department of Genetics, Faculty Veterinary Medicine, University of Liège, Belgium.

ReHal Piétrain was created by introgressing the negative stress gene from the LargeWhite into Piétrain (successive back-cross (BC)).
A total of 5,002 piglets were obtained from commercial sows and boars of different genetic origin, and ReHal (Nn) boars were compared to Landrace and Piétrain pure-bred animals. All the animals were born in 2 farms and fattened on 19 farms. Meat% was analysed with a linear fixed model including fattening farm effect, sow line, sex, boar within breed and breed of the boar. Results indicated that ReHal heterozygote boars performed quite well (58.93%) and that the estimated meat% are closed to the pure Piétrain results (59.48%) and better than Landrace boars (57.99%).
In 1997 and 1998, BC5, BC6 and BC7 generations have been produced. More recently, it has been decided to produce the ReHal homozygote stress negatives called ReHal^{cc}. A large number of ReHal^{cc}, with the meat and carcass performance of Piétrain pigs, is expected to be produced during 1999.

Poster PGN4.25

Effect of RN⁻ gene on the growth rate and carcass quality in crossbreeding of large white sows with P-76 boars
M. Koćwin-Podsiadła*, W. Przybylski, S. Kaczorek, E. Krzęcio. Pig Breeding Department, Agricultural and Pedagogical University, 08-110 Siedlce, 14 Prusa Str., Poland

The aim of study was to analyse the effect of the RN⁻ gene on growth rate and carcass quality in pigs with RN⁻rn⁺ and rn⁺rn⁺ genotypes derived from crossbreeding of polish large white sows with P-76 boars.
The investigations covered 50 animals (28 castrated males and 22 gilts) (25 RN⁻rn⁺ and 25 rn⁺rn⁺) that were originated from crossing 16 polish large white breed sows with 3 crossbreeding French boars P-76 (originated from composite lines Laconie and Penshire from Pen Ar Lan Breeding Company). The genotypes were identified on the basis of glycolytic potential measured in biopsy samples and also verified by technological yield of meat in curing and cooking process called as „Napole yield". Pigs were slaughtered at 100kg live weight. The day after slaughter the chilled half carcasses of all animals were subjected to partial dissection according to the method using in polish Pig Testing Stations for estimation of the carcass quality.
The obtained results shown that RN⁻ gene influenced the grown rate. Pigs with genotypes RN⁻rn⁺ had arround 20 days lower age at slaughter than rn⁺ rn⁺ and also smaller average back fat thickness (at the level of significance $P<0.07$). For other carcass traits there were not found the differences between genotypes.

PIG PRODUCTION   [P]

Poster PGN4.26

Breed and slaughter weight effects on meat quality traits in hal- pig populations
X. Puigvert [1,2], J.Tibau *[2], J. Soler [2], M. Gispert [2], A. Diestre [2]. [1]EPS-Universitat de Girona Avda L.Santaló s/n Girona 17003,Catalunya, Spain, [2]IRTA-CCP-CTC Monells 17121 Girona,Catalunya, Spain

Halotane negative pig populations are the optimal choice to produce female lines in modern pig production schemes but his effect on meat quality of commercial products is still controversial. Several meat characteristics were analyzed on 110 entire males slaughtered at 90 and 110 kg live weight. Animals tested to be non carriers to the halotane gene, of Large White, Landrace and Duroc breeds belongs to the National Pig Breeders Association.

Duroc animals reveals higher pH values (at 45' and 24h *post mortem*). *Longissimus toracis* muscle of Landrace and Large White animales were paler (subjective colour scale) compared with Duroc. The main difference between Duroc and white breeds is its higher intramuscular fat content. Within breeds, the genetic origin (nucleous herds), have a large effect on meat characteristics. Slaughter weight doesn't affects meat quality characteristics. Breed x slaughter weight interactions where not significant in most meat traits analysed.

Poster PGN4.27

Genotypic and allelic frequencies of the RYR1 locus in the Manchado de Jabugo pig breed
A.M. Ramos[1], J.V. Delgado*[2], J. Matos[3], C. Barba[2], F. Simoes[3], R. Sereno[2], A. Clemente[3], T. Rangel-Figueiredo[1] and M. Cumbreras[4]. [1]Div. Fisiologia Animal Departamento Zootecnia Univ. Tras-os-Montes e Alto Douro Apartado 2020 5001 Vila Real Codex Portugal. [2]Departamento de Genética. Universidad de Córdoba. Av. Medina Azahara, 9. 14005 Córdoba. Spain. [3]INETI/IBQTA/DB/BQII Estrada do Paco do Lumiar 1699 Lisboa Codex. Portugal. [4]Servicio de Ganadería Diputación Provincial. Apdo. 26. Huelva . Spain.

Manchado de Jabugo is a breed related to the Iberian Pig branch even though in the origin of this breed have participated with the Iberian Pig other breed from England and Germany, in the first half of the present century. This study is enclosed in an ambiciosus investigation of the implication of the variants of the RYR1 LOCUS in the breeds of the Iberian Peninsula, with a view to determinate the grade of expected affecting of the porcine stress syndrome, on the pig native populations in Spain and Portugal. Manchado de Jabugo because its origin could be a door of entry for high frequencies of the recessive allele from foray breed to the Iberian representative of the Mediterranean pig breed traditionally admitted with a low frequency for PSS. In this study we have employed the PCR methods for genotyping this locus in a sample of animal of both sexes belonging to the breed. The steps of the techniques was of follow : amplification of the locus fragment and digestion with the restriction enzyme HhaI. Here were presenting the genotype and allelic frequencies of the variants of this locus in both sexes, so we can reach important conclusion about the relationship among the diverse Iberian breed populations Spain and Portugal.

PIG PRODUCTION   [P]

Poster PGN4.28

Intramuscular fat content in some native German pig breeds
U. Baulain*[1], P. Köhler[1], E. Kallweit[1] and W. Brade[2], [1]Institute of Animal Science and Animal Behaviour (FAL), Mariensee, 31535 Neustadt, Germany, [2]Chamber of Agriculture Hannover, 30159 Hannover, Germany.

Sows and castrates of some native German pig breeds, Angeln Saddleback (AS), Bentheimer Black Pied (Be) and Swabian Hall Saddleback (SH) and their crosses with Piétrain were fattened at a testing station and compared to Large White x German Landrace (DE*DL), Duroc x German Landrace (Du*DL) and to purebred Piétrain (Pi). Due to the standard feeding scheme at the testing station sows were fed ad libitum but castrates restricted. Fattening, carcass and meat quality data of 399 pigs were registered. Percentage of lean was predicted by FOM. Intramuscular fat content (IMF) of M. long. dorsi was predicted by means of near infrared transmission. The highest IMF was found in SH, Du*DL and AS castrates with 1.64%, 1.58% and 1.42%, respectively. IMF of less than 0.70% was observed in Piétrain sows and castrates as well as in female SH*Pi. Thus IMF in native breeds was less than expected. The lowest pH values (45 min) of M. long. dorsi were found in both sexes of purebred Pi (<5.60) and SH*Pi (<5.93). An over-all correlation of r = 0.46 between IMF and pH was observed. The highest percentage of lean was noticed in female Pi and SH*Pi and in Pi castrates with 61.9%, 57.4% and 57.0%, respectively. Portion of lean was less than 46.5% in Be and SH castrates. An over-all correlation of r = -0.53 between percentage of lean and IMF was determined.

Poster PGN4.29

Fatty acid and cholesterol composition of the lard of different genotypes of swine
J. Csapó, F. Húsvéth[1], Zs. Csapó-Kiss, P. Horn, Z. Házas, É. Varga-Visi. Pannon Univertsity of Agriculture, [1]Faculty of Animal Science Kaposvár, H-7400 Guba S. u. 40. [1]Georgikon Faculty of Agricultural Sciences, H-8360 Keszthely, Deák F. u. 16.

The authors determined the fatty acid composition and the cholesterol content of lard of Mangalica, Hungarian great white x Hungarian Landrace, and Mangalica x Duroc swain. It was established that there are no significant differences among genotypes neither in saturated-, unsaturated- and essential fatty acid content nor cholesterol content of lard. According to their investigations the cholesterol content of the three different swine genotypes was between 71-109 mg/100g. They emphasise the very high concentration of oleic acid (43,57-44,81 relative%), and linoleic acid (10,63-11,47 relative%) content of lard.

PIG PRODUCTION   [P]

Poster PGN4.30

The effect of paternal breed on meat quality of progeny of Hampshire, Duroc and Polish Large White boars
J. Nowachowicz, G. Michalska, B. Rak, W. Kapelanski*. University of Technology and Agriculture, Mazowiecka 28, 85-084 Bydgoszcz, Poland

The study comprised the progeny of Hampshire (H), Duroc (D) and Polish Large White (PLW) boars mated to PLW sows. Each crossbred (HxPLW and DxPLW) and purebred (PLWxPLW) groups consisted of 30 female pigs slaughtered at 185 days of age. The pH_1 value was recorded 45 min postmortem in lumbar region of longissimus dorsi muscle. Colour lightness of meat and soluble protein content were determined 24 h postmortem. There was assumed that the greater value of meat soluble protein content is a reliable indicator of the higher water holding capacity of meat.

The pH_1 values were not significantly differed by the breed of sire (6.32 ± 0.16 in HxPLW; 6.33 ± 0.14 in DxPLW and 6.31 ± 0.39 in pure PLW), similarly as the meat colour lightness (24.6 ± 2.4; 24.5 ± 1.3 and $23.6\pm2.3\%$, resp.). However, the meat protein solubility in HxPLW was the lowest ($7.93\pm0.33\%$) medium in DxPLW ($8.25\pm0.40\%$) and the higtest in purebred PLW ($8.73\pm0.69\%$). The differences were significant ($P<0.05$ for HxPLW vs DxPLW and $P<0.01$ for HxPLW vs pure PLW and for DxPLW vs pure PLW).

It is concluded that Hampshire and Duroc sire effects were unfavourable on meat quality and decreased the water holding capacity of meat.

Poster PGN4.31

Comparison of several pig breeds in fattening and meat quality in some experimental conditions of a Czech region.
T. Adamec*, B. Naděje, J. Laštovková and M. Koucký. Research Institute of Animal Production, Praha 10 - Uhříněves, 104 01 Czech Republic

Experimental fattening of LW, LWxL, (LWxL)xH, (LWxL)x(LW imported), (LWxL)x(BLxD) groups and (LWxL)xCMP (Czech Meat Pig) breed was done in 1992-1998. These groups involved castrates and gilts in more than 2 repetitions. In the last group of the breeds there were tested fattened young boars, too. Conditions of fattening, feed composition, individually boxing, transport, slaughtering and laboratory methods of meat quality were identical.

The (LWxL)xH breed was the best in growing intensity (789g), feed consumption (3.21kg/kg) and in high lean meat (55%), followed by (LWxL)x(BLxD) breed (816g, 3.11kg/kg resp.), with low lean meat of 49.6%, however, which was similar to the original farm conditions. The content of intramuscular fat in m.l.l.t. was lower in hybrids of LW (imported), BL and D breeds (1.3-1.7% unlike 1.7-1.9%), all in father positions. The (LWxL)xCMP breed was best sensorially evaluated in taste and flavour, and the LW and (LWxL)H breeds in texture and humidity of m.l.l.t. The castrates of (LWxL)xCMP breed were absolutely best in daily gains, feed consumption and lean meat, however, the lowest in the content of dry matter, protein and intramuscular fat of m.l.l.t. (1.1%), and also the lowest in the evaluation of all sensorial criteria.

PIG PRODUCTION   [P]

Poster PGN4.32

Early detection of breed differences in fat distribution in pigs measured by CT
K. Kolstad. *Department of Animal Science, Agricultural University of Norway, P.O.Box 5025, N-1432 Ås, Norway.*

The object of this study was to investigate whether genetic differences in fat amounts and distribution at time of slaughter can be detected at an earlier stage of growth. Fat deposition were investigated in 70 pigs of the three genetic groups Norwegian Landrace (L), Duroc (D) and a crossing between Norwegian Landrace and a Norwegian Landrace line selected for increased backfat and slow growth (LP*L). Fat distribution was measured repeatedly by computer tomography (CT). This X-ray based technique allows for non-invasive repeated observations of detailed body components by collecting a series of cross-sectional images with a certain constant distance between them throughout the body. CT at 10, 25, 50, 85 and 105 kg live weight was used to quantify changes in amounts and proportions of the depots subcutaneous, inter/intra muscular and internal fat during growth.

The breeds differed in fat distribution during the growth period. When adjusting to equal fat amounts, LP*L had significantly higher adjusted amounts of subcutaneous fat from 25 kg live weight and through out the experiment, while D had higher adjusted amounts of inter/intra muscular fat during this period. L had significantly higher adjusted amounts of internal fat during the whole experiment. This suggests that breeding for differences in fat distribution can be based on registrations on young pigs.

Poster PGN4.33

Fat score, an index value for fat quality in pigs - its ability to predict properties of backfat differing in fatty acid composition
K. R. Gläser*, M.R.L. Scheeder, C. Wenk, *Swiss Federal Institute of Technology (ETH) Zurich, Institute of Animal Science, Nutrition Biology, CH - 8092 Zurich, Switzerland*

Fat score, a semi-automated determination of double bonds in adipose tissue, is an established method in Swiss slaughterplants to assess backfat quality in pigs. Since body fat composition is directly influenced by dietary fat two feeding trials (i and ii) were conducted. In the first trial i) the effects of monoenoic or polyenoic fatty acids at similar number of double bonds per kg feedstuff (70, 49.5, 31.7 g/kg pork fat, olive oil, soybean oil, respectively) on fat score, consistency and oxidative stability were studied. In a second trial ii) the effects of olein and stearin fraction of pork fat and hydrogenated fat (50 g/kg) were investigated.

Olive and soybean oil impaired fat score and firmness of adipose tissue to a greater extent than pork fat supplemented feed. In contrast, highest induction times and therefore best oxidative stability was measured in fat from pigs fed olive oil. The hydrogenated fat resulted in lower fat scores and higher firmness. Compared to the stearin fraction, the olein fraction led to slightly higher fat scores, softer consistency and reduced oxidative stability, but without reaching statistical significance. Significant correlations were found between fat score and consistency (i: $r^2=0.672$) and oxidative stability (i: $r^2=0.280$, ii: $r^2=0.166$). It may be concluded, that fat score gives an useful at-line estimate of firmness of pig adipose tissue.

PIG PRODUCTION   [P]

Poster PGN4.34

Meat quality with reference to EUROP carcass grading system
W. Kapelanski*, B. Rak, J. Kapelanska, H. Zurawski. University of Technology and Agriculture, Mazowiecka 28, 85-084 Bydgoszcz, Poland

The experiment was conducted on 63 pigs being the progeny of Pietrain boars and crossbred sows (Polish Large White x Polish Landrace). The animals were fed and reared in the same manner and were slaughtered when attained the 105 kg l.w. The carcass lean content (CLC) was measured by UFOM-100 device and classified into EUROP grading system. In particular E,U,R,O class there were 12, 13, 27 and 11 pigs, resp. (P class was absent). The CLC in particular E,U,R,O class attained 58.5±3.2, 52.4±1.4, 47.9±1.3 and 42.2±1.2%. Meat quality generally tended to increase with the decrease of carcass lean content. pH_1 has increased from 5.97 in E through 6.12 and 6.21 in U and R to 6.49 in O class. LF_1 values were the highest in the E class (7.67 vs 4.44, 4.05 and 3.36 mS; P<0.001). Drip loss was the highest in E and the lowest in O class (6.5 vs 2.1%; P<0.001). The WHC expressed as a per cent of the bound water in meat was the lowest in E class (68.5 vs 72.3, 72.2 and 74.8%; P<0.01). Colour lightness was differed in a lower range. The lightest meat was in E class and the darkest was in the O class (25.9 vs 22.7%; P<0.05).
It is concluded that the meat of high graded carcasses, i.e. E class was of inferior quality than in the U, R and O classes.

Poster PGN4.35

The evaluation of pork meat productivity
V.N.Dementyev, I.I. Gudilin, V.G.Pilnikov, Novosibirsk Agrarian University, 160 Dobrolubov Str.,630039 Novosibirsk, Russia

The research was carried out in animal breeds Kemerovskaya and Sibirskaya Severnaya that are common in West Siberia. Fat thickness over the 6 and 7th thoracic ribs and section area of the longest back muscle at the last rib were measured. The pork carcass composition was determined for natural anatomic parts. Preslaughter living weight of young animals was 95-100 kg.
For Kemerovskaya and Sibirskaya Severnaya breeds the correlation between pork portions in a pelvis-thigh part and pork carcass was r=0.61±0.16 (P<0.001) and r=0.71±0.15 (P<0.001), respectively.
By the method of the least squares the equations of straight linear regression were calculated. So, for Kemerovskaya breed the percent of meat in the pork carcass (Y) was calculated for its percent in the pelvis-thigh part (X) and for the section area of the longest back muscle, cm^2 (X), Y=27.41+0.46X (1) and Y=43,97+0.34X (2). respectively. Averaged over groups Y, (1) and (2) it was 58,2, 58,4, and 55,0%, respectively. Hence, the evaluation of pork qualities is done more efficiently for the easily determined composition of the parts than for the measuments.

PIG PRODUCTION   [P]

Poster PGN4.36

The correlations between the fattening and slaughtering performance in pigs.
A. *Pietruszka, R. Czarnecki, Jacyno E. Department of Pig Breeding, University of Agriculture, Dr. Judyma 10 Street, 71-460 Szczecin, Poland.*

The relationship between the fattening performance and slaughter traits of carcass were carried out on 120 fatteners, progeny of Polish Large Whitte x Polish Landrace sows and boars of Pietrain, Pietrain x Polish Synthetic Line 990 and Pietrain x Duroc. The hybrid boars were made using the reciprocal cross. Together there were 5 group. There were after 24 pigs each genotyp (12 gilts and 12 barrows) in the animals experimental group. The animals were fattened from 34 to 100kg of liveweight and slaughtered for testing. Obtained significantly and highsignificantly relationship between daily body weight gain and percentage meat in carcass measured on alive animals $r=0,23^{**}$ and after slaughtered $r=0,14$, between feed conversion per 1kg daily body weight gain and percentage meat in carcass respectivly: $r=-0,18^*$ and $r=-0,17^*$ but feed conversion per 1kg meat gain and percentage meat in carcass respectivly: $r=-0,40^{**}$ and $r=-0,52^{**}$. It was found out that the intake of food for 1kg meat production better correlated with percentage meat in carcass than intake of food for 1kg body weight production. Results of the study clearly show that the pigs with higher geneticaly predisposition of meatness characterized better fattening performance.

Poster PGN4.37

Comparison of fat supplements of different fatty acid profile with growing-finishing swine
J.Gundel*[1] - A.Hermán Ms.[1] - M. Szelényi Ms[1]. - G. Agárdi[2]: [1]Research Institute for Animal Breeding and Nutrition,2053-Herceghalom, Hungary, [2]Debrecen University of Agricultural Sciences, 4015-Debrecen, P.O Box. 36.

Hungarian LW X Dutch LR (n=4x12, individual keeping, 30,5 kg BW), females and barrows were fed by four corn - barley based isocaloric, isoproteic, iso-AA, and iso-lipide rations (ME:13,3 MJ/kg, CP: 16,3 %, LYS: 0,95 %, EE: 4,3 %), using 1) animal fat (AF), 2) sunflower expeller cake(SFEC) 3) fullfat sunflower seed (FFSF) 4) fullfat soybean meal (FFSB) as fat sources of different linoleic acid levels. The linoleic acid levels (%) in the diets were: 1. (AF): 1,5, 2.(SFEC): 2,5, 3. (F. SF): 2,0, 4. (F SB): 2,1.
Slaughtered at a liveweight of 98,1 kg, the following performances were obtained (in the same order as the diets): daily gain: 624, 655, 657, 690 g; feed conversion; 3,55, 3,21, 3,30, 3,14; lean meat, %: 52,1, 52,4, 52,9, 53,5; backfat, fat, %: 13,6, 22,9, 20,0, 17,9.
Comparing fattening performances and carcass quality the vegetable fat sources (partlicularly fullfat soybean) proved to be superior to animal fats. Vegetable fats increased the linoleic acid in the carcass fat related to animal fats (100%) in the following order: FFSF: 168 %, SFEC: 147%, FFSB: 131 %.

PIG PRODUCTION [P] Poster PGN4.38

The pork meat quality in pigs with a different intenzity of nitrogen substances retention
M. Čechová[1], V. Prokop[2], K. Dřímalová[1], Z. Tvrdoň[1], V. Mikule[1]. [1]Department of Animal Breeding, Mendel University of Agriculture and Forestry, Zemědělská 1, Brno 613 00, Czech Republic, [2]Research Institute for Animal Nutrition, Ltd., Pohoøelice 691 23, Czech Republic

In final carcass pig hybrids interbreeding combination No.1(White Improved x Landrase) x Large White and No.2 (White Improved x Landrase) x (Czech Meat Pig x Pietrain) in balance experiment in weight 48 kg there were found limits for nutrigen retention in pig combination No.1 in a level of 147 g per pig per day and in pig combination No.2 in a level of 165 g per pig per day. After following fattening till weigt 100 kg (pigs were fed with the same feed mixture) there were some traits of meat quality determinated in eye muscle: combination No.1 - pH_1 + 6,40, dripping water losses 4,11 and content of intramuscular fat 3,25, combination No.2 - pH_1 + 6,40, dripping water losses 4,35 and content of intramuscular fat 2,53. These results weren't statistically conclusive.

Poster PGN4.39

The effect of dietary Ca-fatty acid salts of linseed oil on cholesterol content in *longissimus dorsi* muscle of finishing pigs*
T. Barowicz*. National Research Institute of Animal Production, 32-083 Balice, Poland

36 Polish Landrace fatteners of both sexes, divided into 3 groups, were fed from 70 to 100 kg liveweight with a complete mixture containing 0 (control), 8 or 15 per cent of Ca salts of fatty acids (CaFAS) made partly from linseed oil and blended fat in the proportion 1:1. The mixture contained 85% UFA, including 30% PUFA. The experiment concluded with slaughter. After dissection, a sample of *longissimus dorsi* muscle was taken to determine total cholesterol after lipid extraction. Total cholesterol was assayed using the enzymatic method.
Total cholesterol in *longissimus dorsi* muscle of experimental fatteners was observed to drop from 55.70 to 52.49 mg/100 g fresh tissue. The differences were particularly noticeable in the meat of sows, amounting to 57.14 (control), 52.68 and 51.11 mg/100 g fresh tissue, respectively. However, the differences were not statistically significant.
It is suggested that CaFAS made partly from linseed oil, especially as a 15 per cent supplement to complete mixture for finishing pigs, may be one of the ways of improving the dietary value of pork.

* The study was financed by Committe for Scientific Research (grant No 5 PO6E 058 14)

PIG PRODUCTION   [P]

Poster PGN4.40

Soybean oil, sex, slaughter weight, cross-breeding - influence on fattening performance and carcass traits of pigs
R. Kratz[1]; E. Schulz[1]; G. Flachowsky[1]; P. Glodek[2]*. [1]Inst. Anim. Nutr., Fed. Agric. Res. Centre, 38116 Braunschweig Bundesalle 50, [2]Inst. Anim. Breed. a. Gen., Univ. Göttingen

High daily live weight gain, protein retention and carcass quality should be obtained by pig fattening. Therefore a fattening trial was started with the factors feed (without / with 2,5 % soybean oil) x sex (castrated male / gilts) x slaughter weight (110 /120 kg) x sire line (PI [nn], PI [NN], PIxHA [NN]) in the live weight range of 30 - 110/120 kg. Feed intake was isoenergetic. Soybean oil had no effect on fattening performance or carcass traits.

sex	castrates						gilts					
slaughter weight	110 kg			120 kg			110 kg			120 kg		
sire line	PI	PI	PIxHA	PI	PI	PIxHA	PI	PI	PIxHA	PI	PI	PIxHA
	[nn]	[NN]	[NN]	[nn]	[NN]	[NN]	[nn]	[NN]	[NN]	[nn]	[NN]	[NN]
ADG[1] [g/day]	819[ab]	829[ab]	823[ab]	808[b]	799[b]	794[b]	876[a]	819[ab]	849[ab]	829[ab]	815[ab]	812[ab]
ECR[2] [MJ ME/kg ADG[1]]	35,4[abc]	34,7[abc]	35,3[abc]	37,1[a]	37,4[a]	37,6[a]	32,5[c]	34,8[abc]	33,7[bc]	35,2[abc]	35,2[ab]	36,4[ab]
lean meat[3] [%]	57,7[cde]	56,7[de]	56,7[de]	57,9[bcde]	58,7[bcd]	56,2[e]	61,1[a]	58,6[bcde]	60,1[ab]	61,2[a]	60,0[abc]	59,7[abc]
ham [kg]	13,2[de]	12,7[ef]	12,3[f]	14,0[abc]	14,2[ab]	13,7[bc]	13,6[cd]	12,9[ef]	12,9[ef]	14,5[a]	14,1[abc]	13,9[bc]
ham, lean[4] [kg]	7,60[bcd]	7,00[de]	6,70[e]	7,78[bc]	7,79[bc]	7,55[bcd]	7,96[ab]	7,29[de]	7,42[bcd]	8,43[a]	7,99[ab]	7,94[ab]

[a,b,c,d,e] means with different superscripts in the same row are significantly different [p ≤ 0,05]; [1] = average daily gain; [2] = energy conversion ratio; [3] = lean meat percentage [Bonner-Formel, 1997]; [4] = ham without skin, bone, separable fat

---

Poster PGN4.41

Transfer of vitamin E supplements from feed into pig tissues
G. Flachowsky [1], H. Rosenbauer [2], A. Berk [1], H. Vemmer [1] and R. Daenicke *[1]. [1]Inst. of Animal Nutrition, FAL, Bundesallee 50, D - 38116 Braunschweig, Germany, [2]Inst. of Chemistry and Physics, BAFF, E.C.Baumann-Str. 20, D - 95326 Kulmbach, Germany

Five feeding experiments with growing-finishing pigs were carried out to investigate the influence of various additional vitamin E levels and application times on vitamin E content of some tissues and the vitamin E transfer into animal body.
Vitamin E content of control diets amounted to 20 mg kg^{-1}. Additional Vitamin E levels (0.5; 1.0; 1.2 g per animal and day) were given 7, 14 or 21 days before slaughtering and compered with 100 or 200 mg vitamin E per kg feed during the total growing-finishing period. Additional vitamin E intake of various groups amounted to 7.0 until 41.6 g per animal. Vitamin E content of tissues increased with higher dosage and longer application time. The vitamin E concentration in liver, muscle and backfat of control group amounted to 4.5; 2.5 and 9.0 mg kg^{-1}. Similar vitamin E intake (~25 g) was achieved in pigs consuming 100 mg vitamin E per kg feed or 1.2 g vitamin E per day during the last 21 days.
Finally additional vitamin E levels increase vitamin E content in foods of pig origin, but its contribution to improve the vitamin E transfer into animal body (about 1% of added vitamin E) and the vitamin E supply of humans are relatively low.

PIG PRODUCTION [P]

Poster PGN4.42

Einfluß der Fütterungsintensität in der Ferkelaufzucht auf die Mast- und Schlachtleistung
W. Wetscherek[1], F. Lettner[1], H. Huber[2], S. Bickel[1] und F. Gaheis[1]. - [1]Abteilung Tier-ernährung, Universität für Bodenkultur Wien, Gregor Mendel Str. 33, A-1180 Wien, Österreich, [2]Landwirtschaftkammer für Oberösterreich, Auf der Gugl 3, A-4010 Linz, Österreich.

In einem Fütterungsversuch wurden 2 Aufzuchtintensitäten mit je 20 Tieren von 10 bis 35 kg Lebendgewicht getestet. Das Futter der Intensivgruppe enthielt 14,0 MJ ME/kg, 20,5 % XP und 1,15 % Lysin. In der Extensivgruppe lag der Nährstoffgehalt nur bei 12,1 MJ ME/kg, 17,5 % XP bzw. 0,92 % Lysin. Auch die übrigen Aminosäuren wurden im gleichen Verhältnis wie das Lysin reduziert. In der darauffolgenden Schweinemast von 35 bis 111 kg Lebendgewicht wurden die Tiere beider Gruppen gleich gefüttert. In der Ferkelaufzucht schnitt die Intensivgruppe mit 506 g Tageszu-wachs bzw. einem Futteraufwand je kg Zuwachs von 1,97 kg deutlich besser ab als die Extensivgruppe mit 473 g Tageszuwachs bzw. einem Futteraufwand je kg Zuwachs von 2,10 kg. Während im ersten Mastabschnitt (35-62 kg Lebendgewicht) die intensiv aufgezogenen Ferkel (INT) noch eine bessere Mastleistung zeigten, änderte sich dies im weiteren Mastverlauf zu gunsten der extensiv aufgezogenen Ferkel (EXT). Insge-samt erreichten die Tiere fast identische Mastleistungen mit 770g (EXT) bzw. 771g (INT) Tageszuwachs und einem Futteraufwand je kg Zuwachs von 3,05 kg (EXT) bzw. 3,09 kg (INT). Im Magerfleischanteil unterschieden sich die Gruppen nicht.

Poster PGN4.43

Effect of healing herb (symhytum peregrinum) fed to pigs on meat quality traits
G. Bee*, G.J. Seewer Lötscher and P.-A. Dufey. Swiss Federal Research Station for Animal Production, 1725 Posieux, Switzerland.

Healing herb is known for the high crude protein content and substances which positively affect growth performance and health status of the pig. However little is known about the influence on determinant parameters of meat quality. In the present study 22 Large White pigs were fed based on live weight either a common growing-finishing diet (C) or the same diet supplemented with 10% leaves from healing herb (H). The animals were slaughtered at an average live weight of 105 kg. Tissue samples of backfat and m. longissimus dorsi were collected 24 h after slaughter. Animals of treatment H had lower stearic and palmitic ($P<.05$), but higher oleic and linolenic acid ($P<.05$) concentration in the adipose tissue than those of treatment C. The differences between treatments were not evident in the muscle lipids. Drip and cooking losses as well as both mean pH values measured 45 min and 24 h post-mortem were not affected by the healing herb supplementation. With respect to color measurements, the a* component (redness) and chroma tended to be higher ($P<.06$) in treatment H compared to C. Furthermore, the taste panel evaluation did not reveal any treatment differences. The data of the present study suggest, that a moderate healing herb supplementation has an impact on the composition of backfat lipids without affecting other meat quality traits.

PIG PRODUCTION [P]

Paper P5.1

Effects of management during the suckling period on post-weaning performance of pigs

S.A. Edwards*[1] and J.A. Rooke[2]. [1]*Department of Agriculture, University of Aberdeen, 581 King Street, Aberdeen AB24 5UA, UK*, [2]*Animal Biology Division, SAC, Craibstone Estate, Bucksburn, Aberdeen AB21 9YA, UK.*

Data from commercial farms and controlled experiments indicate large individual variation in post-weaning growth, even under standard nutritional and environmental conditions, indicating important differences in the predisposition of piglets to cope with the weaning process. The biggest factor influencing this predisposition appears to be litter of origin, which typically explains >50% of variation. This litter influence contains both pre- and post-natal components, indicating genetic and/or developmental factors as well as factors arising from the suckling period. Piglets coming from enriched suckling environments show less growth check, possibly associated with developmental differences in coping strategy to the multiple stressors associated with weaning. Post-weaning performance may be influenced by weight and body composition at weaning, and thus by the quantity and composition of sow milk ingested. Provision of creep feed to enhance digestive enzyme development has limited benefit, and may prime deleterious antigenic responses after weaning if certain raw materials are incorporated, but prior establishment of appropriate gut flora may improve post-weaning performance. The role of the immune system in post-weaning response requires further investigation.

Paper P5.2

Development of the gastrointestinal ecosystem during the weaning period. Need for feed additives?

B.B. Jensen*. *Department of Animal Nutrition and Physiology, Danish Institute of Agricultural Sciences, Research Centre Foulum, 8830 Tjele, Denmark.*

The diverse collection of microorganisms colonising the healthy gastrointestinal tract of pigs, referred to as the microbiota, plays an essential role not only for the well being of the pig but also for animal nutrition and performance and for the quality of animal product. A formidable array of feed additives, claimed to improve pig performance through an effect on the gut ecosystem, is available to the feed compounder and pig producer. There is no doubt that the effect of feed additives is greatest in young pigs. In nature weaning is a slow process, but modern production methods often involve very early and sudden weaning. Several studies have shown that the gut ecosystem is very unstable around weaning. The populations of lactobacilli decline just after weaning, while coliform counts increase, indicating an inverse connection between lactobacilli and coliform bacteria. This tends to make the pig susceptible to scouring and poor growth performance. A number of naturally occurring and artificial factors have been shown to affect the composition and activity of the microbiota around weaning these include: diet composition, growth promoting antibiotics, probiotics, organic acids and fermented liquid feed. The objective of the present paper is to review our current state of knowledge of the development of the microbial ecosystem in the gastrointestinal tract around weaning and how feed composition and feed additive affect this development.

PIG PRODUCTION   [P]

Paper P5.3

Nutritional requirements of weaned pigs and how to meet the demands by appropriate diet formulations
B. Sève and J. LeDividich. INRA Station de Recherches Porcines 35590 Saint-Gilles, France

Pigs are currently weaned at 3-4 weeks of age, whereas the development of the digestive function is not complete, and they are characterised by poor appetite. Therefore, it is important in defining the nutritional requirements to consider, firstly, a period of adaptation of about two weeks and, secondly, a post-adapting period with rapid growth. During the first period, the priority should be given to meeting nutrients requirements for growth of essential tissues, and preservation of further expression of the muscle growth potential. Energy shortage will be tolerated if pigs get sufficient amounts of adequately balanced protein, not only in digestible lysine, but also in digestible secondary-limiting amino acids such as threonine, methionine and tryptophan. Attention should be given to the proper use of industrial amino acids, in order to meet the requirements without any relative excess, which could affect the appetite. Highly digestible ingredients should be preferred for both protein and non-protein energy supply. However a proportion of conventional ingredients, such as vegetal protein and fibre, will help further adaptation of the pig to its diet. At the moment, additives alternative to the antibiotics should be developed for better control of pathogenic microbial flora. In the future, research is still needed on specific nutrients requirements for acquired immunity and on dietary ingredients preventing gastric paresis and more protective of the gastrointestinal mucosa.

Paper P5.4

Strategies and methods for allocation of food and water in the post-weaning period
P.H.Brooks. University of Plymouth, Seale-Hayne Faculty of Agriculture, Food and Land Use, Newton Abbot, Devon, TQ12 6NQ, England.

The newly weaned pig poses the biggest management challenge on many units. Pigs that have been growing at 300 g per day while suckling the sow often grow at half that rate or less in the week following weaning. When suckling the sow the pigs have been used to having their hunger and thirst satisfied by the sow's milk and have been stimulated by the sow to feed at intervals of 40-50 minutes. Post-weaning the piglet has to learn to distinguish between the physiological drives of hunger and thirst and to learn how to satisfy them with water and solid food in order to maintain its homeostatic balance. Work at the University of Plymouth has shown that it can take more than a week for the pig to restore its daily fluid intake to the equivalent of that on the day before weaning. Some pigs take more than 24 hours to take their first feed or drink post-weaning resulting in severe dehydration and hypoglycaemia. Transient inanition results in villous atrophy and this in turn causes impaired nutrient absorption. This paper reviews contemporary research on the behaviour and nutrition of the weaner pig. It considers the ways in which new insights into pig behaviour might be used to inform management strategies for the provision of food and water and thereby ameliorate some of the problems of poor performance experienced by weaner pigs.

PIG PRODUCTION   [P]   Paper P5.5

Environmental requirements of the weaned pigs
J Le Dividich. INRA,Station de RecherchesPorcines,35590 Saint Gilles, France

Environmental conditions have direct effects on the metabolism and performance and are potentially a major risk factor for gastrointestinal disturbances in the weaned pig. This review aims to assess the environmental requirements including ambient temperature, air velocity and relative humidity of the weaned pig. Emphasis is put on ambient temperature, the major component of the climatic environment. Two periods are examined in relation to the level of feed intake. First, during the critical period following immediately weaning, the effects of pronounced underfeeding on the energy metabolism and on the change in the lower critical temperature of the piglets are examined in relation to the age at weaning. During the post-critical period, optimal temperature determination is based on performance. Type of flooring is taken into account. Ways to reduce the heating requirement include provision of a microenvironment (hovers, covers), reduction of the nocturnal temperature and reduction of the overall temperature Their effects on performance and health are analysed. Air velocity and draughts are assessed from their effects on behaviour, temperature requirement and performance of the piglets. Effects of relative humidity are presented in relation with ambient temperature. Finally the effects of environmental conditions on the piglet health are discussed.

Paper P5.6

Housing of weaners - meeting their environmental demands
B. K. Pedersen*, Danish Bacon and Meat Council, Dept. of Housing and Production Systems, Axeltorv 3, DK-1609 Copenhagen V., Denmark.

In nature weaning is a gradual process, which is not completed until 17 weeks postparturition. In modern pig production systems weaning is abrupt and occurs 2-5 weeks after birth. Therefore, meeting the environmental demands of the weaned pig is a challenge comprising several components, such as the sudden separation from the mother, the change in diet and the change of environment. During the last three decades much have been learned about these stressors and how their negative effects might be reduced. Diets meeting the pigs requirement and automatically controlled heated and ventilated nurseries have become common prerequisites of a successful weaning programme. Lately, segregated early weaning has been introduced as a new method of introducing a disease break between mother and off-spring. This procedure has added new possibilities of exploiting the pig's genetic potential for growth without the aid of drugs. However, transport has been added as a new stressor, which needs to be handled appropriately in order not to compromise welfare and production. The public's growing attention of animal well-being has led to new designs of weaner housing, which allow pigs to choose between different thermal zones and flooring within the pen environment. Current Danish research indicate that modern information technology might become an important tool in future management programmes for weaners by electronic monitoring of pigs' eating and drinking patterns.

PIG PRODUCTION [P]

Paper P5.7

Individual feed intake and performance of group-housed weanling pigs
E.M.A.M. Bruininx[*,1], C.M.C. van der Peet-Schwering[1], J.W. Schrama[2], P.C. Vesseur[1], L.A. den Hartog[1], H. Everts[3] and A.C. Beynen[3]. [1]Research Institute for Pig Husbandry, P.O. Box 83, 5240 AB Rosmalen, [2] Department of Animal Science, Wageningen Agricultural University, P.O. Box 338, 6700 AH Wageningen, [3] Department of Nutrition, Utrecht University, P.O. Box 80152, Utrecht, The Netherlands

Feed intake by weanling pigs is often suggested to be related to post weaning problems. IVOG®-feeding stations were used for individual feed intake recordings. The main objective was to describe individual feed intake in relation with performance of group housed weanling (d 28) pigs. At weaning, pigs (n=189) were divided in Light, Middle, and Heavy animals (6.7± .05, 7.9±.04, and 9.3 ±.08 kg). During the first four d after weaning ADFI was unaffected by weight class. Feed intake per kg metabolic weight and ADG differed however between Light, Middle and Heavy pigs. (22, 17, and 15 g/kg$^{.75}$/d: and 86, 56, and 30 g/d, respectively). Averaged over the 34-d experimental period, Light pigs had a lower ADFI and ADG than Heavy pigs. Light pigs ate more within 4 d after weaning than Heavy pigs. Several feed intake traits in relation to performance will be presented. The data indicate that these traits depend on weaning weight.

Paper P5.8

Feeding weaned pigs pellets or meal?
Effects on performance, water intake and eating behaviour
M. Laitat*, M. Vandenheede, A. Désiron, B. Canart, B. Nicks. Faculty of Veterinary Medicine, University of Liège, Belgium

Performance, water intake and feeding behaviour of 2 groups of 30 (test 1), 40 (test 2) or 50 (test 3) weaned pigs fed either pellets or meal of the same formulation were compared. Average daily gains (ADG) were higher for pigs fed pellets in tests 2 (413 vs. 363 g/day, $P<0.001$) and 3 (356 vs. 324 g/day, $P<0.05$). Mean daily water intake (DWI) was higher with meal, but significantly only during test 1 (2.31 vs. 1.65 l/day, $P<0.01$). The higher was the group size, the lower were ADG (both diets) and DWI (only with meal). The occupation time and the number of animals using the feeder simultaneously were higher when pigs were fed meal rather than pellets, whatever the animal density: test 1: 82.6% vs. 69.9% ($P=0.05$) and 3.8 vs. 2.3 ($P<0.01$); test 2: 90.9% vs. 77.9% ($P=0.11$) and 5.2 vs. 3.1 ($P<0.01$); test 3: 96.2 vs. 83.6% ($P<0.05$) and 5.9 vs. 3.8 ($P<0.01$). Pigs showed a preferentially diurnal feeding activity in the 3 tests with pellets ($P<0.01$ or $P<0.001$) but only in test 1 with meal ($P<0.01$). In conclusion, better performance was measured when a low density group was fed pellets. Furthermore, pigs needed more time to eat meal and a high number of pigs/feeder impaired feeding behaviour and eventually welfare. The number of pigs/feeder has thus to be adapted to the food presentation.

PIG PRODUCTION   [P]   Paper P5.9

Restricted feeding for prevention of *E. coli* associated diarrhoea in weaned pigs
L. Jørgensen*, M. Johansen and C. F. Hansen. *The National Committee for Pig Breeding, Health and Production, Axeltorv 3, 1609 Copenhagen V, Denmark*

The aim of this field trial was to compare the effect of restricted feeding with traditional feeding strategy, for prevention of diarrhoea and reduction of mortality in herds with *E. coli* associated diarrhoea. The study was carried out in three herds, where pathogenic types of *E. coli* were identified. There was approximately 1,300 piglets in the test group (restricted feeding) as well as in the control group (*ad libitum* feeding) in each herd. Restricted feeding was defined as 75 % of the amount of feed fed in the control group the first 2 weeks after weaning. Over the next 2 weeks the amount of feed was gradually increased to *ad libitum* feeding. The piglets were weaned at the age of 4 weeks. The piglets in both groups was fed the same feed mixtures. Most of the piglets that died during the trial was submitted to laboratory examination.

The primary effect parameter was mortality the first 4 weeks after weaning. There was distinguished between death caused by diarrhoea and death of other reasons. There was no effect of treatment on mortality. The secondary parameters were number of treatments against diarrhoea, daily gain and feed conversion ratio. Further results regarding the secondary parameters will be presented.

Poster P5.10

Body composition changes in piglets at weaning in response to nutritional modification of sow milk composition and effects on post-weaning performance
G. Jones[1]*, S.A. Edwards[1], S. Traver[1], S. Jagger[2] and S. Hoste[3]. *[1]Scottish Agricultural College, Craibstone Estate, Bucksburn, Aberdeen, AB21 9YA, UK, [2]Dalgety Feed Ltd, Springfield Road, Granthan, NG31 7BG, UK, [3]PIC, Fyfield Wick, Oxford, OX13 5NA, UK*

To determine the effect of body composition on post weaning growth of piglets, 41 sows and their litters were used in a 2x2 factorial experiment in which sows were fed lactation diets differing in energy source (starch v fat added as 0.34 of total DE intake) and lysine: digestible energy ratio (0.43 v 0.64 g/MJ). Sow milk nutrient output was significantly affected by energy source, such that piglets suckling sows fed high starch diets (S) had a higher protein: energy ratio in milk intake ($P< 0.001$) than piglets from sows fed high fat diets (F). Piglet weaning weight did not differ significantly between treatments, but the relationship between body weight and body composition was changed, with S piglets having a higher body protein: lipid ratio ($p< 0.05$). There was a negative relationship between piglet weaning weight and growth rate in the first week post weaning, with a significantly greater slope for F piglets. In the second week, the relationship between growth and weaning weight was significantly positive for both treatments. Therefore, modification of sow milk composition and consequently piglet body composition can influence growth in the immediate post weaning period, and physiological mechanisms require investigation.

PIG PRODUCTION   [P]

Poster P5.11

Effects of sow nutrition and environmental enrichment during the suckling period on post weaning performance of pigs

E. Foster[1], S.A. Edwards*[1], F.M.Davidson[2] and J. Duncan[2]. [1]Department of Agriculture, University of Aberdeen, 581 King Street, Aberdeen AB24 5UA, UK. [2]A Simmers Ltd, Mains of Bogfechel, Whiterashes, Aberdeen AB5 0QU, UK.

Post weaning performance of piglets may be influenced by both physiological and psychological components. An experiment with 3x3 factorial design involving 54 sows and litters investigated the effects of maternal diet (CD = cereal based; UD = 8% added unsaturated fat; SD = 8% added saturated fat) to give different milk composition, and environmental enrichment (CE = fully slatted farrowing pens; PE = physically enriched by provision of novel manipulable objects and substrate; SE = socially enriched by allowing co-mingling of two adjacent litters from 12 days of age). Piglets were weaned at 3 weeks and allocated within litter by weight to one of two post-weaning housing treatments (CP = controlled environment pens or SP = straw court). Weaning weight had a significant ($P<0.001$) negative relationship to week 1 growth rate and a positive relationship to week 2 growth. In CP housing, where treatment groups were separately housed and recorded in detail, week 1 growth rate was higher for CD than for SD and UD ($P<0.001$) and higher for PE than for CE and SE ($p=0.003$). However, these effects did not continue for the second week. Week 1 growth rate was greatly influenced by litter of origin ($P<0.001$), irrespective of treatment or weaning weight.

Poster P5.12

Effects of L-carnitine with different lysine levels on growth and nutrient digestibility in pigs weaned at 21 days of age

W.T. Cho*[1], J.H. Kim[1], In K. Han[1], K.N. Heo[2] and J. Odle[2]. [1]Department of Animal Science & Technology, Seoul National University, Suweon 441-744, Korea, [2]Department of Animal Science, North Carolina State University, NC 27695, USA.

A total of 120 piglets (22±1 d of age 5.9 kg of BW) were allotted into a 3´2 factorial design with three different levels of lysine (1.40%, 1.60% and 1.80%) and two levels of L-carnitine (0 and 1,000 ppm). Growth performance was optimized in pigs fed 1.6% lysine regardless of carnitine addition. Carnitine significantly improved ($p<0.05$) ADG of pigs when the lysine level in the diet was 1.6%. Only in the third week carnitine had a significant influence on growth performance of pigs. Lysine level significantly affected the digestibilities of DM ($p<0.001$), GE ($p<0.001$), CP ($p<0.01$), C. fat ($p<0.05$). Carnitine also significantly improved digestibility of nutrients. Lysine level as well as carnitine level affected the amino acids digestibility. Plasma carnitine content was significant higher ($p<0.05$) in pigs fed L-carnitine. This indicates the increased biological availability of carnitine within the body. L-carnitine supplementation tended to improve feed utilization during the third week ($p<0.10$) and during the entire period ($p=0.10$). Lysine level significantly affected feed utilization of pigs during the third week and entire period ($p<0.05$).

PIG PRODUCTION   [P]   Poster P5.13

Effect of FIX-A-TOX on piglets after weaning
V. Mikule, M. Čechová and Z. Tvrdoň. Department of Animal Breeding, Mendel University of Agriculture and Forestry, Zemědělská 1, Brno 613 00, Czech Republic

FIX-A-TOX is a supplement on mineral base, which desactives mycotoxine in animal organism. The study was carried out on 3786 piglets after weaning (pre-fattening phase - about 40 days in average(32 - 53 days)), where their mothers (F1 - crosses of White Improved x Landrase) were fed with feeding mixtures with FIX-A-TOX all time from the last weaning to clean their organism from mycotoxine. The piglets were carred in five halls and fed with feeding mixtures with FIX-A-TOX. For control there was another group of 2524 piglets without FIX-A-TOX (their mothers weren't fed with mixtures witch included FIX-A-TOX during gravidity and after farrowing) in three halls and we studied percentage of mortality during pre-fattening time. The percentage of mortality in three control groups of piglets was high - 25,81 %, 25,43% and 14,05 %, in average 21,76 %. We have to say, that pig production on that farm is continuous without stop during whole year and a lot of pigs are PRRS positive. The results of five groups with FIX-A-TOX were good, percentage of mortality was lower. In the first hall it was 8,0 %, in the second 3,71 %, in the third 5,78, in the fourth hall 6,54 and in the last hall 6,47 %. The average percentage of mortality was 6,1%.

Poster P5.14

Differentiation of piglets on fear of human
S.Papshev, K.Zhuchaev*, M.Barsukova. Department of Pig Breeding. Novosibirsk State Agrarian University, 630039, Russia.

The fear of human is one of factors negatively effecting the productivity and welfare of animals. The aim of investigations was the comparison of reactions of piglets on the test with unfamiliar man (TUM) and test with nonliving object (TO). The test comprised about 100 piglets 3-month age.
Animals were shared to 3 groups according to results of tests: active (A) - (the time of approach to the stimulus $x < x - 1/2 \delta$), middle (M) - ($x - 1/2 \delta < x > x + 1/2 \delta$), slow (S) - ($x > x + 1/2 \delta$).
Respectively, piglets were distributed after TUM: 44% - 25% - 31%, after TO: 31% - 53% - 12%. The average time of approach on the distance 1 and 1.5 m to the stimulus in TUM was 123.4 and 87.8 s., in TO 34.8 and 19.5 s. The "degree of fear" was evaluated as the difference between results TUM and TO. This trait had strong correlation with results TUM (+ 0.95). Piglets with higher body weight to 2 month age had less fear of human.
The heritability of results TUM was (by different methods) 0,2 - 0,37 (P<0,05). For results of TO and "degree of fear" h^2 was not so high and nonsignifican. According to that, the test with unfamiliar man seems to be quite sufficient to evaluate fear of human.

PIG PRODUCTION [P]

Paper PN6.1

Selective pressure by antibiotic use in food animals
W. Witte, Robert Koch Institute, Wernigerode Branch, D-38855 Wernigerode, Germany

The question whether antibiotic use as feed additives in animal husbandry selects for resistance and whether resistance genes from the animal's intestinal flora reach that of humans is of actual interest and will be elucidated by three examples: -**1**- In 1982 oxytetracycline was replaced as feed additive by the streptothricine antibiotic nourseothricine in former East Germany. Sat-gene (streptothricine acetyltransferase) mediated resistance was first found in 1983 in E. coli from pigs, in 1984 in personnel and family members, in 1985 in uropathogenic E. coli and in 1987 in Shigella sonnei. There is no cross resistance to antibacterials used in humans. -**2**-. The resistance mechanism encoded by the vanA gene cluster mediates cross resistance against vancomycin, teicoplanin and avoparcin which is used as feed additive. Selective pressure avoparcin use in favour of glycopeptide resistant E. faecium (GREF) has been demonstrated by studies in Denmark and in Germany. GREF can be shown in meat products and in the intestinal flora of nonhospitalized humans. Molecular typing of GREF from human and animal sources reveals as polyclonal structure which indicates a frequent transmission among different strains. Also conjugative plasmids from different sources carrying vanA (and ermB) are obviously different. However, as evidenced by overlapping PCR the majority of GREF from humans and animals, isolated in Germany possess a vanA gene cluster of the same configuration. -**3**- We demonstrated resistance to streptogramin A compounds (sat A mediated acetyltransferase) in GREF from humans already before any therapeutic use in Germany. Virginiamycin, however, was used as feed additive since 1974. Sat A mediated resistance was also found in GREF from pigs, chicken and meat products.

Paper PN6.2

Experience with stop for the use of antimicrobial growth promoters in pig production and adjustment of management
S.E. Jorsal*, Department of Pathology and Epidemiology, Danish Veterinary Laboratory, Bülowsvej 27, DK-1790 Copenhagen V., Denmark.

For several decades antimicrobial growth promoters have been added to feed for pigs in most pig-producing countries. In recent years, there has been increasing concern about the risk of selection for antimicrobial resistance related to the use of antimicrobial growth promoters, and documentation for this risk has been presented for instance for avoparcin, virginiamycin and tylosin. A few countries have had a total or limited ban on the use of antimicrobial growth promoters in feed for 10-15 years (Sweden, Norway, Finland), and some countries are presently phasing out the antimicrobial growth promoters (Denmark).

The impact on growth and disease prevalence in pig herds where antimicrobial growth promoters are withdrawn is variable. Experience from the Nordic countries has shown that it is possible to obtain satisfactory production results without growth promotors. In Denmark antimicrobial growth promoters were withdrawn from feed for slaughter pigs in April 1998, apparently without adverse effects in most herds. However, particularly in the postweaning period special concern must be drawn to housing, feeding, water supply, and management in general.

Epidemiological investigations in more than 100 herds in Denmark are carried out in order to evaluate the effect of withdrawal of antimicrobial growth promoters in feed for piglets. Preliminary results as well as a survey of experiences from other sources will be presented at the meeting.

PIG PRODUCTION   [P]   Paper PN6.3

Nutritional and gastrointestinal effects of organic acis supplementation in young pigs
F.X. Roth* and M. Kirchgeßner. Institut für Ernährungswissenschaften, Technische Universität München, 85350 Freising-Weihenstephan, Germany

Experimental data showed a significant improvement of growth rate and feed conversion rate of weanling pigs by the dietary inclusion of organic acids. These ergotropic effects were mainly observed with formic, lactic, sorbic, fumaric, citric and malic acid as well as with different formates and depend for each acid on the optimal dietary dose level. The improvements in performance attained by dietary acidification are mainly considered to be a response to the reduced dietary pH and buffering capacity. However, it has not yet been possible to attain a growth promoting effect in young pigs by lowering the pH value and buffering capacity of the feed with inorganic acids like phosphoric or hydrochloric acid. Studies on the mode of action of organic acids indicated a higher protein and energy digestibility and retention, alteration of bacterial populations and metabolites in the gastrointestinal tract and possibly an effect on metabolism. It seems that the antimicrobial properties of organic acids and of their salts are of great importance for the beneficial effects in weanling pigs.

Paper PN6.4

Current value of alternatives to antimicrobial growth promoters with special emphasis on anti secretory factor
L. Göransson[1] and S. Lange[2]. [1]Swedish Pig Centre Pl 2080, 26890 Svalöv, Sweden, [2]Department of Clinical Bacteriology, University of Gothenburgh, Guldhedsgatan 10, 41346 Gothenburgh, Sweden.

The ban of antimicrobial growth promotors (AGP) in Sweden since 1986 has changed pig production systems. In the search for alternatives to AGP, addition of formic acids has shown positive effects on feed conversion in slaughter pigs. A low crude protein content is critical for designing an optimal diet for weaning pigs. However, several feed additives have been tested without demonstrating consistent effects neither on fattening pigs, nor on post weaning performance. In contrast, the Anti Secretory Factor (AF) concept has proven to significantly decrease the incidence of diarrhoeal disease in pigs, rabbits, calves and humans. AF is an endogenously produced protein, found in all mammals so far investigated, with a critical role for regulation of intestinal water transport. The endogenous synthesis of AF can be significantly stimulated by correctly composed feeds or ORS-solutions. Commercial AF-inducing feeds/ORS-solutions for piglets and slaughter pigs are on the market, and so are human functional food.

PIG PRODUCTION   [P]

Paper PN6.5

Comparison of fumaric acid, calcium formate and mineral levels in diets for newly weaned pigs
P. G. Lawlor*[1,2], P.B. Lynch[1] and P.J. Caffrey[2]. [1]Teagasc, Moorepark Research Centre, Fermoy, Co. Cork, Ireland. [2]Department of Animal Science and Production, University College Dublin, Ireland.

The weaned pig has limited ability to acidify its stomach contents. The objective here was to examine the effect of feeding Fumaric Acid (FA), Calcium Formate (CF) and a diet of low Acid Binding Capacity (ABC) on post-weaning pig performance. Pigs weaned at 19 to 23 days (n=64) were individually fed for 26 days. Treatments were: (1) Control diet, (2) Control + 20g/kg FA, (3) Control + 15g/kg CF, (4) Low Ca (2.8g/kg) and P (5.1g/kg) diet for 7 days followed by Diet 1, (5) Low Ca and P diet for 7 days followed by Diet 2, and (6) Low Ca and P diet for 7 days followed by Diet 3. CF tended to depress intake (DFI) in the final two weeks (691 vs 759 and 749, s.e. 19 g, P=0.07) and overall average daily gain (322 vs 343 and 361 s.e. 11 g, P=0.09) compared with control and FA supplemented diets respectively. Feeding diets with low Ca and P for 7 days post weaning increased DFI (208 vs 178, s.e. 8 g, P<0.01) in week 1 and tended to improve feed conversion efficiency in the first two weeks (1.65 vs 1.85, s.e. 0.10, P=0.09). It is concluded that CF did not improve weaned pig performance but reducing diet ABC or including FA in the diet improved performance after weaning.

Paper PN6.6

Commercial acid products for piglets from 7-30 kg BW
J. Callesen*, H. Maribo and L. Jørgensen. The National Committee for Pig Breeding, Health and Production, Axeltorv 3, 1609 Copenhagen V, Denmark.

The effect of commercial acid products was tested on performance in 14 comparisons with a control group without growth promoters. Each treatment comprised 160 weaned piglets. Identical diets - mainly based on cereal and soybean meal - with and without a product added, were compared in each experiment. The products were dosed due to the recommendations by the supplier (0.2%-1.4%). From the production results a production value (PV) was calculated: PV = (value of weight gained) - (value of feed consumed), followed by an analysis of variance. The cost of each product added was not considered. In 8 out of 14 comparisons the acid products increased the PV (p<.05) compared to the control treatment. The increased PV was in all 8 cases due to an increased ADG (p<.05) and in 4 out of 8 cases also FCR was improved (p<.05). Of the 8 products that increased the PV 3 consisted of organic acids, 4 consisted of organic acids and salts of these and one product consisted of one salt of an organic acid. An analysis across all experiments showed that addition of commercial acid products had the same reducing effect on the number of treatments against diarrhoea as antibiotic growth promoters. The number of treatments was in both cases reduced (p<.05) compared to a control treatment.

PIG PRODUCTION   [P]

Paper PN6.7

Investigations on the effects of dietary nucleotides in weaned piglets
M. Zomborszky-Kovács*[1], S. Tuboly[2], P. Soós[2], G. Tornyos[1], E. Wolf-Táskai[1]. [1]Department of Animal Physiology and Hygiene, Pannon Agricultural University, Faculty of Animal Science, P.O.Box 16, 7401 Kaposvár, Hungary, [2]University of Veterinary Science, P.O.Box 2. 1400 Budapest, Hungary.

Synthetic uracil and adenine (98%, Sigma-Aldrich) were mixed into the diet of weaned pigs at doses of 500 mg/kg diet for each substance. Changes in the animals' daily weight gain, feed consumption, and certain haematological, biochemical and immunological parameters were monitored and recorded during the 3-week period of experiment. Lymphocyte blastogenesis induced by phytohaemagglutinin and concanavalin A, as determined by the lymphocyte stimulation test (LST), increased by 50 and 130%, respectively, due to nucleotide-supplementation. The supplement applied exerted no positive effect on *in vivo* cellular immune response (measured by skin reaction).

The effect of similar supplementation on the intestinal microflora was examined in cannulated weaned pig. The number of lactobacilli and streptococci rose in both the control and treated groups as the experimental period progressed. The number of coliforms and *E. coli* showed significant decrease in the treated group, while it increased in the control animals.

Paper PN6.8

Experiences of the voluntary ban of growth promoters for pigs in Denmark
N. Kjeldsen*, C.F. Hansen and A.Ø. Pedersen. The National Committee for Pig Breeding, Health and Production, Axeltorv 3, 1609 Copenhagen V, Denmark

The National Committee has decided to abolish the use of growth promoters for pigs over 35 kg BW from March 1998 and for piglets from January 2000. 98 per cent of the pig producers have joined the following agreement:
– From March 1998 growth promoters are not used for finishers.
– Pig producers must accept random inspection and analysing of feed samples.
– Not signing or violating the agreement result in a fine of DKK 0.20 per kg carcass weight delivered during a three months period.

After removal of growth promoters from the feed, a period will follow with more cases than usually of loose manure because of changes in the pigs intestinal flora. After about a month, the intestinal flora will have stabilised in some herds, and problems with loose manure will diminish. In general it is important to optimize feeding and housing conditions.

The experiences from 62 herds in 26 weeks after the change are that 60 per cent of the herds had no problems, 30 per cent had a temporary imbalance in the herd, and 10 per cent had problems with lower daily gain or with diarrhoea. The production results after the ban and the experiences of reducing protein level using fermented liquid feed and adding organic acids will be presented.

PIG PRODUCTION [P]

Poster PN6.9

Effect of creep feeding, dietary fumaric acid and whey level on post-weaning pig performance

P.G. Lawlor[*1], P.B. Lynch[1] and P.J. Caffrey[2]. [1]Teagasc, Moorepark, Research Centre, Fermoy, Co. Cork, Ireland, [2]Department of Animal Science and Production, University College Dublin, Ireland.

Legislation on use of feed antibiotics is becoming more restrictive in the EU and alternatives must be evaluated. The objective here was to examine the effect of feeding Fumaric acid (FA) to weaned pigs. In experiment 1, pigs (n = 64, c.6.1 kg) weaned at c.21 days were randomly assigned to the following treatments; (1) no pre-weaning creep, 0 g/kg FA, (2) no creep, 20 g/kg FA, (3) creep, 0 g/kg FA, and (4) creep, 20 g/kg FA. In experiment 2, pigs (n=64, c.5.9 kg) were weaned at c. 21 days, and randomly assigned to the following treatments; (1) 200 g/kg whey in diet, (2) 200 g/kg whey with 20 g/kg FA, (3) 200 g/kg whey with 30 g/kg FA, (4) 50 g/kg whey, (5) 50 g/kg whey with 20 g/kg FA, (6) 50 g/kg whey with 30 g/kg FA. All diets contained barley, wheat, herring meal full fat soya and whey. FA replaced barley. In experiment 1, FA inclusion increased intake (520 vs 469, s.e. 20 g, $p<0.05$), ADG (340 vs 281, s.e. 15 g, $p<0.01$) and improved FCE (1.55 vs 1.70, s.e. 0.05, $p<0.05$) in the first 3 weeks post-weaning. A creep*FA interaction ($p<0.05$) was seen in week 1 where FA increased intake of creep fed pigs but not of pigs that had got no creep. In experiment 2, no effect of treatment was observed ($p>0.05$). In conclusion, FA use can improve post-weaning performance especially in pigs fed creep feed pre-weaning.

Poster PN6.10

Effect of different spray dried plasmas on growth, ileal digestibility and health of early weaned pigs challenged with *E. Coli* k88

P. Bosi[*1], I.K. Han[2], S. Perini[3], L. Casini[1], D. Creston[1], C. Gremokolini[1], S. Mattuzzi[1]. [1]DIPROVAL, University of Bologna, 42100 Reggio Emilia, Italy. [2] Seoul National University, Suweon, 441-744, Korea. [3] Ist. Zooprof. Sperim. della Lombardia e dell'Emilia, 42100 Reggio Emilia, Italy

In two trials, **A** and **B**, a total of 96 pigs weaned at d 13 or d 19 rsp, were fed four diets with: 3 different spray dried plasmas (SDP) or hydrolysed casein (**HC**). SDPs were: **SPP**, from pigs; **SMP** of mixed origin, **SMPIG** of mixed origin and standardised level of immunoglobulins. The diets contained 1,7 % total lysine, 25% of the test protein source, 45% corn starch, 15% lactose, 2% sucrose, 7 % soybean oil. After 2 or 4 days rsp, the piglets were perorally challenged with 10^{10} CFU *E.Coli* K88. In **A**, all the subjects were sacrificed after 14 days, whereas **B** ended after 15 days.
In **A** no clinical sign of infection was detected, no difference for the content of *E.Coli* K88 was found for faeces at 4 and 6 days after the infection and no *E.Coli* K88 was found at the end in the jejunum. In **B**, 3 subjects died in **HC** and 1 in **SPP** group. Deaths were higher ($P<0.01$) for **HC** compared with the others. Feed intake was lower for **HC** in both trials. For **A** and **B** together, daily live weigh gain was 96.2; 106.2; 122.0; 155.2 for **HC, SPP, SMP** and **SMPIG** rsp (**HC** Vs others, $P<0.05$; **SMPIG** Vs other SDP, $P<0.01$). Ileal apparent digestibility of nitrogen in sacrificed piglets was higher for **HC** diet ($P<0.05$). Data on specific immune response in blood and saliva will be presented.

PIG PRODUCTION   [P]

Poster PN6.11

Formi™TLHS - an alternative to antibiotic growth promoters
M. Øverland[1], S. H. Steien[1], G. Gotterbarm[1], T. Granli[1], W. Close[2]*. Hydro Nutrition, Bygdøy Allè 2, N-0240 Oslo, Norway, [2]Close Consultancy, 129 Barkham Road, Workingham, Berkshire, RG41 2RS, UK

The use of antibiotic growth promoters in rearing pigs is being restricted by concerns about the development of resistant bacteria, animal welfare, environmental effects and consumer health. Organic acids have received much attention as an alternative. Formic acid is one of the most effective organic acids against pathogenic bacteria, but has been limited by problems with handling, strong odor, and corrosion during feed processing and on the farm. Formi™TLHS, a dry, odorless, easy to handle potassium di-formate, is an effective growth promoter in diets for weanling pigs and grow-finish pigs. Scientific trials and field trials have shown similar or better growth performance and health status of pigs receiving Formi™TLHS compared to those receiving approved antibiotic growth promoters. Formi™TLHS has also shown to improve carcass quality of pigs. The growth promoting action of Formi™TLHS includes improved feed intake, antibacterial effect in the gastro-intestinal tract, nutrient digestibility and retention of N and minerals. The latter is important from an environmental point of view. The incorporation of Formi™TLHS in a program with proper management and husbandry practices will ensure optimal effect.

Poster PN6.12

Effects of passive protection with spray dried egg protein (PROTIMAX®) against post-weaning diarrhoea and growth check in piglets
W. Krasucki[1]*, E.R. Grela[1], J. Matras[1], and Z. Mroz[2]. [1]Agricultural University of Lublin, Akademicka 13, 20-950 Lublin, Poland. [2] Institute for Animal Science and Health, P.O. Box 65, 8200 AB Lelystad, The Netherlands

In the context of public demands to eliminate in-feed antibiotics, spray dried egg protein (PROTIMAX®) is offered as a natural source of IgG for immunoprophylaxis in piglets. To test its efficacy, in total 339 piglets were used according to 2 x 4 factorial arrangement: PROTIMAX® dose (0.0 vs 1.1 g/kg prestarter or starter diet) and period of its application (22-28, 22-42, 22-56 and 29-56 days of age). The piglets were weaned at 28 days and reared to 56 days of age. Without PROTIMAX®, the rate of clinical $E.\ coli$ diarrhoea was 13.5% (at 2.4% mortality), whereas at presence of this product only less than 4% of piglets suffered of diarrhoea (at no mortality; $P<0.05$). Average diarrhoea-days were 2.1 and 0.9 for piglets fed without and with PROTIMAX®, respectively. The highest efficacy of PROTIMAX® in preventing diarrhoea and in improving piglet performance was noted when applied this product over 22-56 days of age. In this period, the piglets receiving PROTIMAX® grew faster (by 6.7%), consumed more feed (by 1.2%), and had a better feed conversion ratio (by 4.0%; $P<0.05$) as compared to those fed without PROTIMAX®. Also, apparent digestibility of dietary crude protein was improved by up to 2%-units ($P<0.05$). Nutrient digestibility and performance of piglets fed PROTIMAX® over 29-56 days of age was also improved, but to a lesser degree than when applying this product over 22-56 days of age.

PIG PRODUCTION   [P]

Poster PN6.13

The effect of the inclusion of YeaSacc^{1026} at different levels of the concentrate diet on calf performance
R.J. Fallon and B. Earley, Teagasc, Grange Research Centre, Dunsany, Co. Meath, Ireland

The recent banning of a number of in feed antibiotic growth promotors within the EU has generated the need to find suitable non-antibiotic alternatives. The following experiment was undertaken using 80 Friesian male calves (average initial weight of 54 kg) to determine the optimum inclusion rate of YeaSacc^{1026} in a barley soyabean meal ration. Calves were allocated at random to 1 of 4 treatments: 1) 0, 2) 0.625, 3) 1.25 and 4) 2.5 kg YeaSacc per tonne of ration. The concentrate ration was available *ad libitum* throughout the 84 day experimental period and the calves were offered 25 kg of calf milk replacer by bucket over the initial 42 day period. Calf liveweight gain in the period 1-42 days was 0.58, 0.65, 0.65 and 0.68 (SED 0.029) kg/d for treatments 1 through 4, respectively. The corresponding liveweight gain for the period 1 to 84 days were 0.84, 0.92, 0.92 and 0.89 (SED 0.026) kg/d, respectively. Concentrate intake (1 to 42 day) were 25, 29, 30 and 31 (SED 2.1) kg, respectively. It was concluded that the inclusion of 1.25 kg of YeaSacc^{1026} per tonne of ration increased calf liveweight gain by 5 kg in the period 1 to 84 days.

Poster PN6.14

Inclusion of organic chromium in the calf milk replacer: effects on immunological responses of healthy calves and calves with respiratory disease
B. Earley and R. J. Fallon*. Teagasc, Grange Research Centre, Dunsany, Co. Meath, Ireland.

Deficiencies of specific nutrients reduce immune responses and increase disease susceptibility. One hundred Holstein x Friesian calves, approximately 21 days of age at arrival, were used to investigate the effects of supplementation with organic chromium (Cr) on mitogen induced blastogenesis of isolated peripheral blood mononuclear cells. Calves were allocated at random to the following milk replacer treatments; (a). Skim (b). Whey based A, (c), whey based B, (d). Whey based C, (e) Whey based D. Within each treatment, 10 calves received a daily supplementation of organic Cr 250 ppb. The stimulation index (SI) was calculated for all calves. Average daily gain, feed intake, and frequency of antibiotic treatments were not affected by Cr supplementation. Calves supplemented with Cr and having no incidence of respiratory disease had significantly higher blastogenic responses ($P \leq 0.001$). It is concluded that lymphocytes from calves with respiratory disease manifest an impaired capability to blast *in vitro*. Cr supplementation (250ppb) enhanced the blastogenic response in healthy calves.

HORSE PRODUCTION   [H]   Paper H2.1

Genetic parameters of morphofunctional traits in Andalusian horse
A. Molina, M. Valera, R. Dos Santos*, and A. Rodero. Department of Genetic. Veterinary Faculty. University of Cordoba. Avda. Medina Azahara s/n 14005-Córdoba (Spain).

The aim of this study has been to estimate for the first time the genetic parameters of 18 morphofunctional traits in the Andalusian Horse. The data were obtained from the evaluations made on 1273 horses between the years 1991 to 1997. The heritabilities and genetic correlations of these traits were estimated using a BLUP Animal model with REML methodology. The heritabilities obtained for the body measurements were moderate to high (0.35 to 0.95) The estimates obtained for the scored regional conformation show a lower value (0.03-0.50). The heritability for the racial fidelity and overall forms were of 0.58, while the scoring for movements shows a value of 0.15, and 0.08 for temperament, which agrees with the fact that they are the most subjective and complex features being closely tied to the behaviour variables. Lastly, the heritability of the overall evaluation trait, which includes all the previous characters, shows an intermediate magnitude (0.25). The phenotypic and genetic correlations estimated were all positive. The genetic correlations range between 0.11 to 0.94 for zoometric measures and between 0.12 and 0.91 for the regional morphologic evaluations. The punctuation for racial fidelity and overall forms shows a genetic correlation of 0.55 with movement scorings and 0.74 with the overall evaluation of the animal.

Paper H2.2

Genetic study of the Equine population of the Lusitanian Horse
M. Valera*[1], L. Esteves[1], M. M. Oom[3] and A. Molina[2]. [1] Departamento de Genética. Facultad Veterinaria. Universidad Córdoba. Avda. Medina Azahara, s/n 14005- Córdoba (Spain). [2] Departamento de Zoologia e Antropologia. Centro de Biologia Ambiental. Facultade de Ciencias de Lisboa (Portugal).

The population of the Lusitanian horse (PSL) has been characterized from the reproductive point of view using the information that comes from the Breed's Stud Book. This data base included the information of 9486 animals (4245 males and 5241 females), from which a 27.47% correspond to reproductive females (2606) and a 7.42% to stallions (704). The age at first foal/calving obtained was of 82.7 months for the stallions and 73.3 for the females. The age at the last foal/calving was of 125.7 months for stallions and 130,1 for the mares. The average of the reproductive life obtained was of 50.6 months for stallions and of 67.8 for mares. Other analyzed parameters were the calving interval (22.5 months for females), and the number of offspring (12.3 offspring for stallions and 3,9 foals for females).Finally the generation interval for this breed through the four classic ways has been estimated: via father-son (99.64 months), father-daughter (87.19 months), mother-son (109.81 months), mother-daughter (94.63 months). These values have been one of the lowest intervals found in the equine bibliography.

HORSE PRODUCTION [H]

Paper H2.3

The immunogenetic analysis of the Russian trotter mares
M.A.Chechushkova, V.V. Sivtunova, M.L. Kochneva, V.L.Petukhov, Novosibirsk Agrarian University, 160 Dobrolubov Str., 630039 Novosibirsk, Russia

Using erythrocyte antigenes as the genetic markers of the horses labour capacity in seems to be promising. The analysis of the occurrence of the erythrocyte antigenes was carried out in pure mares of Russian trotting breed. The correlations between antigenes and coat color and the speed character were studied.

It was found, that the Aa, Ad, Dc and Dm ahtigenes were characterized by the greatest frequency and the Df and Dh antigenes were rarely observed in the studied population. Dh antigene was found only in bay and black mares, no Df antigen was found in the black ones.

The mares of 2.10 and higher classes were characterized by the lowest frequency of Df antigene (12.5 %) and the highest frequency of Dd and Ka antigenes (70.8 and 79.2 % respectively) comparing with the horses of the other speed classes. Dh antigene was detected only in the mares of 2.10 and higher classes. The significant influence of Ka, Dh and Dg antigenes on the speed of the horses of 2.10 and higher classes was established and the power of the influence constituted 27.9, 12.4 and 3.6 respectively.

Paper H2.4

Damping characteristics of a shock absorbing steel-polyurethane horseshoe
M.A. Weishaupt[1]*, E. Mumprecht[2], C. Lenzlinger[2] and H. Inglin[2]. [1]Department of Veterinary Surgery, University of Zurich, Winterthurerstr. 260, CH-8057 Zurich, Switzerland; [2]Winterthur Polytechnic, P.O. Box 805, CH-8401 Winterthur, Switzerland.

Vertical impact is implicated in the development of various orthopaedic injuries in humans and horses. Nevertheless only a few attempts were made so far to apply shock absorbing materials in horse shoeing techniques. This study compares the damping efficiency of a steel-polyurethane horseshoe (Precotec®) with the one of the traditional steel horseshoe. Field trials were conducted with Warmblood horses adapted over 12 months either to the Precotec® or to the steel horseshoe. Measurements were performed 3 weeks after trimming the hoofs, alternately with the two horseshoe types. Horses were ridden at the walk (1.74±0.04 m/s) and trot (3.24±0.2 m/s) on an asphalt road. Deceleration at impact was assessed using a 3-dimensional, piezoelectrical accelerometer (Type 8694M1, Kistler Instrumente AG, Switzerland) securely mounted laterally on the horseshoe. The damping efficiency was evaluated with mean values ($n=50$ strides) for peak vertical deceleration (Az_{peak}) and with the 'Head Insurance Coefficient' (HIC). The HIC takes into account the total amount and duration of deceleration at impact.

Az_{peak} and the HIC were significantly reduced with the Precotec® horseshoe (in average 2-fold and 20-fold respectively) and therefore their use on horses which are mainly exercising and competing on hard surfaces i.e. endurance horses, can be recommended.

HORSE PRODUCTION [H]   Paper H2.5

The overall performance and prospects of use of Silesian stallions from Ksią Depot in European breeding scheme of heavy warmblood horses
E. Jodkowska*, E. Walkowicz, and H. Geringer de Oedenberg. Department of Horse Breeding, Agricultural University of Wroclaw, Kozuchowska 7, 50-631 Wroclaw, Poland.

Silesian, Oldenburger and Groninger are the contemporary populations of Old-Oldenburger and East Frisian horses. The most numerous is Silesian horse group. The aim of the study was to make overall evaluation of the Silesian stallions in Książ Depot. The study was carried out on 400 stallions used for breeding during the period from 1947 to 1998 year. The detailed analyses of conformation and genetic structure were done. Reproductive performance was analyzed due to number of matings. Differences in biometrical traits values of Silesian stallions due to pedigree were found. Prospects of Silesian male lines for future improvement of the breed are discussed. Obtained results are related to aim of International Horse Breeders Association Heavy Warmblood (IHW) to use Silesian horse in European breeding scheme.

Paper H2.6

The share of Arabian and Thoroughbred horses blood in the Małopolska horses breeding
M. Kulisa, M. Pieszka* and G. Ciuraszkiewicz. Department of Horse Breeding, Agricultural University, Al. Mickiewicza 24/28, 30-059 Kraków, Poland

The Małopolska breed belongs to the part of Anglo-Arabian group of horses. Their most valuable traits (dryness and beauty, gentle character and high-courage temperament, strength and willingness to work) were taken after Arabian and Thoroughbred horses.
The aim of this study was to define the percentage of Arabian and Thoroughbred blood in pedigrees of Małopolska mares bred in six Polish studs as well as to investigate the Arabian and Thoroughbred horses blood effects on point-scale estimation and colours of those horses. Data concerning 323 Małopolska mares were considered.
It was stated that in the above-mentioned population the average percentage of Arabian blood in pedigrees amounted to 12.049% and Thoroughbred blood to 42.298%. No influence based on point-scale estimation was stated but biometrical indices were dependent on the percentage of Arabian blood in pedigrees of Małopolska mares. In the group with higher share of Arabian horses blood more grey mares were observed.

HORSE PRODUCTION [H] Paper H2.7

Evaluation of horse population of Shagya-Arab breed bred in Czech Republic
J. Navrátil. Department of Cattle Breeding and Dairying, Czech University of Agriculture Prague, 165 21 Prague 6, Czech Republic

The Shagya-Arab breed has been recognized since the year 1978 by WAHO (World Arabian Horse Organization) as an independent breed of purebred Arabian horses. Its breeding in CR takes place on the basis of 13 internationally recognized families, to them at present belong 111 breeding mares, 8 breeding stallions and 47 young horses. Pure Bred Shagya-Arab Society CR (established 1993) organizes every year National Breeding Shows where the independent comission of the International Union ISG evaluates by ten-point scale the type of exterior (head, neck, body conformation) and mechanics of movement (pace and trot). The contribution is engaged in analysis of point values of characteristics observed, reached on the last 3 shows in the years 1996-98. At present is being completed standardization of basic body measurements and evaluation of performance testing of stallions included into breeding linking up the similar system of testing the mare performance.

Poster H2.8

The variation of Russian Trotter coat colors
V.V. Sivtunova, M.A. Chechushkova, M.L. Kochneva, Novosibirsk Agrarian University, 160 Dobrolubov Str., 630039 Novosibirsk, Russia

The retrospective analysis of horse coat colors changes in Russian Trotters from "Chiksky" stud was carried out in 1968, 1988-1997. The observations in the changes showed that in 1968 the horses were of bay and grey colors. Most horses (60 %) were of grey color. It might be attributed to the influence of the Orlovsky Trotter. From 1988 to 1997 the portion of bay horses increased, at present they make up 79,4 %. The quantity of the grey horses has reduced, now they total 6.6 %. The rare of the black-brown horses has also restricted to 6.7 and 2.2 % in 1988 and 1997, respectively.
For nine years the ratio of the black and bay horses changed. The portion of the black and bay horses constituted 16.4 and 4.9 %, 1.5 and 10.3 % in 1988 and 1997, respectively. Thus, the bay color is the main coat color of the Russian trotters.

HORSE PRODUCTION   [H]

Poster H2.9

Percentage distribution of the main whey proteins in milk of Italian Saddle Horse nursing mares during the first two lactation months
F. Martuzzi*[1], A. Tirelli[2], A. Summer[1], A.L. Catalano[1] and P. Mariani[1]. [1]Istituto di Zootecnica Alimentazione e Nutrizione, Facoltà di Medicina Veterinaria, 43100 Parma, Italy. [2]DISTAM sez. Industrie Agrarie, via Celoria 2, 20133 Milano, Italy

The aim was to study the distribution of the mare milk main whey proteins separated by HPLC. 42 individual milks of Italian Saddle Horse nursing mares reared in 14 farms in Northern Italy were analysed. Mares, 500÷650 kg live weight and 3÷22 years old (1÷16 parities), were fed with hay *ad libitum* (~10-13 kg) and 4-5 kg of concentrate. Milk was taken by total manual milking of one gland, while the foal was suckling the other. Nitro-gen compounds were determined by Kjeldahl; whey proteins were separated by reversed-phase HPLC. Data were analysed by ANOVA. For the nitrogen fractions the following mean values (±sd) were obtained: crude protein 2.34±0.31; casein 1.24±0.20; true whey protein 0.87±0.15; non-protein N (x6.38) 0.24±0.03 g per 100 g of milk. For the whey proteins the following percentage distribution was observed: β-lactoglobulin 41.80±6.48 %; α-lactalbumin 33.02±6.42%; blood serum albumin 4.97± 2.37%; immunoglobulin 20.21±6.21%. The milk produced in the 2nd month of lactation (n=19) contains less protein and less whey protein *in toto* compared with the milk produced from the 5th to the 30th lactation day (n=23). The percentage distribution values of the whey protein indivi-dual fractions, instead, show moderate differences statistically not significant (P>0.05).

Poster H2.10

The relationships between horse gray coat color and prenatal vitality of Orlov trotters
S.P. Knyazev*[1], N.V. Gutorova[1], S.V. Nikitin[2], O.I. Staroverova[3], R.M. Dubrovskaya[4]. [1]Novosibirsk State Agrarian University, Novosibirsk 630039, Russia, [2]Institute of Cytology and Genetics, Novosibirsk 630090, Russia, [3]Novotomnikovo Stud Farm, Tambov, Russia, [4]All-Russian Institute of Horse Breeding, Ryazan, Russia.

The variability of prenatal vitality of descendants of 137 Orlov Trotter's mars (Novotomnikovo Stud Farm) depends on dam's gray or non-gray coat color was investigated under the parentage control by means of blood groups systems and bio-chemical polymorphisms. As a criterea of prenatal vitality the ratio of alive born foals number to the total number of matings of these mars was used. The phenotype of gray coat color is genetically heterogene ous for *G* locus (consists of 2 genotypes: *G/G* and *G/g*); non gray colors is a homogenous phenotype for it (genotype *g/g*).
Our investigation revealed that mars of gray and non gray phenotypes were characterised by different variability of their offspring: the significant (P<0,01) heterogeneity of gray dams and uniformity of non gray ones. The concordance of homo- and heterogeneity of mar phenotypes for *G* locus with homo- and heterogeneity of their offspring prenatal vitality points on relationships between dominant epistatic *Gray* gene and studied performance. More than, it was observed the decreasing of prenatal vitality among *G*-heterozygotes. The relative vitality of foals with different *G*-genotypes was determined using the specially constructed model; homozygotes (*G/G* and *g/g*) and heterozygous foals (*G/g*) have been vitality 1,0 and 0,7 respectively.

HORSE PRODUCTION   [H]

Poster H2.11

Tyrosinase activity and ocular characteristics in Asinara white donkeys
W. Pinna*[1], P.Todde[2], G. M. Cosseddu[1], G. Moniello[1] and A. Solarino[1]. [1]University of Sassari, Via Vienna, 2 - 07100 Sassari, Italy, [2]Anatomia ed Istologia Patologica,Ospedale S.Michele - Brotzu - 09123 Cagliari, Italy.

The population of white asses of the Asinara isle (Italy) is characterised by white hair and coat, pink colour of glabrous areas of the body and skin, partial pigmentation of the iris. Our previous studies allowed to classify this phenotype as the expression of an oculocutaneous albinism. Aiming to go deeper into the knowledge of this kind of albinism in the population, the ophthalmic examination was performed and the tyrosinase activity of the hair bulb was studied in 5 animals of different age. The *fundus* of eye was observed an photographed with the portable hand held *fundus* camera (Kowa) and it showed a complete lack of pigment both in the retinal pigment epithelium and in the choroid. The lack of pigment makes transparent these two layers thus the vascular structure of the choroid becomes visible, both in the non tapetal *fundus* and in correspondence of the tapetal *fundus*, which is transparent too. Specific biochemical tests with L-tyrosine and L-DOPA, were carried aiming to evaluate the tyrosinase activity of the hair bulb, a marked pigmentation of the bulb was obtained when using DOPA, but the melanin was not produced at all when using the tyrosine. This means that the synthesis of melanin has a metabolic block of the conversion from tyrosine to DOPA which is the not active the next reaction which leads to the production of melanin.

Poster H2.12

Genetic analysis of pedigree in the Andalusian Horse
M. Valera*, A. Molina and A. Rodero. Dpto. Genética. Facultad de Veterinaria. Universidad de Córdoba. Avda. Medina Azahara s/n 14005-Córdoba (Spain)

In this work the genetic analysis of pedigree has been analysed in 33486 PRE horses (Andalusian horses) registered in the Stud Book. According to the results obtained, the level of completeness of the genealogical trees of the animals born in the last generation period analysed (1980-1991), in the Andalusian Horse is very high, because they present a greater IC of the 95% for 7th generation. The IC in the 15th generation is 55%. The coefficient of inbreeding and the effective size of the population, in order to characterise the PRE population. This analysis has been carried out along 10 generations that has been constituted for intervals of 10 years. Likewise the inbreeding effect over reproductive (number of the descendants, age to the first birth, calving interval), morphologic (morphologic evaluation) and productive (useful productive life characters). The average inbreeding obtained in the P.R.E. population could be considered within acceptable limits ($F=5.05\%$), not existing a significant relationship with reproductive, morphologic or productive characters. Nevertheless it is very high the percentage of consanguineous animals (64.9% of the population with a average F of 8.64%). Approximately 85% of the group of 37 sires qualify as more influential into the breed (with a $F_1 >=100$ or $F_2 >=500$ or $F_3 >=1000$ or $F_4 >=5000$ or $F_5 >=7000$) belong to the Carthusian Strain.

HORSE PRODUCTION [H]

Paper H3.1

Pferdezucht - und haltung in der Schweiz
Rudolf Schatzmann, Verband Schweizerischer Pferdezuchtorganisationen, Les Longs Prés, CH-1580 Avenches
Der gesamte Pferdebestand in der Schweiz umfasst rund 65'000 Pferde. Das Pferd ist auch in der Schweiz ein bedeutender Wirtschaftsfaktor, steuert es doch rund 600 Mio. Schweizerfranken zum Bruttosozialprodukt bei. Es schafft 10'000 Vollarbeitsplätze und ca. 150'000 Menschen beschäftigen sich regelmässig mit dem Pferd. Die Pferdezucht wird von den verschiedenen Rassenorganisationen betrieben. Sie arbeiten nach eigenen Zuchtprogrammen und Reglementen. Der Staat unterhält ein eigenes Gestüt und für anerkannte Rassenorganisationen beteiligt er sich finanziell an verschiedenen züchterischen Massnahmen. Den grössten Zuchtbestand weist der Freiberger auf, gefolgt vom Warmblut, dem Haflinger, den Kleinpferden und Ponys, dem Araber sowie den Rennpferden. Die verschiedenen Rassenorganisationen sind im Dachverband der Schweizerischen Pferdezuchtorganisationen zusammengefasst, der die Interessen der Pferdezucht im In- und Ausland wahrnimmt.(Vertretungen in Organisationen, politischer Ansprechpartner für übergeordnete Fragen der Pferdezucht, Förderung des Absatzes von Schweizer Produkten im Ausland und eine in der Landwirtschaft eingebundene Pferdehaltung). Gleichzeitig bietet er den verschiedenen Rassenorganisationen Dienstleistungen an. (Herdebuchführung, Zuchtberatung, Geschäftsführungen, Organisation von Finalprüfungen etc.). Der Pferdesport wird von einem eigenständigen Verband, der gleichzeitig Mitglied der FEI ist, betrieben. Er regelt das gesamte Spektrum des Leistungssportes, die Koordinierung der Pferdesportdaten, der Ausbildung von Reiter und Sportfunktionären sowie die Beschickung an Internationale Veranstaltungen. Es besteht eine enge Zusammenarbeit zwischen dem Verband für Pferdesport und dem Dachverband der Zucht.

Paper H3.2

Die Freibergerzucht in der Schweiz - population, zuchtziele und selektion
A. Lüth, Schweizerischer Freibergerzuchtverband, Postfach 190, CH-1580 Avenches
Durch die Anpaarung von kleineren, in der Erscheinung bescheidenen Landesstuten mit kompakten Warmbluthengsten englischer und französischer Herkunft entstand um 1880 die Freibergerrasse. Sie hat ihren Ursprung in den Freibergen im Jura an der nordwestlichen Grenze des Landes. Damit ist der Freiberger das einzige echte Schweizer Pferd. Im Jahr 1998 waren rund 5000 zuchtaktive Stuten und 170 Hengste im Herdebuch des Schweizerischen Freibergerzuchtverbandes registriert. Gezüchtet wird ein sehr ausdrucksvolles, mittelrahmiges, robustes und korrektes sowie leistungsstarkes Pferd im mittelschweren Typ mit schwungvollen, elastischen Bewegungen und trittsicheren Gängen. Hervorstechende Eigenschaft des Freibergers ist sein ausgeprägt sehr guter Charakter. Gegenwärtig vollzieht sich die Umzüchtung vom Gebrauchspferd der Landwirtschaft und Armee zum beliebten, vielseitig einsetzbaren Freizeitpferd für alle Arten des Pferdesportes. Die Widerristhöhe beträgt 150 bis 160 cm. Das Zuchtziel wird heute mit der Methode der Reinzucht angestrebt. Um im Umzüchtungsprozess die typischen Merkmale des Freibergers zu erhalten, wurde das Reinzuchtregister am 1.1.1998 geschlossen. Im Alter von drei Jahren absolvieren Stuten und Wallache den Feldtest. Geprüft werden die Fahr- und Reiteignung sowie der Charakter. Alle vorgestellten Tiere werden in 24 Exterieurmerkmalen linear beschrieben. Die Resultate dienen neben den Ergebnissen aus den Promotions- und Freizeitprüfungen der Stutenkategorisierung und der Nachzuchtbeurteilung der Hengste. Die dreijährigen Hengstanwärter werden im Januar anlässlich einer zentralen Exterieurbeurteilung vorselektiert. Die besten erhalten die Zulassung zum 40-tägigen Stationstest im Nationalen Gestüt in Avenches. Neben der Beurteilung der Fahr- und Reitmerkmale wird besonders der Charakter geprüft.

HORSE PRODUCTION [H] Paper H3.3

Der fremdblutanteil beim Freiberger pferd
H. Binder[*1] *und P.-A. Juillerat*[2]. *Veterinäramt Zürich, Culmannstrasse 1, 8090 Zürich, Schweiz, ESIA Zollikofen, Zollikofen, Schweiz*

Der Einsatz rassefremder Hengste in der Freiberger Zucht zum raschen Umsetzen sich wandelnder Zuchtziele hat Tradition. Seit 1930 wurden 4 Hengste mit Araber-Abstammung sowie 13 Hengste verschiedener Warmblutrassen eingesetzt (4 Anglonormänner, 2 Schweden, 3 Franzosen, 2 Holsteiner, 1 Schweizer). Die genetische Prägung der Rasse durch die Fremdhengste wurde anhand der Verwandtschaftsbeziehungen zu den Fohlen der Jahrgänge 1977 (333 Fohlen) und 1997 (3357 Fohlen) analysiert.

Im Fohlenjahrgang 1977 wurde durchschnittlich 2.5 % Fremdblutanteil gefunden, wobei 64.6 % der Fohlen reine Freiberger-Abstammung aufwiesen. Im Fohlenjahrgang 1997 erscheinen noch 10.1 % der Fohlen originaler Abstammung, während 11.5 % über 25 % Fremdblutanteil aufwiesen und 0.5 % gar über 50 %. In 30 % der Fohlen liegt der Fremdblutanteil zwischen 10 % und 25 % und in mehr als der Hälfte der Fohlen (57.6 %) unter 10 %. Der durchschnittliche Fremdblutanteil betrug 12.1 %. Als bedeutenster Vererber tritt der ab 1964 eingesetzte Schwedenhengst ALADIN in Erscheinung ($\varnothing$ 5.2 % Blutanteil) insbesondere durch seinen Sohn ALSACIEN. Von 5 der Fremdhengste konnten keine Nachkommen mehr gefunden werden.

Paper H3.4

Die Warmblutzucht in der Schweiz - Population, Zuchtziele und Selektion
A. Lüth, Zuchtverband des CH-Warmblut-Sportpferdes, Postfach 190, CH-1580 Avenches

Durch den Import von Hengsten und Stuten aus den wichtigsten europäischen Zuchtgebieten wird die Schweizer Warmblutzucht seit den 60er Jahren von einer grossen Blutlinienvielfalt geprägt. Gegenwärtig ist man bemüht, in einer Phase der Konsolidierung die besten Merkmale in der CH-Sportpferdezucht zu verankern. Im Jahr 1998 waren rund 1800 zuchtaktive Stuten und 440 Hengste im Herdebuch des Zuchtverbandes des CH-Warmblut-Sportpferdes registriert. Rund ein Drittel der zuchtberechtigten Hengste steht im Ausland. Gemäss Zuchtziel wird ein Sportpferd gezüchtet, das Leistungen auf höchstem Niveau in allen Disziplinen des Pferdesportes erbringt, gesund ist und ein funktionelles und ästhetisches Exterieur besitzt. Im Alter von drei Jahren absolvieren die Stuten und Wallache den Feldtest Reiten. Anlässlich dieser Prüfung werden die Merkmale des Freispringens und die Grundgangarten unter dem Reiter geprüft. Gleichzeitig werden alle vorgestellten Tiere in 24 Exterieurmerkmalen linear beschrieben. In allen Sportdisziplinen werden Promotionsprüfungen für 4- bis 6jährige CH-Warmblutpferde durchgeführt. Nach regionalen Qualifikationen werden die besten Pferde anlässlich eines Finales ermittelt. Die Resultate der Feldtests und der Promotionsprüfungen dienen der Stutenkategorisierung und der Nachzuchtbeurteilung der Hengste. Die Prüfung der Hengste erfolgt nach einer Vorselektion im Exterieur und in der Gesundheit ausschliesslich über den Sport. Um das Stuten-Hengst-Verhältnis und damit die Grundlagen für eine aussagekräftige Nachzuchtbeurteilung zu verbessern, wurden die Anforderungen für die Körung von Hengsten 1998 verschärft. Ziel ist es dabei, die inländische Zucht zu fördern, gleichzeitig aber die besten ausländischen Leistungsträger nicht auszuschliessen.

HORSE PRODUCTION [H]

Paper H3.5

Der feldtest beim CH-warmblutpferd
L. Egli*, Les Vieux Chênes, CH-1583 Villarepos

Seit 1991 werden in der Schweiz sogenannte Feldtests durchgeführt. Dabei handelt es sich um Zuchtprüfungen für dreijährige CH-Pferde. Beurteilt werden das Exterieur, das Freispringen und die drei Grundgangarten unter dem Sattel. Ziel dieser Prüfung ist es, die potentielle Leistungsveranlagung unter möglichst standardisierten Bedingungen schon beim jungen Pferd sichtbar zu machen. Es wurden nun die Zusammenhänge zwischen dem Feldtest und der Promotion "Springen" (= Springprüfungen für vier- bis sechsjährige CH-Pferde) untersucht, um zu prüfen, ob von den Zuchtwerten des Feldtests direkt auf die genetische Veranlagung als Springpferd geschlossen werden kann. Insgesamt wurden die Zuchtwerte von 7103 Pferden untersucht. Mittels multipler Regressionsanalyse wurden die Korrelationen ermittelt. Erwartungsgemäss ist die Korrelation zwischen Freispringen und Promotion Springen am grössten. Es konnten Werte zwischen 0.388 und 0.471 gefunden werden. Der Zusammenhang zwischen dem Feldtestmerkmal Schritt und der Promotion Springen ist vernachlässigbar. Die ebenfalls durchgeführte Varianzanalyse zeigte die Unterschiede zwischen schlechten, mittelmässigen und gute Springpferden auf. Es kam zum Vorschein, dass Pferde mit einem hohen Zuchtwert in der Promotion "Springen" signifikant höhere Zuchtwerte in den Feldtestmerkmalen (Ausnahme Schritt) haben als mittelmässige und schlechte. Analog haben Pferde mit tiefem Zuchtwert in der Promotion "Springen" signifikant tiefere Zuchtwerte in den Feldtestmerkmalen als mittelmässige und gute Tiere.

Paper H3.6

Die haflingerzucht in der Schweiz - population, zuchtziele und selektion
A. Lüth, Schweizerischer Haflingerverband, Postfach 190, CH-1580 Avenches

Ende der 50er Jahre begann man vor allem in der Ostschweiz mit der gezielten Zucht des Haflingers. Durch Importe von Zuchttieren vor allem aus Nordtirol wurde der Grundstein für die heutige Schweizer Haflingerzucht gelegt. Neben dem Freiberger, dem Schweizer Warmblut-Sportpferd und dem Maultier zählte der Haflinger bis Ende 1998 zu den geförderten Rassen in der Schweiz. Das heisst seine Zucht und Verwendung im Militär wurde durch direkte finanzielle Subventionen von Seiten des Staates gefördert. Im Jahr 1998 waren rund 620 zuchtaktive Stuten und 40 Hengste im Herdebuch des Schweizerischen Haflingerverbandes registriert. Das Zuchtziel lehnt sich an die Vorgaben es Welthaflinger-Verbandes an und fordert neben den rassetypischen Farbmerkmalen einen unkomplizierten Charakter, ein ausgeglichenes Temperament, Leistungsbereitschaft sowie eine gute Fahr- und Reiteignung. Das Zuchtziel wird mit der Methode der Reinzucht angestrebt. Die dreijährigen Stuten und Wallache absolvieren den Feldtest. Neben den Merkmalen der Fahr- und Reiteignung wird auch der Charakter überprüft. Alle vorgestellten Pferde werden in 24 Exterieurmerkmalen linear beschrieben. Die Resultate dienen neben den Ergebnissen aus Promotions- und Freizeitprüfungen der Stutenkategorisierung sowie der Nachzuchtbeurteilung der Hengste. Zusätzlich werden die besten dreijährigen Stuten anlässlich einer zentralen Elitschau aufgrund ihrer Exterieurbeurteilung rangiert. Die Hengstanwärter werden im Exterieur vorselektiert. Die besten erhalten die Zulassung zum 40-tägigen Stationstest. Es werden die Fahr- und Reiteignung sowie der Charakter der Hengste beurteilt. Bei der Organisation der Leistungsprüfungen (Feld- und Stationstest) nutzt man die vorhandenen Synergien mit der Freibergerzucht.

HORSE PRODUCTION [H]

Paper H5.1

Effect of insemination timing and dose on pregnancy rates in mares bred with frozen-thawed equine semen

J. Knaap[2], J.L. Tremoleda, A. van Buiten[2] and B. Colenbrander[1], [1]Department of Equine Sciences, Faculty of Veterinary Medicine, Utrecht University, Yalelaan 12, 3584 CM Utrecht, The Netherlands, [2]Research station for Cattle, Sheep and Horse Husbandry, P.O. Box 2176, 8203 AD Lelystad, The Netherlands.

A breeding trial was conducted to evaluate the effect of insemination timing and dose on mares bred with frozen-thawed semen. Seventeen stallions of known fertility and 61 fertile mares were included in the study. During two breeding seasons mares were teased and scanned every 48 hours to predict insemination time as close to ovulation as possible. Mares were inseminated with either 100TNM (Total number of motile morphologically normal spermatozoa X10^6) (n:50) or 200TNM (n:54) frozen-thawed semen. Inseminations were performed within 48 hours intervals if ovulation had not yet occurred. Frozen-thawed sperm samples were analysed for motility, viability and acrosome integrity. The number of inseminations (mean) per mare per cycle was 1.5, with an average number of cycles per mare of 1.9. The total pregnancy rate was 77%, with rates of 82% and 71% for mares inseminated with 100TNM and 200TNM, respectively. There was no significant difference in the pregnancy rate between the 2 insemination doses ($P>0.05$). We have shown that low insemination doses of 100 and 200 TNM would yield an adequate pregnancy rate. We demonstrated that it is possible to obtain good pregnancy rates with frozen-thawed semen following less intensive protocols.

Paper H5.2

Studies on possibilities of stallion semen selection for deep freezing and artificial insemination (AI)

K. Kosiniak-Kamysz, Z. Podstawski*. Department of Horse Breeding, Agriculture University, Al. Mickiewicza 24/28, 30-059 Kraków, Poland

Analysis of lactic dehydrogenase (LDH) activity and total protein contents in fresh semen plasma of 211 ejaculates collected from 54 stallions revealed a significant correlation $r= -0.25$ ($P \leq 0.01$) between examined components and semen freezability assessed on the basis of spermatozoa motility and the time of spermatozoa survival at the temperature of 4^0C. While analysing these parameters in the ejaculates qualified for freezing on the basis of motility assessment and spermatozoa concentration it has been determined that if total protein contents did not exceed 26 mg/cm^3 or LDH activity was higher than 100mU/cm^3, between 85-90% of those ejaculates fulfilled criteria for artificial insemination (AI) after thawing. Besides, an evaluation of examined semen performed using filtration method through G-15 sephadex columns glass wool revealed conformity of thawed semen value determination using method of motility and the time of spermatozoa survival at 4^0C assessment to filtration method up to 93%. However, no usability of filtration method for fresh stallion semen qualification for deep freezing was noticed.

HORSE PRODUCTION [H]

Paper H5.3

Successful low-dose insemination by hysterscopy in the mare
Lee H-A. Morris[1], R.H.F. Hunter[2] and W.R. Allen[1]. [1]*University of Cambridge, Department of Clinical Veterinary Medicine Equine Fertility Unit, Mertoun Paddocks, Woodditton Road, Newmarket, Suffolk, CB8 9BH, UK.* [2]*Department of Clinical Studies - Reproduction, Royal Veterinary University, Copenhagen, Denmark.*

In two successive breeding seasons a hysteroscopic intrauterine insemination procedure was used to deposit doses of 10, 5, 1, 0.5, 0.1, or 0.001 x 10^6 spermatozoa onto the utero-tubal papilla at the tip of the uterine horn ipsilateral to the ovary containing a preovulatory follicle of >35mm in, respectively, 10, 8, 26, 10, 11 and 10 oestrous mares. A 1.5ml aliquot of fresh semen had been centrifuged through a Percoll density gradient to provide a concentrated motile fraction of spermatozoa which was resuspended in 59-141 µl Tyrode's medium to give the required dose. This was aspirated into an equine GIFT catheter which was passed through the working channel of a Pentax EPM 3000 videoendoscope which, in turn, was passed up the ipsilateral uterine horn for deposition onto the uterine papilla of the utero-tubal junction. Conception rates of 60%, 75%, 62%, 30%, 22% and 10% were achieved in the 6 groups of mares. The simplicity and success of the technique establishes a practical method to exploit the low numbers of viable spermatozoa made available after pre-fertilisation gender selection or cryopreservation of stallion semen.

Paper H5.4

Variation in fertility of warmblood stallions
T. Dohms and E. Bruns*. *Institute for Animal Breeding and Husbandry, University of Goettingen, Albrecht-Thaer-Weg 3, D 37075 Goettingen, Germany.*

The objective of this study was to analyse the development and environmental effects on male reproduction of horses. Furthermore, it was intended to estimate the repeatability for conception rate. The study included annual observations of 720 stallions of the Hanoverian state stud of Celle and the Westphalian state stud of Warendorf between 1980 and 1996. Altogether these stallions covered nearly 250.000 mares. The analyses of data showed that the Hanoverian stallions have reached the maximum of fresh semen artificial insemination with more than 90% of coverings since 1993. In Warendorf the technique of artificial insemination has been more and more applied; 24% of coverings were through artificial insemination in 1996. The fertility of stallions described by their conception rate is defined as the number of fertilised mares (mares with foals born alive or dead and abortions) divided by the number of covered minus non-proved mares. The average conception rate reached 72%. Only stallions with at least ten mares covered per year are considered in the analyses. The conception rate substantially differed between stallions; 25% of the stallions fertilised less than 67% of covered mares and 25% of stallions showed a conception rate higher than 77%. The estimated coefficients of repeatability were 22% (Hanoverian stallions) and 27% (Westphalian stallions). The stallions significantly influenced the conception rate, but also environmental effects (year of covering, type of covering, age of stallion) were responsible for yearly variations in fertility of stallion. In future the effects on female reproduction will be analysed.

HORSE PRODUCTION   [H]   Paper H5.5

The gestation length of the Andalusian and Arabian Horse
F. Blesa[1], M. Valera*[1], M. Vinuesa[2] and A. Molina[1]. [1]*Departamento de Genética. Facultad Veterinaria. Universidad Córdoba. Avda. Medina Azahara, s/n 14005- Córdoba (Spain).* [2]*Cría Caballar. Yeguada Militar de Jerez de la Frontera, Cádiz (Spain).*

The duration of the gestation, considering it as the period between the last service and the day of foaling (when the product is viable), is a physiological variable of great economical importance in all the domestic species, presenting the completed gestations a duration within a short range of days. Although a certain variability is appreciated, justifiable in a genetic and in an environmental point of view. In this work, data from 757 length of gestations, belonging to 143 mares of Andalusian horses (P.R.E.) and to 73 mares of Arabian horses (P.R.Á.) were obtained from the *Yeguada Militar* stud of Jerez de la Frontera (Spain). This data were collected during 13 reproductive years. The average of the gestation length in the P.R.E. mares has been 336.5± 0.67 (c.v. 3.65%) days and 341.87± 0.78 (c.v. 2.65%) days for P.R.Á. mares. The 50% of the gestations are included between the interval of 332-346 days. The influence of the month and year of the service, breed, age of the mare, sex of the product, coat colour of the mare and product and the effect of the stallion have been analysed. Of these factors the breed, age of the mother, year of service, and month of service have resulted statistically significant. This last parameter explains more than an 80% of the variability.

Paper H5.6

Genetic determination of the period between consecutive foalings in East Bulgarian mares
I. I. Sabeva, Research Buffalo Institute, 9700 Shoumen, Bulgaria.

Non orthogonal set of data was analysed containing information of 2221 mating seasons of mares from East Bulgarian Riding breed, acted in the studs after 1958. Two periods of time were studied, and each of them comprised two generation intervals. The components of variance and heritability were computed through the mixed model methodology.
During the years 1958 - 1977 the heritability of the duration of the period between consecutive foalings, computed by really measured quantities has been 0.056 and after 1978 - 0.508. After logarithmic transformations of the scale of measuring the heritability coefficients have been respectively 0.065 and 0.367. During the last two generation intervals the population genetic variance exceeds the lines' one, so the lines have been equalized regarding the studied trait and have stud character. Repeated increase has been established of the additive conditioned variance for a relatively short time.

HORSE PRODUCTION   [H]   Paper H5.7

Influence of maternal size on fetal and postnatal development in the horse
W.R. Allen[1], Francesca Stewart[1], Caroline Turnbull[1], Melanie Ball[1], Abigail Fowden[2], Jennifer Ousey[3] and P.D. Rossdale[3]. University of Cambridge Departments of [1]Clinical Veterinary Medicine and [2]Physiology, Cambridge, and 3Beaufort Cottage Stables, Newmarket, Suffolk; UK.

The effects of uterine environment on development of the placenta, fetus and growing foal were examined in 7 Pony-in-Thoroughbred (P-inTb), 8 Thoroughbred-in-Pony (Tb-in-P), 7 Pony-in-Pony (P-in-P) and 7 Thoroughbred-in-Thoroughbred (Tb-in-Tb) pregnancies established by between-breed embryo transfer. The mares foaled spontaneously at term (days 322-348) and they showed good correlations between maternal weight and foal weight, placental weight and foal weight, and placental area versus foal weight across the 4 types of pregnancy. The P-in-Tb foals were much heavier at birth (37±5kg) than their P-in-P counterparts (24±4kg) and they reached mean weight and height advantages over the latter of 38 kg and 6 cm at weaning. In contrast, the Tb-in-P foals were lighter at birth (33±7kg) than their Tb-in-Tb counterparts (53±6kg) and they grew slower to register mean weight and height disadvantages of 23 kg and 5 cms at weaning. These results showed that, in the horse, maternal size controls fetal growth by limiting the area of uterine endometrium available for attachment of the diffuse epitheliochorial placenta.

Paper H5.8

Influence of maternal age on placental size and function in the mare
Sandra Wilsher and W.R. Allen. University of Cambridge Department of Clinical Veterinary Medicine Equine Fertility Unit, Mertoun Paddocks, Woodditton Road, Newmarket, Suffolk CB8 9BH, UK.

Normal, term placentae were recovered at third stage labour from 4 groups, each of 8 Thoroughbred mares, that included primiparous mares aged 4-7 years and multiparous mares aged 5-9 years, 10-15 years and ≥16 years. Each allantochorion was weighed and its gross area and volume were measured. Ten randomly selected biopsies were prepared histologically and submitted for computer assisted morphometric measurement of the microscopic area of contact between trophoblast and maternal endometrial epithelium.
The mean weights, volumes and gross and microscopic areas of the placentae were all less in the primiparous mares than the other 3 groups and the mean microscopic, but not the gross, area and the volume of the placentae from the ≥16 y.o. mares were also reduced. These age and parity related differences were closely reflected by similar differences in the mean birth weights of the foals.

HORSE PRODUCTION   [H]

Paper H5.9

Superovulation treatment is associated with endogenous LH depression in mares
C. Briant* and D. Guillaume. Haras Nationaux INRA, PRMD, 37380, Nouzilly, France.

Superovulation in mares with equine FSH increases the number of ovulations but not the number of embryos.
During the breeding season, 6 pony mares were followed in a Latin square design, during 2 treated and 1 control follicular phases. Treatments were performed with daily doses of partially purified eFSH (0,75 mg of eFSH) by IM (1 injection/day) or IV (3 injections/day). Follicles were measured by rectal ultrasonography. Plasma LH, progesterone (P_4) and total estrogens were assayed by RIA.

Results: mean ± SEM	IV treatment	IM treatment	Control
Number of preovulatory follicles >28 mm	4,33 ± 1,89	1,5 ± 0,8	1 ± 0
Number of ovulations	0,67 ± 0,9	1,5 ± 0,8	1 ± 0
eLH (ng/ml) during 3,5 days before ovulation	24,5 ± 4	30 ± 6	42,5 ± 8
P_4 (ng/ml) during follicular phase	0,4 ± 0,06	0,4 ± 0,07	0,4 ± 0,06
Estrogens (ng/ml) during follicular phase	1,3 ± 0,33	0,39 ± 0,16	0,30 ± 0,09

IV treatment significantly increased the number of preovulatory follicles but not the number of ovulations. This could be explained by the significant depression of LH after the end of treatment (IV and IM). This depression cannot be associated with P_4 levels but with a significant increase in estrogens.
The interaction between superovulation treatments and LH levels could be an explanation for disappointing results of these treatments in mares.

Paper H5.10

Treatment of endometrosis in infertile mares by intra-uterine infusion of kerosene
Verena Bracher, Anke Neuschaefer and W.R. Allen. Thoroughbred Breeders' Association Equine Fertility Unit, Mertoun Paddocks, Woodditton Road, Newmarket, Suffolk CB8 9BH, U.K.

A 50ml aliquot of the commercially available petroleum spirit, kerosene, was infused into the uteri of 4 young fertile mares and 24 older, infertile mares exhibiting varying degrees of aged-related fibrous degenerative changes to their endometrium (endometrosis). Videoendoscopic inspections of the endometrium and the recovery of endometrial biopsies for microscopic examination were carried out on 4 occasions between 1 and 21 days after treatment.
Diffuse oedema and intense inflammation of the endometrium on day 1 after treatment subsided by days 4 and 7 when tall, active-looking columnar epithelial cells were the dominant histological feature in both the lumenal epithelium and the apical portions of the endometrial glands. By days 14 and 21 both the gross and the histological appearance of the endometrium had returned to normal. Nine of the 11 mares which had exhibited severe endometrosis prior to treatment conceived when they were mated after treatment but only 5 of them carried their pregnancies to term.

HORSE PRODUCTION [H]

Paper H5.11

Immunisation of recurrently aborting mares with stallion lymphocytes
Susanna Mathias and W.R. Allen. University of Cambridge Department of Clinical Veterinary Medicine Equine Fertility Unit, Mertoun Paddocks, Woodditton Road, Newmarket, Suffolk CB8 9BH, UK

Recurrent spontaneous abortion (RSA) of unknown aetiology affects some 2 - 5 % of human couples and a small population of Thoroughbred mares exhibit a similar condition of repeated early pregnancy losses. In one uncontrolled trial (n=24 mares), and in a second controlled trial (n=31 mares), Thoroughbred mares with histories of RSA were immunised during oestrus, and again between 16 and 22 days of gestation, with 2 - 5 x 108 lymphocytes isolated from the peripheral blood of the mating stallion. In the controlled trial, approximately half the mares, selected randomly, were injected with their own (autologous) lymphocytes.
In Trial 1, 31 of the 32 (97%) pregnancies established in the 24 immunised mares resulted in the birth of live foals. In Trial 2, 15 of the 17 mares (88%) immunised with stallion lymphocytes similarly carried their pregnancies to term. However, 10 of the 13 control mares (77%) injected with autologous lymphocytes also produced live foals. This unexpectedly high foaling rate in the control mares negated any significance of the results in the actively immunised animals and it cast doubt on the parameters used to recruit mares to the trial.

Paper H5.12

Maternal recognition of pregnancy in the mare
T.A.E. Stout and W.R. Allen. University of Cambridge Department of Clinical Veterinary Medicine, Madingley Road, Cambridge CB3 0ES, UK.

The developing equine conceptus appears to achieve luteostasis in the mare by completely suppressing uterine $PGF_{2\alpha}$ secretion during the period when luteolysis would otherwise occur. The blastocyst signal which governs this process has yet to be identified although it is likely that the conceptus exerts its effect, at least in part, by inhibiting the cyclical upregulation of endometrial receptors for oxytocin. However, the scale of the decrease in oxytocin binding capacity between pregnant and dioestrous endometrium is not sufficient to account for the dramatic inhibition of $PGF_{2\alpha}$ secretion that occurs in gestation. Therefore, it is probable that additional mechanisms are involved in bringing about luteostasis.
$PGF_{2\alpha}$ is not detectable in uterine lumenal flushings recovered from pregnant mares during days 12-16 after ovulation. However, $PGF_{2\alpha}$ concentrations rise again sharply to cyclical dioestrous levels in uterine flushings recovered between days 18 and 30 of gestation indicating that, in the mare, maternal recognition of pregnancy involves a delay in, rather than a total abolition of, the development of uterine oxytocin responsiveness. Thus, during days 10-16 after ovulation in the pregnant mare, the migrating conceptus must prevent the endometrium from releasing $PGF_{2\alpha}$ in response to oxytocin challenge. But beyond this time, a second strategy must be employed to prevent an apparently still functional $PGF_{2\alpha}$ release pathway from being triggered.

HORSE PRODUCTION   [H]

Paper H6.1

How to breed and train a showjumper?
A. Barneveld[1], R. van Weeren[1], J. Knaap*[2]
[1]Department of Equine Sciences, Faculty of Veterinary Medicine, Utrecht University, Yalelaan 12, 3584 CM Utrecht, The Netherlands, [2]Research station for Cattle, Sheep and Horse Husbandry, P.O. Box 2176, 8203 AD Lelystad, The Netherlands.

Rearing young horses is a time and money-consuming experience. The results of this large investment remain uncertain for a long time and may often be disappointing as in many cases the athletic performance will not come up to the (high) expectations. The goal of this project is to develop criteria for the selection and effective and injury-free training of showjumpers. 42 Warmblood foals are reared and trained according to two different systems under controlled circumstances. One group will be raised following an extensive, low-input system. The other group will be trained twice a week in free jumping. At the age of three the second phase of the project starts when all horses are broken and further trained up to a level at which they will be able to participate successfully in medium-class competition at the age of 5 years. Performance of both groups will be objectively compared. During the years of training jumping performance is scored by two professional international riders/trainers. Jumping technique is measured by using a Proreflex® kinematic analysis system. Muscle composition is determined by microscopical and biochemical analysis of biopsies. Also the mobility of the spinal column will be a specific subject of study.

Paper H6.2

A new type of early performance test: gait and conformation measurements in 3 years old horses
E. Barrey *[1], M. Holmström [2], S. Biau [3], D. Poirel, B. Langlois [1]. [1] INRA, SGQA, F-78352 Jouy-en-Josas; [2] Horse Evaluation system, S-24495 Dösjehro; [3] ENE, F-49411 Saumur.

The purpose of this study was to measure gait and conformation variables in 3 years-old horses presented in a performance test. The relationships between these measurements and the judge scores were analyzed.
Materials and Methods: The measures were done on 39 horses (3 years old) presented at the national final « Dressage Plus » in Fontainebelau 1997 & 1998. The gaits, conformation traits and ability (3 scores + total) of each horse was evaluated by 4 experts. Stride variables (2x11) were measured on walking and trotting horses in hand using an accelerometric device attached on to the sternum (Equimétrix). The femur length, pelvis and femur inclination, hock and stiffle joint angles were measured using digitized image of the left lateral view of the horse (Horse Evaluation System).
Results: At the walk, the scores were significantly correlated with the stride length (0.31) and vertical displacement (0.33). At trot, significant correlations were found between the scores and speed (0.32 to 0.45), stride length (0.33 to 0.56) and symmetry (-0.35). The stiffle angle (-0.36) and conformation index (0.33) combining the main conformation variables were correlated with the scores. No significant correlation was found between conformation and gait variables. The 3 scores were correlated (0.57 to 0.67).

HORSE PRODUCTION [H]

Paper H6.3

Influence of early exercise on de locomotor system
A. Barneveld[1], R. van Weeren[1], J. Knaap*[2]
[1]Department of Equine Sciences, Faculty of Veterinary Medicine, Utrecht University, Yalelaan 12, 3584 CM Utrecht, The Netherlands, [2]Research station for Cattle, Sheep and Horse Husbandry, P.O. Box 2176, 8203 AD Lelystad, The Netherlands.

The influence of exercise during the first months of life on the development of the equine musculoskeletal system was studied with special attention to the development of osteochondrosis. Forty-three Dutch Warmblood foals were randomly allotted to 3 groups at one week of age. Group "box" was housed 24 h/day in a 3 x 3,5 m box stall. Group "training" was housed likewise, but was subjected to an exercise regime consisting of an increasing number of gallop sprints. Group "pasture" was kept at pasture 24 h/day and used as a reference group. All foals were weaned at 5 months and part of them where euthanised. The remaining foals were housed together and given similar exercise in order to verify if any exercise induced effects were reversible. The foals where euthanised at 11 months. At regular intervals the foals were radiographed, muscle and tendon biopsies were taken and the locomotion was analysed. After euthanasia, samples from a large number of muscular, tendnous, bony and articular components were analysed. Some conclusions: Osteochondrosis is a dynamic process during the first year of life. Pasture exercise decreases the risk for osteochondrosis. Exercise has a positive influence on the development of the musculoskeletal system, but overtraining can cause irreversible damage.

Paper H6.4

The rider effect in a genetic evaluation of showjumping performance of horses.
S. Janssens*[1], D. Geysen[1] and W. Vandepitte[1]. [1]Centrum voor huisdierengenetica en selectie, Department Animal Production, K.U.Leuven, Minderbroederstraat 8, 3000 Leuven, Belgium

Performance of horses in sport competition is affected by many factors and the objective is to include them in the (mixed) model for genetic evaluation.
The impact of the rider on the performance of the horse can hardly be underestimated, but many models do not include this effect. Reasons are that information on the identity of the rider is missing or perfomance is measured cumulatively and cannot be linked to one rider. More important however is that the rider- and the direct genetic effect are (statistically) confounded. Individual performance data on jumping (n=113,380, 1995-1998) were obtained from L.R.V. (Landelijke Rijverenigingen). 4 "skill" levels (B-L-M-Z) are present and combinations (=rider*horse) can rise to a higher level by accumulating credit points. About 75% of the 4330 riders mounts at least 2 horses (mean=4.8, max.=59) and about 64% of all horses has at least 2 drivers. Variance components were estimated by combining a direct genetic, a permanent environmental and a rider-effect in 4 different models. Fixed effects of age, sexe and single competition were always included. Heritability estimates for showjumping ranged from 0.37 to 0.11. The rider effect accounted for 14 % of the variance. Correlations between estimated breeding values ranged from 0.81 to 0.95 (all horses, n=22,093) and 0.85 to 0.96 (stallions, n=530). Models are compared using the AIC.

HORSE PRODUCTION [H]　　　　　　　　　　　Poster H6.5

Genetic correlations in performance test results among Haflinger mares and stallions
M. Zeiler, H. Hamann and O. Distl*. Department of Animal Breeding and Genetics, School of Veterinary Medicine Hannover, P.O. Box 711180, 30545 Hannover, Germany.

Haflinger mares and stallions are tested on station for temperament, performance readiness, jumping, riding, cross country and carting abilities. The data analysis included 60 stallions and 600 mares being performance tested on four different stations in the years 1991 to 1995. Pedigree information was built up for eight generations. The objective of our study is to develop a procedure to predict breeding values using information from performance tests on stations. Genetic parameters and breeding values were estimated employing a linear multitrait animal model including the fixed effects for test station, year and season of test, age at test and the covariates Arab blood percentage and inbreeding coefficient. Records from stallions and mares were separately used to predict breeding values. Genetic correlations were analysed in a combined dataset. Analysis of correlations among mares and stallions using breeding values revealed for most traits rather small correlations and even negative values. Riding ability showed a consistent rather high positive genetic correlation among sexes, whereas trotting ability seems to be negatively genetically correlated. However, in the case of top ranking sires all breeding values predicted from test results of both sexes were highly positive. Problems in selection arise in lower ranking sires due to contradictory breeding values between mares and stallions.

Poster H6.6

Les trends genetiques des caracterès utilitaires des étalons demi-sang dans les stations d'entraînement polonaises
M.Kaproń, M. Łukaszewicz, G. Zięba. Department of Horse Breeding and Use, Agricultural University, Akademicka 13, 20-950 Lublin, Poland.

Les études menées ont porté sur 1865 étalons demi-sang, entraînes dans les stations d'entraînement polonaises dans les années 1973-1992. Dans le cadre du système I on a constaté la plus haute héritabilité dans le cas de l'épreuve de course (steeple chase-h^2 =0,568), de l'appréciations générale pour la totalité des épreuves (h^2 =0,436) et l'appréciations des caractères généraux des étalons par le chef d'entraînement (h^2 =0,336). Le trend génétique positive essentiel a été constaté dans le cas de l'appréciation faite par le chef de l'entraînement ($P<0,05$), cependant les trends négátifs essentiels ont été constatés pour l'essai de vitesse au pas ($P<0,05$), steeple chase ($P<0,01$) et l'appréciation de l'état d'avancement de l'entraînement ($P<0,01$). Dans le système II la plus haute héritabilité a été constatée pour la vitesse au pas (h^2 =0,564), pour le style du galop au cross country (h^2 =0,464) et pour l'appréciations de chef d'entraînement (h^2 =0,368). Les trends génétiques positifs et essentiels ont été constatés dans le cas de l'appréciation générale, dans le cas de la totalité des épreuves de performances ($P<0,01$), l'essai d'envie à tirer ($P<0,01$), l'appréciation des caractères généraux par le chef d'entraînement ($P<0,05$) et le style du galop au cross country ($P<0,05$), cependant un trend négatif insignifiant a apparu dans le cas de l'essai de vitesse au pas.

HORSE PRODUCTION   [H] Poster H6.7

Corrélation entre les indices choisis de l'appréciation de l'entraînement des chevaux de course et la vitesse et distance du galop préparatoire
M.Kaproń, I. Janczarek, H. Kaproń. Department of Horse Breeding and Use, Agricultural University, Akademicka 13, 20-950 Lublin, Poland.

Les présentes recherches ont porté sur 10 chevaux pur sang arabe de trois ans (5 juments et 5 étalons), que l'on appréciait dans le cadre de 40 tests d'entraînement. Le paramètre analysé c'était le pouls enregistré à l'aide d'un appareil électronique de télémesure "Sport-Tester 4000". Les enregistrements du pouls ont servi à déterminer les indices suivants: le pouls moyen observé au cours de l'effort [*Pm*] et l'indice de réaction à l'effort (le rapport du pouls moyen à l'effort au pouls au repos multiplié par le pouls moyen à l'effort - [*Irel*]). L'influence de distance du galop préparatoire sur le niveau des indices analysés s'est montrée différenciée. Dans le cas du pouls moyen [*Pm*] on a déterminé les corrélations suivantes: juments - 0,16 à (-0,46^{xx}), (ensemble - 0,24^{xx}); étalons - 0,17 à 0,37^x Par rapport à l'indice de réaction à l'effort [*Ire*] les coefficients de corrélation se sont montrés insignificatifs staststiquement. Dans le cas de la vitesse du galop préparatoire les coefficients de corrélation [R_{xy}] entre le parametre énuméré et le pouls moyen [*Pm*] s'enfermaient dans les limites: juments - 0,29 à 0,61^{xx}; étalons - 0,14 à 0,61^{xx}. Dans le cas [*Ire*] les valeurs R_{xy} se présentaient: juments - 0,13 à 0,48^{xx}; étalons - 0,28 à 0,55^{xx}.

Poster H6.8

Relationship between the racing performance of pure - bred Arabian horses and their coat colour
R. Pikuła[1], M. Smugała[1], D. Gronet[1], W. Grzesiak[2]. [1]Department of Horse Breeding. Agricultural University, Doktora Judyma Str. 24, 71 - 460 Szczecin, Poland, [2]Department of Genetics and Animal Breeding. Agricultural University, Doktora Judyma Str. 6, 71 - 460 Szczecin, Poland.

Studies included pure - bred Arabian horses from National Studs, that run a race on the Służewiec Horseracing course. The investigations covered 657 three - year - old horses (I season of racing test) and 329 four - year - old horses (II season of racing test). For every horse the following data, were determined coefficient of success and coefficient of persistence. All calculations were made using the Statistica 5.1 computer programme, considering influence of coat colour of horses to their racing performance (in dependence of sex).
Statistical differences were found between coat colour of horse and their racing performance.

HORSE PRODUCTION [H] Poster H6.9

Characteristics of racing performance of Thoroughbreds in Korea
K. J. Lee*[1], K, D, Park[2]. *Department of Dairy Science, Konkuk University. 93-1 Mojin-Dong, Kwangjin-Gu, Seoul Korea, 143-701*

A total of 100,839 racing records for Thoroughbreds in Korea was collected from 1990 to 1997 and was utilized for this study. Average racing times for 1,000m, 1,700m and 2,000m were 65.66±1.46 seconds, 116.54±2.01, and 136.49±2.21 seconds, respectively. As the racing distances were increasing, standard deviations showed the increasing trend, while coefficients of variation were decreased. The average generation intervals of sire of son, sire of daughter, dam of son, and dam of daughter were 10.84, 10,39, 10.16, and 10.39 years, respectively. And the average generation interval of parant of offspring was 10.51 years which were similar to 10.5 years reported by Langlois(1983). Data for the annual earnings was standardized by standard deviation of age-birth year groups. The phenotypic and genetic correlations between annual earnings and standardized annual earnings were 0.676 and 0.817, respectively. Mean months of age at the first start, last start, and racing longevity were 45.88, 72,40, and 26.51. Estimates of heritabilities for racing longevity, number of starts and earnings in lifetime were 0.184, 0,216, and 0,178, respectively. The genetic trend for racing time was estimated to be 0.02 seconds per year which was very low, comparing with that of other countries.

Poster H6.10

Racing performances of Anglo-Arab-Sardinian horses analysed by mixed linear models
S.P.G. Rassu[1], N.P.P. Macciotta[1], A. Cappio-Borlino[1], S. Delogu[2], G. Enne[1]. *[1]Dipartimento di Scienze Zootecniche, Università degli Studi di Sassari, Via E. De Nicola 07100 Sassari, Italy; [2]Istituto Incremento Ippico della Sardegna, P.zza D. Borgia 4, 07014 Ozieri (SS), Italy.*

The performances of 87 Anglo-Arab-Sardinian horses in the first racing season (three years of age) were analysed by mixed linear models in order to evaluate the incidence of individual variability on racing times, corrected for the effects of the main factors that influence racing performances. Trainer, Arab blood percentage and total endowment affected significantly racing times, whereas the sex of the horse was not an important discriminating factor.
Repeatability of racing performances, as the ratio between the variance component associated to the horse effect and the sum of this component plus the residual variance was 0.25. Such a low value, already found in other studies, evidences the main limits in the use of racing horses in Sardinia: a scarce specialisation, especially regarding the length of the race, and a general attitude of owners to exploit horses also in an early stage of physical preparation. Moreover, being the value of repeatability rather small (0.34) also when the model takes account of the different physical preparation stages, the low correlation value among racing performances can be mainly ascribed to the irrational use of the horses during the racing season.

HORSE PRODUCTION   [H]   Poster H6.11

A human encounter test to assess young horses' reaction to humans
E. Søndergaard[*1], J. Ladewig[2] and C.C. Krohn[1], [1]Danish Institute of Agricultural Sciences, P.O. Box 50, 8830 Tjele, Denmark, [2]The Royal Veterinary and Agricultural University, Grønnegårdsvej 3, 1870 Frederiksberg C, Denmark.

The aim of the experiment was to assess horses' reaction to humans. Nineteen young horses were housed in boxes, 7 of them singly and 12 in groups of three. Three of the single housed and six of the group housed horses were handled 3 times per week, for 10 minutes. After weaning at approx. 4 months of age the colts were blocked according to sire and age. The human encounter test was conducted after one winter period of handling. During the test the horses were left alone for 3 minutes where after a person entered and stood by the wall for another 3 minutes. Behaviour was registered for both periods by scan sampling every 10 sec. Additionally, average heart rate in the two periods, latency to first contact with person, number of contacts with person and time from end of the test until the horse was caught were recorded. Heart rate was lower, frequency of standing was higher and frequency of walking and defecation were lower when the person was present. Group housed horses were standing less and walking more than single housed horses, indicating that they were more nervous by the isolation. Handled horses walked less than non-handled horses and made faster and more frequent contact to the person. The results show that by combining recordings of behaviour, latencies and heart rate it is possible to assess the reaction of horses to humans. To improve the test, also position in the arena should be recorded.

Poster H6.12

Individual differences in temperamental traits of horses
E.K. Visser[*1], H.J. Blokhuis[1], J. Knaap[2], and A. Barneveld[3]. [1]Institute for Animal Science and Health ID-DLO, P.O.Box 65, 8200 AB Lelystad, The Netherlands, [2]Institute for Horse Husbandry "De Waiboerhoeve", Wisentweg 55, 8219 PL Lelystad, The Netherlands, [3]Faculty of Veterinary Medicine, Utrecht University, P.O.Box 80.153, 3508 TD Utrecht, The Netherlands.

The interest in behavioural research on temperamental traits in animals has grown since we know that they are relevant to the science of animal welfare. It places a particular emphasis on the way in which individuals react to environmental change and challenge. With a behavioural test one can objectively assess the individual response on an environmental change or challenge.
The goal of this study is to develop and execute behavioural tests to assess individual characteristics in horses at an early age. 42 Dutch warm-blood horses will go through several behavioural tests. To test the consistency these tests will be repeated twice a year with a four weeks interval. Behavioural tests that are executed between 5 months and 13 months include an arena test, a novel object test and a handling test. During these tests behaviour is recorded with real time video camera and later analysed with the Observer system (Noldes Information Technology). During the tests heart rate is measured with Polar Horse Trainer transmitters. Before and after the tests saliva is collected to analyse cortisol levels. Behavioural data will be linked to physiological data. Responses in different tests will be compared between and within horses.

HORSE PRODUCTION   [H]

Behaviour performance of half bred horses tested on race track
H.K. Geringer and J. Kasprzak. Department of Horse Breeding, Agricultural University, 51-631 Wrocław, 7 Kożuchowska Str. Poland.

Poland is the only country where half bred horses pass performance test also in racetrack at the age of 3 years. Stallions nonqualified for the performance test station and mares are entering races. Horses participate in hurdle race and steeple chase in the second half of race season. Some horses continue steeple chase in the age of 4 years and more, some are destined to sport or (and) breeding after first racing season. Our study concern 93 half bred, 3- years old horses which took part in flat race, hurdle race and steeple chase on Wrocław Park course in 1997. Behaviour of horses for 4 elements were judged [the scale from 2 pts (the worst) to 5 pts (the best)]during: presentation on the paddock (mean- 4,63 pts, SD 0,59), saddling (mean- 4,08 pts, SD 1,11), mounting (mean- 4,72 pts, SD 0,66), entering the starting-gate (mean-4,51 pts,SD 0,90). Highly significant (p<0,01) correlations between analyzed elements were stated. Significant correlations (p<0,05) between individual succes coefficient and presentation and mount scores were observed. Influence of number of starts, sex of horse and stable-trainer were analysed. Low (0,05) and negativ (-0,06) correlations were stated between behaviour in starting gate, behaviour in the time of saddling and individual succes coefficient.

CATTLE PRODUCTION   [C]   Paper C5

Management systems to improve the production and financial performance of dairy farms
R. W. Palmer[*]. Department of Dairy Science, University of Wisconsin-Madison, 1675 Observatory Drive, Madison, WI 53705-1284, USA.

The role of the dairy manager is to plan strategically and to direct resources in a way that leads to a profitable and sustainable dairy enterprise. Management is the process of decision making and has three major functions: planning, implementation and control. Tremendous development, adoption and management of new production enhancing technologies over the past few decades has led to rapid increases in herd size and milk production levels. The challenge to the managers of these modern dairy herds is to economically achieve high milk yield without sacrificing animal health and welfare, deterioration of the environment or human safety. The complexity and scope of the decisions required of the dairy manager requires the creation of a new level of decision support systems which can collect, organize and analyze pertinent information on a timely basis. These new support systems should include farm monitoring and control procedures which compare real time performance values with benchmark information and be able to identify when corrective actions need to be taken.

AUTHORS INDEX

A

	Page		Page
Abbas, S.	248	Arana, A.	239
Abd Alla, M.M.	229	Archer, J.A.	62
Abd El-Hamid, I.A.	70	Archibald, A.L.	1
Abd El-Moty, A.K.I.	79	Arendonk, J.A.M. van	3, 5, 19, 35,
Abd El-Razek, S.T.	225		36, 38, 83
Abd-El-Hakeam, AA.A.	79	Arieli, A.	95
Abijaoudé, J.A.	258	Arnaud, F.	208
Abramson, S.M.	95	Arthur, P.F.	62
Abrutat, D.J.	286	Aşkın, Y.	233
Acatincăi, S.	185	Augustini, C.	198
Adamec, T.	145, 316	Averdunk, G.	4, 207
Adamović, M.	80	Azeroglu, F.	139
Ådnøy, T.	221, 264		
Agárdi, G.	319		

B

	Page		Page
Aggrey, S.E.	37	Bach Knudsen, K.E.	268
Ahmed, S.	50	Baiomy, A.A.	79
Aigner, B.	6	Bali Papp, A.	126
Akçay, H.	189	Balicka-Ramisz, A.	145
Akulich, E.G.	129	Balika, S.	195, 201
Al-Bayati, H.	41	Ball, M.	349
Al-Khbeer, A.M.S.	231	Balogh, R.	175
Albano, P.	305	Bán, B.	7
Albina, E.	135	Banda, J.W.	288
Alcaide, B.	141, 142	Barłowska, J.	192, 200
Alderson, L.	46	Barajas, F.	227
Algers, B.	137, 155	Baran, M.	250
Allain, D.	263	Bárány, I.	202, 203
Allen, W.R.	347, 349, 349, 350, 351, 351	Barba, C.	227, 281, 314
Alpan, O.	139	Barbat, A.	27, 60
Alps, H.	61	Barbato, O.	166
Álvarez, F.	194	Barbieri, S.	112
Amarger, V.	312	Barcikowski, B.	250
Ambrosjeva, E.D.	47	Bareille, N.	138
Amer, P.R.	260	Barile, V.L.	166
Amigues, Y.	310	Barneveld, A.	352, 353, 357
Ananjev, V.N.	48	Barowicz, T.	320
Andersen, H.J.	301	Barrey, E.	352
Andersen, S.	308	Bartenschlager, H.	4
Andersson, L.	312	Baruskova, M.	329
Ando, S.	140	Bata, Á	136
Andres, M.	144	Bauer, C.	261
Anglada, M.	141, 141	Baulain, U.	309, 315
Aniołowski, K.	224	Baumung, R.	207, 209
Antczak, K.	102	Bava, L.	257
Antropov, V.G.	81	Bazard, C.	83
Apeldoorn, E.J.	54	Beauchamp, C.J.	276

	Page		Page
Beaudeau, F.	138	Bodó, I.	7, 202
Becková, R.	312	Boer, G. van de	215
Bedhiaf, S.	228	Böhm, J.	97
Bedö, S.	195	Böhme, H.	99, 289
Bee, G.	322	Boichard, D.	27, 59, 60
Beeckmann, P.	4	Boidi, G.	143
Beek, S. van der	42	Boland, M.P.	124
Behr, V. de	179, 190, 197, 202	Bölcskey, K.	182, 184, 202, 203
Beilharz, R.	54	Bolet, G.	161
Beiu, F.	86, 92, 180	Bolocan, E.	180
Bek, Y.	32	Bonaïti, B.	27
Bekenev, V.A.	291	Bonneau, M.	161
Ben Gara, A.	228	Bonomi, A.	143
Ben Hamouda, M.	228	Borba, A.E.S.	104
Benito, J.	285	Borba, A.F.R.S.	104
Berg, P.	23, 33, 36	Borell, E. von	113, 137, 238
Berg, Th. van den	118	Borghese, A.	166
Berk, A.	321	Borka, G.	182, 184
Berry, N.	77	Borys, B.	232, 236, 240
Bertschinger, H.U.	9	Bosch, J.C.	118
Besbes, B.	29	Boscher, M.Y.	46
Beuing, R.	90, 261	Bosi, P.	334
Beynen, A.C.	326	Bouix, J.	227, 259
Bhuiyan, A.K.F.H.	212	Boulanger, R.	276
Biagini, D.	179	Bouraoui, R.	217
Biau, S.	352	Bouska, J.	176
Bickel, S.	322	Bovenhuis, H.	36, 38
Bidanel, J.-P.	15, 302	Boyazoglu, S.	169
Bielen, A.	197	Boykov, Yu.V.	80
Biereder, S.	162	Bozó, S.	174, 202, 203
Bijma, P.	34	Bracher, V.	350
Billon, Y.	310	Brade, W.	44, 315
Binder, H.	344	Bradshaw, R.H.	294
Bingöl, M.	233	Brandt, H.	35
Bink, M.C.A.M.	38	Brascamp, E.W.	3
Birchmeier, A.N.	45	Bray, A.R.	228
Bishop, S.C.	5, 260, 263	Brem, G.	6
Blanco Roa, E.N.	213	Brenig, B.	3, 4, 10, 10, 37, 40, 41, 51
Blasco, A.	22	Briant, C.	350
Blaszczýk, B.	250, 251, 251	Briend, M.	60
Blesa, F.	348	Brink, D. van den	199
Bloc, N.	208	Brits, M.	103
Blokhuis, H.J.	114, 357	Brka, M.	64
Blum, J.W.	77, 152, 158	Brocard, V.	77
Blümel, J.	4, 37	Brockmann, G.	4, 37, 41
Bluzmanis, J.	122	Brooks, P.	324
Bocian, M.	310, 311	Broom, D.M.	113, 294
Bódis, K.	218	Brotherstone, S.	55

Bruckental, I.	95	Chagunda, M.G.	216
Bruckmaier, M.	77	Chardon, P.	312
Bruininx, E.	326	Charon, K.M.	234, 262
Bruns, E.W.	216, 347	Chechushkova, M.A.	338, 340
Brzostowski, H.	237, 243, 244	Chevalet, C.	46
Bugnacka, D.	289	Chládek, G.	247
Buiten, A. van	346	Chmielnik, H.	196
Buleca, J.	127, 205, 284	Cho, W.T.	328
Bulla, J.	40	Chrenek, P.	40
Bünger, A.	12	Chrenková, M.	108
Bünger, L.	53	Christensen, L.G.	13, 33
Bunghiuz, C.	86, 92	Christensen, M.	149
Bureau, F.	57	Chudoba, K.	219
Burger, J.	40	Ciobanu, D.	47
Burkert, O.	84	Ciuraszkiewicz, G.	339
Busk, H.	309	Ciuruś, J.	223
		Çivi, A.	233

C

		Claeys, E.	199
Caffrey, P.	93, 332, 334	Clemente, A.	8, 234, 311, 314
Caja, G.	132	Clinquart, A.	75, 190
Callesen, J.	332	Close, W.	335
Campo, J.L.	65	Codjo, B.	155
Campo, M.M.	197	Coghe, J.	57
Canart, B.	326	Cole, D.J.A.	273
Canavesi, F.	17	Colenbrander, B.	346
Cannas, A.	94	Conill, C.	132
Cantarero, J.	141, 142	Coppieters, W.	36
Cantet, R.J.C.	45	Corning, S.	136
Cappai, P.	169	Cosseddu, G.M.	342
Cappelli, K.	2	Coudurier, B.	46
Cappio Borlino, A.	18, 356	Creston, D.	334
Carballo, J.A.	206	Crimella, C.	112
Cariolet, R.	135	Cromie, A.	74
Caritez, J.C.	310	Cronin, E.J.	93
Carta, A.	18	Cropper, M.	131
Casanova, L.	11	Crosby, E.J.	229
Casciotti, D.	2	Crosse, S.	91
Casini, L.	334	Crovetto, M.G.	257
Catalano, A.L.	341	Crowley, B.A.	261
Catillo, G.	126	Cruz-Sagredo, R.	197
Cavani, C.	112	Csapó, J.	110, 293, 315
Cechová, M.	320, 329	Csapó-Kiss, Zs.	293, 315
Cengiz, F.	233	Cumberas, M.	314
Cepica, S.	4	Cuypers, M.	131
Ceresnáková, Z.	108	Cwikla, A.	224, 225
Chacko, C.T.	217	Cytowski, J.	206
Chadd, S.	273	Czaplicka, M.	187, 193
Chadio, S.	256	Czarnecki, R.	279, 282, 300, 300, 301, 319

	Page		Page
Czarniawska-Zajaç, S.	237	Dohms, T.	347
Cziszter, L.T.	185, 188	Dohy, J.	43, 126, 176, 190
Czudy, W.	193	Doll, K.	213
		Donovan, A.	124
D		Dore, M.G.	277
Daccord,	255	Döring, L.	139
Daelemans, J.	174	Dorner, Cs.	188
Daenicke, R.	94, 99, 289, 321	Dos Santos, R.	337
Dahlstedt, L.	108	Doskočil, M.	96
Damgaard, L.H.	13	Dourmad, J.-Y.	273
Dandrifosse, G.	150	Dragos, M.	4
Danell, B.	46	Dréau, D.	154
Danielsen, V.	270	Dřevo, V.	246
Danshin, V.	192	Dřimalová, K.	320
Dauncey, M.J.	151	Drögemüller, C.	67
David, P.	312	Drozd-Janczak, A.	88
Davidson, F.M.	328	Druet, T.	208
Davoli, R.	46	Dublecz, K.	110
De Marchis, F.	2	Dubrovskaya, R.M.	341
Deaville, E.	76	Duchamp, G.	165
Debenedetti, A.	166	Ducrocq, V.	12, 29
Delgado, F.	194, 227	Duda, J.	4
Delgado, J.V.	46, 281, 314	Dufey, P.A.	198, 322
Deligeorgis, S.	293	Duffy, P.	124
Delikator, B.	282, 300, 301	Dufrasne, I.	75, 159, 202
Dellal, I.	266	Dunayeva, G.S.	267
Delogu, s.	356	Duncan, J.	328
Deloyer, P.	150	Dündar, S.	257, 258
Dementieva, N.V.	72	Duru, M.	74
Dementjev, V.N.	318	Dymnicki, E.	177, 195
Demeyer, D.	199, 199	Dzapo, V.	9
Denisenko, V.Y.	168	Dziadek, K.	282, 300, 301
Depuydt, J.	239		
Désiron, A.	326	**E**	
Detilleux, J.	21, 214, 226	Eady, S.	260
Dieguez, E.	281	Early, B.	336, 336
Diepen, J. Th. M. van	275	Edwards, S.A.	116, 136, 140, 270,
Dierkes, B.	51		323, 327, 328
Diestre, A.	314	Eenaeme, C. van	75, 159, 190
Dietl, G.	212	Egger-Danner, C.	209
Diez, M.	179	Egli, L.	345
Dijkhuizen, A.A.	121	Egri, Z.	7
Distl, O.	66, 67, 210, 213, 214, 354	Eguinoa, P.	239
Dividich, J. Le	151, 154, 155, 160, 296, 324, 325	Eik, L.O.	221, 264
		Eilart, K.	292
Djemali, M.	217, 228	Eissen, J.J.	54
Dobicki, A.	219	Èrtek, J.	280, 283
Dodds, K.G.	8	El Fadili, M.	21, 226

	Page		Page
El Hakaam, A.A.	229	Félix, A.	214
El Shahat, Aly Ahmed	253	Felska, L.	144
El-Barody, M.A.A.	79, 225	Fernandez, X.	303
El-Feel, F.M.R.	79	Fernández, J.A.	95
El-Ghamry, A.A.	97, 106	Fernández Barranquero, C.	142, 142
El-Kaiaty, A.M.	157	Fernando, R.L.	28, 45
El-Mallah, G.M.	106	Ferrera, J.L.	285
El-Sayed, T.M.	70	Ferris, C.P.	74
Elsasser, T.H.	158	Février, C.	271
Elsen, J.M.	227, 259, 310	Ficco, G.	126
Emler, K.	224	Fiedler, J.	281
Ender, B.	178	Filozof, A.	88
Ender, K.	178, 304	Filya, I.	232
English, P.R.	140	Fischer, E.	84
Enne, G.	356	Fisker, B.N.	274, 274
Enser, M.	240	Flachowsky, G.	321, 321
Erasmus, G.J.	20, 43	Florek, M.	88
Erdin, D.	73, 82	Florescu, E.	89
Erhardt, G.	4, 37, 90, 261, 262	Förster, M.	4, 37
Erkus, A.	266	Fortun-Lamothe, L.	161, 161
Erokhin, A.S.	128	Foster, E.	328
Ertugrul, O.	139	Fourichon, C.	138
Eryomenko, V.V.	184	Fowden, A.	349
Espejo, M.	197	Freyer, G.	121
Esslemont, R.J.	215	Friedli, K.	173
Esteves, L.	337	Fries, R.	9, 41
Eszes, F.	242	Frileux, Y le	254, 255
Euw, D. von	73	Frost, A.R.	286
Eveno, E.	135	Fuerst, C.	207
Everts, H.	326	Fulka, J.	169
Evrard, M.	190, 197, 202	Furukawa, T.	39
Ewen, M.	116	Futerová, J.	63
Eychenne, F.	227		

F

G

Fahr, R.-D.	86, 139	Gabel, M.	178
Failla, S.	204	Gábor, G.	124, 182, 184
Falcão e Cunha, L.	98	Gädeken, D.	94
Falcinelli, M.	2	Gaheis, F.	322
Falkowski, J.	289	Gajic, Z.	307
Fallon, R.J.	336, 336	Galassi, G.	257
Fandrejewski, H.	96, 162	Galasso, A.	166
Fantova, M.	231	Galesloot, P.	215
Farnir, F.	21	Galieva, L.D.	169
Faucitano, L.	302	Gall, C.F.	83
Fechner, H.	299	Gálová, Z.	108
Fedele, V.	254	Gamarnic, N.G.	81
Fedoskov, E.D.	169	García Casco, J.M.	285
		Garcia-Cachan, M.D.	197

Gardzińska, A.	278	Gravillon, G.	296
Gardzina, E.	66, 187, 191	Grela, E.R.	335
Gáspárdy, A.	174, 188, 218, 242	Gremokolini, C.	334
Gauly, M.	90, 261, 262	Greyling, J.P.C.	249
Gawęcki, J.	225	Grigoriadis, D.F.	140
Geldermann, H.	4, 9	Grindflek, E.	303
Gengler, N.	208	Grochowska, R.	177
Georges, M.	227	Groen, A.	33
Georgescu, D.	86, 92	Groenen, M.A.M.	3, 38, 46
Gérard, N.	165	Groeneveld, E.	15, 44
Gérard, O.	159, 179, 202	Gronet, D.	355
Geringer, H.K.	358	Groot, P.N. de	3
Geringer de Oedenberg, H.	339	Grozea, A.	127
Geysen, D.	308, 353	Grubić, G.	80
Ghassan Agil,	241	Gruner, L.	259
Ghişe, Gh.	127	Grupe, S.	4, 37
Ghodratnama, A.	153	Grzesiak, W.	355
Ghsnem, Y.S.	253	Gudilin, I.I.	318
Gianola, D.	25, 29, 30, 59	Gudmundson, O.	110
Gidenne, T.	98	Guérin, B.	119, 120
Gierzinger, E.	209	Guienne, B. Le	120
Gigli, S.	76, 204	Guilhermino, M.M.	218
Ginja, C.	8	Guillaume, D.	350
Ginste, J. vande	105	Guler, A.	127
Gispert, M.	314	Gulyás, L.	201
Gitsakis, N.	18	Gundel, J.	319
Givens, I.	76	Günzburg, W.H.	6
Givens, M.D.	119	Guo, Z.	216
Gjerde, B.	33	Guturova, N.V.	341
Gläser, K.	317	Gutzwiller, A.	156
Glodek, P.	35, 46, 321	Györkös, I.	174, 182, 184, 202, 203, 218
Gnyp, J.	188, 210		

H

Goe, M.R.	217	Haaksma, J.	271
Goff, G. Le	269	Haley, C.	46
Gökdal, Ö.	233	Hamann, H.	210, 354
Gonçalves, P.M.M.O.	104	Hamdy, A.M.M.	170
Gondret, F.	161	Hamman, H.	152, 214
González, J.	65	Hammond, K.	46
González-Rodríguez, A.	172	Hamon, L.	135
Góralczyk, M.	233	Han, In K.	328, 334
Göransson, L.	331	Hanenberg, E.H.A.T.	16
Gordon, F.J.	74	Hanrahan, J.P.	124, 261
Gotterbarm, G.	335	Hansen, B.K.	23
Goulas, C.	256	Hansen, C.F.	327, 333
Grajewska, S.	310	Hansen, L.L.	134
Granciu, I.	89	Harazim, J.	99, 108
Grandfils, C.	150	Harlizius, B.	67
Granli, T.	335		

	Page		Page
Hartog, L.A. den	326	Huba, J.	40, 212, 213
Hashish, S.M.	97, 106	Huber, H.	322
Hasselmann, L.	276	Hucko, B.	84, 96, 104, 109, 111
Hatzikas, A.	248	Huërou-luron, I. Le	154, 155
Haumann, P.	245	Huirne, R.	121
Hauser, R.	173	Hunter, R.H.F.	347
Hayes, J.F.	37	Husson, C.	57
Házas, Z.	293, 315	Húsvéth, F.	315
Hecht, W.	2		
Hedemann, M.S.	151	**I**	
Heetkamp, M.J.W.	271	Iacurto, M.	76, 179, 204
Heger, J.	104	Ikonen, T.	62, 64
Heinrich, H.	79, 90, 138, 171	Ilaslan, M.	189
Henckel, P.	149	Inglin, H.	338
Henning, M.	309	Iolchiev, B.S.	49
Henryon, M.	36	Isakov, V.	307
Heo, K.N.	328	Ishii, K.	39
Herd, R.M.	62	Istasse, L.	75, 159, 179, 190, 197, 202
Heringstad, B.	59	Ivancsics, J.	126
Hermán, A.	319	Ivanova, O.A.	204
Herrmann, R.	213	Iwańczuk-Czernik, K.	143
Hertel, S.H.	309	Izquierdo, M.	197
Hervieu, J.	259		
Herzig, I.	108	**J**	
Hetényi, L.	15, 40, 116, 212, 213	Jacyno, E.	319
Heydarpour, M.	19	Jagger, S.	327
Heylen, K.	245	Jaguelin-Peyraud, Y.	271
Hibner, A.	189	Jagusiak, W.	67
Hiendleder, S.	2, 4, 8, 37	Jamison, W.	306
Hill, W.G.	53	Janczarek, I.	355
Hirabayashi, M.	109	Jandurova, O.	63, 63
Hoeschele, I.	28	Janicki, B.	240
Hofer, A.	14, 15, 30, 304	Jankowski, P.	205
Hoffmann, K.	90, 262	Jánosa, Á.	190
Holcman, A.	106	Jansen, M.B.	130
Holló, G.	176	Jansen, S.	10
Holmström, M.	352	Janss, L.L.G.	3
Holzer, Z.	90, 138	Janssens, S.	308, 353
Homolka, P.	99, 103	Jarczyk, A.	111
Honing, Y. van der	275	Jávor, A.	265
Honkatukia, M.S.	38	Jávorka, L.	242
Horn, P.	293, 315	Jędrychowski, L.	111
Hornick, J.-L.	75, 159, 179, 190, 197	Jędryczko, R.	111
Horvai Szabó, M.	176, 201	Jemeljanovs, A.	122, 122
Hošek, M.	247	Jeneckens, I.	10
Hoste, S.	327	Jensen, B.B.	151, 323
Houška, L.	281	Jensen, J.	13, 23, 33
Hrouz, J.	57	Jeon, J.T.	312

Jestin, A.	135	Karus, A.	168, 181
Jezewska, G.	117	Karus, V.	181
Ježková, A.	85	Kasprzak, J.	358
Jodkowska, E.	339	Kato, S.	105
Johansen, M.	327	Kaufmann, D.	15
Jones, G.	327	Kaulfuß, K.-H.	249
Jong, G. de	42, 215	Kawai, Y.	165
Jongbloed, A.W.	275	Kawęcka, M.	279, 282, 300, 300, 301
Jonins, V.	122	Kaya, O.	139
Jordaan, G.F.	31	Keane, M.G.	75
Jorsal, S.E.	330	Kędzierska, M.	200
Jovanović, R.	80	Keeling, L.	114
Juillerat, P.-A.	344	Kelleher, D.L.	229
Junge, N.	64	Kellems, R.O.	175, 180
Just, A.	95	Kelly, D.	153
Juszczak, J.	189	Kempers, A.	67
Jørgensen, H.	95	Kenawy, M.N.	146
Jørgensen, L.	327, 332	Kerouanton, J.	77
Jørgensen, P.F.	149	Kerrour, M.	190
		Keszthelyi, T.	242

K

		Kharchenko, P.G.	283
Kaam, J.B.C.H.M. van	38	Khidr, R.	97
Kaart, T.	279	Khimich, N.G.	183
Kaasiku, U.	194	Kiiman, H.	185
Kaczorek, S.	313	Kim, S.W.	328
Kadokawa, H.	165	Kirchgeßner, M.	198, 331
Kahi, A.K.	83	Kirukova, Ju. S.	169
Kahl, S.	158	Kjeldsen, N.J.	333
Kaljo, A.	168	Kjos, N.-P.	305
Kallweit, E.	315	Klapil, L.	98
Kalm, E.	4, 37, 41, 64, 280, 312	Klautschek, G.	212
Kampf, D.	289	Klecker, D.	57, 107, 107
Kamyczek, M.	279	Klemetsdal, G.	59
Kamyczek, M.	300	Klindtworth, M.	132
Kamyk, P.	299	Klocek, C.	278, 290, 298
Kanitz, E.	166	Klont, R.E.	303
Kanitz, W.	162	Klooster, C.E. van 't	131
Kann, G.	163	Kluciński, Wł	234
Kapelanska, J.	318	Knaap, J.H.	346, 352, 353, 357
Kapelanski, W.	307, 310, 311, 316, 318	Knap, P.W.	53
Kaproń, H.	355	Knight, C.H.	147, 167
Kapron, M.	354, 355	Knížková, I.	186
Karabulut, A.	232	Knizkova, N.	186
Karamucki, T.	311	Knol, E.F.	16, 282
Karasov, E.A.	235	Knyazev, S.P.	341
Karaszewska, A.	163, 233	Koç, A.	189
Karlikov, D.V.	89	Kochnev, N.N.	68, 173, 184
Karlsson, A.	303, 309	Kochneva, M.L.	70, 71, 71, 338, 340

	Page		Page
Koćwin-Podsiadła, M.	313	Kuchtík, J.	247
Koczanowski, J.	278, 290, 298	Kudrna, V.	101, 211
Kodes, A.	96, 104, 109, 111	Kuhn, G.	304
Kögel, J.	81	Kühn, C.	4, 37
Köhler, P.	315	Kuhnlein, U.	37
Kohn, G.	309	Kukovics, S.	265
Kollers, S.	41	Kulikova, S.G.	70, 71, 71
Kolstad, N.	317	Kulisa, M.	339
Kondo, D.	105	Kumar, J.	168
Koning, D-J. de	3, 38	Kumar, M.-A.	168
Konskich, N.V.	81	Kunc, P.	186, 186
Koops, W.J.	19	Künzi, N.	14, 15, 28, 30, 73, 82, 224
Kopecny, J.	63	Kurt, E.	240
Korn, C.	131	Kuryło, B.	205
Korotkevich, O.S.	285, 296, 297, 297	Kuryl, J.	310, 311
Korotkova, G.N.	70, 71, 297	Kuzmina, T.I.	69, 164, 164, 168, 169
Korsgaard, I.R.	13, 30	Kuznetsov, V.G.	297, 298
Kortz, J.	310, 311	Kyriakou, K.	256
Koshel, V.	294, 295		
Kosiniak-Kamysz, K.	346	**L**	
Kosolapikov, A.V.	173	Labroue, F.	14
Kostomakhin, N.M.	128	Ladewig, J.	357
Kotomin, K.N.	298	Lærke, H.N.	151
Koubková, M.	186, 186	Lafuente, M.J.	154
Koucký, M.	316	Lagant, H.	310
Koutník, V.	247	Lahucky, R.	309
Koutsenko, N.	72	Laitat, M.	326
Koutsotolis, K.	163	Lallès, J.-P.	154
Kovac, M.	35, 106, 135	Lambooy, E.	131
Kovacs, A.	126, 195, 201	Lamminger, A.	210, 214
Kovács, G.	110	Lang, P.	101, 211
Kovács, K.	182, 184	Lange, S.	331
Kowalski, P.	188	Langhammer, M.	58, 125
Kozera, W.	289	Langholz, H.J.	61, 226
Krása, A.	85	Langlois, B.	352
Krasucki, W.	335	Lantier, F.	259
Kratz, R.	321	Larroque, H.	12
Kraus, M.	90, 261	Larsen, M.	17
Kräußlich, H.	210	Lasota, B.	250
Kremer, V.D.	35	Laštovková, J.	21, 316
Krempler, A.	10, 41, 51	Látits, G.	218
Kreuzer, M.	77, 100	Latouche, K.	115
Kriegesmann, B.	51	Lawlor, P.	332, 334
Krinskiy, Y.A.	68	Lazzaroni, C.	112, 179, 204
Krohn, C.C.	357	Lebas, F.	161
Krzęcio, E.	313	Lebedev, V.A.	164
Kubouškova, M.	21	Lebedeva, I.Yu.	164
Küchenmeister, U.	304	Lebengartz, Ya. Z.	182

	Page		Page
Lebzien, P.	94, 99	Lynch, B.	332, 334
Lechevestrier, Y.	271		
Ledoux, D.R.	287, 287	**M**	
Lee, K.J.	356	Małyska, T.	188
Leek, A.	277, 286	Macarie, G.	92
Leenhouwers, J.I.	282	Macarie, H.	92
Leiding, C.	118	Macciotta, N.P.P	18, 356
Leitgeb, R.	97	Machacová, E.	102
Lekeux, P.	57	MacKenzie, K.	5
Lende, T. van der	35, 282	Madec, F.	135
Lengerken, G. von	86, 139, 162, 245	Madjoub, A.	217
Lengyel, A.	244	Madsen, P.	17, 30
Lenzlinger, C.	338	Magic, D.	205
Leotta, R.	134	Magistrini, M.	165
Leroy, P.L.	21, 57, 214, 226, 227, 313	Mahé, D.	135
Lesniewska, V.	151	Mahrous, K.F.	70
Lettner, F.	322	Mahrous, U.E.	238, 238
Leuenberger, H.	224	Makeeva, T.V.	183
Lew, H.	97	Mäki, K.	22
Leyhe, B.	4, 37	Mäki-Tanila, A.	34, 38
Liamadis, D.	235, 248	Makulska, J.	78, 187, 191
Lichovníková, M.	107, 107	Malecki, J.	145
Lien, S.	303	Malfatti, A.	166
Liinamo, A.-E.	22, 60, 61	Malinovskiy, A.M.	130
Lin, C.Y.	37	Malovrh, S.	35, 135
Lindqvist, A.	155	Mandersloot, F.	216
Litwińczuk, A.	88, 192, 200	Mantey, C.	41
Litwińczuk, Z.	192, 200, 205	Mäntysaari, E.A.	26
Loi, L.	169	Mäntysaari, P.	171
Looft, C.	4, 37, 312	Mantzios, A.	163
López-Gatius, F.	123, 123	Mantzke, V.	257, 258, 276, 299
Lorenz, D.	290, 304	Marcq, F.	227
Loret, S.	150	Mareček, E.	108
Loucka, R.	102	Marenkov, V.G.	49, 68
Louda, F.	87, 87, 88, 125	Marguerat, C.	224
Louveau, I.	151	Mariani, P.	341
Lovas, B.	116	Maribo, H.	332
Lüchinger-Wüest, R.	222	Marie, M.	83
Łuczak, W.	225	Marounek, M.	21
Lugasi, A.	202	Marquardt, V.	117
Luiting, P.	53, 54	Marschall, B.	10
Lukács, P.	32	Martelli, G.	272
Łukaszewicz, M.	354	Martínez, A.M.	46
Łukaszewski, Z.	102	Martins, A.	234
Lund, M.S.	30	Martuzzi, F.	341
Lunden, A.	177	Marx, G.	117
Lüth, A.	343, 344, 345	Marzanov, N.S.	49
Luzi, F.	112	Marzouk, K.M.	229, 231

Masun, S.R.	71	Møller, P.D.	110
Mathiak, H.	90, 262	Molnár, B.	236
Mathias, S.	351	Moniello, G.	342
Matlova, V.	223, 231	Monserrat, L.	206
Matos, J.	8, 234, 311, 314	Morand-Fehr, P.	255, 258, 259
Matras, J.	335	Mörchen, F.	86
Matsui, T.	109, 248	Mordenti, A.	272
Matthes, H. -D.	79, 90, 138, 171	Morel, I.	157
Mattová, J.	205, 284	Moreno, C.	259, 310
Mattuzzi, S.	334	Moretti, C.	2
McEwan, J.C.	260	Moritz, S.	249
McKeon, M.P.	277	Moroz, T.A.	130
Medjugorac, I.	4, 37	Morris, C.A.	45
Meier, W.	115	Morris, L.H.-A.	347
Meijerink, E.	9	Morvan, P.	135
Mele, M.	134	Moser, G.	4, 9
Meloni, U.	131	Moskwa, B.	262
Menaya, C.	285	Mozgis, V.	122
Mendizabal, J.A.	239	Mroczkowski, S.	236
Menegatos, J.	256	Mroz, Z.	275, 335
Merchant, M.	266	Mtimuni, J.P.	288
Merks, J.W.M.	16	Mudřik, Z.	84, 96
Meschy, F.	252, 254	Mueller, E.	4
Meunier-Salaün, M.C.	270	Muir, D.D.	167
Meur, D. Le	77	Mulder, H.	217
Meuwissen, T.H.E.	26	Müller, M.	6
Meyer, K.	14, 25	Mulligan, F.J.	93
Michalska, G.	307, 316	Mulume, C.G.	288
Michaux, C.	21, 57, 226	Mumprecht, E.	338
Mieleńczuk, G.	144	Münchow, H.	257, 258, 276, 299
Mielenz, N.	44	Muratovic, S.R.	100
Miesenberger, J.	207	Murphy, O.J.	229
Migdal, W.	278, 290, 298		

N

Miglior, F.	11, 17
Mihina, S.	116
Mihók, S.	7
Mikula, I.	284
Mikule, V.	320, 329
Milan, D.	46, 312
Milenković, I.	80
Milgen, J. van	273
Milioudis, S.	248
Milis, Ch.	235, 248
Mirzabekov, S.Sh.	230, 230
Mlázovská, P.	101
Mølbak, L.	110
Molina, A.	194, 227, 337, 337, 342, 348
Moll, J.	11

Naděje, B.	316
Nagy, A.	47
Nagy, SZ.	126
Naitana, S.	169
Nakamura, A.	39
Näsholm, A.	241
Nasserian, A.A.	92
Nassiry, M.R.	49, 235, 242, 243
Navrátil, J.	340
Neacsu, C.	256
Neander, S.	67
Neckrasova, N.N.	164
Negovanović, D.	101
Nehring, R.	132

	Page		Page
Nemec, Z.	84, 96, 109	Oliván, M.	197
Németh, T.	293	Oliver, M.A.	197
Nephawe, K.A.	43	Olivier, J.J.	20
Neser, F.W.C.	20, 43	Olleta, J.L.	197
Nešetřilová, H.	203	Ollivier, L.	46
Nest, M. van der	249	Olori, V.	229
Nesterenko, N.N.	183	Oom, M.M.	337
Neuenschwander, S.	9	Oprzadek, J.	177, 195
Neuschaefer, A.	350	Oravcová, M.	40, 212
Nezamzadeh, R.	10	Orzechowska, B.	284
Nezavitin, A.G.	196, 204	Osako, H.	105
Nicks, B.	326	Osikowski, M.	232, 240
Nielen, M.	121	Osoro, K.	197
Nijssen, J.M.A.	216	Otsuki, K.	140
Nikitin, S.	341	Otten, W.	166
Nikokyris, P.N.	293	Ousey, J.	349
Nikolaeva, T.N.	285, 298	Øverland, M.	305, 335
Nikolaou, E.	163	Owsianny, J.	282
Nitter, G.	83	Ożgo, M.	156, 167
Niznikowski, R.	223, 265		
Noblet, J.	269, 273	**P**	
Noguera, J.L.	13	Pakulski, T.	232
Nohejlová, L.	246	Palic, D.	103
Noia, G.	126	Palmer, R.W.	359
Nosenko, N.A.	277	Pambalk, K.	6
Nová, V.	87, 87, 246	Panea, B.	197
Novicov, A.A.	47, 48	Panicke, L.	84, 121
Nowachowicz, J.	307, 316	Panov, B.L.	71
Nowakowski, P.	78, 224, 225	Papshev, S.	329
Nürnberg, K.	304	Papstein, H.J.	178
Nute, G.R.	240	Paramonova, T.B.	70
Nuutila, E-M	264	Pareek, C.S.	69
Nylander, A.	264	Park, K.D.	356
		Parkkonen, P.	60
O		Pärna, E.	42
O'brien, B.	91	Parsini, P.	272
O'Connell, D.	228	Paschma, J.-M.	290
O'Doherty, J.V.	277, 286	Pastushenko, V.	79, 90, 138, 171
O'Mara, F.P.	93	Pászthy, G.	244
O'Riordan, E.G.	75	Pataki, B.	7
Obadálek, J.	96	Paul, S.	312
Obermaier, A.	81	Pavel, Ü.	168
Odle, J.	328	Pavelek, L.	99, 108
Odnoshevsky, D.A.	297	Pavelková, L.	99
Ojala, M.	22, 60, 61, 62, 64	Pavie, J.	208
Okan, A.E.	189	Pavlicevic, A.	101
Oksbjerg, N.	149, 149	Pavlík, J.	205, 281
Olesen, I.	33	Peclaris, G.M.	163

	Page		Page
Pecsi, T.	126	Pour, M.	280, 283
Pedersen, A.O.	333	Prange, H.	117
Pedersen, B.	308	Pribyl, J.	31, 211
Pedersen, B.K.	325	Pribylová, J.	31
Peet, G.F.V. van der	114	Prokop, V.	320
Peet-Schwering, C.M.C. van der	326	Proshin, S.N.	69
Perini, S.	334	Proshina, O.V.	80
Permyakov, A.A.	196, 204	Protais, M.	29
Peskovicová, D.	15, 40, 212, 213	Prunier, A.	161
Petkov, K.	102	Przybylski, W.	313
Petrović, M.P.	101	Puchajda, Z.	187, 193
Petukhov, V.L.	49, 68, 70, 71, 173, 285, 297, 298, 338	Puigvert, X.	314
		Pujol, M.R.	16
Peulen, O.	150	Pulkrábek, J.	203, 281
Pezotti, M.	2	Puntila, M.-J.	264
Pfeifer, I.	40, 51	Purroy, A.	239
Phoya, R.K.D.	288, 288	Purup, S.	160
Phua, S.H.	2	Püski, J.	174
Pickl, M.	81	Pyś, J.B.	278
Piedrafita, J.	16, 197	Pyzhov, A.P.	89
Piekarzewska, A.	58		
Pierzynowski, S.G.	151	**Q**	
Pieszka, M.	339	Quiniou, N.	273
Pieterse, C.	131	Quintanilla, R.	16, 197
Pietruszka, A.	279, 300, 319		
Pikula, R.	355	**R**	
Piles, M.	22	Raboy, V.	287, 287
Pilnikov, V.G.	291, 318	Rădulescu, L.	86, 180
Pinheiro, V.	98	Radzik-Rant, A.	265
Pinna, W.	342	Radzka, E.	187
Pires da Costa, J.S.	305	Ragab, F.A.	157
Pirkelmann, H.	132, 133	Raj, St.	96, 162
Piwczyński, D.	236	Rak, B.	307, 316, 318
Pla, M.	22	Ramisz, A.	145
Plastow, G.S.	1, 46, 47, 302	Ramos, A.M.	234, 311, 314
Połoszynowicz, J.	58	Rangel-Figueiredo, T.	8, 234, 311, 314
Podstawski, Z.	346	Rapetti, L.	257
Pohlenz, J.	67	Rassu, S.P.G.	356
Poirel, D.	352	Rath, M.	91, 93
Polák, P.	212, 213	Rattink, A.P.	3
Põldvere, A.	292	Rauw, W.M.	54
Polgár, J.P.	65, 191	Rebsamen, A.	290, 304
Polgár, P.J.	32	Reeg, K.J.	261
Pollmann, U.	133	Reents, R.	4
Pomykala, D.	117	Rehfeldt, C.	147
Pool, M.H.	26	Reimann, W.	132
Porceddu, A.	2	Reine, A.	122
Poucet, A.	131	Reiner, G.	4, 9

Reinhardt, F.	84	Rózycki, M.	279, 282, 300, 301
Reinsch, N.	4, 37, 41, 64, 312	Ružić, D.	101
Rekaya, R.	29, 59	Ruban, J.D.	52
Reklewska, B.	163, 233	Ruban, S.	193
Remeikiene, J.	278	Rubino, R.	255
Renne, U.	58, 125	Rumsey, T.N.	158
Renner, M.	6	Rundgren, M.	155
Reuillon, J.L.	208	Ruottinen, O.	62
Rey, V.	312	Rupp, R.	59
Ribó, O.	131, 132	Russ, I.	4, 37
Richard, A.	259	Russel, A.J.F.	263, 266
Richardson, E.C.	62	Russo-Almeida, P.	234
Richardson, J.M.	93	Rutkowski, R.	262
Richtrová, A.	57	Ryan, G.	91
Riis, B.	148	Rybarczyk, A.	310
Rijnen, M.M.J.A.	271	Ryniewicz, Z.	233
Rinaldo, D.	296		
Ringdorfer, F.	222	**S**	
Ritter, D.	7	Saage, H.	139
Rózycki, M.	284	Sabbioni, A.	143
Robert, S.	270	Sabeva, I.I.	348
Robert-Granié, C.	27	Sáblíková, L.	63, 63
Robic, A.	312	Šáda, I.	241
Rodero, A.	46, 48, 194, 227, 337, 342	Šafus, P.	211
Rodero, E.	141, 141, 142, 142	Şahin, M.	172, 220
Rodrigáñez, J.	45	Sakowski, T.	206, 213
Rodríguez, M.C.	45	Sallam, M.T.	229
Roeder, L.	180	Saloniemi, H.	155
Rogel Gaillard, C.	312	Saminadin, G.	296
Roguet, J.M.	263	Samoré, A.B.	17
Röhe, R.	280	Sanchez, L.	206
Rohr, P. von	28	Sandrucci, A.	257
Röhrmoser, G.	210	Sangild, P.T.	152
Rom, H.B.	275	Sanna, S.	18
Romaneko, N.I.	48	Santolaria, P.	123
Romanowicz-Barcikowska, K.	250	Santos e Silva, J.	305
Romé, V.	154, 155	Sañudo, C.	197
Rooke, J.A.	116, 323	Sardi, L.	272
Rosenbauer, H.	321	Sárdi, J.	202, 203
Rosenberger, E.	210, 213	Sartin, J.L.	159
Rosochacki, S.	58	Sasaki, O.	39
Rossdale, P.D.	349	Satoh, M.	39
Roth, F.X.	331	Saveli, O.	185, 194
Rothschild, M.F.	1	Sawa, A.	196
Roughsedge, T.	55	Schäffer, D.	117, 137
Rouw, O.L.A.M. de	35	Schahidi, R.	146
Roux, C.Z.	43	Schaik, G. van	121
Roy, P. le	310, 312	Schattschneider, T.	168

Schatzmann, R.	343	Seynaeve, M.	199
Schaub, J.	173	Shabi, Z.	95
Scheeder, M.R.L.	100, 226, 317	Shabla, V.P.	181
Schelling, M.	28	Shaker Momani, M.	241
Scheppingen, A.T.J. van	216	Shayakhmetov, D.	72
Schmeiserová, L.	98	Shekalova, V.P.	235, 242, 243
Schmidely, Ph.	258	Shelouchina, T.V.	164
Schneeberger, M.	222	Shikalova, V.P.	49
Schneider, F.	162, 217	Shipilin, N.N.	71
Schneider, P.	11	Shores, M.A.	159
Schnyder, U.	14, 30	Short, T.H.	1
Schoeman, S.J.	31	Sierra, A.C.	227
Schofield, C.P.	286	Silió, L.	45
Schougaard, H.	134	Silvestrelli, M.	2
Schrama, J.W.	271, 326	Šimeček, K.	98
Schrijver, R. de	105	Siminski, E.	240
Schroeffel, J.	4	Simões, F.	234, 311, 314
Schrooten, C.	36	Simoongwe, V.	288
Schukken, Y. H.	121, 220	Sinclair, L.A.	93, 240
Schüler, L.	44	Siuta, A.	278
Schulman, N.	38	Sivtunova, V.V.	338, 340
Schulz, E.	321	Skiba, G.	96
Schwager-Suter, R.	73, 82	Skřivan, M.	21
Schwarz, F.J.	198	Skrivanová, V.	21
Schwerin, M.	4, 37, 69	Skrzypczak, W.F.	156, 167
Schwörer, D.	290, 304	Slawon, J.	117
Scipioni, R.	272, 272	Sloth, N.M.	275
Scott, S.L.	153	Słowiński, M.	206
Scurtu, I.	252	Słoniewski, K.	213
Šeba, K	31	Smaragdov, M.	72
Sebezhko, O.I.	297	Smet, S. de	199, 199, 239
Secchiari, P.L.	134	Smirnova, T.A.	69
Seegers, H.	138	Smits, A.C.	131
Seeland, G.	66	Smugala, M.	355
Seewer Lötscher, J.	322	Snell, H.	245, 247
Sejrsen, K.	160	Snochowski, M.	177
Şekerden, Ö.	172, 175, 220	Socha, S.	24, 117
Sellier, P.	158, 310	Soest, P.J. van	94
Semak, M.S.	48	Sokol, J.	116
Seoane, J.R.	153	Solarino, A.	342
Seremak, B.	250	Soler, J.	314
Serenius, T.	34	Sölkner, J.	207, 208, 209, 209
Sereno, J.R.B.	141, 141, 142, 142	Sommer, A.	108
Sereno, R.	314	Sonck, B.	174
Serra, X.	197	Søndergaard, E.	134, 357
Settar, P.	20	Soós, P.	333
Sève, B.	148, 160, 324	Sorensen, A.	167
Sevón-Aimonen, M.-L.	34	Sorensen, D.	30

	Page		Page
Sørensen, G.	274, 274	Superchi, P.	143
Sørensen, M.K.	33	Sürie, C.	226
Sørensen, M.T.	149	Suriyasathaporn, W.	220
Soret, B.	239	Süss, R.	238, 238
Sormunen-Cristian, R.	246	Šustala, M.	103
Sosnicki, A.	302	Sutter, F.	77, 100
Souffrant, W.B.	269	Sutton, J.	253
Southwood, O.	1	Swalve, H.	12
Sowińska, J.	143, 237, 243, 244	Świderek, W.P.	234, 262
Spoolder, H.	136	Szabó, F.	32, 43, 65, 191
Šprysl, M.	280, 283	Szarek, J.	66, 78, 127, 187,
Sraïri, M.T.	219		191, 200, 205, 284
Sretenović, Lj.	80	Szatkowski, R.	187
St-Laurent, G.	153, 276	Szczepański, W.	237
Stádník, L.	88, 125	Szelényi, M.	319
Stanciu, G.	185	Szentpáli, K.	43
Standal, N.	264	Sztych, D.	265
Stanek, P.	205, 210	Szücs, E.	174, 182, 184, 188, 218
Staroverova, O.I.	341	Szüts, G.	110
Starozhilova, T.	72	Szyda, J.	303
Stasiak, A.	299	Szymańska, A.	193
Staufenbiel, R.	84	Szyszkowska, A.	78
Staun, H.	134		
Stear, M.J.	260	**T**	
Steenbergen, E.J. van	55	Tafta, V.	256
Steien, S.H.	335	Tahtalỳ, Y.	32
Steinhardt, M.	146	Taibi, L.	126
Steinhart, H.	198	Takács, E.	7
Steinheim, G.	221	Talmant, A.	310
Stekar, J.	106	Tamburini, A.	257
Stenske, M.	66	Tänavots, A.	279
Stern, S.	292	Tański, Z.	237, 243, 244
Stewart, F.	349	Tapkı, I.	172
Štípková, M.	63, 63, 176	Tari, J.	43
Stoićević, Lj.	80	Tarkowski, J.	210
Štolc, L.	87, 87, 246	Tawfik, E.S.	245, 247
Stout, T.A.E.	351	Teichmann, S.	86
Straková, E.	108	Telezhenko, E.V.	173, 184
Stranzinger, G.	9	Terzano, G.M.	126, 166
Stratil, A.	4	Tessier, J.	258, 259
Stricker, C.	27, 28, 82	Thénard, V.	83
Stringfellow, D.	119	Therkildsen, M.	177
Stuhec, I.	135	Theron, H.E.	43
Stupka, R.	280, 283	Thibier, M.	119, 120
Suchỳ, P.	108	Thielscher, H.H.	146
Suess, R.	245	Thieven, u.	67
Sulik, M.	144	Thomas, F.	154, 155
Summer, A.	143, 341	Thomsen, H.	4, 37

	Page		Page
Thorpe, W.	83	Vandehoek, J.E.D.	118
Tibau, J.	314	Vandenheede, M.	326
Timm, M.	198	Vandepitte, W.	308, 353
Tirelli, A.	341	Vangen, O.	52, 54
Tkáčiková, L.	284	Varela, A.	206
Tkalčic, E.	106	Varga-Visi, É.	315
Toborowicz, R.	127	Vargas, B.	19
Todde, P.	342	Varona, L.	13
Togashi, K.	39	Värv, S	51
Tornsten, A.	312	Vázquez Cisneros, C.	285
Tornyos, G.	136, 333	Vázquez Yáñez, O.P.	172
Toro, M.A.	45	Veerkamp, R.	73
Toscano Pagano, G.	179	Vega-Pla, J.L.	46, 48
Tóth, Á	136	Vemmer, H.	321
Toušová, R.	88, 125	Ventura, J.	305
Tözsér, J.	195, 201, 218	Verini Supplizi, A.	2
Trautman, J.	188, 210	Verleyen, V.	313
Traver, S.	327	Vernin, P.	310
Tremoleda, J.L.	346	Verstegen, M.W.A.	54, 271
Tribout, T.	302	Vervaeke, V.	174
Třináctý, J.	85, 98 99, 103	Vesseur, P.C.	326
Truong, C.	135	Vestergaard, M.	149, 160, 177, 270
Tsvetkova, O.G.	89	Vetési, F.	136
Tuboly, S.	333	Veum, T.L.	287, 287
Tuchscherer, M.	166	Viinalass, H.	51
Tuiskula-Haavisto, M.	38	Vilinská, Z.	205
Tumienie, M.	192	Vilkki, J.	38
Tùmová, E.	21	Vincze, L.	110
Turnbull, C.	349	Vintila, I.	127
Turner, S.P.	116	Vinuesa, M.	348
Tvrdoň, Z.	320, 329	Visscher, A.H.	55
Tyrisevä, M.	64	Visscher, P.M.	55
Tzortzios, S.	18	Visser, E.K.	357
		Vodehnal, D.	104
U		Vögeli, P.	9
Udała, J.	250, 250, 251, 251, 279, 300	Vogrin-Bračič, M.	135
Ufimtseva, N.S.	183	Vohradsky, F.	241
Unal, N.	139	Voigt, K.	198
Urban, F.	176	Völgyi Csík, J.	182
Ustinova, V.I.	183	Voorde, G. van de	199, 199, 239
Utz, J.	213	Vörösmarthy, E.	32
Uystepruyst, Ch.	57	Vouzela, C.F.M.	104
		Vreman, K.	275
V		Vries, A.G. de	302
Vajda, L.	32	Vrzalová, D.	85
Valenta, H.	99	Vu Tien Khang, J.	259
Valera, M.	194, 337, 337, 342, 348	Vukasinovic, N.	11
Valizadeh, R.	82		

	Page		Page
W		Wood, J.D.	240, 286
Waaij, E.H. van der	5	Woolliams, J.	34
Wagenhoffer, Zs.	32, 65	Wrathall, A.E.	120
Wágner, L.	110	Wróblewska, B.	111
Walawski, K.	69	Wülbers-Mindermann, M.	137
Walczak, J.	295	Wyk, J. van	20
Wales, R.	47		
Walkiewicz, A.	299	**X**	
Walkowicz, E.	339	Xu, N.	4, 37
Wall, E.E.	229	Xu, R.J.	150
Wanke, R.	213		
Wassell, B.R.	215	**Y**	
Wassell, T.R.	215	Yakovlev, A.F.	69
Wassmuth, R.	8, 61	Yamada, Y.	165
Waterhouse, A.	221	Yamamoto, N.	39
Weber, R.	178	Yániz, J.	123, 123
Wechsler, B.	173, 178	Yano, F.	105
Weerdt, M. -L. van de	57	Yano, H.	109, 248
Weeren, R. van	352, 353	Yerle, M.	312
Węglarz, A.	66, 78, 187, 191, 200	Yermekov, M.A.	230
Weigel, K.A.	29, 56, 59	Yi, Z.	180
Weishaupt, M.A.	338	Young, S.R.	228
Weller, J.I.	20	Yue, G.	4
Wendl, G.	132		
Wenk, C.	268, 317	**Z**	
Wentink, G.H.	118	Zachwieja, A.	189
Weremko, D.	96	Zadworny, D.	37
Westhuizen, J. van der	43	Zakharov, N.B.	196, 201, 204
Wetscherek, W.	97, 322	Zalewski, W.	205
Wettstein, H.R.	100	Zamorano, M.J.	48
White, R.P.	286	Zamwel, S.	95
Wicke, M.	162	Zapletal, P.	66, 200
Wieser, M.	217	Zarnecki, A.S.	67
Wigren, I.	292	Zaugg, A.	222
Wilkinson, R.G.	93, 240	Zdziarski, K.	233
Willam, A.	207, 209	Zecchini, M.	112
Wilsher, S.	349	Zeiler, M.	354
Winne, P. de	131	Zeman, L.	107, 107
Winnicka, A.	234, 262	Zervas, G.P.	256
Wirth-Dzięciołowksa, E.	163	Zheltikov, F.I.	49
Witte, W.	330	Zhuchaev, K.	294, 295, 329
Wittlingerová, Z.	111	Ziaei, A.	82
Wójcik, A.	143	Zięba, G.	354
Wolf, E.	6	Ziemiński, R.	189
Wolf, J.	15	Zikova, E.	145
Wolf-Táskai, E.	333	Zimmermann, M.	306
Wollny, C.B.A.	216, 288, 288	Zollitsch, W.	97
Wolski, K.	78, 224	Zomborszky-Kovács, M.	136, 333

	Page
Zurawski, H.	*318*
Zwierzchowski, L.	*177, 195*
Żychlińska, J.	*127*